Holt Pre-Algebra

Assessment Resources

HOLT, RINEHART AND WINSTON

A Harcourt Education Company

Orlando • **Austin** • New York • San Diego • Toronto • London

Copyright © by Holt, Rinehart and Winston

All rights reserved. No part of this publication may be reproduced or transmitted in any form or by any means, electronic or mechanical, including photocopy, recording, or any information storage and retrieval system, without permission in writing from the publisher.

Teachers using HOLT PRE-ALGEBRA may photocopy complete pages in sufficient quantities for classroom use only and not for resale.

Printed in the United States of America

ISBN 0-03-069613-5

10   1409   09

CONTENTS

- **Introduction** v
  - Assessment in *Holt Pre-Algebra* v
  - Assessing Prior Knowledge v
  - Progress Monitoring v
  - Summative Evaluation vi
  - Test Preparation vi
- **Assessment Options at a Glance** vii
- **Preparing Students for Success** viii
- **Formal Assessment** ix
- **How to Use the Inventory Test** x
  - Inventory Test Profiles xi
- **Daily Assessment** xiii
- **Performance Assessment** xiv
- **Student Self-Assessment** xv
- **Management Forms** xvi
  - Individual Record Forms xvii
  - Class Record Forms xxix
- **Checklists, Rubrics, and Surveys**
  - Project Scoring Rubric xxxi
  - Group Project Evaluation xxxii
  - Individual Group Member Evaluation xxxiii
- **Portfolio Assessment** xxxiv
  - Guide to My Math Portfolio xxxv
  - Portfolio Evaluation xxxvi
  - Family Response to Portfolio xxxvii
- **Assessing Problem Solving** xxxviii
  - Think Along, Oral Response xxxix
  - Oral Response, Scoring Guide xl
  - Think Along, Written Response xli
  - Written Response Scoring Guide xlii
- **Test-Taking Tips** xliii
- **Test Answer Sheets** xlv

Assessment Resources   iii

TESTS

▶ INVENTORY TEST ... 1

▶ Quizzes
- Chapter 1 ... 5
- Chapter 2 ... 7
- Chapter 3 ... 9
- Chapter 4 ... 11
- Chapter 5 ... 13
- Chapter 6 ... 15
- Chapter 7 ... 17
- Chapter 8 ... 19
- Chapter 9 ... 21
- Chapter 10 ... 23
- Chapter 11 ... 25
- Chapter 12 ... 27
- Chapter 13 ... 29
- Chapter 14 ... 31

▶ Chapter Tests
- Chapter 1 ... 33
- Chapter 2 ... 39
- Chapter 3 ... 45
- Chapter 4 ... 51
- Chapter 5 ... 57
- Chapter 6 ... 63
- Chapter 7 ... 69
- Chapter 8 ... 75
- Chapter 9 ... 81
- Chapter 10 ... 87
- Chapter 11 ... 93
- Chapter 12 ... 99
- Chapter 13 ... 105
- Chapter 14 ... 111

▶ Performance Assessment
- Chapter 1 ... 117
- Chapter 2 ... 119
- Chapter 3 ... 121
- Chapter 4 ... 123
- Chapter 5 ... 125
- Chapter 6 ... 127
- Chapter 7 ... 129
- Chapter 8 ... 131
- Chapter 9 ... 133
- Chapter 10 ... 135
- Chapter 11 ... 137
- Chapter 12 ... 139
- Chapter 13 ... 141
- Chapter 14 ... 143

▶ Cumulative Tests
- Chapter 1 ... 145
- Chapter 1-2 ... 157
- Chapter 1-3 ... 169
- Chapter 1-4 ... 181
- Chapter 1-5 ... 193
- Chapter 1-6 ... 205
- Chapter 1-7 ... 217
- Chapter 1-8 ... 229
- Chapter 1-9 ... 241
- Chapter 1-10 ... 253
- Chapter 1-11 ... 265
- Chapter 1-12 ... 277
- Chapter 1-13 ... 289
- Chapter 1-14 ... 301

▶ END-OF-YEAR TEST ... 313

▶ ANSWER KEY ... 317

Assessment Resources

INTRODUCTION

Assessment is aligned with instruction. Students are assessed frequently to determine progress. The results of the assessment can be used to determine future instruction.

Assessment in Holt Pre-Algebra

Holt Pre-Algebra provides a wide range of assessment tools to measure student achievement before, during, and after instruction. These tools include:

- Assessing Prior Knowledge
- Progress Monitoring
- Summative Evaluation
- Test Preparation

Assessing Prior Knowledge

Inventory Test—This test, provided on pages 1–4, can be administered at the beginning of the school year to determine a baseline for student mastery of major objectives. The baseline may also be used to evaluate a student's growth.

Are You Ready?—This feature appears at the beginning of every chapter in *Holt Pre-Algebra* student book. It can be used before chapter instruction begins to determine if students possess crucial prerequisite skills. Tools for intervention are included in "Are You Ready? Intervention and Enrichment".

Progress Monitoring

Daily Assessment—Obtain daily feedback on students' understanding of concepts by using the Spiral Review and Test Prep that is on last page of each lesson in the student book. A Warm Up, Problem of the Day, and Lesson Quiz are on the first page of each lesson in the teacher's edition and also on transparencies in the Chapter Resource Books.

Student Self-Assessment—Students can evaluate their own work using the Group Project Evaluation on page xxxii, Individual Group Member Evaluation on page xxxiii, the Portfolio Guide on page xxxv, and the journal which is in the TE on the last page of each lesson.

Introduction   **Assessment Resources**   v

Summative Evaluation

Formal Assessment—Several options are provided to determine whether students have mastered the goals of given section or chapter. They include

- Section Quizzes in this book
- Mid-Chapter Quizzes in the student book
- Chapter Tests in the student book
- Chapter Tests (Levels A, B, C) in this book
- Cumulative Tests (Levels A, B, C) in this book
- Standardized Test Prep in the student book
- End-of-Year Test in this book

Performance Assessment—Assess students' understanding of chapter concepts and combined problem-solving skills. These options include

- Performance Assessment in the student book. The teacher's edition includes the scoring rubric for this.
- Performance Assessment and Teacher Support for it are in this book.

Portfolio Assessment—Portfolio opportunities appear throughout the student book and teacher's edition. Suggested work sample include

- Problem Solving Project in the teacher's edition and Chapter Resource Book
- Performance Assessment in student book
- Portfolio Guide in this book
- Journal in the teacher's edition
- Write About It in the student book

Test Preparation

Review and practice for chapter and standardized tests include

- Standardized Test Prep in the student book
- Spiral Review with Test Prep in the student book
- Study Guide and Review in the student book
- Reviews in the Chapter Resource Book
- Test Prep Took Kit
- Technology: Test and Practice Generator CD-ROM
- Technology: State Specific Test Practice Online KEYWORD: MP4 TestPrep

ASSESSMENT OPTIONS AT A GLANCE

ASSESSING PRIOR KNOWLEDGE

Inventory Test, pages 1–4
Are You Ready?, student book

DAILY ASSESSMENT

Spiral Review and Test Prep, student book
Warm Up, TE and Daily Transparency in CRB
Problem of the Day, TE and Daily Transparency in CRB
Lesson Quiz, TE and Daily Transparency in CRB

STUDENT SELF-ASSESSMENT

Group Project Evaluation, page xxxii
Individual Group Member Evaluation, page xxxiii
Portfolio Guide, page xxxv
Journal, TE

FORMAL ASSESSMENT

Section Quizzes, AR
Mid-Chapter Quizzes, student book
Chapter Tests, student book
Chapter Tests (Levels A, B, C) AR
Cumulative Tests (Levels A, B, C) AR
Standardized Test Prep, student book
End-of-Year Test, AR

PERFORMANCE ASSESSMENT

Performance Assessment, student book
Performance Assessment, AR

PORTFOLIO ASSESSMENT

Problem Solving Project, TE and CRB
Performance Assessment, student book
Portfolio Guide, AR
Journal, TE
Write About It, student book

TEST PREPARATION

Standardized Test Prep, student book
Spiral Review with Test Prep, student book
Study Guide and Review, student book
Reviews, CRB
Test Prep Took Kit

Technology: Test and Practice Generator CD-ROM
Technology: State specific Test Practice Online KEYWORD: MS4 TestPrep

Key: AR = *Assessment Resources*, TE = *Teacher's Edition*,
CRB = *Chapter Resource Book*

Assessment Options **Assessment Resources** vii

▶ PREPARING STUDENTS FOR SUCCESS

Assessing Prior Knowledge

Assessment of prior knowledge is essential to planning mathematics instruction and to ensure students' progress from what they already know to higher levels of learning. In *Holt Pre-Algebra*, each chapter begins with "Are You Ready?", a tool to assess prior knowledge that can be used to determine whether students have the prerequisite skills to move on to the new skills and concepts of the chapter.

If students are found lacking in some skills or concepts, appropriate strategies are suggested. The "Are You Ready? Intervention and Enrichment" ancillary provides additional options for intervention. The teacher's edition provides suggestions for reaching students with a variety of learning needs and the chapter resource books provide reteaching, practice (A, B, C), and challenge masters.

Test Preparation

With increasing emphasis today on standardized tests, many students feel intimidated and nervous as testing time approaches. Whether they are facing teacher-made tests, program tests, or state-wide standardized tests, students will feel more confident with the test format and content if they know what to expect in advance.

Holt Pre-Algebra provides multiple opportunities for test preparation. At the end of each lesson there is a Spiral Review and Test Prep, which provides items in a standardized-test format. A Standardized Test Prep page at the end of each chapter provides more practice with test preparation. There are test-taking tips on pages xliii and xliv of this book.

▶ FORMAL ASSESSMENT

Formal assessments in *Holt Pre-Algebra* consist of a series of reviews and tests that assess how well students understand concepts, perform skills, and solve problems related to program content. Information from these measures (along with information from other kinds of assessment) is needed to evaluate student achievement and to determine grades. Moreover, analysis of results can help determine whether additional practice or reteaching is needed.

Formal assessment in *Holt Pre-Algebra* includes the following:

- Section Quizzes in this book
- Mid-Chapter Quizzes in the student book
- Chapter Tests in the student book
- Chapter Tests (Levels A, B, C) in this book
- Cumulative Tests (Levels, A, B, C) in this book
- Standardized Test Prep in the student book
- End-of-Year Tests in this book

A **Section Quiz** and a **Mid-Chapter Quiz** assess just the objectives for the section.

There is a **Chapter Test** at the end of each chapter in the student book. It can be used to determine whether there is a need for more instruction or practice. Three levels (A, B, C) of free-response tests are available for every chapter. They can be found in this book.

Three levels (A, B, C) of multiple-choice **Cumulative Tests** are available for every chapter. These can be found in this book.

An **End-of-Year Test** in multiple choice format is available to assess the main skills and concepts covered in the year's curriculum.

The **Answer Key** in this book provides reduced replications of the tests with answers.

Two **Record Forms** are available for formal assessment—the individual record form (pages xvii–xxviii) and class record form (pages xxix–xxx).

Students can record their answers directly on the test sheets. However, for multiple-choice tests, you may choose to have students use one of the **Answer Sheets,** similar to the "bubble form" used for standardized tests, that is located on pages xlv and xlvi.

Formal Assessment   **Assessment Resources   ix**

Interpreting the Inventory Test Results

The scores for an individual student can be summarized for each of the 14 chapters.

CHAPTER	TEST ITEMS	PROFICIENCY LEVEL	STUDENT SCORE	PROFICIENCY SCORE (%)
Algebra Toolbox	1-3	2/3		
Integers and Exponents	4-6	2/3		
Rational and Real Numbers	7–10	3/4		
Collecting, Displaying, and Analying Data	11-14	3/4		
Plane Geometry	15-17	2/3		
Perimeter, Area, and Volume	18-21	3/4		
Ratios and Similarity	22-25	3/4		
Percents	26-29	3/4		
Probability	30-31	2/3		
More Equations and Inequalities	32-34	2/3		
Graphing Lines	35-37	2/3		
Sequences and Functions	38-40	2/3		

By referring to the Proficiency Score column, you can quickly determine whether the student has failed to demonstrate proficiency in one or more chapters.

Using the Class Profile

The Class Profile can be used to summarize the results of the Inventory Test. By using the Class Profile, you can make better use of classroom instruction time. For example, if the test results indicate that most students are proficient in decimals, you may decide to present a brief unit on decimals and provide remedial work for weak students. However, if the results also indicate that most students are not proficient in fractions, you may choose to spend more time teaching that chapter.

Class Profile for Inventory Test

STUDENT NAME	Ch 1	Ch 2	Ch 3	Ch 4	Ch 5	Ch 6	Ch 7	Ch 8	Ch 9	Ch 10	Ch 11	Ch 12	Ch 13	Ch 14
1.														
2.														
3.														
4.														
5.														
6.														
7.														
8.														
9.														
10.														
11.														
12.														
13.														
14.														
15.														
16.														
17.														
18.														
19.														
20.														
Total Number Proficient														
% of Students Proficient														

Class Profile for Inventory Test

STUDENT NAME	Ch 1	Ch 2	Ch 3	Ch 4	Ch 5	Ch 6	Ch 7	Ch 8	Ch 9	Ch 10	Ch 11	Ch 12	Ch 13	Ch 14
21.														
22.														
23.														
24.														
25.														
26.														
27.														
28.														
29.														
30.														
Total Number Proficient														
% of Students Proficient														

The Class Profile will help you look for patterns of performance among students.

Some ways you may choose to use the Class Profile include:

- grouping students with similar instructional needs
- identifying content for extra teaching
- identifying skills that need regular reinforcement
- noting whether students' class performance verifies strengths and weaknesses identified on the test

▶ DAILY ASSESSMENT

Daily assessment is assessment embedded in daily instruction. Students are assessed as they learn and learn as they are assessed. First you observe and evaluate your students' work on an informal basis, and then you seek confirmation of those observations through other program assessments.

Holt Pre-Algebra offers the following resources to support informal assessment on a daily basis:

- Spiral Review and Test Prep that is on the last page of each lesson in the student book.
- Warm Up in the teachers edition and also on transparencies in the Chapter Resource Books.
- Problem of the Day in the teacher's edition and also on transparencies in the Chapter Resource Books.
- Lesson Quiz in the teacher's edition and also on transparencies in the Chapter Resource Books.

The **Warm Up** allows you to adjust instruction so that all students are progressing toward mastery of skills and concepts.

Problem of the Day can be used any time during the class. It relates to the content of the lesson and the students' world.

The **Lesson Quiz** is a quick check of students' mastery of lesson skills. Depending on what you learn from students' responses to this quiz, you may wish to use Problem Solving, Reteach, Practice (A, B, C), or Challenge copying masters found in the Chapter Resource Book before starting the next lesson.

Daily Assessment   **Assessment Resources**

▶ PERFORMANCE ASSESSMENT

Performance assessment can help reveal the thinking strategies students use to work through a problem. Students usually enjoy doing the performance tasks.

Holt Pre-Algebra offers the following assessment measures, scoring instruments, and teacher observation checklists for evaluating student performances.

- Project Scoring Rubric, in this book
- Portfolio Evaluation in this book
- Problem Solving Think Along Response Sheets and Scoring Guides in this book
- Scoring Rubric in teacher's edition for the Performance Assessment in the student book
- Teacher Support and scoring rubric for the Performance Assessment in this book

The **Performance Assessment** in the student book includes one task per chapter. These tasks can help you assess students' ability to use what they have learned to solve everyday problems. The Performance Assessment in this book has a different task for the chapter.

The **Problem Solving Think Along** is a performance assessment that is designed around the problem-solving method used in *Holt Pre-Algebra.* You may use either the Oral Response or Written Response form to evaluate the students. For more information see pages xl–xli.

Portfolios can also be used to assess students' mathematics performance. For more information, see pages xxxiv–xxxvii.

▶ STUDENT SELF-ASSESSMENT

Research shows that self-assessment can have significant positive effects on students' learning. To achieve these effects, students must be challenged to reflect on their work and to monitor, analyze, and control their learning. Their ability to evaluate their behaviors and to monitor them grows with their experience in self-assessment.

Holt Pre-Algebra offers the following self-assessment tools:

- Group Project Evaluation on page xxxii
- Individual Group Member Evaluation on page xxxiii
- Portfolio Guide on page xxxv
- Journal in the teacher's edition

The **Group Project Evaluation** ("How Did Our Group Do?") is designed to assess and build up group self-assessment skills. The **Individual Group Member Evaluation** ("How Well Did I Work in My Group?") helps the student evaluate his or her own behavior in and contributions to the group. Discuss directions for completing these forms with the students. Tell them there are no "right" responses to the items. Discuss reasons for various responses.

The **Journal** is a collection of student writings that may communicate feelings, ideas, and explanations as well as responses to open-ended problems. it is an important evaluation tool in math even though it is not graded. Use the journal to gain insights about student growth that you cannot obtain from other assessments.

▶ MANAGEMENT FORMS

This Assessment Resource contains two types of forms to help you manage your record keeping and evaluate students in various types of assessment. On pages xvii–xxviii, you will find Individual Record Forms that contain all of the objectives for each chapter. After each objective are correlations to the items in Level A, B, and C of the Chapter Tests. Criterion scores for each objective are given. The form provides a place to enter a single student's scores on the chapter tests and to indicate the objectives he or she has not met. A list of review options is also included. The options include lessons in the student book and teacher's edition and ancillaries in the Chapter Resource Books that you can assign to a student in need of addition practice.

The Class Record Form on pages xxix and xxx makes it possible to record the test scores of an entire class on a single form.

Individual Record Form

Pre-Algebra • Chapter 1 Algebra Toolbox

Student Name _____ Test Level | A | B | C |

	OBJECTIVES	CHAPTER TEST			REVIEW OPTIONS						
Lesson	Learning Goal	Test Items	Criterion Score	Student's Score	SE/TE Lesson	Pages in CRB					
						P(A)	P(B)	P(C)	R	C	PS
1-1	Learn to evaluate algebraic expressions.	1-3	2/3								
1-2	Learn to write algebraic expressions.	4-6	2/3								
1-3	Learn to solve equations using addition and subtraction.	7-10	3/4								
1-4	Learn to solve equations using multiplication and division.	11-14	3/4								
1-5	Learn to solve and graph inequalities.	15-18	3/4								
1-6	Learn to combine like terms in an expression.	19-24	4/6								
1-7	Learn to write solutions of equations in two variables as ordered pairs.	25-26	2/2								
1-8	Learn to graph points and lines on the coordinate plane.	27-28	2/2								
1-9	Learn to interpret information given in a graph or table and to make a graph to solve problems.	29-30	2/2								

Key: P-Practice, **R**-Reteach, **C**-Challenge, **PS**-Problem Solving, **CRB**-Chapter Resource Book

Individual Record Form   **Assessment Resources   xvii**

Individual Record Form

Pre-Algebra • Chapter 2 Integers and Exponents

Student Name _____ Test Level | A | B | C |

Lesson	OBJECTIVES Learning Goal	CHAPTER TEST Test Items	Criterion Score	Student's Score	REVIEW OPTIONS SE/TE Lesson	P(A)	P(B)	P(C)	R	C	PS
2-1	Learn to add integers.	1-6	4/6								
2-2	Learn to subtract integers.	7-12	4/6								
2-3	Learn to multiply and divide integers.	13-18	4/6								
2-4	Learn to solve equations with integers.	19-21	2/3								
2-5	Learn to solve inequalities with integers.	22-24	2/3								
2-6	Learn to evaluate expressions with exponents.	25-33	6/9								
2-7	Learn to apply the properties of exponents and to evaluate the zero exponent.	34-36	2/3								
2-8	Learn to evaluate expressions with negative exponents.	37-42	4/6								
2-9	Learn to express large and small numbers in scientific notation.	43-47	4/5								

Key: P-Practice, **R**-Reteach, **C**-Challenge, **PS**-Problem Solving, **CRB**-Chapter Resource Book

xviii   Assessment Resources   Individual Record Form

Individual Record Form

Pre-Algebra • Chapter 3 Rational and Real Numbers

Student Name _____   Test Level | A | B | C |

	OBJECTIVES	CHAPTER TEST			REVIEW OPTIONS						
Lesson	Learning Goal	Test Items	Criterion Score	Student's Score	SE/TE Lesson	\<Pages in CRB\>					
						P(A)	P(B)	P(C)	R	C	PS
3-1	Learn to write rational numbers in equivalent forms.	1-7	5/7								
3-2	Learn to add and subtract decimals and rational numbers with like denominators.	8-11	3/4								
3-3	Learn to multiply fractions and mixed numbers.	12-17, 43-44	6/8								
3-4	Learn to divide fractions.	15-23	4/6								
3-5	Learn to add and subtract fractions with unlike denominators.	24-27	3/4								
3-6	Learn to solve equations with rational numbers.	28-30	2/3								
3-7	Learn to solve inequalities with rational numbers.	31-33	2/3								
3-8	Learn to find square roots.	34-36	2/3								
3-9	Learn to estimate square roots to a given number of decimal places and solve problems using square roots.	37-39	2/3								
3-10	Learn to determine if a number is rational or irrational.	40-42	2/3								

Key: P-Practice, **R**-Reteach, **C**-Challenge, **PS**-Problem Solving, **CRB**-Chapter Resource Book

Individual Record Form

Pre-Algebra • Chapter 4 Collecting, Analyzing, and Displaying

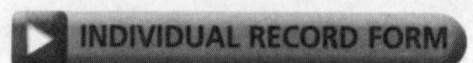

Student Name _____   Test Level | A | B | C |

	OBJECTIVES	CHAPTER TEST			REVIEW OPTIONS					
Lesson	Learning Goal	Test Items	Criterion Score	Student's Score	SE/TE Lesson	\multicolumn{6}{c}{Pages in CRB}
						P(A)	P(B)	P(C)	R	C	PS
4-1	Learn to recognize biased samples and to identify sampling methods.	1-3	2/3								
4-2	Learn to organize data in tables and stem-and-leaf plots.	4	1/1								
4-3	Learn to find appropriate measures of central tendency.	5-7, 12-14	4/6								
4-4	Learn to find measures of variability.	9-11, 16-18	4/6								
4-5	Learn to display data in bar graphs, histograms, and line graphs.	19-20	2/2								
4-6	Learn to recognize misleading graphs and statistics.	21	1/1								
4-7	Learn to create and interpret scatter plots.	22-24	2/3								

Key: P-Practice, **R**-Reteach, **C**-Challenge, **PS**-Problem Solving, **CRB**-Chapter Resource Book

Individual Record Form

Pre-Algebra • Chapter 5 Plane Geometry

Student Name _____   Test Level | A | B | C |

OBJECTIVES		CHAPTER TEST			REVIEW OPTIONS						
Lesson	Learning Goal	Test Items	Criterion Score	Student's Score	SE/TE Lesson	Pages in CRB					
						P(A)	P(B)	P(C)	R	C	PS
5-1	Learn to classify and name figures.	1-5	4/5								
5-2	Learn to identify parallel and perpendicular lines and the angles formed by a transversal.	6-8	2/3								
5-3	Learn to find unknown angles in triangles.	9-10	2/3								
5-4	Learn to classify and find angles in polygons.	12-14	2/3								
5-5	Learn to identify polygons in the coordinate plane.	15-16	2/2								
5-6	Learn to use properties of congruent figures to solve problems.	17-18	2/2								
5-7	Learn to transform plane figures using translations, rotations, and reflections.	19-20	2/2								
5-8	Learn to identify symmetry in figures.	21-22	2/2								
5-9	Learn to predict and verify patterns involving tessellations.	23	1/1								

Key: P-Practice, R-Reteach, C-Challenge, PS-Problem Solving, CRB-Chapter Resource Book

Individual Record Form   **Assessment Resources   xxi**

Individual Record Form

Pre-Algebra • Chapter 6 Perimeter, Area, and Volume

Student Name _____   Test Level | A | B | C |

	OBJECTIVES	CHAPTER TEST			REVIEW OPTIONS					
Lesson	Learning Goal	Test Items	Criterion Score	Student's Score	SE/TE Lesson	\multicolumn{6}{c}{Pages in CRB}
						P(A)	P(B)	P(C)	R	C	PS
6-1	Learn to find perimeter and area of rectangles and parallelograms.	1-3	2/2								
6-2	Learn to find the area of triangles and trapezoids.	2-4	2/2								
6-3	Learn to use the Pythagorean Theorem and its converse to solve problems.	5-6	2/2								
6-4	Learn to find the area and circumference of circles.	7-8	2/2								
6-5	Learn to draw and identify parts of three-dimensional figures.	9-10	2/2								
6-6	Learn to find the volume of prisms and cylinders.	11-12	2/2								
6-7	Learn to find the volume of pyramids and cones.	13-14	2/2								
6-8	Learn to the surface area of prisms and cylinders.	15-16	2/2								
6-9	Learn to the surface area of pyramids and cones.	17-18	2/2								
6-10	Learn to find the volume and surface area of spheres.	19-20	2/2								

Key: P-Practice, **R-**Reteach, **C-**Challenge, **PS-**Problem Solving, **CRB-**Chapter Resource Book

Individual Record Form

Pre-Algebra • **Chapter 7** Ratios and Similarity

Student Name _____   Test Level | A | B | C |

	OBJECTIVES	CHAPTER TEST			REVIEW OPTIONS					
Lesson	Learning Goal	Test Items	Criterion Score	Student's Score	SE/TE Lesson	P(A)	P(B)	P(C)	R	C	PS
7-1	Learn to find equivalent ratios to create proportions.	1-7	5/7								
7-2	Learn to work with rates and ratios.	8-9	2/2								
7-3	Learn to use one or more conversion factors to solve problems.	10-13	3/4								
7-4	Learn to solve proportions.	14-16	2/3								
7-5	Learn to identify and create dilations of plane figures.	17-18	2/2								
7-6	Learn to determine whether figures are similar, to use scale factors, and to find missing dimensions in similar figures.	19-20	2/2								
7-7	Learn to make comparisons between and find dimensions of scale drawings and actual objects.	21-22	2/2								
7-8	Learn to make comparisons between and find dimensions of scale models and actual objects.	23-25	2/3								
7-9	Learn to makes scales models of solid figures.	26-27	2/2								

Key: P-Practice, **R**-Reteach, **C**-Challenge, **PS**-Problem Solving, **CRB**-Chapter Resource Book

Individual Record Form   **Assessment Resources   xxiii**

Individual Record Form

Pre-Algebra • Chapter 8 Percents

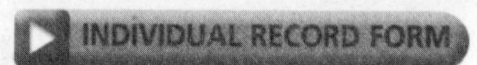

Student Name _____   Test Level | A | B | C |

	OBJECTIVES	CHAPTER TEST			REVIEW OPTIONS					
Lesson	Learning Goal	Test Items	Criterion Score	Student's Score	SE/TE Lesson	\multicolumn{6}{c}{Pages in CRB}
						P(A)	P(B)	P(C)	R	C	PS
8-1	Learn to relate decimals, fractions, and percents.	1-5	4/5								
8-2	Learn to find percents.	6-7, 12	2/3								
8-3	Learn to find a number when the percent is known.	8-9, 13-15	4/5								
8-4	Learn to find the percent increase and decrease.	16-18	2/3								
8-5	Learn to estimate with percents.	19	1/1								
8-6	Learn to find commission, sales tax, and withholding tax.	20-21	2/2								
8-7	Learn to compute simple interest.	23-25	2/3								

Key: P-Practice, **R-**Reteach, **C-**Challenge, **PS-**Problem Solving, **CRB-**Chapter Resource Book

Individual Record Form

Pre-Algebra • Chapter 9 Probability

Student Name _____ Test Level | A | B | C |

OBJECTIVES		CHAPTER TEST			REVIEW OPTIONS						
Lesson	Learning Goal	Test Items	Criterion Score	Student's Score	SE/TE Lesson	\multicolumn{6}{c|}{Pages in CRB}
						P(A)	P(B)	P(C)	R	C	PS
9-1	Learn to find the probability of an event by using the definition of probability.	1-2	2/2								
9-2	Learn to estimate probability using experimental methods.	3-4	2/2								
9-3	Learn to use a simulation to estimate probability.	5-6	2/2								
9-4	Learn to estimate probability using theoretical methods.	7-8	2/2								
9-5	Learn to find the number of possible outcomes in an experiment.	9-10	2/2								
9-6	Learn to find permutations and combinations.	11-12	2/2								
9-7	Learn to find the probabilities of independent and dependent events.	13-14	2/2								
9-8	Learn to convert between probabilities and odds.	15-17	2/3								

Key: P-Practice, **R**-Reteach, **C**-Challenge, **PS**-Problem Solving, **CRB**-Chapter Resource Book

Individual Record Form Assessment Resources xxv

Individual Record Form

Pre-Algebra • Chapter 10 More Equations and Inequalities

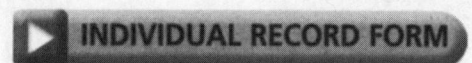

Student Name _____   Test Level | A | B | C

Lesson	OBJECTIVES — Learning Goal	CHAPTER TEST — Test Items	Criterion Score	Student's Score	REVIEW OPTIONS — SE/TE Lesson	P(A)	P(B)	P(C)	R	C	PS
10-1	Learn to solve two-step equations.	1-4, 28	4/5								
10-2	Learn to solve multistep equations.	5-8	3/4								
10-3	Learn to solve equations with variables on both sides of the equal sign.	9-12	3/4								
10-4	Learn to solve two-step inequalities and graph the solution of an inequality on a number line.	13-15, 29	3/4								
10-5	Learn to solve an equation for a variable.	16-21	4/6								
10-6	Learn to solve a system of linear equations.	22-27, 30	5/7								

Key: P-Practice, R-Reteach, C-Challenge, PS-Problem Solving, CRB-Chapter Resource Book

xxvi   Assessment Resources   Individual Record Form

Individual Record Form

Pre-Algebra • Chapter 11 Graphing Lines

Student Name _____   Test Level | A | B | C |

OBJECTIVES		CHAPTER TEST			REVIEW OPTIONS						
Lesson	Learning Goal	Test Items	Criterion Score	Student's Score	SE/TE Lesson	\multicolumn{6}{c}{Pages in CRB}
						P(A)	P(B)	P(C)	R	C	PS
11-1	Learn to identify and graph linear equations.	1-2	1-2								
11-2	Learn to find the slope of a line and use the slope to understand and draw graphs.	3-4	3-4								
11-3	Learn to use slopes and intercepts to graph liner equations.	5-8	5-8								
11-4	Learn to find the equation of a line given one point and the slope.	9-10	9-10								
11-5	Learn to recognize direct variation by graphing tables of data and checking for constant ratios.	11-12	11-12								
11-6	Learn to graph inequalities on the coordinate plane.	13-14	13-14								
11-7	Learn to recognize relationships in data and find the equation of a line of best fit.	15-16	15-16								

Key: P-Practice, **R**-Reteach, **C**-Challenge, **PS**-Problem Solving, **CRB**-Chapter Resource Book

Individual Record Form — Assessment Resources

Individual Record Form

Pre-Algebra • Chapter 12 Sequences and Functions

Student Name _____   Test Level | A | B | C |

	OBJECTIVES	CHAPTER TEST			REVIEW OPTIONS					
Lesson	Learning Goal	Test Items	Criterion Score	Student's Score	SE/TE Lesson	\multicolumn{6}{c}{Pages in CRB}
						P(A)	P(B)	P(C)	R	C	PS
12-1	Learn to find terms in an arithmetic sequence.	1-2	2/2								
12-2	Learn to find terms in a geometric sequence.	3-4	2/2								
12-3	Learn to find patterns in sequences.	5-6	2/2								
12-4	Learn to represent functions with tables, graphs, and equations.	7-8	2/2								
12-5	Learn to identify linear functions.	9-10	2/2								
12-6	Learn to identify and graph exponential functions.	11, 16	2/2								
12-7	Learn to identify and graph quadratic functions.	12-13	2/2								
12-8	Learn to recognize inverse variation by graphing tables of data.	14-15	2/2								

Key: P-Practice, **R**-Reteach, **C**-Challenge, **PS**-Problem Solving, **CRB**-Chapter Resource Book

Individual Record Form

Pre-Algebra • Chapter 13 Algebra Toolbox

Student Name _____   Test Level | A | B | C |

	OBJECTIVES	CHAPTER TEST			REVIEW OPTIONS					
Lesson	Learning Goal	Test Items	Criterion Score	Student's Score	SE/TE Lesson	\multicolumn{5}{c}{Workbooks}
						P(A)	P(B)	P(C)	R	C	PS
13-1	Learn to classify polynomials by degree and by the number of terms.	1-8									
13-2	Learn to simplify polynomials.	9-13									
13-3	Learn to add polynomials.	14-20									
13-4	Learn to subtract polynomials.	21-29									
13-5	Learn to multiply polynomials by monomials.	30-34									
13-6	Learn to multiply binomials.	35-40									

Key: P-Practice, **R**-Reteach, **C**-Challenge, **PS**-Problem Solving

Individual Record Form   **Assessment Guide   xxix**

Individual Record Form

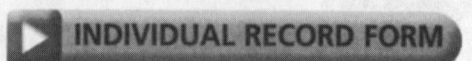

Pre-Algebra • Chapter 14 Integers and Exponents

Student Name _____   Test Level A B C

Lesson	OBJECTIVES - Learning Goal	CHAPTER TEST - Test Items	Criterion Score	Student's Score	SE/TE Lesson	P(A)	P(B)	P(C)	R	C	PS
14-1	Learn to understand mathematical sets and set notation.	1-4									
14-2	Learn to describe the intersection and union of sets.	5-8									
14-3	Learn to make and use Venn diagrams.	9-11									
14-4	Learn to differentiate between conjunctions and disjunctions and make truth tables.	12-13									
14-5	Learn to understand conditional statements and reason deductively.	14-16									
14-6	Learn to find Euler circuits.	17, 19									
14-7	Learn to find and use Hamiltonian circuits.	20-21									

REVIEW OPTIONS — Workbooks

Key: P-Practice, **R-**Reteach, **C-**Challenge, **PS-**Problem Solving

xxx **Assessment Guide**   **Individual Record Form**

Formal Assessment
Class Record Form

CHAPTER TESTS

School											
Teacher											
NAMES	Date										

Formal Assessment
Class Record Form

CUMULATIVE TESTS

School									
Teacher									
NAMES	Date								

Student's Name _____ Date _____ ▶ **PROJECT SCORING RUBRIC**

Project Scoring Rubric

Check the indicators that describe a student's or group's performance on a project. Use the check marks to help determine the individual's or group's overall score.

Score 3 Indicators: The student/group
_____ makes outstanding use of resources.
_____ shows thorough understanding of content.
_____ demonstrates outstanding grasp of mathematics skills.
_____ displays strong decision-making/problem-solving skills.
_____ exhibits exceptional insight/creativity.
_____ communicates ideas clearly and effectively.

Score 2 Indicators: The student/group
_____ makes good use of resources.
_____ shows adequate understanding of content.
_____ demonstrates good grasp of mathematics skills.
_____ displays adequate decision-making/problem-solving skills.
_____ exhibits reasonable insight/creativity.
_____ communicates most ideas clearly and effectively.

Score 1 Indicators: The student/group
_____ makes limited use of resources.
_____ shows partial understanding of content.
_____ demonstrates limited grasp of mathematics skills.
_____ displays weak decision-making/problem-solving skills.
_____ exhibits limited insight/creativity.
_____ communicates some ideas clearly and effectively.

Score 0 Indicators: The student/group
_____ makes little or no use of resources.
_____ fails to show understanding of content.
_____ demonstrates little or no grasp of mathematics skills.
_____ does not display decision-making/problem-solving skills.
_____ does not exhibit insight/creativity.
_____ has difficulty communicating ideas clearly and effectively.

Overall score for the project. _____

Comments: _____
_____

Project _____ Date _____
Group members _____

How Did Our Group Do?

Discuss the question. Then circle the score your group thinks it earned.

SCORE

How well did our group	Great Job	Good Job	Could Do Better
1. share ideas?	3	2	1
2. plan what to do?	3	2	1
3. carry out plans?	3	2	1
4. share the work?	3	2	1
5. solve group problems without seeking help?	3	2	1
6. make use of resources?	3	2	1
7. record information and check for accuracy?	3	2	1
8. show understanding of math ideas?	3	2	1
9. demonstrate creativity and critical thinking?	3	2	1
10. solve the project problem?	3	2	1

Write your group's answer to each question.

11. What did our group do best? _____
_____
_____

12. How can we help our group do better? _____
_____
_____

xxxiv   Assessment Resources   Group Project Evaluation

Name _____ Date _____  INDIVIDUAL GROUP MEMBER EVALUATION

Project _____

How Well Did I Work in My Group?

Circle **yes** if you agree. Circle **no** if you disagree.

1. I shared my ideas with my group.   yes   no
2. I listened to the ideas of others in my group.   yes   no
3. I was able to ask questions of my group.   yes   no
4. I encouraged others in my group to share their ideas.   yes   no
5. I was able to discuss opposite ideas with my group.   yes   no
6. I helped my group plan and make decisions.   yes   no
7. I did my fair share of the group's work.   yes   no
8. I understood the problem my group worked on.   yes   no
9. I understood the solution to the problem my group worked on.   yes   no
10. I can explain to others the problem my group worked on and its solution.   yes   no

Individual Group Member Evaluation Assessment Resources xxxv

▶ PORTFOLIO ASSESSMENT

A portfolio is a collection of each student's work gathered over an extended period of time. A portfolio illustrates the growth, talents, achievements, and reflections of the learner and provides a means for the teacher to assess the student's performance and progress.

Building a Portfolio

There are many opportunities to collect students' work throughout the year as you use *Holt Pre-Algebra*. Suggested portfolio items are found throughout the *Teacher's Edition*. Give students the opportunity to select some work samples to be included in the portfolio.

- Provide a folder for each student with the student's name clearly marked.
- Explain to students that throughout the year they will save some of their work in the folder. Sometimes it will be their individual work; sometimes it will be group reports and projects or completed checklists.
- Have students complete "A Guide to My Math Portfolio" several times during the year.

Evaluating a Portfolio

The following points made with regular portfolio evaluation will encourage growth in self-evaluation:

- Discuss the contents of the portfolio as you examine it with each student.
- Encourage and reward students by emphasizing growth, original thinking, and completion of tasks.
- Reinforce and adjust instruction of the broad goals you want to accomplish as you evaluate the portfolios.
- Examine each portfolio on the basis of individual growth rather than in comparison with other portfolios.
- Use the Portfolio Evaluation sheet for your comments.
- Share the portfolios with families during conferences or send the portfolio, including the Family Response form, home with the students.

Name _____
Date _____

PORTFOLIO GUIDE

A Guide to My Math Portfolio

What is in My Portfolio	What I Learned
1.	
2.	
3.	
4.	
5.	

I organized my portfolio this way because _____
_____
_____

Portfolio Guide (Student)   **Assessment Resources**

Name _____
Date _____

▶ **PORTFOLIO EVALUATION**

Evaluating Performance	**Evidence and Comments**
1. What mathematical understandings are demonstrated?	
2. What skills are demonstrated?	
3. What approaches to problem solving and critical thinking are evident?	
4. What work habits and attitudes are demonstrated?	

Summary of Portfolio Assessment

For This Review			Since Last Review		
Excellent	Good	Fair	Improving	About the Same	Not as Good

xxxviii   Assessment Resources

Portfolio Evaluation (Teacher)

PORTFOLIO FAMILY RESPONSE

Date _____

Dear Family,

    This is your child's math portfolio. It contains work samples that your child and I have selected to show how his or her abilities in math have grown. Your child can explain what each sample shows.

    Please look over the portfolio with your child and write a few comments in the blank space at the bottom of this sheet about what you have seen. Your child has been asked to bring the portfolio with your comments included back to school.

    Thank you for helping your child evaluate his or her portfolio and for taking pride in the work he or she has done. Your interest and support is important to your child's success in school.

Sincerely,

_____
(Teacher)

Response to Portfolio:

_____
(Family member)

▶ ASSESSING PROBLEM SOLVING

Assessing a student's ability to solve problems involves more than checking the student's answer. It involves looking at how students process information and how they work at solving problems. The problem solving method used in *Holt Pre-Algebra*—Understand the Problem, Make a Plan, Solve, and Look Back—guides the student's thinking process and provides a structure within which the student can work toward a solution. The following instruments can help you assess students' problem solving abilities.

- Think Along Oral Response Form, page xxxix
- Oral Response Scoring Guide, page xl
- Think Along Written Response Form, page xli
- Written Response Scoring Guide, page xlii

The **Oral Response Form** (page xxxix) can be used by a student or group as a self-questioning instrument or as a guide for working through a problem. It can also be an interview instrument you can use to assess students' problem solving skills.

The analytic **Scoring Guide for Oral Responses** (page xl) has a criterion score for each chapter. It may be used to evaluate the oral presentation of an individual or group.

The **Think Along Written Response Form** (page xli) provides a recording sheet for a student or group to record their responses as they work through each section of the problem solving process.

The analytic **Written Response Scoring Guide** (page xlii), which gives a criterion score for each section, will help you pinpoint the parts of the problem solving process in which your students need more instruction.

Name _____
Date _____

▶ ORAL RESPONSE FORM

Problem-Solving Think Along: Oral Response Form

Solving problems is a thinking process. Asking yourself questions as you work through the steps in solving a problem can help guide your thinking. These questions will help you understand the problem, plan how to solve it, solve it, and then look back and check your solution. These questions will also help you think about other ways to solve the problem.

Understand the Problem

1. What is the problem about?

2. What is the question?

3. What information is given in the problem?

Make a Plan

4. What problem-solving strategies might I try to help me solve the problem?

5. What is my estimated answer?

Solve

6. How can I solve the problem?

7. How can I state my answer in a complete sentence?

Look Back

8. How do I know whether my answer is reasonable?

9. How else might I have solved this problem?

Think Along, Oral Response Form

Name _____

Date _____

▶ ORAL RESPONSE GUIDE

Problem-Solving Think Along:
Scoring Guide • Oral Responses

Understand the Problem Criterion Score 4/6 Pupil Score _____

_____ 1. Restate the problem in his or her own words.
 2 points Complete problem restatement given.
 1 point Incomplete problem restatement.
 0 points No restatement given.

_____ 2. Identify the question.
 2 points Complete problem restatement of the question given.
 1 point Incomplete problem restatement of the question given.
 0 points No restatement of the question given.

_____ 3. State list of information needed to solve the problem.
 2 points Complete list given.
 1 point Incomplete list given.
 0 points No list given.

Make a Plan Criterion Score 3/4 Pupil Score _____

_____ 1. State one or more strategies that might help solve the problem.
 2 points One or more useful strategies given.
 1 point One or more strategies given but are poor choices.
 0 points No strategies given.

_____ 2. State reasonable estimated answer.
 2 points Reasonable estimate given.
 1 point Unreasonable estimate given.
 0 points No estimated answer given.

Solve Criterion Score 3/4 Pupil Score _____

_____ 1. Describe a solution method that correctly represents the information in the problem.
 2 points Correct solution method given.
 1 point Incorrect solution method given.
 0 points No solution method given.

_____ 2. State correct answer in complete sentence.
 2 points Complete sentence given; answer to question is correct.
 1 point Sentence given does not answer the question correctly.
 0 points No sentence given.

Look Back Criterion Score 3/4 Pupil Score _____

_____ 1. State sentence explaining why the answer is reasonable.
 2 points Complete and correct explanation given.
 1 point Sentence given with incomplete or incorrect reason.
 0 points No solution method given.

_____ 2. Describe another strategy that could have been used to solve the problem.
 2 points Another useful strategy described.
 1 point Another strategy described, but strategy is a poor choice.
 0 points No other strategy described.

TOTAL 13/18 Pupil Score _____

xlii Assessment Resources Oral Response Scoring Guide

Name _____
Date _____

WRITTEN RESPONSE FORM

Problem Solving

Understand the Problem

1. Retell the problem in your own words. _____
   _____
   _____
   _____

2. Restate the question as a fill-in-the-blank sentence. _____
   _____

3. List the information needed to solve the problem. _____
   _____

Make a Plan

4. List one or more problem-solving strategies that you can use. _____
   _____

5. Predict what your answer will be. _____
   _____

Solve

6. Show how you solved the problem. _____
   _____

7. Write your answer in a complete sentence. _____
   _____

Look Back

8. Tell how you know your answer is reasonable. _____
   _____

9. Describe another way you could have solved the problem. _____
   _____

Think Along, Written Response   **Assessment Resources   xliii**

> **WRITTEN RESPONSE GUIDE**

Assessing Problem Solving
Scoring Guide • Written Responses

Understand the Problem
Indicator 1:
Student restates the problem in his or her own words.

Criterion Score 4/6
Scoring:
- 2 points Complete problem restatement written.
- 1 point   Incomplete problem restatement written.
- 0 points No restatement written.

Indicator 2:
Student restates the question as a fill-in-the-blank statement.

- 2 points Correct restatement of the question.
- 1 point   Incorrect or incomplete restatement.
- 0 points No restatement written.

Indicator 3:
Student writes a complete list of the information needed to solve the problem.

- 2 points Complete list made.
- 1 point   Incomplete list made.
- 0 points No list made.

Make a Plan
Indicator 1:
Student lists one or more problem-solving strategies that might be helpful in solving the problem.

Criterion Score 3/4
Scoring:
- 2 points One or more useful strategies listed.
- 1 point   One or more strategies listed, but strategies are poor choices.
- 0 points No strategies listed.

Indicator 2:
Student gives a reasonable estimated answer.

- 2 points Reasonable estimate given.
- 1 point   Unreasonable estimate given.
- 0 points No estimated answer given.

Solve
Indicator 1:
Student shows a solution method that correctly represents the information in the problem.

Criterion Score 3/4
Scoring:
- 2 points Correct solution method written.
- 1 point   Incorrect solution method written.
- 0 points No solution method written.

Indicator 2:
Student writes a complete sentence giving the correct answer.

- 2 points Sentence has correct answer and completely answers the question.
- 1 point   Sentence has an incorrect numerical answer or does not answer the question.
- 0 points No sentence written.

Look Back
Indicator 1:
Student writes a sentence explaining why the answer is reasonable.

Criterion Score 3/4
Scoring:
- 2 points Gives a complete and correct explanation.
- 1 point   Gives an incomplete or incorrect reason.
- 0 points No sentence written.

Indicator 2:
Student describes another strategy that could have been used to solve the problem.

- 2 points Another useful strategy described.
- 1 point   Another strategy described, but it is a poor choice.
- 0 points No other strategy described.

TOTAL 13/18

▶ Test-Taking Tips

Being a good test taker is like being a good problem solver. When you answer test questions, you are solving problems. Remember to **UNDERSTAND THE PROBLEM, MAKE A PLAN, SOLVE,** and **LOOK BACK**.

―――――――――――――――――― **Understand the Problem** ――――――――――――――――――

Read the problem.
- Look for math terms and recall their meanings.
- Reread the problem and think about the question.
- Use the details in the problem and the question.
- Each word is important. Missing a word or reading it incorrectly could cause you to get the wrong answer.
- Pay attention to words that are in **bold** type, all CAPITAL letters, or *italics*.
- Some other words to look for are round, about, only, best, or least to greatest.

―――――――――――――――――― **Plan** ――――――――――――――――――

Think about how you can solve the problem.
- Can you solve the problem with the information given?
- Pictures, charts, tables, and graphs may have the information you need.
- Sometimes you may need to remember some information that is not given.
- Sometimes the answer choices have information to help you solve the problem.
- You may need to write a number sentence and solve it to answer the question.
- Some problems have two steps or more.
- In some problems you may need to look at relationships instead of computing an answer.
- If the path to the solution isn't clear, choose a problem-solving strategy.
- Use the strategy you chose to solve the problem.

Follow your plan, working logically and carefully.
- Estimate your answer. Look for unreasonable answer choices.
- Use reasoning to find the most likely choices.
- Make sure you solved all the steps needed to answer the problem.
- If your answer does not match one of the answer choices, check the numbers you used. Then check your computation.

Solve

Solve the problem.
- If your answer still does not match one of the choices, look for another form of the number such as decimals instead of fractions.
- If answer choices are given as pictures, look at each one by itself while you cover the other three.
- If you do not see your answer and the answer choices include NOT HERE, make sure your work is correct and then mark NOT HERE.
- Read answer choices that are statements and relate them to the problem one by one.
- Change your plan if it isn't working. You may need to try a different strategy.

Look Back

Take time to catch your mistakes.
- Be sure you answered the question asked.
- Check that your answer fits the information in the problem.
- Check for important words you may have missed.
- Be sure you used all the information you needed.
- Check your computation by using a different method.
- Draw a picture when you are unsure of your answer.

Don't forget!

Before the Test

- Listen to the teacher's directions and read the instructions.
- Write down the ending time if the test is timed.
- Know where and how to mark your answers.
- Know whether you should write on the test page or use scratch paper.
- Ask any questions you have before the test begins.

During the Test

- Work quickly but carefully. If you are unsure how to answer a question, leave it blank and return to it later.
- If you cannot finish on time, look over the questions that are left. Answer the easiest ones first. Then go back to the others.
- Fill in each answer space carefully and completely. Erase completely if you change an answer. Erase any stray marks.

Name _____   Date _____

Test Answer Sheet

Test Title _____

1. Ⓐ Ⓑ Ⓒ Ⓓ
2. Ⓕ Ⓖ Ⓗ Ⓙ
3. Ⓐ Ⓑ Ⓒ Ⓓ
4. Ⓕ Ⓖ Ⓗ Ⓙ
5. Ⓐ Ⓑ Ⓒ Ⓓ

6. Ⓕ Ⓖ Ⓗ Ⓙ
7. Ⓐ Ⓑ Ⓒ Ⓓ
8. Ⓕ Ⓖ Ⓗ Ⓙ
9. Ⓐ Ⓑ Ⓒ Ⓓ
10. Ⓕ Ⓖ Ⓗ Ⓙ

11. Ⓐ Ⓑ Ⓒ Ⓓ
12. Ⓕ Ⓖ Ⓗ Ⓙ
13. Ⓐ Ⓑ Ⓒ Ⓓ
14. Ⓕ Ⓖ Ⓗ Ⓙ
15. Ⓐ Ⓑ Ⓒ Ⓓ

16. Ⓕ Ⓖ Ⓗ Ⓙ
17. Ⓐ Ⓑ Ⓒ Ⓓ
18. Ⓕ Ⓖ Ⓗ Ⓙ
19. Ⓐ Ⓑ Ⓒ Ⓓ
20. Ⓕ Ⓖ Ⓗ Ⓙ

21. Ⓐ Ⓑ Ⓒ Ⓓ
22. Ⓕ Ⓖ Ⓗ Ⓙ
23. Ⓐ Ⓑ Ⓒ Ⓓ
24. Ⓕ Ⓖ Ⓗ Ⓙ
25. Ⓐ Ⓑ Ⓒ Ⓓ

26. Ⓕ Ⓖ Ⓗ Ⓙ
27. Ⓐ Ⓑ Ⓒ Ⓓ
28. Ⓕ Ⓖ Ⓗ Ⓙ
29. Ⓐ Ⓑ Ⓒ Ⓓ
30. Ⓕ Ⓖ Ⓗ Ⓙ

31. Ⓐ Ⓑ Ⓒ Ⓓ
32. Ⓕ Ⓖ Ⓗ Ⓙ
33. Ⓐ Ⓑ Ⓒ Ⓓ
34. Ⓕ Ⓖ Ⓗ Ⓙ
35. Ⓐ Ⓑ Ⓒ Ⓓ

36. Ⓕ Ⓖ Ⓗ Ⓙ
37. Ⓐ Ⓑ Ⓒ Ⓓ
38. Ⓕ Ⓖ Ⓗ Ⓙ
39. Ⓐ Ⓑ Ⓒ Ⓓ
40. Ⓕ Ⓖ Ⓗ Ⓙ

41. Ⓐ Ⓑ Ⓒ Ⓓ
42. Ⓕ Ⓖ Ⓗ Ⓙ
43. Ⓐ Ⓑ Ⓒ Ⓓ
44. Ⓕ Ⓖ Ⓗ Ⓙ
45. Ⓐ Ⓑ Ⓒ Ⓓ

46. Ⓕ Ⓖ Ⓗ Ⓙ
47. Ⓐ Ⓑ Ⓒ Ⓓ
48. Ⓕ Ⓖ Ⓗ Ⓙ
49. Ⓐ Ⓑ Ⓒ Ⓓ
50. Ⓕ Ⓖ Ⓗ Ⓙ

HOLT PRE-ALGEBRA

Test Answer Sheet

Name _____ Date _____ **HOLT PRE-ALGEBRA**

Test Answer Sheet for Form A Tests

Test Title _____

1. Ⓐ Ⓑ Ⓒ Ⓓ		26. Ⓐ Ⓑ Ⓒ Ⓓ	
2. Ⓐ Ⓑ Ⓒ Ⓓ		27. Ⓐ Ⓑ Ⓒ Ⓓ	
3. Ⓐ Ⓑ Ⓒ Ⓓ		28. Ⓐ Ⓑ Ⓒ Ⓓ	
4. Ⓐ Ⓑ Ⓒ Ⓓ		29. Ⓐ Ⓑ Ⓒ Ⓓ	
5. Ⓐ Ⓑ Ⓒ Ⓓ		30. Ⓐ Ⓑ Ⓒ Ⓓ	
6. Ⓐ Ⓑ Ⓒ Ⓓ		31. Ⓐ Ⓑ Ⓒ Ⓓ	
7. Ⓐ Ⓑ Ⓒ Ⓓ		32. Ⓐ Ⓑ Ⓒ Ⓓ	
8. Ⓐ Ⓑ Ⓒ Ⓓ		33. Ⓐ Ⓑ Ⓒ Ⓓ	
9. Ⓐ Ⓑ Ⓒ Ⓓ		34. Ⓐ Ⓑ Ⓒ Ⓓ	
10. Ⓐ Ⓑ Ⓒ Ⓓ		35. Ⓐ Ⓑ Ⓒ Ⓓ	
11. Ⓐ Ⓑ Ⓒ Ⓓ		36. Ⓐ Ⓑ Ⓒ Ⓓ	
12. Ⓐ Ⓑ Ⓒ Ⓓ		37. Ⓐ Ⓑ Ⓒ Ⓓ	
13. Ⓐ Ⓑ Ⓒ Ⓓ		38. Ⓐ Ⓑ Ⓒ Ⓓ	
14. Ⓐ Ⓑ Ⓒ Ⓓ		39. Ⓐ Ⓑ Ⓒ Ⓓ	
15. Ⓐ Ⓑ Ⓒ Ⓓ		40. Ⓐ Ⓑ Ⓒ Ⓓ	
16. Ⓐ Ⓑ Ⓒ Ⓓ		41. Ⓐ Ⓑ Ⓒ Ⓓ	
17. Ⓐ Ⓑ Ⓒ Ⓓ		42. Ⓐ Ⓑ Ⓒ Ⓓ	
18. Ⓐ Ⓑ Ⓒ Ⓓ		43. Ⓐ Ⓑ Ⓒ Ⓓ	
19. Ⓐ Ⓑ Ⓒ Ⓓ		44. Ⓐ Ⓑ Ⓒ Ⓓ	
20. Ⓐ Ⓑ Ⓒ Ⓓ		45. Ⓐ Ⓑ Ⓒ Ⓓ	
21. Ⓐ Ⓑ Ⓒ Ⓓ		46. Ⓐ Ⓑ Ⓒ Ⓓ	
22. Ⓐ Ⓑ Ⓒ Ⓓ		47. Ⓐ Ⓑ Ⓒ Ⓓ	
23. Ⓐ Ⓑ Ⓒ Ⓓ		48. Ⓐ Ⓑ Ⓒ Ⓓ	
24. Ⓐ Ⓑ Ⓒ Ⓓ		49. Ⓐ Ⓑ Ⓒ Ⓓ	
25. Ⓐ Ⓑ Ⓒ Ⓓ		50. Ⓐ Ⓑ Ⓒ Ⓓ	

Assessment Resources

Name _____ Date _____ Class _____

Inventory Test

Select the best answer for questions 1–40.

1. Twice a number, increased by 28 equals 48 plus the number. What is the number?
   - **A** 21
   - **B** 20
   - **C** 18
   - **D** −18

2. An airplane flies at a cruising altitude of 2900 ft. It descends 1200 ft as it begins to reach its destination. As it approaches the airport, it descends an additional 1400 ft. What is the new altitude of the airplane?
   - **F** 1700 ft
   - **G** 1500 ft
   - **H** 300 ft
   - **J** 150 ft

3. If you can type 1,240 words in 40 minutes, how many words per minute can you type?
   - **A** 180 words/min
   - **B** 31 words/min
   - **C** 27 words/min
   - **D** 9 words/min

4. Solve the inequality $-5x < -45$.
   - **F** $x > 9$
   - **G** $x < -5$
   - **H** $x < 9$
   - **J** $x < -9$

5. Evaluate the expression $-|7 - y|$ for $y = -6$.
   - **A** −1
   - **B** −13
   - **C** 1
   - **D** 13

6. Which is equivalent to $(-4)^4(-4)^9$.
   - **F** $(16)^{36}$
   - **G** $(-4)^{13}$
   - **H** $(16)^{13}$
   - **J** $(-4)^{36}$

7. Express 680,000 using scientific notation.
   - **A** 6.8×10^5
   - **B** 6.8×10^4
   - **C** 6.8×10^{-5}
   - **D** 6.8×10^{-4}

8. Simplify $6 \times 1\frac{9}{10}$.
   - **F** $6\frac{9}{10}$
   - **G** $\frac{25}{11}$
   - **H** $11\frac{2}{5}$
   - **J** $\frac{79}{10}$

9. Divide $\frac{6}{7} \div \frac{4}{5}$.
   - **A** $\frac{3}{28}$
   - **B** $\frac{24}{35}$
   - **C** $\frac{14}{17}$
   - **D** $\frac{15}{14}$

10. Which is equivalent to $9\sqrt{36 + 64}$?
    - **F** 90
    - **G** 126
    - **H** 30
    - **J** 23

11. A potential candidate for mayor wants to know whether she is positioned to launch a successful campaign. 25 residents chosen at random from each of 10 randomly identified city neighborhoods are asked about their opinions on 10 key issues. Identify the sampling method used.
    - **A** biased
    - **B** random
    - **C** systematic
    - **D** stratified

12. To the nearest tenth, find the mean weight of a group of apples with individual weights of 4.7, 4.0, 6.2, 6.5, 6.1, 4.7, 4.0, 6.2, 6.3, 6.5, 4.7, 6.2, 6.5, and 6.0 ounces.
    - **F** 4.7 oz
    - **G** 6.5 oz
    - **H** 5.6 oz
    - **J** 6.2 oz

Holt Pre-Algebra

Name _____ Date _____ Class _____

Inventory Test

13. Six employees of a small company were asked the distance they traveled to work, to the nearest tenth of a mile. Find the range of the distances: 2.7, 5.9, 1.7, 4.7, 6.9, 3.5.
    A 0.8 mi   C 5.2 mi
    B 1.7 mi   D 5.9 mi

14. Identify the pair of angles that are complementary.

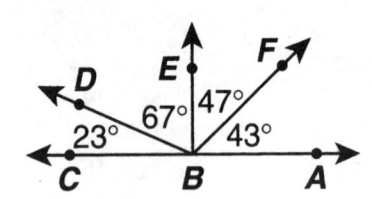

    F ∠CBE & ∠CBD
    G ∠ABF & ∠CBD
    H ∠DBE & ∠EBF
    J ∠ABF & ∠FBE

15. Which angle is congruent to ∠5?

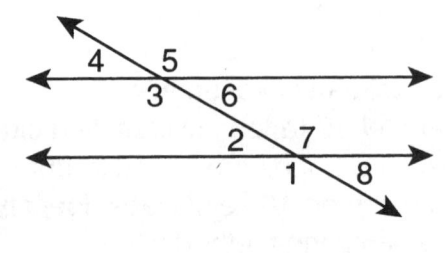

    A ∠6   C ∠7
    B ∠8   D ∠2

16. A decagon has ten sides. What is the sum of the interior angles in a decagon?
    F 1,440°
    G 1,800°
    H 2,160°
    J Unknown. The interior angle formula only applies to regular polygons.

17. The diameter of an ice-hockey puck is 3.0 inches. To the nearest tenth, what is the area of the flat upper surface? Use 3.14 for π.
    A 3.5 in^2 C 7.1 in^2
    B 9.4 in^2 D 28.3 in^2

18. Find the missing length in the right triangle.

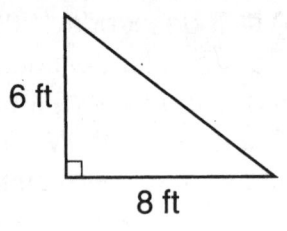

    F 7 ft   H 9 ft
    G 8 ft   J 10 ft

19. Find the volume of the pyramid.

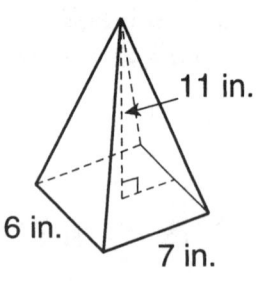

    A 462 in^3 C 154 in^3
    B 242 in^3 D 14 in^3

20. To estimate the number of fish in a lake, biologists catch, mark, and release 50 fish. One week later, they catch 33 fish and find that 15 of them are marked. What is their best estimate for the total number of fish in the lake?
    F 110   H 83
    G 73   J 60

Copyright © by Holt, Rinehart and Winston.
All rights reserved.

Holt Pre-Algebra

Name _____ Date _____ Class _____

Inventory Test

21. Convert 8 weeks to minutes.
  A 80,640 min   **C** 40,320 min
  B 57,600 min   **D** 10,080 min

22. Find the missing value in the proportion $\frac{4}{8} = \frac{x}{64}$.
  F $x = 4$ **H** $x = 32$
  G $x = 12$ **J** $x = 256$

23. An interior decorator is making a scale drawing of a room. If the scale is 2 in. : 3 ft, how wide is the drawing of 7.5-ft bay window?
  A 5 in.   **C** 15 in.
  B 11.25 in. **D** 2.5 in.

24. What is $6\frac{1}{5}\%$ of 8000?
  F 500   **H** 4,960
  G 496   **J** 49,600

25. 56 is 140% of what number?
  A 40   **C** 400
  B 78.4   **D** 19,600

26. A computer, priced at $1,380, is marked down to $1,057. What is the rate of markdown?
  F 22.7%   **H** 29.6%
  G 23.4%   **J** 30.6%

27. A six-sided die is rolled twice. What is the probability of rolling the same number twice in a row?
  F $\frac{1}{6}$ **H** $\frac{2}{3}$
  G $\frac{1}{36}$ **J** $\frac{1}{3}$

28. Find the value of $_{10}C_2$.
  A 90   **C** 40,320
  B 45   **D** 80,640

29. In a certain town, 4% of the population commutes to work by bicycle. If a person is randomly selected from the town, what are the odds against selecting someone who commutes by bicycle?
  A 24:25   **C** 1:25
  B 24:1   **D** 1:24

30. Solve $-7c + 6 = 2 + 8c$.
  F $c = \frac{15}{4}$ **H** $c = \frac{4}{15}$
  G $c = -\frac{15}{4}$ **J** $c = 8$

31. Find the inequality that corresponds to the given number line.

  A $2x + 3 < 11$ **C** $2x + 3 \geq 11$
  B $2x + 3 > 11$ **D** $2x + 3 \leq 11$

32. Solve the formula $A = \frac{bh}{2}$ for b.
  F $b = \frac{A}{2h}$ **H** $b = \frac{2A}{h}$
  G $b = \frac{Ah}{2}$ **J** $b = \frac{h}{2A}$

33. Solve the system of equations:
  $2x + 2y = 18$
  $6x + y = 39$
  A $(-6, 2)$ **C** $(7, 2)$
  B $(6, 3)$ **D** $(-2, 11)$

Holt Pre-Algebra

Name _____ Date _____ Class _____

Inventory Test

34. Find the rule for the linear function.

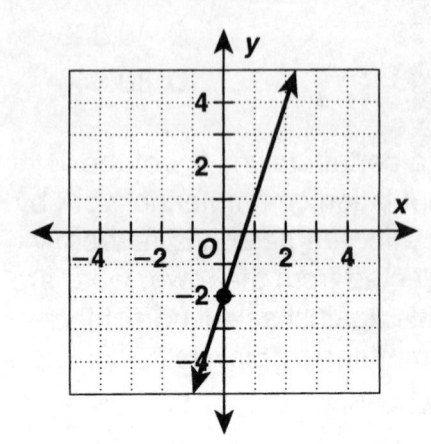

F $f(x) = 3x - 2$ **H** $f(x) = -3x - 2$
G $f(x) = 3x + 2$ **J** $f(x) = x - 2$

35. What is the point-slope form of the equation for the line with a slope of $\frac{2}{3}$ that passes through (0, 2).

A $y - 2 = \frac{2}{3}x$ **C** $y - 3 = \frac{3}{2}x$
B $y + 2 = \frac{2}{3}x$ **D** $y + 2 = -\frac{2}{3}x$

36. Given that y varies directly with x, find the equation of direct variation if y is 10 when x is 5.

F $y = x + 5$ **H** $y = \frac{50}{x}$
G $y = 2x$ **J** $y = 0.5x$

37. Find the 12th term in the arithmetic sequence 2, 7, 12, 17, ...

A 62   **C** 57
B 60   **D** 55

38. Complete the ordered pairs in the table for the equation $y = -x + 8$.

x	3	8	0.5
y	??	??	??

F 15, 0, 8.5   **H** 3, 8, 0.5
G 5, −2, 2.5   **J** 5, 0, 7.5

39. A colony of bacteria triples in size every 4 days. If there were originally 27 cells, how many cells are there after 4 weeks?

A 729   **C** 19,683
B 2187   **D** 59,049

40. Find $f(4)$ for the quadratic function $f(x) = 3x^2 - x - 5$.

F 3   **H** 39
G 15   **J** 135

Name _____ Date _____ Class _____

CHAPTER 1 Quiz
Section A

Choose the best answer.

1. Evaluate $2x + 8$ for $x = 5$.
   - **A** 10
   - **B** 18
   - **C** 28
   - **D** 32

2. Evaluate $2.5r + 12$ for $r = 6$.
   - **A** 24
   - **B** 162
   - **C** 27
   - **D** 44

3. Evaluate $4a + 7c$ for $a = 5$ and $c = 3$.
   - **A** 118
   - **B** 41
   - **C** 30
   - **D** 19

4. Which algebraic expression represents "4 times the sum of 12 and b"?
   - **A** $4 + 12 + b$
   - **B** $4 + (12 + b)$
   - **C** $4(12 + b)$
   - **D** $4(12 - b)$

5. Which algebraic expression represents "3 less than the sum of 5 and r"?
   - **A** $(5 \cdot r) - 3$
   - **B** $(5 + r) - 3$
   - **C** $3 - (5 + r)$
   - **D** $(5 - r) - 3$

6. Which value of z is the solution for the equation $43 - z = 18$?
   - **A** $z = 25$
   - **B** $z = 15$
   - **C** $z = 13$
   - **D** $z = 61$

7. Which value of t is the solution for the equation $9.45 = t + 3.7$?
   - **A** $t = 13.15$
   - **B** $t = 6.2$
   - **C** $t = 5.75$
   - **D** $t = 6.12$

8. What is the value of k for this equation: $\frac{k}{8} = 12$?
   - **A** $k = 96$
   - **B** $k = 20$
   - **C** $k = 66$
   - **D** $k = \frac{3}{2}$

9. What is the value of m for this equation: $4m - 15 = 33$?
   - **A** $m = 48$
   - **B** $m = 52$
   - **C** $m = 60$
   - **D** $m = 12$

10. Which inequality is represented by this graph?

    7 8 9 10 11 12 13 14 15 16 17

    - **A** $x + 3 > 10$
    - **B** $t - 3 \leq 22$
    - **C** $y + 5 > 15$
    - **D** $5 > \frac{w}{2}$

11. Simplify $3(2x - 5) + 2x$.
    - **A** $6x - 15 + 2x$
    - **B** $8x - 15$
    - **C** $7x - 5$
    - **D** $6x - 15$

12. Solve $6d + 4 + 5d - 2d = 58$.
    - **A** $d = 13d$
    - **B** $d = 10$
    - **C** $d = 6$
    - **D** $d = 13d + 4$

Holt Pre-Algebra

Name _____ Date _____ Class _____

CHAPTER 1 Quiz
Section B

Choose the best answer.

1. Which ordered pair is a solution for $y = 2x + 8$?
   A (1, 20)   C (3, 15)
   B (4, 12)   D (4, 16)

2. Which ordered pair is a solution for $y = 5x - 3$?
   A (17, 4)   C (4, 17)
   B (5, 28)   D (3, 18)

3. The price of a 5-line newspaper ad is $8, plus $1.50 for each additional line. The equation for price p for buying an ad is $p = 8 + 1.50l$. What is the cost of an ad with 4 extra lines?
   A $14   C $11.50
   B $13   D $9.50

4. What ordered pair is missing from the table below for $y = 2x - 5$?

x	2x	y	(x, y)
8	2(8)	11	(8, 11)
9	2(9)	13	(9, 13)
10	2(10)	15	(,)
11	2(11)	17	(11, 17)

   A (10, 15)   C (10, 14)
   B (10, 12)   D (15, 11)

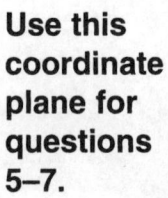

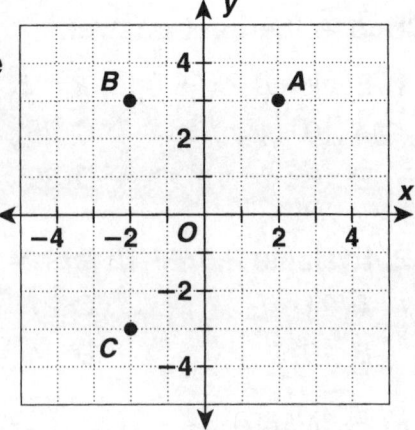

Use this coordinate plane for questions 5–7.

5. Identify the coordinates for point B.
   A (−2, −3)   C (−2, 3)
   B (2, 3)   D (2, 3)

6. Identify the coordinates for point A.
   A (3, 3)   C (2, 2)
   B (2, 3)   D (−2, −3)

The table shows how many seconds it takes for a garage door to open.

Time (seconds)	Door Opening (feet)
0	0
2	1
4	2
6	3
8	4
10	5
12	6
14	7
16	8

7. How many seconds does it take for the garage door to be half-way up?
   A 4 seconds   C 6 seconds
   B 8 seconds   D 16 seconds

Name _____ Date _____ Class _____

CHAPTER 2 Quiz
Section A

Add.

1. $6 + (-9)$
   - A −15
   - B −3
   - C 3
   - D 15

Evaluate the expression.

2. $11 + d + (-4)$ for $d = -6$
   - A −9
   - B −1
   - C 1
   - D 9

Subtract.

3. $-12 - (-4)$
   - A −16
   - B −8
   - C 8
   - D 16

Multiply or divide.

4. $-7(-2)$
   - A −14
   - B −5
   - C 5
   - D 14

5. $\dfrac{-9(11)}{-3}$
   - A −33
   - B 27
   - C 33
   - D 99

Solve.

6. $-17 + v = 3$
   - A $v = -14$
   - B $v = 3$
   - C $v = 14$
   - D $v = 20$

7. $8r = -48$
   - A $r = -8$
   - B $r = -6$
   - C $r = 6$
   - D $r = 384$

8. $\dfrac{h}{6} = -7$
   - A $h = -42$
   - B $h = -13$
   - C $h = 13$
   - D $h = 42$

9. $4 - p = 12$
   - A $p = -16$
   - B $p = -8$
   - C $p = 8$
   - D $p = 24$

10. $\dfrac{k}{5} < -2$
    - A $k < -10$
    - B $k < -7$
    - C $k < 7$
    - D $k < 10$

Copyright © by Holt, Rinehart and Winston.
All rights reserved.

Holt Pre-Algebra

Name _____ Date _____ Class _____

CHAPTER 2 **Quiz**
Section B

Write using exponents.

1. $n \cdot n \cdot n \cdot n \cdot n$
   - **A** n^3
   - **B** n^4
   - **C** n^5
   - **D** n^6

Evaluate.

2. $(-3)^4$
   - **A** -81
   - **B** -12
   - **C** 12
   - **D** 81

Simplify.

3. $14 + (-2 - (-2)^3)$
   - **A** -4
   - **B** 4
   - **C** 20
   - **D** 24

4. $(6 \cdot 2)^2 + 4$
   - **A** 28
   - **B** 148
   - **C** 164
   - **D** 576

Multiply or divide. Write the product as one power.

5. $4^3 \cdot 4^0 \cdot 4^2 \cdot 4^5$
   - **A** 4^9
   - **B** 4^{10}
   - **C** 4^{11}
   - **D** 256^{10}

6. $\dfrac{f^7}{g^3}$
   - **A** fg^4
   - **B** fg^{10}
   - **C** $\dfrac{f^4}{g}$
   - **D** Cannot simplify

Simplify.

7. 10^{-4}
   - **A** 0.00001
   - **B** 0.0001
   - **C** 1000
   - **D** $10,000$

8. $\dfrac{6^5}{6^3}$
   - **A** 1
   - **B** 12
   - **C** 36
   - **D** 216

Write the number in standard notation.

9. 1.42×10^{-2}
   - **A** 0.00142
   - **B** 0.0142
   - **C** 1.42
   - **D** 142

Write the number in scientific notation.

10. 0.000721
    - **A** 0.721×10^{-5}
    - **B** 7.21×10^{-5}
    - **C** 7.21×10^{-4}
    - **D** 721×10^{-4}

Holt Pre-Algebra

Name _____ Date _____ Class _____

CHAPTER 3 Quiz
Section A

Choose the best answer.

1. Simplify the fraction $\frac{42}{90}$.

   A $\frac{22}{45}$ C $\frac{9}{15}$

   B $\frac{7}{15}$ D $\frac{16}{30}$

Add or subtract. Express in simplest form.

2. $\frac{7}{10} + \frac{9}{10}$

   F $\frac{1}{5}$ H $1\frac{1}{3}$

   G $1\frac{1}{5}$ J $1\frac{3}{5}$

3. $\frac{8}{9} - \frac{2}{9}$

   A $\frac{6}{9}$ C $\frac{10}{9}$

   B $\frac{2}{3}$ D $1\frac{1}{9}$

4. Multiply $-0.02(32.7)$.

   F -0.654 H 0.654

   G -6.54 J 6.54

5. Evaluate $3\frac{1}{5}n$ for $n = \frac{3}{4}$.

   A $1\frac{5}{9}$ C $3\frac{1}{9}$

   B $2\frac{2}{5}$ D $3\frac{3}{20}$

6. Divide $3\frac{3}{5} \div 6$. Express in simplest form.

   F $\frac{3}{5}$ H $\frac{108}{5}$

   G $18\frac{3}{5}$ J $6\frac{11}{5}$

Add or subtract.

7. $\frac{6}{7} - \frac{4}{5}$

   A $1\frac{23}{25}$ C 1

   B $\frac{1}{12}$ D $\frac{2}{35}$

8. $\frac{3}{8} + \frac{1}{7}$

   F $\frac{29}{56}$ H $\frac{18}{56}$

   G $-\frac{4}{15}$ J $-\frac{3}{56}$

9. Solve $x + \frac{4}{15} = \frac{2}{3}$.

   A $x = \frac{14}{15}$ C $x = \frac{2}{5}$

   B $x = -\frac{6}{15}$ D $x = -\frac{14}{15}$

10. Solve $3.7 + y > 4.9$.

    F $y < -1.2$ H $y < 1.2$

    G $y > -1.2$ J $y > 1.2$

Holt Pre-Algebra

Name _____ Date _____ Class _____

CHAPTER 3 Quiz
Section B

Choose the best answer.

Simplify each expression.

1. $\sqrt{9+16}$
   - A 4
   - B 5
   - C 7
   - D 25

2. $\sqrt{64} - \sqrt{36}$
   - F 10
   - G 0
   - H 2
   - J 5.29

3. $11 + \sqrt{25}$
   - A 36
   - B 6
   - C 12
   - D 16

Use a calculator to find each square root. Round to the nearest tenth.

4. $\sqrt{132}$
   - F 11.2
   - G 11.5
   - H 11.8
   - J 14.9

5. $-\sqrt{47}$
   - A −6.9
   - B −6.2
   - C 6.4
   - D 6.9

6. $\sqrt{56.4}$
   - A 5.7
   - B 6.2
   - C 7.5
   - D 8.4

7. What kind of number is 0?
   - F rational
   - G irrational
   - H not a real number
   - J negative number

8. What kind of number is $-\sqrt{2}$?
   - A rational
   - B irrational
   - C not a real number
   - D natural number

Name _____ Date _____ Class _____

CHAPTER 4 Quiz
Section A

Choose the best answer.

1. Identify the population and sample in the following example. A convenience store owner wants to know which soda-pop is the most popular in his store. He surveys 30 random customers and notes which soda they buy.

   A All people; customers who buy soda
   B Customers who buy soda; people who shop at the store
   C People who shop at the store; customers who are surveyed
   D People who shop at the store; customers who buy soda

2. Identify the sampling method used in the following example. A door-to-door poll-taker stops at every third house.

   F Random
   G Systematic
   H Stratified
   J Not a sample method

3. Which set of numbers represent the data values in the stem-and-leaf plot?

Stem	Leaves
5	1 4
6	0 5 8
7	
8	7 7 9

   A 51, 54, 60, 64, 68, 87, 87, 89
   B 51, 54, 60, 65, 68, 68, 87, 87
   C 50, 54, 60, 65, 68, 87, 87, 89
   D 51, 54, 60, 65, 68, 87, 87, 89

4. Find the mean, median, and mode of the data set: 15, 7, 9, 12, 21, 11, 13, 12, 18

   F mean: 12, median: 12, mode: 13.1
   G mean: 13.1, median: 12, mode: 12
   H mean: 12, median: 13.1, mode: 12
   J mean: 13.5, median: 12, mode: 12

5. Find the range and the first and third quartiles of the data set:
   42, 33, 47, 50, 37, 51, 49, 35, 52, 48, 53

   A range: 18, 1st: 42, 3rd: 50
   B range: 20, 1st: 40, 3rd: 49
   C range: 20, 1st: 37, 3rd: 51
   D range: 18, 1st: 38, 3rd: 51

6. Use the box-and-whiskers graph to find the third quartile.

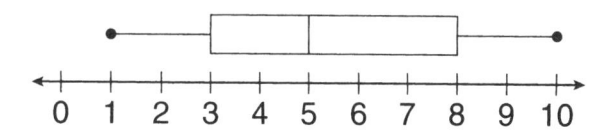

   F 5
   G 8
   H 10
   J 9

Name _____ Date _____ Class _____

CHAPTER 4 Quiz
Section B

Choose the best answer.

1. What is the frequency of the data value 8 in the data set below?
   9, 7, 8, 6, 7, 6, 5, 9, 8, 6, 9, 8, 7
   A 0   C 1
   B 2   D 3

2. Use the following line graph to estimate the number of units sold for the month of August.

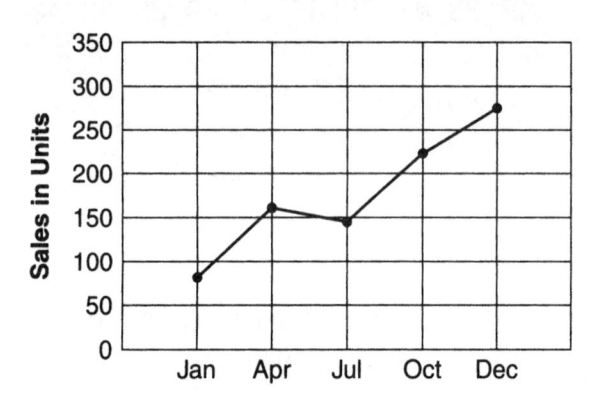

   F 150   H 140
   G 175   J 200

3. What is the interval used in the following histogram?

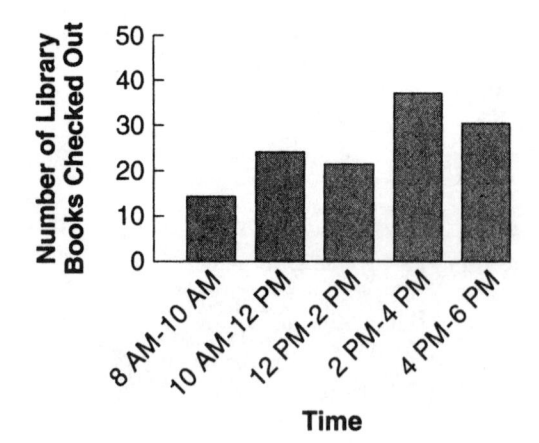

   A 1 day   C 10 hours
   B 2 hours   D 1 week

4. Which of the following statements is not misleading?
   F A little league baseball player hits the ball 55% percent of the time, while a professional player hits the ball 33% of the time. Little-leaguers are better hitters than professionals.
   G A basketball player makes 75% of her shots during a single game. She is obviously a great shooter.
   H A hockey team scores no points during one game. The best the team can do is tie the game.
   J The kicker for the school football team scored more points than any other player on the team for the season. He is the best player on the team.

5. Which of the following data sets has a negative correlation?
   A The total distance driven and the amount of tread on the tires.
   B The number of miles driven and the amount of gas used.
   C The number of passengers in a car and the number of driver licenses.
   D The speed of the car and the rotation rate of its tires.

Copyright © by Holt, Rinehart and Winston.
All rights reserved.

Holt Pre-Algebra

Name _____ Date _____ Class _____

CHAPTER 5 Quiz
Section A

Choose the best answer.

1. Which of the following is not true for the figure below?

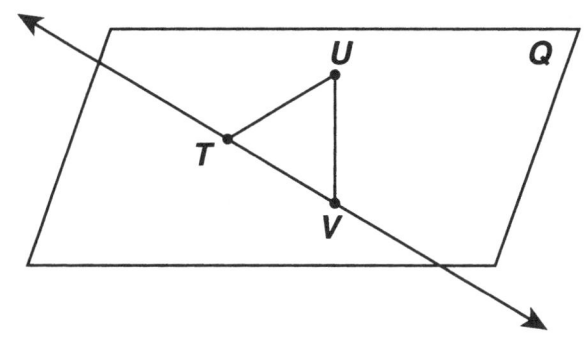

   A plane TUV, point T, point U, point V
   B plane Q, point T, point U, point V
   C $\overleftrightarrow{TQ}$, $\overline{TU}$, $\overline{UV}$
   D $\overleftrightarrow{TV}$, $\overline{TU}$, $\overline{UV}$

2. If ∠F and ∠G are vertical angles and m∠F = 63°, find m∠G.
   F 180°   H 117°
   G 63°   J 100°

3. Which of the following is true for the figure below where line e ∥ line f?

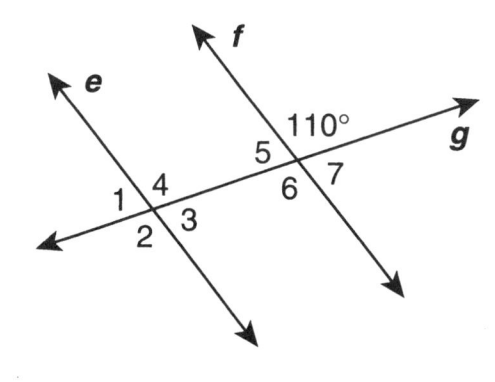

   A ∠2 ≅ ∠7   C line f ⊥ line g
   B m∠3 = 70°   D m∠6 = 70°

4. If the congruent angles of an isosceles triangle each measure 41°, what is the measure of the third angle?
   F 41°   H 59°
   G 98°   J 118°

5. A regular polygon that has six sides is a hexagon. What is the angle measure of a hexagon?
   A 120°   C 720°
   B 60°   D 90°

6. Which set of terms applies to the figure below?

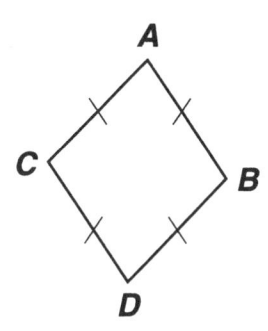

   F quadrilateral, parallelogram, square
   G quadrilateral, parallelogram, rhombus
   H parallelogram, trapezoid, rectangle
   J quadrilateral, trapezoid, rhombus

7. What is the slope of the line that passes through the points (−3, 7) and (5, −1)?
   A −1   C 1
   B 3   D −3

Holt Pre-Algebra

Name _____ Date _____ Class _____

CHAPTER 5 Quiz
Section B

Choose the best answer.

1. Identify the correct congruence statement for the pair of olygons shown below.

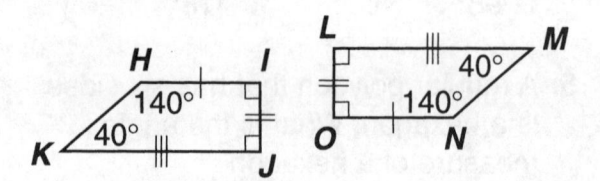

    A trapezoid HIJK ≅ trapezoid LMNO
    B trapezoid KJIH ≅ trapezoid MLON
    C trapezoid IJKH ≅ trapezoid ONML
    D trapezoid JIHK ≅ trapezoid ONML

2. triangle ABC ≅ triangle EDF
   Find f.

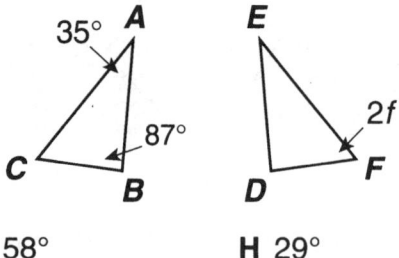

    F 58°
    G 87°
    H 29°
    J 35°

3. Identify which transformation is represented in the following figure.

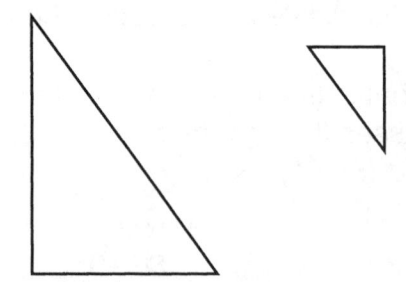

    A translation
    B reflection
    C rotation
    D none of these

4. A square has vertices $T(1, 2)$, $U(4, 2)$, $V(4, 5)$, $W(1, 5)$. If the square is rotated clockwise 90 degrees around $(0, 0)$, what is the coordinate of W?

    F $(5, -1)$
    G $(5, -4)$
    H $(2, -4)$
    J $(2, -1)$

5. Which of the following is not true about tessellations?

    A The angles at the vertices add to 360°.
    B Tessellations cover an entire plane.
    C A regular tessellation uses two or more regular polygons.
    D There are no gaps or overlaps in a tessellation.

Holt Pre-Algebra

Name _____ Date _____ Class _____

CHAPTER 6 Quiz
Section A

Choose the best answer.

1. Find the perimeter of the figure below.

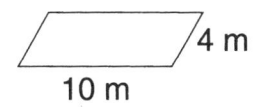

   4 m / 10 m (parallelogram)

   A 40 m C 28 m
   B 10 m D 4 m

2. Find the area of the figure with vertices $(-1, -1)$, $(3, -1)$, $(3, 4)$, and $(-1, 4)$.
   F 18 units^2 H 10 units^2
   G 20 units^2 J 9 units^2

3. Find the perimeter of the figure below.

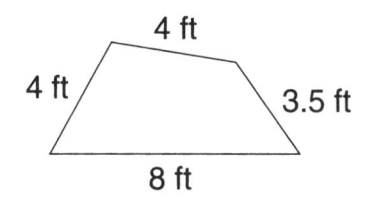

   4 ft (top), 4 ft (left), 3.5 ft (right), 8 ft (bottom)

   A 19.5 ft   C 24 ft
   B 21 ft   D 18 ft

4. Find the area of the figure with vertices $(0, 3)$, $(3, -1)$, and $(-2, -1)$.
   F 20 units^2 H 5 units^2
   G 10 units^2 J 4 units^2

5. Find the length of the hypotenuse of a right triangle that has legs of lengths 12 in. and 9 in.
   A 17 in.   C 21 in.
   B 13 in.   D 15 in.

6. Find the circumference of a circle that has a radius of 7 cm. Use 3.14 for π.
   F 153.86 cm   H 43.96 cm
   G 114.74 cm   J 21.98 cm

7. Find the area of a circle with center $(2, 3)$ that passes through $(2, 7)$. Use 3.14 for π.
   A 50.24 units^2 C 25.12 units^2
   B 12.56 units^2 D 41.36 units^2

8. A merry-go-round that has a radius of 20 ft has a duration per ride of 15 revolutions. How far does a person travel who rides on the merry-go-round? Use 3.14 for π.
   F 600 ft   H 942 ft
   G 18,840 ft   J 1884 ft

Copyright © by Holt, Rinehart and Winston.
All rights reserved.

Holt Pre-Algebra

Name _____ Date _____ Class _____

Chapter 6 Quiz
Section B

Choose the best answer.

1. A flat surface of a three-dimensional figure is a(n) _____.
   A edge
   B vertex
   C point
   D face

2. Find the volume of a rectangular prism with base 2 units by 3 units and height 5 units.
   F 15 units^3
   G 30 units^3
   H 62 units^3
   J 20 units^3

3. Find the volume of the figure below to the nearest tenth of a unit using 3.14 for π.

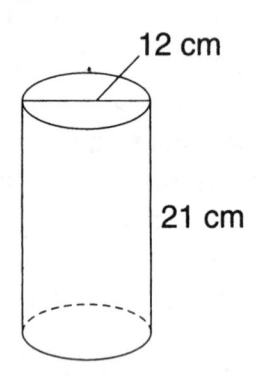

   12 cm
   21 cm

   A 2373.8 cm^3
   B 1017.4 cm^3
   C 791.3 cm^3
   D 9495.4 cm^3

4. Find the volume of the figure below.

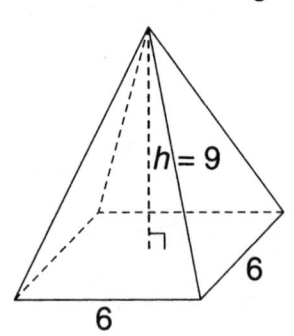

   $h = 9$
   6
   6

   F 324 units^3
   G 144 units^3
   H 108 units^3
   J 236 units^3

5. Find the surface area of the figure below. Use 3.14 for π.

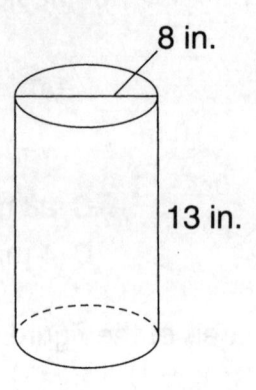

   8 in.
   13 in.

   A 1055.04 in^2
   B 427.04 in^2
   C 326.56 in^2
   D 653.12 in^2

6. A regular square pyramid has a base with sides that measure 4 ft and a slant height of 7 ft. Find the surface area of the pyramid.
   F 37.3 ft^2
   G 56 ft^2
   H 128 ft^2
   J 72 ft^2

7. Find the surface area of a cone with diameter 6 cm and slant height 12 cm. Use 3.14 for π.
   A 339.12 cm^2
   B 141.30 cm^2
   C 113.04 cm^2
   D 452.16 cm^2

8. Find the volume of a sphere with radius 11 m to the nearest tenth of a unit using 3.14 for π.
   F 1519.8 m^3
   G 2696.5 m^3
   H 3271.1 m^3
   J 5572.5 m^3

9. Find the surface area of a sphere with diameter 34 in. to the nearest tenth of a unit using 3.14 for π.
   A 20,569.1 in^2
   B 3629.8 in^2
   C 14,519.4 in^2
   D 6112.96 in^2

Copyright © by Holt, Rinehart and Winston.
All rights reserved.

Holt Pre-Algebra

Name _____ Date _____ Class _____

Quiz
CHAPTER 7 — **Section A**

Choose the best answer.

1. Find two ratios that are equivalent to the ratio $\frac{15}{6}$.

   A $\frac{30}{10}, \frac{10}{3}$ **C** $\frac{30}{18}, \frac{5}{2}$

   B $\frac{45}{18}, \frac{5}{2}$ **D** $\frac{45}{10}, \frac{10}{3}$

2. Which conversion is correct?

   F 3 yd = 36 in.
   G 1 day = 1440 sec
   H 2 mi = 10,560 ft
   J 200 km = 2,000,000 cm

3. A thunder storm produced 114 lightning strikes in $1\frac{1}{2}$ hours. What was the unit rate of lightning strikes?

   A 76 strikes per hour
   B 228 strikes per hour
   C 114 strikes per hour
   D 57 strikes per hour

4. The typical speed of sound is 760 miles per hour. If a plane is traveling exactly at the speed of sound, it is said to be at Mach 1. What is the Mach of a plane, to the nearest tenth, traveling 960 miles per hour?

   F 1.5   **H** 1.3
   G 1.7   **J** 0.8

5. If an object is traveling 30 feet per second, what is its speed in miles per hour to the nearest whole number? There are 5280 feet in a mile.

   A 20 mph   **C** 25 mph
   B 55 mph   **D** 32 mph

6. What is the value of *a* in the proportion $\frac{a}{24} = \frac{6}{16}$?

   F $a = 64$ **H** $a = 8$
   G $a = 3$ **J** $a = 9$

Name _____ Date _____ Class _____

CHAPTER 7 **Quiz**
Section B

1. A rectangle with dimensions 7 × 13 is enlarged by 15%. What are the new dimensions?
   A 10.5 × 19.5   C 8.05 × 14.95
   B 1.05 × 1.95   D 5.95 × 11.05

2. A figure has vertices (−13, 13), (26, 52), (39, 39). What would be the new coordinates of the vertices if the image were reduced by a scale factor of 1.3 with the origin as the center of dilation?
   F (−16.9, 16.9), (33.8, 67.6), (50.7, 50.7)
   G (−10, 10), (20, 40), (30, 30)
   H (10, 10), (−20, 40), (−30, 30)
   J (16.9, 16.9), (33.8, 67.6), (50.7, 50.7)

3. A picture that is originally 820 pixels long and 410 pixels tall is to be scaled to 600 pixels long. To the nearest pixel, how tall is the new picture?
   A 560   C 190
   B 1200   D 300

4. An advertisement on a billboard measures 22 ft long and 8 ft high. If the ad is transferred to the side of a bus and is 30 in. long, how tall is the new ad, to the nearest inch?
   F 11 in.   H 10 in.
   G 9 in.   J 12 in.

5. If two towns are 3 inches apart on a map and in real dimensions they are 15 miles apart, what is the scale of the map?
   A 1 in.:15 mi   C 3 in.:1 mi
   B 1 in.:5 mi   D 1.5 in.:3 mi

6. What scale factor would reduce an object by 33%?
   F 24 in.:1 yd   H 10 mm:1 cm
   G 1 ft:66 ft   J 2 m:5 m

7. A common scale for do-it-yourself airplane models is 1:48. The F-117A Stealth Fighter is 63 feet, 9 inches long. To the nearest inch, how long would a model of this plane be?
   A 9 in.   C 16 in.
   B 12 in.   D 13 in.

Name _____ Date _____ Class _____

CHAPTER 8 Quiz
Section A

Choose the best answer.

1. What is the percent of the unknown value *p* represented on the number line below?

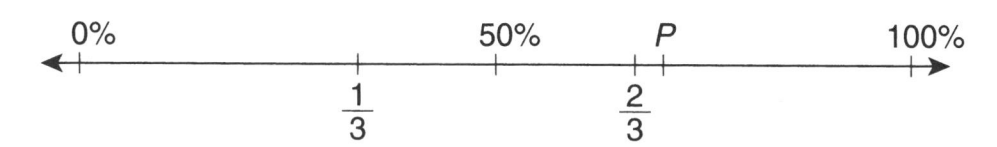

   ```
   0%           50%      P      100%
   ←——+——————+———+——+———————→
          1/3         2/3
   ```

   A 66%   **C** 70%
   B 25%   **D** 60%

2. What is 40% written as a fraction?
   F $\frac{1}{25}$ **H** $\frac{3}{8}$
   G $\frac{2}{5}$ **J** $\frac{3}{5}$

3. What is $\frac{5}{8}$ written as a decimal?
   A 0.625   **C** 0.5
   B 0.375   **D** 0.58

4. 54 is what percent of 150?
   F 50%   **H** 42%
   G 54%   **J** 36%

5. Find 5% of 356.
   A 71.2   **C** 35.6
   B 17.8   **D** 20

6. 33 is 220% of what number?
   F 15   **H** 22
   G 72.6   **J** 100

7. The length of the Nile river is 6693 km. The Amazon river is 96% as long as the Nile. To the nearest kilometer, what is the length of the Amazon?
   A 6693 km   **C** 6425 km
   B 6256 km   **D** 6021 km

Name _____ Date _____ Class _____

CHAPTER 8 Quiz
Section B

Choose the best answer.

1. What is the percent increase from 19 to 27, to the nearest percent?
   A 30%   C 22%
   B 42%   D 46%

2. If a stock started the week at $23.04 per share, and at the end of the week it had lost 8.5% of its value, to the nearest penny what was the final price of the stock?
   F $1.96   H $21.08
   B $20.96   J $27.56

3. Estimate 19% of 25.
   A about 5   C about 6
   B about 4   D about 3

4. 80 is about what percent of 228?
   F 66%   H 80%
   G 33%   J 50%

5. A car salesman was able to sell a car for $12,500, earning a commission of 5%. How much was his commission?
   A $716   C $500
   B $423   D $625

6. The sales tax in Alex's city is 7.33%. He bought a video game system for $299, and 2 games at $49 apiece. What was his total bill, to the nearest dollar?
   F $397   H $426
   G $335   J $468

7. A computer program used to require 450 clock cycles to process a certain input file. After the programmer optimized the program, it only took 60 clock cycles to process the same file. To the nearest percent, what was the percent decrease?
   A 87%   C 60%
   B 72%   D 89%

Name _____ Date _____ Class _____

CHAPTER 9 Quiz
Section A

Choose the best answer.

1. A recent insurance industry survey discovered that 25% of teenage drivers will have an accident within 6 months of receiving their license. What is the probability that a teenage driver will not get into an accident within 6 months of receiving his or her license?

   A 60%   C 75%
   B 25%   D 33%

2. Eight teams meet annually for a basketball tournament. The table below shows each teams' probability of winning. Team 6 is from River City North; team 7 is from River City South; and team 8 is from River City East. What is the probability that a team not from River City will win the tournament?

Team	Probability
1	0.07
2	0.13
3	0.05
4	0.14
5	0.16
6	0.18
7	0.21
8	0.06

   F 37.5%   H 45%
   G 55%   J 62.5%

3. There are five finalists at the Edgewater Kennel Show. The German Shepherd has a $\frac{1}{6}$ chance of winning the show. The Boxer is $\frac{1}{2}$ as likely to win as the German Shepherd, and the Black Lab, Poodle, and Border Collie have an equal chance of winning. What is the probability of the Border Collie's winning the show?

   A $\frac{1}{3}$ C $\frac{1}{5}$
   B $\frac{1}{4}$ D $\frac{1}{6}$

4. Kitty watches the birds fly by the windows. She notices that out of 52 birds, 17 are finches. Estimate to the nearest percent the probability that the next bird flying by will be a finch.

   F 25%   H 50%
   G 33%   J 75%

5. A baseball player's chance of getting a hit is 0.29. Using the following table, estimate the chances the baseball player has of getting a hit at least 2 out of 5 times.

   86 58 52 79 19 86 31 87 72 91
   47 94 18 39 47 32 66 67 89 93
   26 44 61 27 34 22 43 54 56 12
   74 83 21 96 11 25 16 52 23 30
   52 73 74 02 98 58 28 19 32 68

   A 40%   C 33%
   B 25%   D 50%

Name _____ Date _____ Class _____

Chapter 9 Quiz
Section B

Choose the best answer.

1. What is the probability when rolling two dice that the total shown on the dice is 2 or 3?

   A $\frac{1}{6}$ C $\frac{1}{36}$

   B $\frac{1}{4}$ D $\frac{1}{12}$

2. In a weekend bingo game Agnes has a 5% chance of winning the pot. Her friend Dorothy bought more cards than her and has an chance 8% of winning. What is the probability that Agnes or Dorothy will win the pot?

   F 13%   H 8%

   G 5%   J 3%

3. In computer languages, an identifier is a word the programmer uses to name a variable, function, or label. Some languages allow all letters, the underscore character '_', and the digits 0-9 to be used in identifiers, but the first character cannot be a digit. If the language is not case sensitive, i.e., upper and lower case letters are the same, how many possible 4-character identifiers are there?

   A 1,874,161   C 1,367,631

   B 1,213,056   D 531,441

4. Simplify 7!

   F 5040   H 40,320

   G 720   J 120

5. In a yacht race with 12 boats, how many ways out of the total number of boats can participants finish 1st, 2nd, and 3rd?

   A 220   C 1320

   B 479,001,600   D 12

6. Three cards are chosen at random from a standard deck of 52 cards. If each card is replaced before the next card is drawn, are the events dependent or independent? And, what is the probability of drawing a five, then a red card, and then a face card?

   F dependent; $\frac{3}{338}$

   G dependent; $\frac{42}{2197}$

   H independent; $\frac{3}{338}$

   J independent; $\frac{42}{2197}$

7. The probability of a student passing a certain multiple-choice test if he or she guesses at all of the answers is $\frac{3}{20}$. What are the odds that the student will pass?

   A 3:20   C 17:3

   B 3:17   D 3:23

Name _____ Date _____ Class _____

CHAPTER 10 Quiz
Section A

Choose the best answer.

1. Chris has a job selling newspapers every morning. He receives a base pay of $10 per day and also gets 15 cents for every paper he sells. One morning he made $16.45. How many papers did he sell?
   - **A** 52
   - **B** 43
   - **C** 64
   - **D** 39

2. Solve $18s + 22 = -14$.
   - **F** $s = -2$
   - **G** $s = -3$
   - **H** $s = 3$
   - **J** $s = 2$

3. Solve $\frac{n}{2} + 7 = 22$.
   - **A** $n = 58$
   - **B** $n = 30$
   - **C** $n = 26$
   - **D** $n = 15$

4. Solve $9y - 6 + y + 8 = 42$.
   - **F** $y = 10$
   - **G** $y = 5$
   - **H** $y = 4$
   - **J** $y = 8$

5. Solve $\frac{h}{2} + \frac{h}{5} = 14$.
   - **A** $h = 16$
   - **B** $h = 38$
   - **C** $h = 24$
   - **D** $h = 20$

6. Solve $12x + 15 = 24 - 6x$.
   - **F** $x = 6\frac{1}{2}$
   - **G** $x = \frac{1}{2}$
   - **H** $x = 2\frac{1}{6}$
   - **J** $x = 1\frac{1}{2}$

7. Solve $\frac{3w}{2} + \frac{1}{2} = w + 4$.
   - **A** $w = 7$
   - **B** $w = 4$
   - **C** $w = 8$
   - **D** $w = 3$

Holt Pre-Algebra

Name _____ Date _____ Class _____

CHAPTER 10 Quiz
Section B

Choose the best answer.

1. A school booster club is selling T-shirts. The silk-screening company is selling shirts to the boosters for $3 each, plus a one-time fee of $150 to cover supply costs. If the boosters want to make at least $300 in profits and they sell a shirt for $12, how many shirts do they need to sell?
   - **A** 35 or more
   - **B** 60 or more
   - **C** 25 or more
   - **D** 50 or more

2. Solve $5x - 2 < 13$.
   - **F** $x < 2$
   - **G** $x < 7$
   - **H** $x < 3$
   - **J** $x < 15$

3. Solve $-14 > 2x + 4$.
   - **A** $x < -5$
   - **B** $x < -9$
   - **C** $x > 6$
   - **D** $x > 8$

4. Solve $8x - 5 - 4x \le 3$.
   - **F** $x \le 2$
   - **G** $x \ge 2$
   - **H** $x \le -2$
   - **J** $x \ge -\frac{1}{2}$

5. Solve $\frac{3x}{4} - \frac{1}{2} \ge \frac{5}{8}$.
   - **A** $x \ge \frac{5}{3}$
   - **B** $x \ge \frac{3}{2}$
   - **C** $x \ge 2$
   - **D** $x \ge 3$

6. Solve $t - 3u + 5 = v$ for t.
   - **F** $t = v - 3u + 5$
   - **G** $t = v + 3u + 5$
   - **H** $t = v - 3u - 5$
   - **J** $t = v + 3u - 5$

7. Solve $A = \pi r^2$ for r.
   - **A** $r = \frac{\sqrt{A}}{\pi}$
   - **B** $r = \frac{A}{\pi}$
   - **C** $r = \sqrt{\frac{A}{\pi}}$
   - **D** $r = \frac{A}{\sqrt{\pi}}$

8. Identify the solution for the system
   $x + 2y = 11$
   $3x + y = 8$.
   - **F** (1, 2)
   - **G** (1, 5)
   - **H** (3, 7)
   - **J** (2, 2)

Name _____ Date _____ Class _____

CHAPTER 11 **Quiz**
Section A

Choose the best answer.

1. Which of the following equations is not linear?
   A $y = x + 6$ **C** $y = 2x^2$
   B $y = -8$ **D** $y = 9x$

2. The formula for converting temperature from Celsius to Fahrenheit is $f = \frac{9}{5}c + 32$ where c is the temperature in Celsius. If the temperature is 21° Celsius, what is the temperature in degrees Fahrenheit?
   F 45.5°F   **H** 21.0°F
   G 72.3°F   **J** 69.8°F

3. What is the slope of the line that passes through the two points (1, 7) and (4, 6)?
   A $-\frac{1}{3}$ **C** $\frac{3}{2}$
   B $-\frac{2}{3}$ **D** $\frac{1}{2}$

4. What is the slope of the line in the figure below?

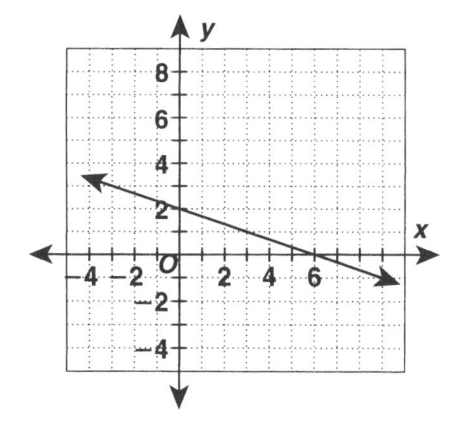

   F $-\frac{1}{3}$ **H** -3
   G $\frac{1}{2}$ **J** -2

5. Identify the line that is perpendicular to the line passing through the points (−5, 0) and (9, −7).
   A $5x - y = 7$ **C** $-2x + y = -6$
   B $x + 6y = 4$ **D** $-3x + 4y = 8$

6. Identify the x- and y-intercepts for the line $8x + 4y = 16$.
   F $x = 1, y = 5$ **H** $x = 8, y = 4$
   G $x = 2, y = 4$ **J** $x = 0, y = 0$

7. Express the equation $2x - 6y = 8$ in slope-intercept form.
   A $y = 2x - 6$ **C** $y = \frac{1}{2}x - \frac{3}{2}$
   B $y = \frac{1}{3}x - \frac{4}{3}$ **D** $y = 3x - 4$

8. Find the equation that has a slope of $-\frac{1}{8}$ and passes through the point (0, −4).
   F $y = -\frac{1}{4}x - 8$ **H** $y = \frac{1}{4}x - 4$
   G $y = \frac{1}{8}x$ **J** $y = -\frac{1}{8}x - 4$

Copyright © by Holt, Rinehart and Winston.
All rights reserved.

Holt Pre-Algebra

Name _____ Date _____ Class _____

CHAPTER 11 Quiz
Section B

Choose the best answer.

1. The data set below represents a direct variation. Identify the constant of proportionality.

x	1	2	3	4	5	6	7
y	$\frac{9}{2}$	9	$\frac{27}{2}$	18	$\frac{45}{2}$	27	$\frac{63}{2}$

   A $\frac{2}{9}$ **C** $\frac{9}{2}$
   B 9   **D** 2

2. Find the equation of direct variation, given that y varies directly with x, and x is 36 when y is 99.

   F $y = 99x$ **H** $y = 36x$
   G $y = \frac{11}{4}x$ **J** $y = \frac{4}{11}x$

3. Which inequality is represented in the graph below?

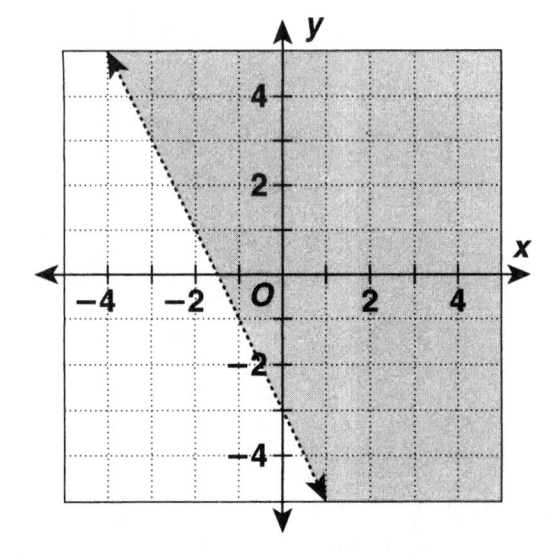

   A $y > -2x - 3$ **C** $y < -2x - 3$
   B $y \leq -2x - 3$ **D** $y \geq -2x - 3$

4. Which inequality is represented in the graph below?

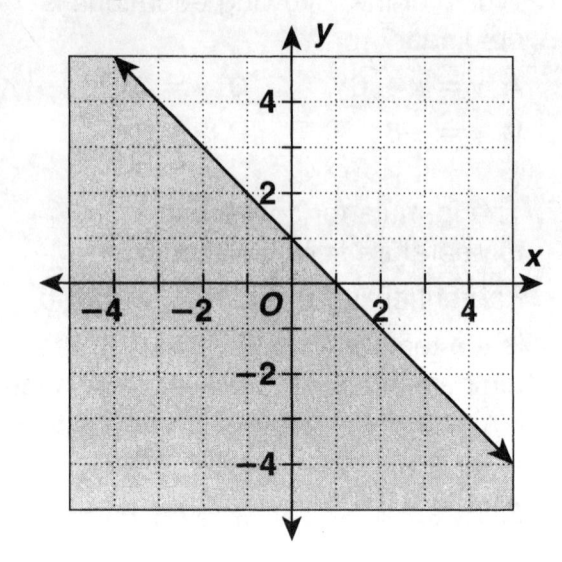

   F $y \geq -x + 1$ **H** $y < 2x + 1$
   G $y < -x - 1$ **J** $y \leq -x + 1$

5. Which inequality represents the following scenario? A newspaper stand in a city sells paper A for 35 cents and paper B for 50 cents. To be profitable, the stand needs to sell at least $50 worth of newspapers per day.

   A $35A + 50B > 50$
   B $7A + 10B \geq 1000$
   C $35A + 50B \geq 500$
   D $7A + 10B > 50$

6. Find a line of best fit for the data in the table below.

x	−5	−3	0	1	3	5	9
y	1	3	2	3	5	6	6

   F $y = -\frac{1}{2}x + 3$ **H** $y = 2x + 3$
   G $y = \frac{1}{2}x + 3$ **J** $y = 2x - 3$

Name _____ Date _____ Class _____

CHAPTER 12 Quiz
Section A

Choose the best answer.

1. Identify the common difference for the following arithmetic sequence:
   52, 49, 46, 43, ...
   A 3   **C** −3
   B −5 **D** 7

2. Find the 18th term for the arithmetic sequence where $a_1 = 6$ and $d = 3$.
   F 60   **H** 54
   G 57   **J** 51

3. Find the next three terms for the sequence 7, 20, 33, 46, 59, ...
   A 75, 91, 107   **C** 64, 69, 74
   B 66, 73, 80   **D** 72, 85, 98

4. Identify the common ratio for the following geometric sequence:
   $\frac{1}{3}, \frac{2}{9}, \frac{4}{27}, \frac{8}{81}, \ldots$
   F $\frac{2}{3}$ **H** $\frac{1}{3}$
   G $\frac{1}{6}$ **J** $\frac{2}{9}$

5. Find the 6th term for the following geometric sequence:
   1, 3, 9, 27, ...
   A 81   **C** 30
   B 243   **D** 729

6. Find the next three terms for the sequence −1, 2, −4, 8, −16, ...
   F 32, 64, 128   **H** −32, 64, −128
   G 18, 20, 22   **J** 32, −64, 128

7. Find the next three terms for the sequence 0, 2, 6, 12, 20, ...
   A 25, 30, 40   **C** 30, 42, 56
   B 22, 26, 32   **D** 28, 30, 32

8. Find the first three terms of the sequence for which the first term is 2 and each successive term is 1 more than twice the previous term.
   F 2, 4, 8   **H** 2, 4, 6
   G 2, 5, 11   **J** 2, 3, 5

9. Find the first three terms of the sequence defined by $a_n = \frac{n-1}{n}$.
   A $\frac{1}{2}, \frac{2}{3}, \frac{3}{4}$ **C** $\frac{1}{2}, \frac{3}{4}, \frac{5}{6}$
   B $0, 2, \frac{3}{2}$ **D** $0, \frac{1}{2}, \frac{2}{3}$

Name _____ Date _____ Class _____

CHAPTER 12 Quiz
Section B

Choose the best answer.

1. Which relationship below is not a function?
   A $y = 5x$
   B $y = x^2$
   C $x = 3$
   D $y = 6$

2. Evaluate the function $f(x) = -x + 4$ for $x = -6$.
   F -6
   G -2
   H 10
   J 4

3. Identify the equation for the function represented in the graph below.

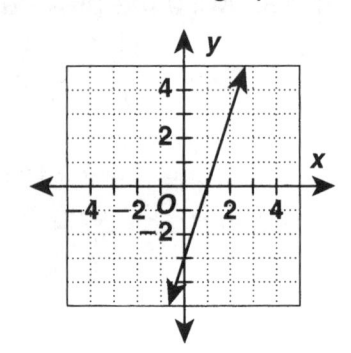

   A $y = x + 4$
   B $y = \frac{1}{3}x - 3$
   C $y = x^2 - 5$
   D $y = 3x - 3$

4. Identify the equation for the function represented in the graph below.

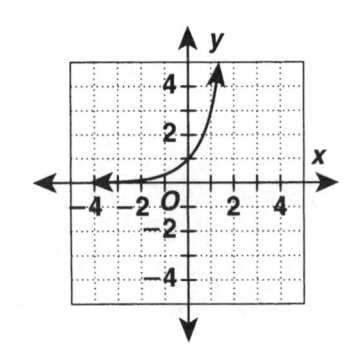

   F $y = 3^x$
   G $y = x^3$
   H $y = 2^x$
   J $y = 3x$

5. The radioactive isotope carbon-14 is the element used in the common technique of carbon dating to determine the age of an organic material. The half-life of carbon-14 is about 5500 years. If a sample contains 30 mg of carbon-14, how much would remain after 16,500 years?
   A 3 mg
   B 2.75 mg
   C 3.75 mg
   D 2 mg

6. Identify the equation for the function represented in the graph below.

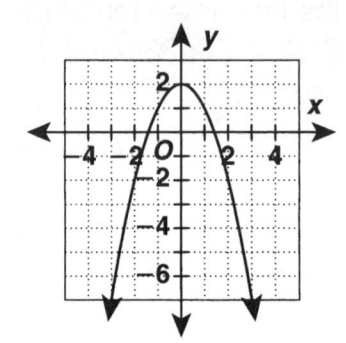

   F $y = -x + 2$
   G $y = x + 2$
   H $y = x^2 + 2$
   J $y = -x^2 + 2$

7. Find the constant of proportionality for the inverse variation represented in the table below.

x	−9	−5	−1	1	3	5	9
y	$-\frac{1}{3}$	$-\frac{3}{5}$	-3	3	1	$\frac{3}{5}$	$\frac{1}{3}$

   A 3
   B −2
   C −3
   D 2

Name _____ Date _____ Class _____

CHAPTER 13 **Quiz**
Form A

Choose the best answer.

1. Classify $6y^5 + 5x$.
   A monomial
   B binomial
   C trinomial
   D none of the above

2. Classify $7mn^{0.6}$.
   A monomial
   B binomial
   C trinomial
   D none of the above

3. Find the degree of $5y^4 + y^2 + 7$.
   A 4   C 7
   B 5   D 12

4. Which polynomial has a degree of 7?
   A $2x^5 + 4x^2$ C $2x + 5x^4$
   B $4f^6 + 3f^7 + f$ D $4c + 20c^{0.7}$

5. Identify like terms in the polynomial $6x + 5y^3 - 6 + 2y^3$.
   A 6, 6x
   B y^3
   C $5y^3$; $2y^3$
   D no like terms

6. Simplify $7n^2 + 7p^2 - 4n^2 + p^2$.
   A $11n^2 - 8p^2$ C $3n^2 + 7n^2$
   B $3n^2 + 8p^2$ D $11n^2 - 7p^2$

7. Simplify $9(5x^4 + 7x)$.
   A $45x^4 + 7x$ C $45x^4 + 63x$
   B $45x^{36} + 63$ D $108x^5$

8. Simplify $3(3x^2 + 5y) - 7x^2$.
   A $2x^2 + 5y$ C $2x^2 + 15y$
   B $16x^2 + 15y$ D $30x^2 + 15y$

29

Holt Pre-Algebra

Name _____ Date _____ Class _____

CHAPTER 13 Quiz Form B

Choose the best answer.

1. Add $(7a^2 - 5b) + (-4a^2 + 2b + 8)$.
   A $11a^2 + 7b + 8$ C $3a^2 + 3b^2 + 8$
   B $3a^2 - 3b + 8$ D $3a^2 + 7b + 8$

2. Add $(5xy^2 + 3x - 2y) + (3xy^2 + y - 3)$.
   A $8xy^2 + 3x + y - 3$
   B $8xy^2 + 3x - y - 3$
   C $8xy^4 + 3x - 2y - 3$
   D $15xy^2 + 3x - 2y - 3$

3. Find the opposite of $-3hj^3 + 5hj - 4$.
   A $-3hj^3 - 5hj + 4$
   B $3hj^3 + 5hj + 4$
   C $3hj^3 - 5hj + 4$
   D $4 - 5hj + -3hj^3$

4. Subtract $(x^4 - 4x^2) - (4x - x^2 + 7)$.
   A $x^4 + 4x^2 - 4x - x^2 - 7$
   B $x^4 + 5x^4 - 4x - 7$
   C $x^4 + 4x^2 + 4x + 7$
   D $x^4 - 3x^2 - 4x - 7$

5. Multiply $(4m^4n^4)(3m^3n)$.
   A $7m^{12}n^4$ C $12m^7n^4$
   B $12m^{12}n^4$ D $12m^7n^5$

6. Multiply $3y^3z(6y^4 - 5z^2)$.
   A $9y^7z - 2y^3z^3$
   B $18y^7z - 15y^3z^3$
   C $18y^7z^5 - 15z^2$
   D $18y^{12} - 15y^3z^2$

7. Multiply $(a + 5)(a + 7)$.
   A $a^2 + 7a + 5a + 12$
   B $a^2 + 35a + 35$
   C $a^2 + 12a^2 + 35$
   D $a^2 + 12a + 35$

Copyright © by Holt, Rinehart and Winston.
All rights reserved.

Holt Pre-Algebra

Name _____ Date _____ Class _____

CHAPTER 14 Quiz
Form A

Choose the correct answer.

1. cats ☐ {pets}
   - A ∉
   - B ∈
   - C ⊂
   - D ∅

2. $3x + \left(\dfrac{12}{x}\right) + y^3$ ☐ {polynomials}
   - A ∉
   - B ∈
   - C ⊂
   - D ∩

3. Describe the set: H = {odd integers}
   - A infinite
   - B finite
   - C ∉ {all numbers}
   - D ⊄ {all numbers}

4. Which of the following is true about sets M and N? $M = \{0, 4, 8, 9\}$ $N = \{-4, 4, 7\}$
   - A $M \cap N = \{\ \}$
   - B $M \cup N = \{4\}$
   - C $M \cap N = \{4\}$
   - D $M \cap N = \{-4\}$

5. What is the union of sets T and R?
   $T = \{3, 7, 11\}$
   R = {prime numbers}
   - A $T \cup R = T$
   - B $T \cup R = R$
   - C $T \cup R = \{\ \}$
   - D $T \cap R = R$

6. What is the intersection of sets E and F? $E = \{2, 8, 104\}$
   F = {even numbers}
   - A $E \cup F = F$
   - B $E \cap F = F$
   - C $E \cup F = \{\ \}$
   - D $E \cap F = E$

7. Identify $X \cap Y$ in the Venn diagram.

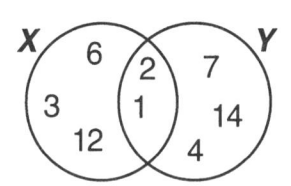

   X: 6, 3, 12 | 2, 1 | Y: 7, 14, 4

   - A {3, 6, 12}
   - B {4, 7, 14}
   - C {1, 2}
   - D { }

Name _____ Date _____ Class _____

Chapter 14 Quiz Form B

Choose the best answer.

For 1 and 2, use the following disjunction:
P: The person must be 8 years old.
Q: The person must be at least 44 inches tall.

1. James is 7 years old and 45 inches tall.
   A P: False; Q: True; P or Q: False
   B P: True; Q: False; P or Q: True
   C P: False; Q: True; P or Q: True
   D P: False; Q: False; P or Q: False

2. Star is 9 years old and 48 inches tall.
   A P: True; Q: False, P or Q: False
   B P: True; Q: True; P or Q: True
   C P: False; Q: True; P or Q: True
   D P: True; Q: False; P or Q: True

3. In the statement, "You will be healthier if you exercise every day," what is the term that describes, "exercise every day?"
   A hypothesis   C conjunction
   B conclusion   D deduction

Use the figure for 4 and 5.

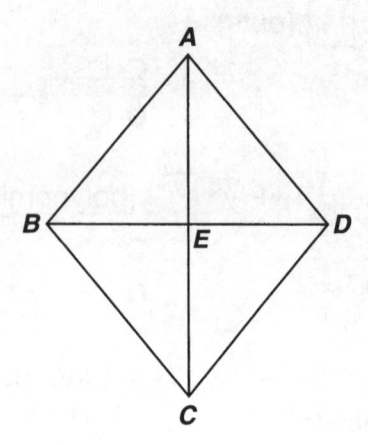

4. What is the degree of point E?
   A 1 degree   C 3 degrees
   B 2 degrees   D 4 degrees

5. Name the figure.
   A Euler circuit
   B Königsberg Bridge
   C Hamiltonian circuit
   D Euler path

Name _____ Date _____ Class _____

CHAPTER 1 Chapter Test
Form A

Evaluate each expression for the given values of the variables.

1. $3x + 2y$ for $x = 8$ and $y = 6$

   _____

2. $13m - 2n$ for $m = 3$ and $n = 4$

   _____

3. $5(k + 8) - 2m$ for $k = 6$ and $m = 3$

   _____

Write an algebraic expression for each word phrase.

4. 5 more than twice a number p

   _____

5. 7 times the sum of h and 12

   _____

6. 4 less than the sum of g and 15

   _____

Solve.

7. $z + 18 = 54$

   _____

8. $m - 4.5 = 12$

   _____

Write an equation, then solve.

9. The depth of Lake Superior is 1330 feet. This is 407 feet deeper than Lake Michigan. How deep is Lake Michigan?

   _____

10. The length of Lake Ontario is 193 miles, which is 48 miles less than the length of Lake Huron. How many miles long is Lake Huron?

    _____

Solve and check.

11. $\dfrac{n}{8} = 12$

    _____

12. $7k = 91$

    _____

Write an equation, then solve.

13. The Smith family spends an average of $450 monthly for groceries. Groceries account for $\dfrac{1}{8}$ of their monthly costs. How much are their monthly costs?

    _____

14. Janice spent $325 on clothes. This was 5 times the amount spent for her 5-year-old brother. How much money was spent on clothes for her brother?

    _____

Solve and graph.

15. $x + 8 > 12$ _____

    ←—+—+—+—+—+—+—+—+—→

16. $z - 5 \leq 15$ _____

    ←—+—+—+—+—+—+—+—+—→

Name _____ Date _____ Class _____

Chapter 1 Chapter Test
Form A, continued

Solve and graph.

17. $4x \geq 52$ _____

18. $\dfrac{k}{5} < 12$ _____

Simplify.

19. $4(z + 8) - z$ _____

20. $2(4y + 6) - 3y$ _____

21. $3(x - 5) + 8x$ _____

Solve.

22. $4y + 3y = 63$

23. $10g - 4g = 54$

24. $12t - 4t = 96$

Determine whether the ordered pair is a solution of the given equation.

25. $y = x + 9$; (7, 16)

26. $y = 2x + 8$; (5, 28)

Give the coordinates of each point identified on the coordinate plane.

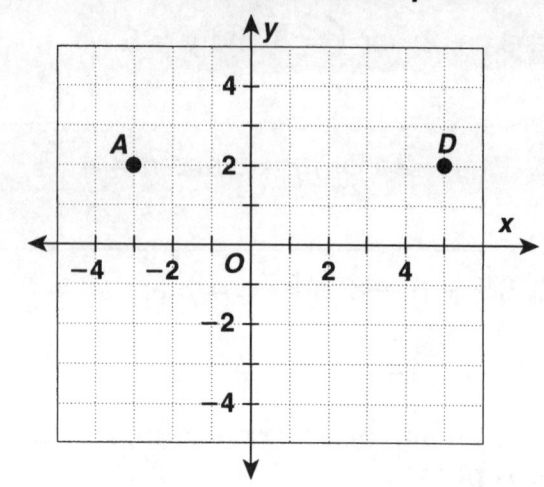

27. A _____

28. D _____

Study the table and use it to answer the following questions.

Rocket Experiment with Water Pressure

Water (ounces)	Height of Rocket (feet)
4	22
8	50
10	60
16	91

29. How much water was used to launch the rocket 50 feet?

30. If twice as much water is used, will the rocket soar twice as far? Explain.

34 Holt Pre-Algebra

Name _____ Date _____ Class _____

Chapter 1 Chapter Test
Form B

Evaluate each expression for the given values of the variables.

1. $6x + 3y$ for $x = 7$ and $y = 8$

2. $1.8s - 5p$ for $s = 9$ and $p = 2$

3. $j(5 + t) - 8$ for $j = 8$ and $t = 6$

Write an algebraic expression for each word phrase.

4. 8 more than the product of z and 8

5. 4 less than the quotient of x and 8

6. 7 less than the sum of m and 12

Solve and check.

7. $n + 36 = 154$

8. $4.9 = m - 2.6$

Write an equation, then solve.

9. The average weight of an elephant is 5450 kg. This is 4270 kg more than the average weight of giraffe. What is the average weight of a giraffe?

10. The giraffe sleeps 180 minutes per day, which is 60 minutes less than an elephant sleeps each day. How long does the elephant sleep each day?

Solve and check.

11. $\frac{m}{15} = 18$

12. $23y = 92$

Write an equation, then solve.

13. A koala bear eats about 2.5 pounds of eucalyptus leaves each day. This is about $\frac{1}{10}$ of his total body weight. What does a koala bear weigh?

14. The hummingbird beats its wings 5400 beats per minute. This is 30 times faster than a stork's wing beats per minute. How many wing beats per minute does a stork make?

Solve and graph.

15. $12 + x \geq 20$

16. $m - 12 \leq 8$

Name _____ Date _____ Class _____

CHAPTER 1 Chapter Test
Form B, continued

Solve and graph.

17. $13y \leq 104$ _____

18. $6 \geq \dfrac{x}{2}$ _____

Simplify.

19. $7(2b - 8)$

_____

20. $4(4x - 4) + 2x$

_____

21. $3(4b + 3) - 5$

_____

Solve.

22. $24k - 7k = 51$ _____

23. $45 = 3y + 6y$ _____

24. $\dfrac{z}{12} = 9$ _____

Determine whether the ordered pair is a solution of the given equation.

25. $y = 7x + 7;\ (4, 30)$

_____

26. $y = 4x + 12;\ (3, 24)$

_____

Give the coordinates of each point identified on the coordinate plane.

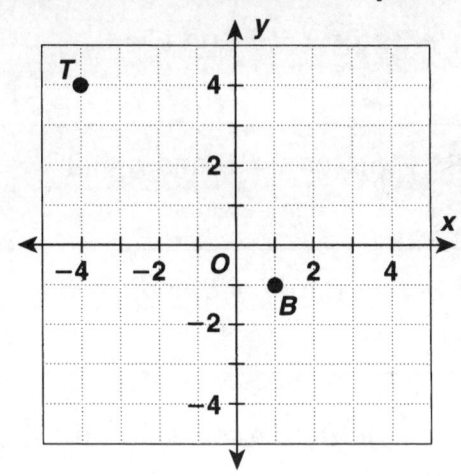

27. T _____

28. B _____

Study the table and use it to answer the following questions.

Filling the Bath Tub

Time (minutes/seconds)	Height of Bath Water
1:00	4 inches
2:00	7.5 inches
3:00	10.2 inches
4:00	11.6 inches
4:23	12 inches

29. During which minute does the most water enter the bathtub as it is being filled?

_____

30. After 2 minutes of filling the tub, how many more inches of water must be added for the tub to be filled?

_____

Copyright © by Holt, Rinehart and Winston.
All rights reserved.

36

Holt Pre-Algebra

Name _____ Date _____ Class _____

Chapter Test
Chapter 1 Form C

Evaluate each expression for $x = 2.5$, $y = 12$, and $z = 4$

1. $2y - 2x$

2. $z(4 + x) + 8$

3. $5xyz$

Write an algebraic expression for each word phrase.

4. 5 less than the product of 8 and k

5. twice the sum of k and 8

6. half the sum of 12 and t

Solve and check.

7. $2.8 + t = 9.4$

8. $18 = d - 5$

Write an equation, then solve.

9. The world's largest cherry pie weighs 37,740 pounds. This is 7625 pounds heavier than the largest apple pie. How heavy is the largest apple pie?

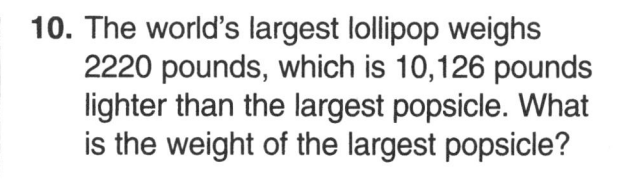

10. The world's largest lollipop weighs 2220 pounds, which is 10,126 pounds lighter than the largest popsicle. What is the weight of the largest popsicle?

Solve and check.

11. $\dfrac{n}{6} - 4 = 8$

12. $13x + 14 = 40$

Write an equation, then solve.

13. The population of Nevada is close to 1.6 million, which is about $\dfrac{1}{5}$ of Michigan's population. What is the estimated population of Michigan?

14. There are about 1.2 million people in the U.S. who speak Chinese. This is about 3 times the number of people who speak Greek. How many people speak Greek?

Solve and graph.

15. $3.2 + x \geq 9$ _____

16. $8 < x - 3$ _____

Name _____ Date _____ Class _____

Chapter Test
Form C, continued

Solve and graph.

17. $3x + 8 \geq 41$ _____

18. $6 < \dfrac{a}{4}$ _____

Simplify.

19. $6(3h + 9) - 4h$

_____

20. $6(5y - 7) + 8y$

_____

21. $4(2y - 3) + 5y$

_____

Solve.

22. $12g + 13g = 450$ _____

23. $45 = 3y + 6y$ _____

24. $42 = 9k - 2k$ _____

Determine whether the ordered pair is a solution of the given equation.

25. $y = 17x - 3$; (2, 31)

_____

26. $y = 4x - 12$; (8, 4)

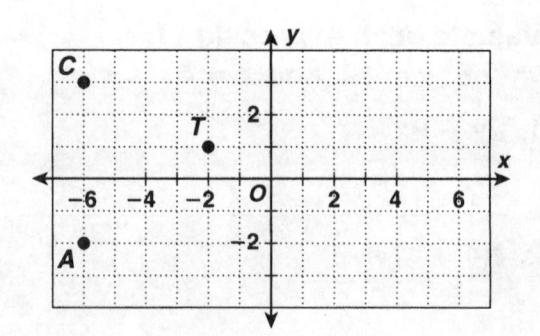

_____

27. Give the coordinates point C, point T and point A, on the coordinate plane.

_____

28. Connect all 3 points to form a geometric figure. Identify the figure formed.

_____

Office Deliveries 5-Story Building

Floor #	# of Deliveries	Time (minutes)
1	2	11
2	6	20
3	3	13
4	1	4
5	3	15

29. On which floors were 3 deliveries made?

_____

30. Do 6 deliveries take twice as long as 3 deliveries? Why or why not?

_____

Name _____ Date _____ Class _____

CHAPTER 2 Chapter Test
Test A

Add.

1. 6 + (−1) _____

2. −9 + (−5) _____

3. −6 + 6 _____

Evaluate each expression for the given value of the variable.

4. 10 + j for j = −3 _____

5. n + (−2) for n = 5 _____

6. s + 6 for s = −6 _____

Subtract.

7. −9 − 2 _____

8. 2 − (−7) _____

9. −12 − (−5) _____

Evaluate each expression for the given value of the variable.

10. 4 − g for g = −2 _____

11. −16 − t for t = 4 _____

12. n − (−6) for n = 5 _____

Multiply or divide.

13. 5(−7) _____

14. −2(−12) _____

15. $\dfrac{2^5}{2^2}$ _____

Simplify.

16. 8(−10 + 7) _____

17. 9(−1 + 4) _____

18. $\dfrac{-4(5)}{-4}$ _____

Solve.

19. −9 + d = 23 _____

20. 4h = −28 _____

21. $\dfrac{k}{9} = -3$ _____

Solve and graph. (2-5)

22. t − 1 ≤ 4 _____

←―+―+―+―+―+―→

Name _____ Date _____ Class _____

CHAPTER 2 Assessment
Test A, continued

Solve and graph.

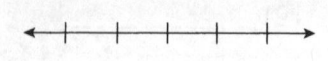

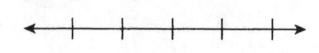

23. $\dfrac{y}{4} < -2$ _____

<-+--+--+--+--+--+->

24. $4c \geq 12$ _____

<-+--+--+--+--+--+->

Write using exponents.

25. $3 \cdot 3 \cdot 3$ _____

26. $(-n) \cdot (-n)$ _____

27. $k \cdot k \cdot k \cdot k \cdot k$ _____

Simplify.

28. 2^3 _____

29. $(-4)^3$ _____

30. 5^4 _____

31. $(8 - 7)^2$ _____

32. $9 + (-5)^2$ _____

33. $(1 \cdot 6)^2$ _____

Multiply or divide. Write the product or quotient as one power.

34. $n^1 \cdot n^4 \cdot n^7$ _____

35. $4^3 \cdot 4^2 \cdot 4^0 \cdot 4^1$ _____

36. $\dfrac{u^7}{u^2}$ _____

Simplify the powers of 10.

37. 10^{-3} _____

38. 10^{-2} _____

39. 10^{-6} _____

Simplify.

40. $10^{-3} \cdot 10^5$ _____

41. 1.6×10^{-2} _____

42. $\dfrac{2^3}{2^5}$ _____

Write each number in standard notation.

43. 3.2×10^3 _____

44. 1.75×10^{-3} _____

45. 9.46×10^0 _____

Write each answer in scientific notation.

46. In one year, Americans made 445 billion phone calls. Write this number in scientific notation.

_____

47. A bookstore orders a shipment of books. The books weigh 3.2 lb each. How much will the shipment of 100 books weigh?

_____

Name _____ Date _____ Class _____

CHAPTER 2 Assessment
Test B

Add.

1. $7 + (-8)$ _____

2. $-18 + (-5)$ _____

3. $-6 + 15$ _____

Evaluate each expression for the given value of the variable.

4. $14 + j$ for $j = -7$ _____

5. $n + (-9)$ for $n = 15$ _____

6. $s + 7 - (-3)$ for $s = -6$ _____

Subtract.

7. $-9 - 22$ _____

8. $10 - (-7)$ _____

9. $-18 - (-6)$ _____

Evaluate each expression for the given value of the variable.

10. $9 - g$ for $g = -15$ _____

11. $-26 - t$ for $t = 9$ _____

12. $n - (-14)$ for $n = 1$ _____

Multiply or divide.

13. $25(-7)$ _____

14. $-28(-12)$ _____

15. $\dfrac{8^9}{8^3}$ _____

Simplify.

16. $8(-13 - 7)$ _____

17. $-9(-16 + 4)$ _____

18. $\dfrac{-3(14)}{-6}$ _____

Solve.

19. $19 + d = 53$ _____

20. $14h = -98$ _____

21. $\dfrac{k}{81} = -3$ _____

Solve and graph.

22. $t - 15 \leq -9$ _____

←—+—+—+—+—+—→

Holt Pre-Algebra

CHAPTER 2 Assessment
Test B, continued

Solve and graph.

23. $\frac{y}{8} < -2$ _____

<—+—+—+—+—+—>

24. $7c \geq 49$ _____

<—+—+—+—+—+—>

Write using exponents.

25. $7 \cdot 7 \cdot 7 \cdot 7$ _____

26. $(-k) \cdot (-k) \cdot (-k)$ _____

27. $t \cdot t \cdot t \cdot t \cdot t$ _____

Simplify.

28. 7^4 _____

29. $(-9)^3$ _____

30. 12^2 _____

31. $(8 + (-7))^8$ _____

32. $19 + (-5) + (-3)^3$ _____

33. $(5 + 7 \cdot 4)^2 + 7$ _____

Multiply or divide. Write the product or quotient as one power.

34. $n^{11} \cdot n^{11} \cdot n^{11}$ _____

35. $4^6 \cdot 4^2 \cdot 4^0 \cdot 4^1$ _____

36. $\frac{u^{16}}{u^{14}}$ _____

Simplify the powers of 10.

37. 10^{-6} _____

38. 10^{-4} _____

39. 10^{-10} _____

Simplify.

40. $10^{-5} \cdot 10^{11}$ _____

41. 2.49×10^{-2} _____

42. $\frac{6^3}{6^5}$ _____

Write each number in standard notation.

43. 1.42×10^6 _____

44. 3.56×10^{-4} _____

45. 5.12×10^0 _____

Write each answer in scientific notation.

46. The thinnest commercial glass is 0.000984 in. thick. The glass on an aquarium is 1000 times as thick. How thick is the glass?

_____

47. A pet store buys a truckload of dog food. The bags weigh 50 lb each. How much will all 1000 bags ordered weigh?

_____

Name _____ Date _____ Class _____

CHAPTER 2 Assessment
Test C

Add.

1. $27 + (-38)$ _____

2. $-48 + (-25)$ _____

3. $-46 + 25$ _____

Evaluate each expression for the given value of the variable.

4. $84 + j$ for $j = -28$ _____

5. $n + (-69)$ for $n = 75$ _____

6. $s + 27 - (-93)$ for $s = -16$ _____

Subtract.

7. $-89 - 32$ _____

8. $46 - (-15)$ _____

9. $-58 - (-66)$ _____

Evaluate each expression for the given value of the variable.

10. $35 - g$ for $g = -26$ _____

11. $-86 - t$ for $t = 45$ _____

12. $n - (-64)$ for $n = 29$ _____

Multiply or divide.

13. $525(-4)$ _____

14. $-108(-14)$ _____

15. $\dfrac{17^8}{17^3}$ _____

Simplify.

16. $26(-23 - 19)$ _____

17. $-14(-36 + 14)$ _____

18. $\dfrac{-13(20)}{4}$ _____

Solve.

19. $-126 = 39 - d$ _____

20. $74h = -518$ _____

21. $\dfrac{4h}{-7} = -64$ _____

Solve and graph.

22. $t - 25 \leq -19$ _____

←—+—+—+—+—+—→

Copyright © by Holt, Rinehart and Winston.
All rights reserved.

Holt Pre-Algebra

Name _____ Date _____ Class _____

CHAPTER 2 Assessment
Test C, continued

Solve and graph.

23. $\dfrac{-y + 12}{4} > 8$ _____

24. $16c \geq 112$ _____

Write using exponents.

25. $17 \cdot 17 \cdot 17 \cdot 17$ _____

26. $(-t) \cdot (-t) \cdot (-t)$ _____

27. $-5 \cdot 5 \cdot 5 \cdot s \cdot s \cdot b \cdot b \cdot b \cdot b$ _____

Simplify.

28. 34^3 _____

29. $(-12)^3$ _____

30. 18^4 _____

31. $(9 + (-6))^8$ _____

32. $19 + (-5)^3$ _____

33. $(13 \cdot 2)^2 + 56$ _____

Multiply or divide. Write the product or quotient as one power.

34. $(n^2)^2 \cdot n^{15} \cdot n^{11} \cdot n^{19}$ _____

35. $7 \cdot 7^2 \cdot 7 \cdot 6^0$ _____

36. $\dfrac{u^{45}}{u^{14}}$ _____

Simplify the powers of 10.

37. 10^{-11} _____

38. 10^{-7} _____

39. 10^{-4} _____

Simplify.

40. $(13 - 3)^{15} (6 + 4)^{-9}$ _____

41. 3.567×10^{-3} _____

42. $\dfrac{12^3}{12^5}$ _____

Write each number in standard notation.

43. 5.234×10^6 _____

44. 1.9×10^{-5} _____

45. 6.48×10^0 _____

Write each answer in scientific notation.

46. The mass of a small insect is 0.0000569 g. How much would 100 of the insects weigh?

_____

47. A pet store buys a truckload of dog food. The bags weigh 25 lb each. How much will all 10,000 bags weigh?

_____

Name _____ Date _____ Class _____

Chapter Test
Chapter 3 Form A

Simplify.

1. $\dfrac{6}{9}$ _____

2. $-\dfrac{25}{75}$ _____

3. $\dfrac{12}{28}$ _____

Write each decimal as a fraction in simplest form.

4. 0.39 _____

5. 0.24 _____

Write each fraction as a decimal.

6. $\dfrac{7}{10}$ _____

7. $\dfrac{1}{5}$ _____

Add or subtract.

8. $\dfrac{3}{5} - \dfrac{1}{5}$ _____

9. $-1.4 + 1.9$ _____

Evaluate each expression for the given value of the variable.

10. $2.9 + j$ for $j = 1.3$ _____

11. $4 + n$ for $n = -\dfrac{2}{3}$ _____

Multiply. Write each answer in simplest form.

12. $\dfrac{3}{5}\left(\dfrac{1}{2}\right)$ _____

13. $5.3(4.1)$ _____

14. $-0.2(2.7)$ _____

Evaluate $\dfrac{1}{6}y$ for each value of y.

15. $y = 4$ _____

16. $y = \dfrac{1}{9}$ _____

17. $y = \dfrac{2}{13}$ _____

Divide. Write each answer in simplest form.

18. $\dfrac{9}{16} \div \dfrac{1}{2}$ _____

19. $9 \div \dfrac{3}{4}$ _____

Divide.

20. $0.36 \div 0.24$ _____

21. $9.12 \div 0.5$ _____

Copyright © by Holt, Rinehart and Winston.
All rights reserved.

Holt Pre-Algebra

Name _____ Date _____ Class _____

CHAPTER 3 Chapter Test
Form A, continued

Evaluate $\frac{2}{x}$ for each value of x.

22. $x = 0.4$ _____

23. $x = 0.2$ _____

Add or subtract.

24. $\frac{1}{2} + \frac{1}{6}$ _____

25. $3\frac{1}{6} + 1\frac{7}{12}$ _____

Evaluate each expression for the given value of the variable.

26. $\frac{1}{8} + x$ for $x = \frac{1}{4}$ _____

27. $x - \frac{4}{5}$ for $x = \frac{1}{5}$ _____

Solve.

28. $m - 1.6 = 5.2$ _____

29. $\frac{2}{9}m = \frac{2}{3}$ _____

30. $3.0h = 12.0$ _____

Solve.

31. $y - \frac{1}{8} \geq \frac{5}{8}$ _____

32. $7 + y > 8.2$ _____

33. $3a < \frac{1}{6}$ _____

Simplify each expression.

34. $\sqrt{2} + 2$ _____

35. $\sqrt{36} - \sqrt{4}$ _____

36. $\sqrt{16} + 2$ _____

Use a calculator to find each value. Round to the nearest tenth.

37. $\sqrt{196}$ _____

38. $\sqrt{12}$ _____

30. $-\sqrt{29}$ _____

State if the number is rational, irrational, or not a real number.

40. 0.91 _____

41. -5 _____

42. $0.\overline{3}$ _____

43. Jessie bought 6 carnations that were $0.39 each. How much did she spend?

_____

44. There is a fourth of a pie left. Your mom says you can have half of it. How much of the pie is that?

_____

Copyright © by Holt, Rinehart and Winston.
All rights reserved.

Holt Pre-Algebra

Name _____ Date _____ Class _____

Chapter Test
Chapter 3 Form B

Simplify.

1. $\dfrac{21}{30}$ _____

2. $-\dfrac{45}{120}$ _____

3. $\dfrac{15}{17}$ _____

Write each decimal as a fraction in simplest form.

4. -3.46 _____

5. -0.07 _____

Write each fraction as a decimal.

6. $\dfrac{7}{25}$ _____

7. $-\dfrac{25}{5}$ _____

Add or subtract.

8. $\dfrac{7}{23} - \dfrac{9}{23}$ _____

9. $-1.4 + 0.72$ _____

Evaluate each expression for the given value of the variable.

10. $8 + n$ for $n = -\dfrac{1}{4}$ _____

11. $\dfrac{4}{12} + f$ for $f = -\dfrac{9}{12}$ _____

Multiply. Write each answer in simplest form.

12. $\dfrac{3}{4}\left(\dfrac{1}{8}\right)$ _____

13. $-5\left(1\dfrac{5}{6}\right)$ _____

14. $4.73(3.1)$ _____

Evaluate $5\dfrac{2}{3}y$ for each value of y.

15. $y = 4$ _____

16. $y = -\dfrac{1}{9}$ _____

17. $y = \dfrac{7}{17}$ _____

Divide. Write each answer in simplest form.

18. $\dfrac{5}{6} \div \dfrac{9}{16}$ _____

19. $6\dfrac{1}{3} \div 3\dfrac{1}{2}$ _____

Divide.

20. $1.9 \div 0.05$ _____

21. $7.15 \div 1.3$ _____

Copyright © by Holt, Rinehart and Winston.
All rights reserved.

Holt Pre-Algebra

Name _____ Date _____ Class _____

CHAPTER 3 Chapter Test
Form B, continued

Evaluate $\frac{8}{x}$ for each value of *x*.

22. $x = 2.5$ _____

23. $x = 0.04$ _____

Add or subtract.

24. $\frac{2}{3} + \frac{4}{5}$ _____

25. $2\frac{5}{6} - 1\frac{3}{10}$ _____

Evaluate each expression for the given value of the variable.

26. $\frac{1}{8} + x$ for $x = -4\frac{1}{8}$ _____

27. $n - 3\frac{5}{6}$ for $n = -\frac{1}{15}$ _____

Solve.

28. $m - 5.6 = -0.9$ _____

29. $\frac{3}{7}m = -\frac{6}{7}$ _____

30. $3.7h = 0.74$ _____

Solve.

31. $y - \frac{1}{8} \geq \frac{3}{4}$ _____

32. $5.5 + y > 2.3$ _____

33. $6a < -\frac{5}{8}$ _____

Simplify each expression.

34. $\sqrt{7 + 9}$ _____

35. $\sqrt{36} - \sqrt{81}$ _____

36. $\sqrt{25} + 24$ _____

Use a calculator to find each value. Round to the nearest tenth.

37. $\sqrt{136}$ _____

38. $-\sqrt{34}$ _____

39. $\sqrt{82.9}$ _____

State if the number is rational, irrational, or not a real number.

40. $\sqrt{27}$ _____

41. $-\sqrt{\frac{81}{16}}$ _____

42. $\sqrt{\frac{1}{9}}$ _____

43. If you buy 2.5 pounds of hamburger and it costs $2.10 per pound, how much does the package cost?

_____

44. A book of stamps contains 20 stamps. If you used one fourth of them, how many did you use?

_____

Name _____ Date _____ Class _____

Chapter 3 Chapter Test
Form C

Simplify.

1. $\dfrac{75}{204}$ _____

2. $-\dfrac{108}{320}$ _____

3. $\dfrac{14}{53}$ _____

Write each decimal as a fraction in simplest form.

4. -4.3125 _____

5. $0.0\overline{3}$ _____

Write each fraction as a decimal.

6. $\dfrac{9}{16}$ _____

7. $-\dfrac{19}{40}$ _____

Add or subtract.

8. $\dfrac{5}{136} - \dfrac{9}{136}$ _____

9. $-17.67 + 25.13$ _____

Evaluate each expression for the given value of the variable.

10. $54 + n$ for $n = -\dfrac{7}{16}$ _____

11. $\dfrac{2}{9} + f$ for $f = -\dfrac{8}{9}$ _____

Multiply. Write each answer in simplest form.

12. $13\dfrac{4}{7}\left(\dfrac{3}{5}\right)$ _____

13. $11\left(4\dfrac{5}{12}\right)$ _____

14. $-0.47(92.8)$ _____

Evaluate $16\dfrac{2}{5}y$ for each value of y.

15. $y = 4$ _____

16. $y = \dfrac{1}{16}$ _____

17. $y = \dfrac{5}{29}$ _____

Divide. Write each answer in simplest form.

18. $13\dfrac{3}{4} \div 1\dfrac{2}{3}$ _____

19. $6\dfrac{2}{3} \div 7\dfrac{1}{2}$ _____

Divide.

20. $0.512 \div 0.08$ _____

21. $2002 \div 5.2$ _____

Name _____ Date _____ Class _____

CHAPTER 3 Chapter Test
Form C, continued

Evaluate $\frac{24}{x}$ for each value of x.

22. x = 0.9 _____

23. x = 0.012 _____

Add or subtract. Write each answer in simplest form.

24. $\frac{4}{5} + \frac{3}{8}$ _____

25. $6\frac{5}{8} - 1\frac{9}{10}$ _____

Evaluate each expression for the given value of the variable.

26. $\frac{2}{3} + x$ for $x = \frac{7}{12}$ _____

27. $n - 5\frac{5}{6}$ for $n = -4\frac{7}{8}$ _____

Solve.

28. $m + 25.6 = -0.19$ _____

29. $\frac{4}{17}m = -\frac{8}{17}$ _____

30. $23.7h = 16.59$ _____

Solve.

31. $y - \frac{1}{2} \geq \frac{7}{8}$ _____

32. $43.6 + y > 32.3$ _____

33. $-3a < -13\frac{4}{5}$ _____

Simplify each expression.

34. $\sqrt{460} + 24$ _____

35. $\sqrt{121} - \sqrt{196}$ _____

36. $\sqrt{324} + 324$ _____

Use a calculator to find each value. Round to the nearest tenth.

37. $\sqrt{1436}$ _____

38. $-\sqrt{284}$ _____

39. $\sqrt{982.9}$ _____

State if the number is rational, irrational, or not a real number.

40. $\sqrt{11}$ _____

41. $\sqrt{576}$ _____

42. $\sqrt{5\frac{1}{16}}$ _____

43. The area of a rectangle is its length times its width. Find the area if the length is 2.3 feet and the width is 0.7 foot.

_____

44. Find the area of a rectangle if the length is $2\frac{2}{3}$ inches and the width is $\frac{4}{9}$ inches.

_____

Name _____ Date _____ Class _____

CHAPTER 4 Chapter Test
Form A

Identify the population, sample, and sampling method.

1. Every 15th student eating in the school cafeteria was asked his or her favorite dessert.

   _____
   _____
   _____

2. 300 households in Sylvania were selected by random digit dialing to respond to a survey.

   _____
   _____
   _____

3. Ten video stores in a large city were randomly selected and 50 customers were randomly selected from each store and asked what was their favorite video.

   _____
   _____
   _____

Organize the data to make a stem-and-leaf plot.

4. The following are test scores for a math class.

Test Scores			
98	85	70	93
85	74	98	81
89	78	100	85

Use the data. 10, 25, 18, 15, 21, 20, 12 Round to the nearest tenth.

5. Find the mean. _____
6. Find the median. _____
7. Find the mode. _____
8. Find the range. _____
9. Find the third quartile. _____
10. Find the first quartile. _____
11. Make a box-and-whisker plot.

Use the data. Round to the nearest tenth. 8, 5, 5, 8, 6, 5, 3, 4

12. Find the mean. _____
13. Find the median. _____
14. Find the mode. _____
15. Find the range. _____
16. Find the third quartile. _____
17. Find the first quartile. _____
18. Make a box-and-whisker plot.

Name _____ Date _____ Class _____

Chapter Test
Chapter 4 Form A, continued

Use the data.
21, 21, 19, 22, 24, 19, 21, 24, 21, 24, 24, 22, 19, 22, 21, 19, 24, 21

19. Make a frequency table.

20. Make a bar graph.

Explain why the graph is misleading.

21.

(Bar graph: 70–79: 30, 80–89: 40, 90–99: 20; y-axis 15 to 45)

_____
_____
_____
_____

Use the data to work with scatter plots.

Test Score	70	85	60	95	70	80	75	85	90
Hours Studied	5	6	3	7	4	5	5	5	7

22. Use the data to construct a scatter plot.

23. Draw a line of best fit on the scatter plot you drew for problem 22.

_____
_____

24. Does the data set have a positive, negative, or no correlation? Explain.

_____
_____
_____
_____

Copyright © by Holt, Rinehart and Winston.
All rights reserved.

Holt Pre-Algebra

Name _____ Date _____ Class _____

CHAPTER 4 Chapter Test
Form B

Identify the population, sample, and sampling method.

1. 250 students in an elementary school were randomly selected and asked what was their favorite color.

   _____

   _____

2. Every twentieth visitor to the local zoo was asked a series of questions about the animal exhibits.

   _____

   _____

3. Five grocery stores in a city were randomly selected and 100 customers were randomly surveyed from each store as to their favorite flavor of ice cream.

   _____

   _____

4. Make a back-to-back stem-and-leaf plot.

Test Scores for 2 Math Classes

Class #1	Class #2
98 85 70 93	88 94 82 100
85 74 98 81	88 100 84 100
89 78 100 85	88 94 94 71
89 93 85 81	94 82 82 94

Use the data. 98, 95, 89, 85, 89, 90

5. Find the mean. _____

6. Find the median. _____

7. Find the mode. _____

8. Find the range. _____

9. Find the third quartile. _____

10. Find the first quartile. _____

11. Make a box-and-whisker plot.

Use the data. Round to the nearest tenth. 28, 25, 28, 22, 28, 29, 23, 24

12. Find the mean. _____

13. Find the median. _____

14. Find the mode. _____

15. Find the range. _____

16. Find the third quartile. _____

17. Find the first quartile. _____

18. Make a box-and-whisker plot.

Copyright © by Holt, Rinehart and Winston.
All rights reserved.

Holt Pre-Algebra

Name _____ Date _____ Class _____

CHAPTER 4 Chapter Test
Form B, continued

Use the test score data to answer the questions.
38,15,18,17,28,29,22,24,34,20,35,31,25, 33,14

19. Make a frequency table with an interval of 10.

20. Make a histogram.

Explain why the graph is misleading.

21.   Bookings

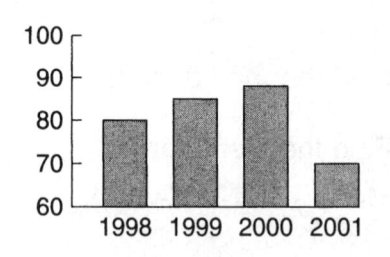

Use the data to work with scatter plots.

Test Score	55	75	50	80	95	90	85	80	85	90
Hours Watching TV per Week	22	15	25	10	7	9	10	6	15	7

22. Use the data to construct a scatter plot. Draw a line of best fit.

23. Draw a line of best fit on the scatter plot you drew for problem 22.

24. Does the data set have a positive, negative, or no correlation? Explain.

Copyright © by Holt, Rinehart and Winston.
All rights reserved.

Holt Pre-Algebra

Name _____ Date _____ Class _____

Chapter 4 Test Form C

Identify the population, sample, and sampling method.

1. Seven elementary schools in a large city were randomly selected and then 100 students were randomly selected from each school. Each student was asked how much time each week he spends on homework.

   _____
   _____
   _____
   _____

2. A manufacturer of automobile tires wants to check his best brand of tires for wear. He randomly selects 10 tires and puts them on the testing machine.

   _____
   _____

3. A local grocery store wants to know from how far away people come to shop at the store. Every 5th customer is asked how far he or she drove to get to the store.

   _____
   _____

4. Make a back-to-back stem-and-leaf plot.

   Test Scores for 2 Math Classes

Class #1	Class #2
78 75 76 97	85 92 87 95
81 84 78 83	84 94 87 100
82 98 95 85	87 82 94 94
89 96 75 83	74 100 82 77

Use the data. 198, 195, 189, 185, 189, 190

5. Find the mean. _____

6. Find the median. _____

7. Find the mode. _____

8. Find the range. _____

9. Find the third quartile. _____

10. Find the first quartile. _____

11. Make a box-and-whisker plot.

Chapter 4 Test
Form C, continued

Use the data. 1128, 507, 1634, 989, 1350, 1275, 1647, 1301, 1035

12. Find the mean. _____

13. Find the median. _____

14. Find the mode. _____

15. Find the range. _____

16. Find the third quartile. _____

17. Find the first quartile. _____

18. Make a box-and-whisker plot.

Use the test score data.
138, 115, 128, 117, 128, 129, 122, 124, 134, 120, 135, 131, 125, 133, 114

19. Make a frequency table with an interval of 10.

20. Make a histogram

21. The graph shows the number of cars and trucks on a highway. Explain why the graph is misleading.

= 50 cars
= 50 trucks
100 cars   150 trucks

_____
_____
_____

Test Scores	85	80	67	89	74	91	81	79	96	79
Hours of Exercise	3.0	8.0	3.0	8.5	1.0	8.0	5.5	9.5	2.0	2.5

22. Use the data above to construct a scatter plot. Draw a line of best fit.

23. Draw a line of best fit on the scatter plot you drew for problem 22.

_____
_____

24. Does the data set have a positive, negative, or no correlation? Explain.

_____
_____

Name _____ Date _____ Class _____

CHAPTER 5 Chapter Test
Form A

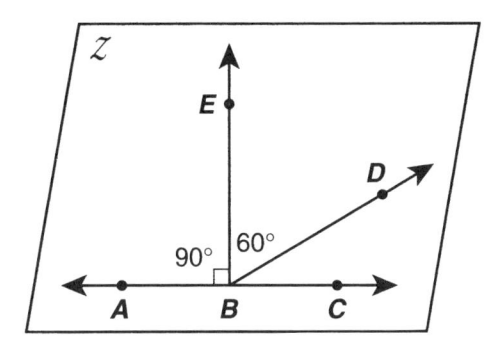

1. Name one point in the figure.

   _____

2. Name a line in the figure.

   _____

3. Name a plane in the figure.

   _____

4. Name one line segment in the figure.

   _____

5. Name one ray in the figure.

   _____

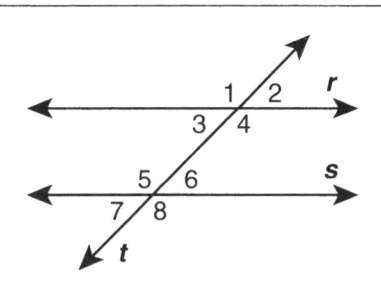

6. Name one angle congruent to ∠7.

   _____

7. Which line is the transversal?

   _____

8. If m∠6 is 40°, what is m∠5?

   _____

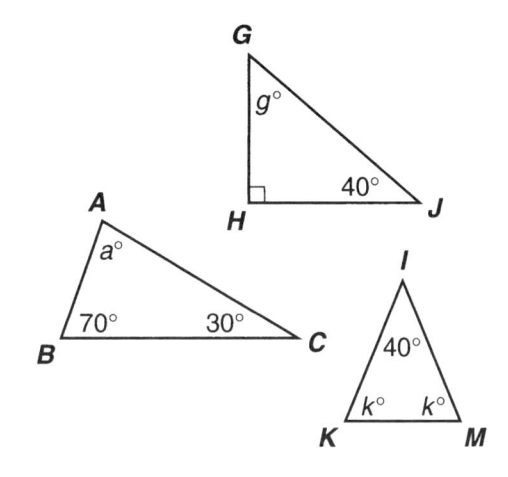

9. Find g in the right triangle.

   _____

10. Find a in the acute triangle.

    _____

11. Which triangle is an isosceles triangle?

    _____

Find the sum of the angle measures.

12. pentagon

    _____

13. triangle

    _____

14. rectangle

    _____

Copyright © by Holt, Rinehart and Winston.
All rights reserved.

Holt Pre-Algebra

Name _____ Date _____ Class _____

CHAPTER 5 **Chapter Test**
Form A, continued

Graph the quadrilaterals with the given vertices. Write all names.

15. (1, 1), (1, 3), (3, 3), (3, 1)

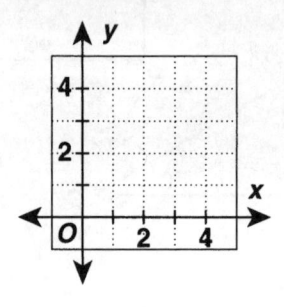

16. (2, 3), (6, 3), (6, 0), (2, 0)

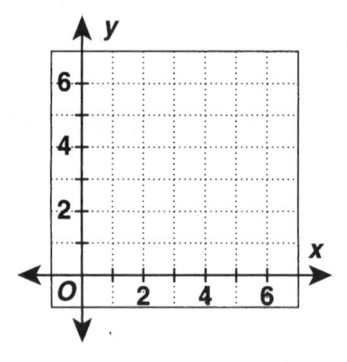

Write a congruence statement.

17.

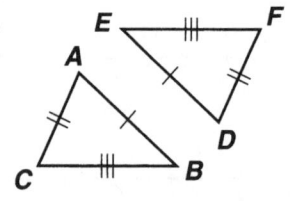

18.

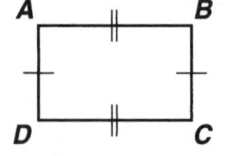

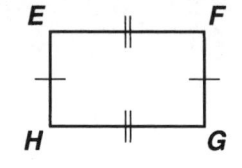

Identify each as a translation, rotation, reflection, or none of these.

19.

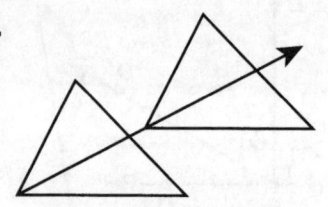

20.

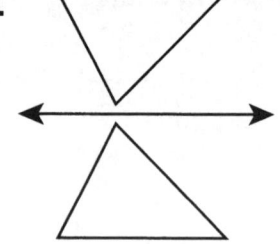

Complete the figure. The dashed line is the line of symmetry.

21.

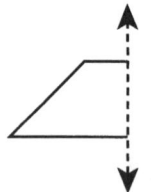

22.

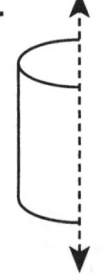

Create a tessellation with the given figure.

23.

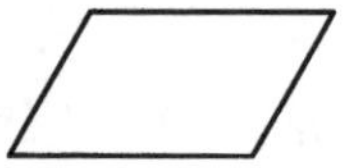

Copyright © by Holt, Rinehart and Winston.
All rights reserved.

Holt Pre-Algebra

Name _____ Date _____ Class _____

Chapter 5 Chapter Test
Form B

1. Name three points in the figure.

2. Name a line in the figure.

3. Name a plane in the figure.

4. Name two line segments in the figure.

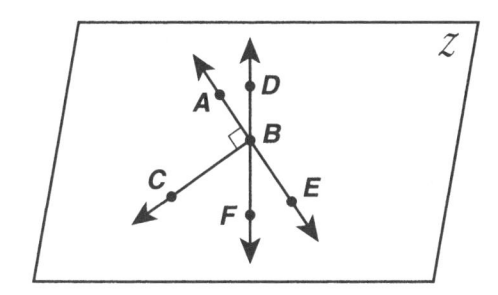

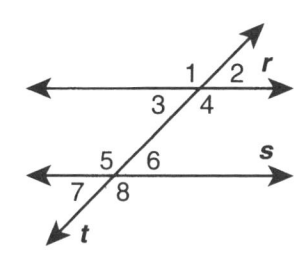

5. Name two rays in the figure.

6. Name two angles congruent to ∠3.

7. Which line is the transversal?

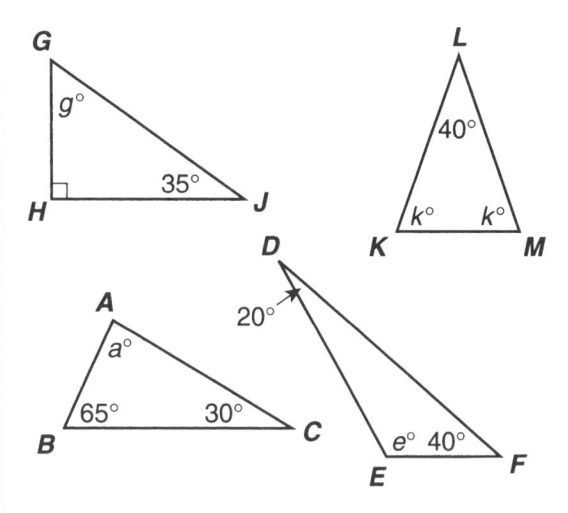

8. If m∠6 is 40°, what is m∠7?

9. Find e in the obtuse triangle.

10. Find the unknown angle measures in the isosceles triangle.

11. Name one acute triangle.

Find the sum of the angle measures.

12. hexagon

13. heptagon

14. octagon

Holt Pre-Algebra

Name _____ Date _____ Class _____

Chapter Test
Chapter 5 Form B, continued

Graph the quadrilaterals with the given vertices. Write all names.

15. (2, 2), (6, 2), (6, −1), (2, −1)

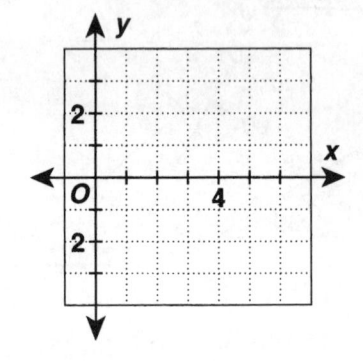

16. (−2, 4), (−3, 1), (1, 1), (2, 4)

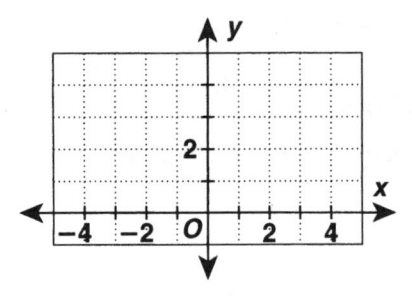

Write a congruence statement.

17.

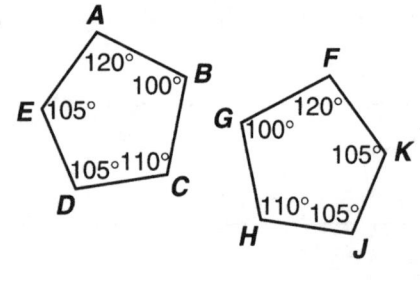

18.

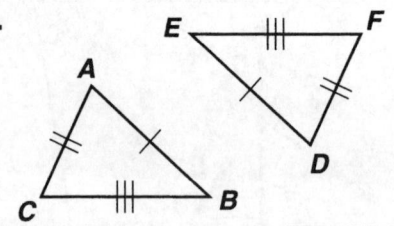

Identify each as a translation, rotation, reflection, or none of these.

19.

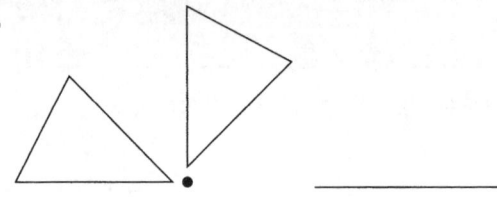

20.

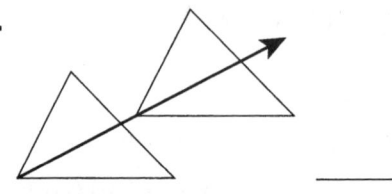

Complete the figure. The dashed line is the line of symmetry.

21.

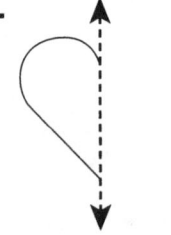

22.

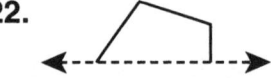

Create a tessellation with the given figure.

23.

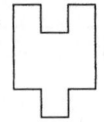

Name _____ Date _____ Class _____

Chapter Test
Form C

CHAPTER 5

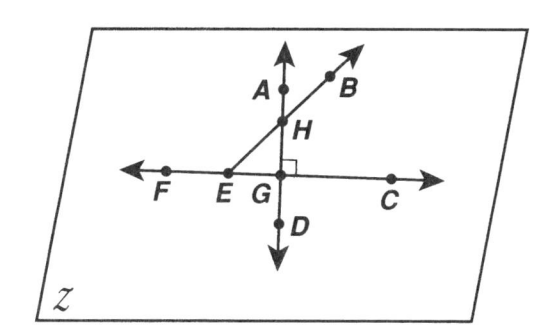

1. Name all points in the figure.
   _____
   _____

2. Name all lines in the figure.
   _____

3. Name a plane in the figure.
   _____

4. Name eight line segments in the figure.
   _____
   _____

5. Name all acute angles in the figure.
   _____

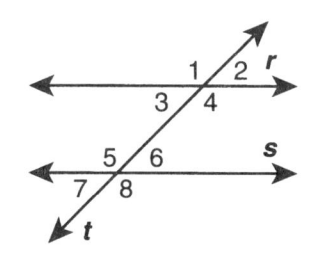

6. Name all angles congruent to ∠8.
   _____

7. Which line is the transversal? _____

8. If m∠7 is 35°, what is m∠4? _____

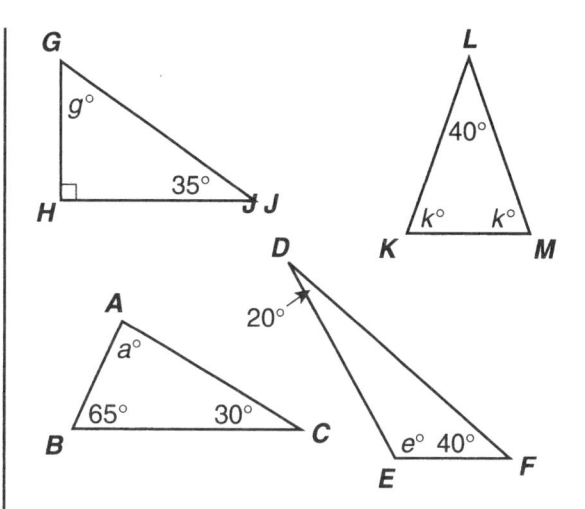

9. Find *a* in the acute triangle.
   _____

10. Find *e* in the obtuse triangle.
    _____

11. Which triangles are scalene triangles?
    _____

Find the sum of the angle measures.

12. hexagon
    _____

13. heptagon
    _____

14. octagon
    _____

Copyright © by Holt, Rinehart and Winston.
All rights reserved.

Holt Pre-Algebra

Name _____ Date _____ Class _____

Chapter Test
CHAPTER 5 Form C, continued

Graph the quadrilaterals with the given vertices. Write all names.

15. (−3, 0), (−2, −2), (−3, −4), (−4, −2)

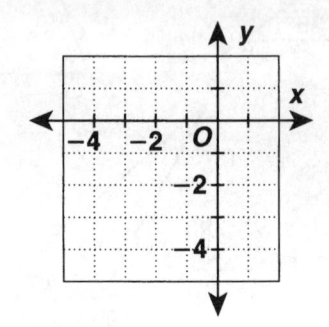

_____

16. (−4, 3), (0, 3), (1, 1), (−3, 1)

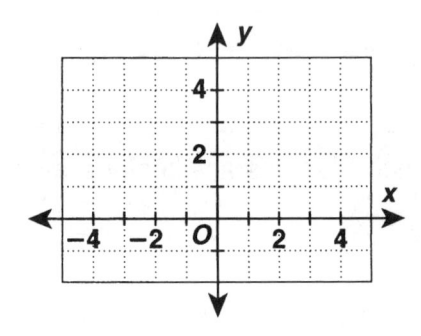

_____

Write a congruence statement.

17.

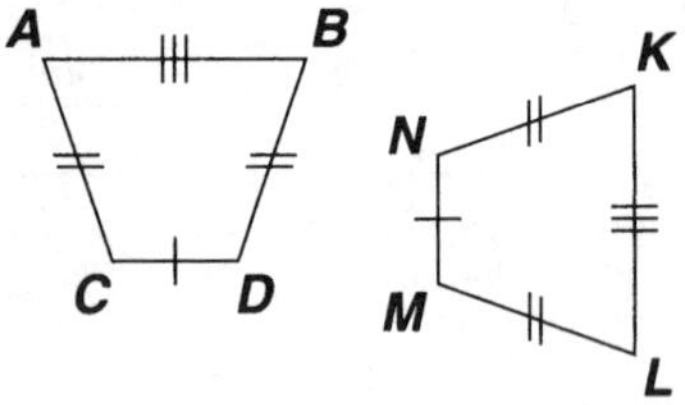

18.

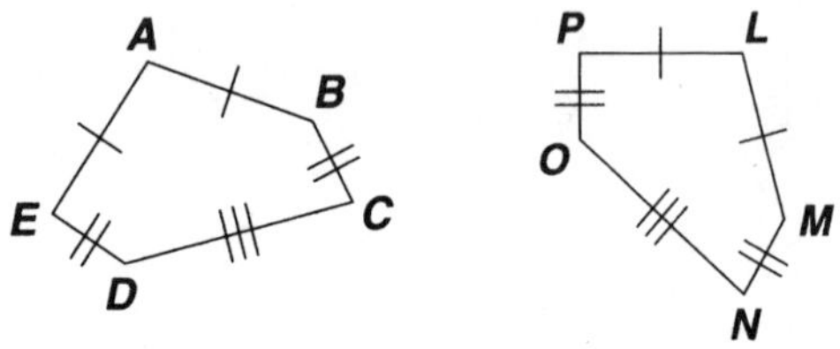

Identify each as a translation, rotation, reflection, or none of these.

19.

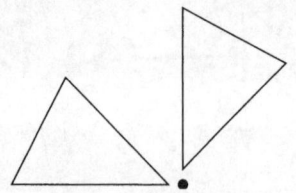

20.

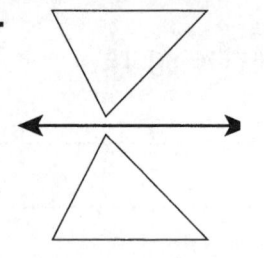

Complete the figure. The dashed line is the line of symmetry.

21.

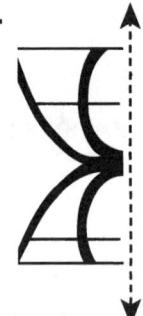

22.

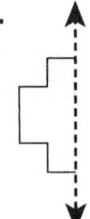

Create a tessellation with the given figure.

23.

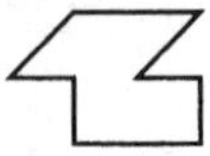

62 Holt Pre-Algebra

Name _____ Date _____ Class _____

Chapter Test
Chapter 6 — Form A

Find the perimeter of each figure.

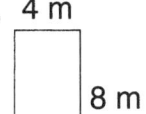

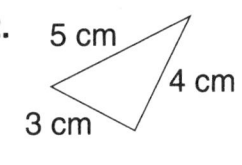

1. 4 m, 8 m (rectangle)

2. 5 cm, 4 cm, 3 cm (triangle)

_____ _____

Graph and find the area of each figure with the given vertices.

3. (0, 0), (0, 4), (3, 4), (3, 0)

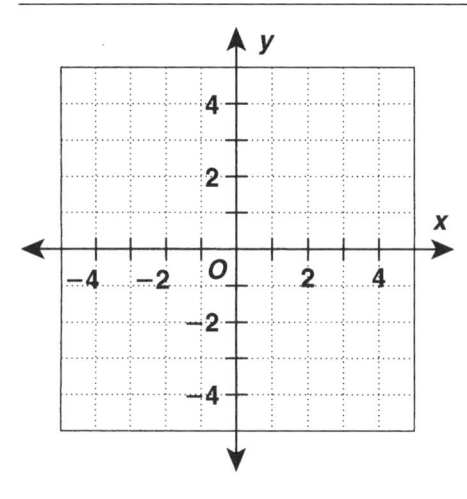

_____

4. (1, 0), (3, 4), (5, 0)

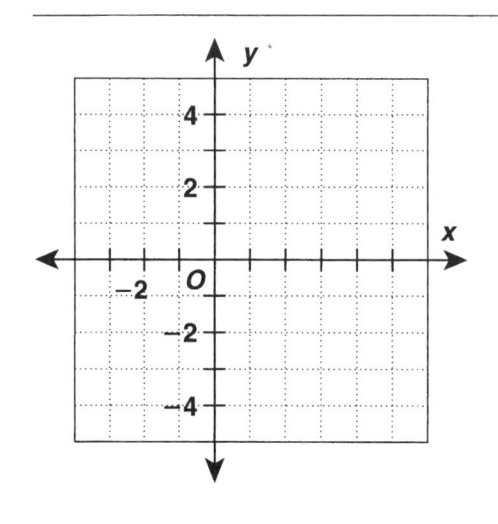

_____

5. Find the length of the hypotenuse.

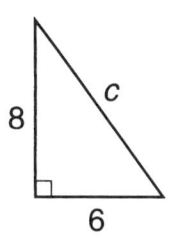

(right triangle with legs 8 and 6, hypotenuse c)

_____

6. Find the unknown side.

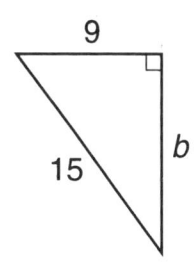

(right triangle with sides 9, 15, and b)

_____

Find the circumference and area of each circle, both in terms of π and to the nearest tenth of a unit using 3.14 for π.

7. circle with radius 7 in.

_____
_____

8. circle with diameter 16 cm

_____
_____

9. Use isometric dot paper to sketch a rectangular box that is 4 units long, 2 units wide, and 3 units tall.

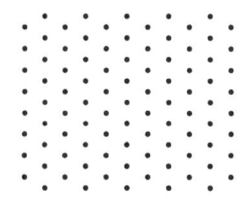

Copyright © by Holt, Rinehart and Winston.
All rights reserved.

Holt Pre-Algebra

Name _____ Date _____ Class _____

CHAPTER 6 Chapter Test
Form A, continued

10. Sketch a one-point perspective drawing of a cube.

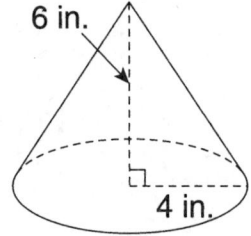

Find the surface area of each figure to the nearest tenth. Use 3.14 for π.

15. the figure in Exercise 11

_____

16. the figure in Exercise 12

_____

17. 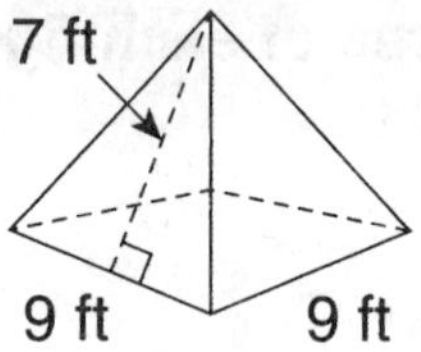 7 ft, 9 ft, 9 ft

_____

18. 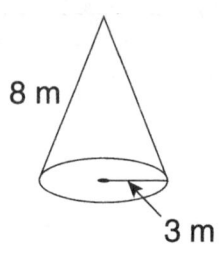 8 m, 3 m

_____

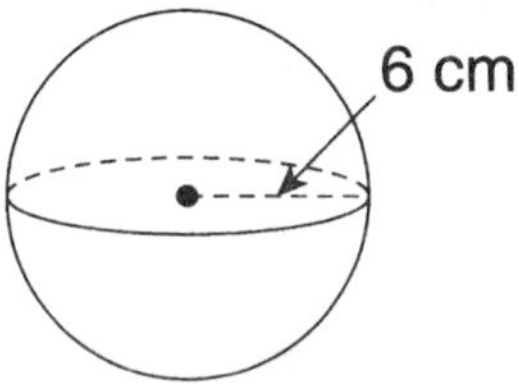

Find the volume of each figure to the nearest tenth. Use 3.14 for π.

11. 8 cm, 3 cm _____

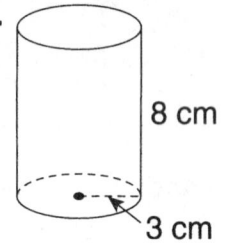

12. 4 in., 2 in., 6 in. _____

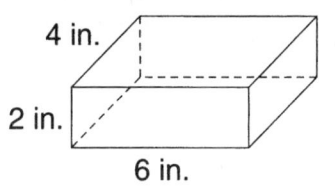

13. 5 cm, 3 cm, 4 cm _____

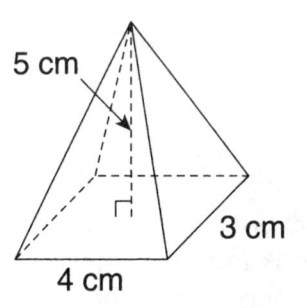

14. 6 in., 4 in. _____

6 cm (sphere)

19. Find the surface area of the sphere, both in terms of π and to the nearest tenth of a unit using 3.14 for π.

_____

20. Find the volume of the sphere, both in terms of π and to the nearest tenth of a unit using 3.14 for π.

_____

Holt Pre-Algebra

Name _____ Date _____ Class _____

CHAPTER 6 Chapter Test
Form B

Find the perimeter of each figure.

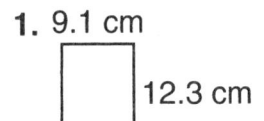

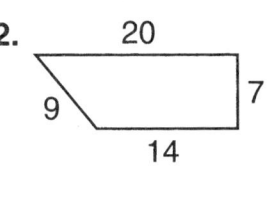

1. 9.1 cm, 12.3 cm (rectangle)

2. Trapezoid with sides 20, 7, 14, 9

_____   _____

Graph and find the area of each figure with the given vertices.

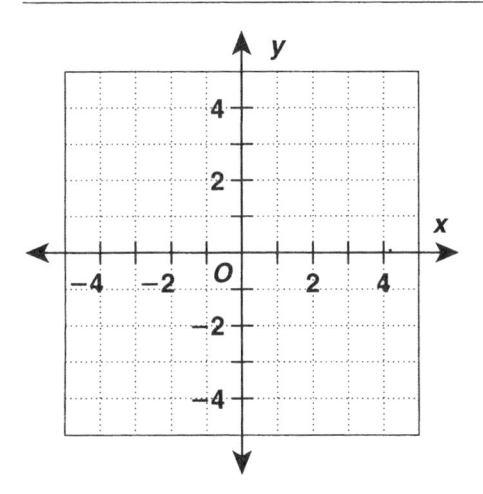

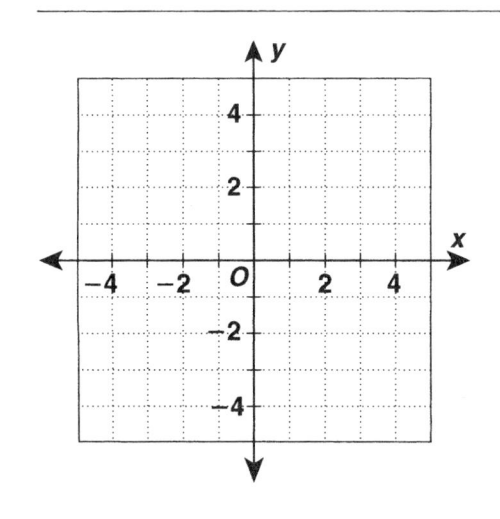

3. $(-2, 1), (0, 4), (5, 4), (3, 1)$

_____

4. $(-3, 1), (-1, 5), (2, 1)$

5. Find the length of the hypotenuse to the nearest tenth.

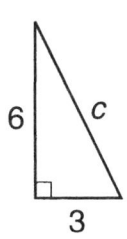

(right triangle with legs 6 and 3, hypotenuse c)

_____

6. Find the unknown side to the nearest tenth.

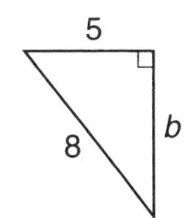

(right triangle with hypotenuse 8, leg 5, leg b)

_____

Find the circumference and area of each circle, both in terms of π and to the nearest tenth of a unit using 3.14 for π.

7. circle with radius 15.2 in.

_____

_____

8. circle with diameter 28.6 cm

_____

_____

9. Use isometric dot paper to sketch a cube 5 units on each side.

Name _____ Date _____ Class _____

Chapter Test
Chapter 6 Form B, continued

10. Sketch a one-point perspective drawing of a triangular box.

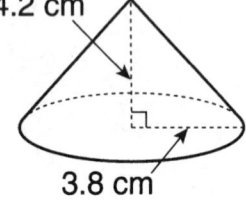

Find the surface area of each figure to the nearest tenth. Use 3.14 for π.

15. the figure in Exercise 11

_____

16. a rectangular prism 7 in. by 4 in. by 5 in.

_____

Find the volume of each figure to the nearest tenth. Use 3.14 for π.

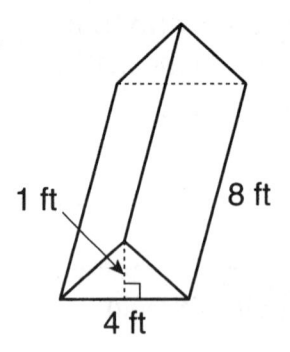

11.
16 cm
7 cm

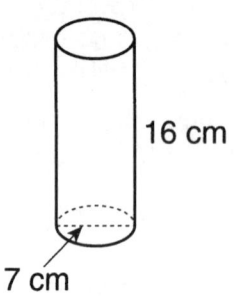

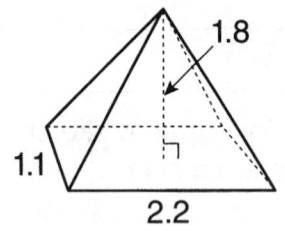

17. 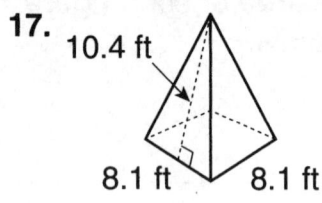
10.4 ft
8.1 ft   8.1 ft

_____

12.
1 ft   8 ft
4 ft

18. 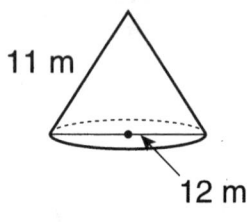
11 m
12 m

_____

13.
1.8
1.1
2.2

9 cm

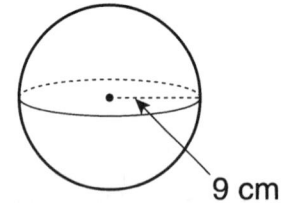

19. Find the surface area of the sphere, both in terms of π and to the nearest tenth of a unit using 3.14 for π.

_____

20. Find the volume of the sphere, both in terms of π and to the nearest tenth of a unit using 3.14 for π.

14. 4.2 cm
3.8 cm

_____

Copyright © by Holt, Rinehart and Winston.
All rights reserved.

Holt Pre-Algebra

Name _____ Date _____ Class _____

CHAPTER 6 Chapter Test
Form C

Find the perimeter of each figure.

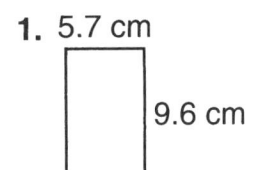

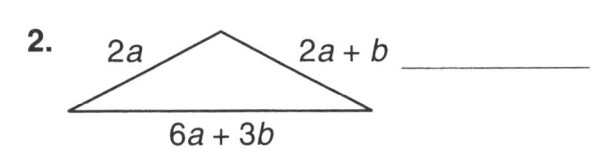

1. 5.7 cm, 9.6 cm (rectangle) _____

2. Triangle with sides 2a, 2a + b, 6a + 3b _____

Graph and find the area of each figure with the given vertices.

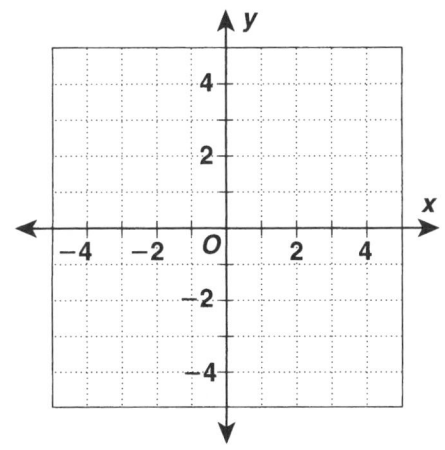

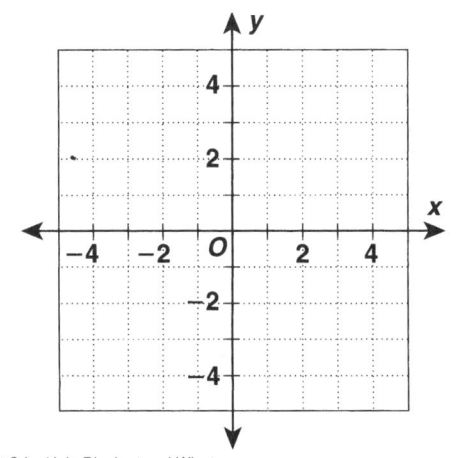

3. (−3, 3), (−3, 0), (−2, 0), (−2, −2), (1, −2), (1, 0), (3, 0), (3, 2), (0, 2), (0, 3)

_____

4. (−3, −2), (−2, 2), (3, 2), (4, −2)

5. Find the length of the hypotenuse to the nearest hundredth.

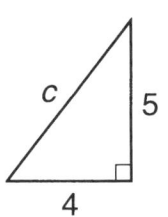

Triangle with legs 4 and 5, hypotenuse c _____

6. Find the unknown side to the nearest hundredth.

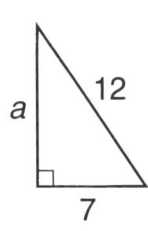

Triangle with side a, hypotenuse 12, other leg 7 _____

Find the circumference and area of each circle, both in terms of π and to the nearest tenth using 3.14 for π.

7. circle whose circumference is one-sixth the circumference of a circle with radius 18 in.

_____

8. circle whose area is four times the area of a circle with diameter 10 cm

_____

9. Use isometric dot paper to sketch a rectangular box with a base 5 units long by 3 units wide and a height of 2 units.

Copyright © by Holt, Rinehart and Winston.
All rights reserved.

Holt Pre-Algebra

Name _____ Date _____ Class _____

Chapter Test
Form C, continued

10. Sketch a two-point perspective drawing of a square box.

14.
20.1
18.6

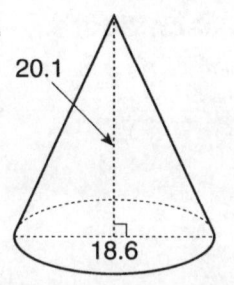

Find the surface area of each figure to the nearest tenth. Use 3.14 for π.

15. the figure in Exercise 11 _____

16. the figure in Exercise 12 _____

Find the volume of each figure to the nearest tenth. Use 3.14 for π.

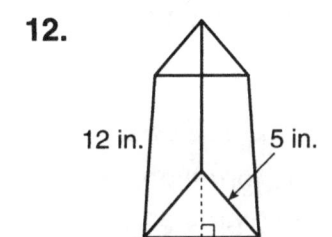

11. _____
18.4 cm
12.2 cm

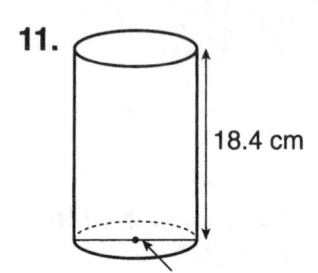

17. 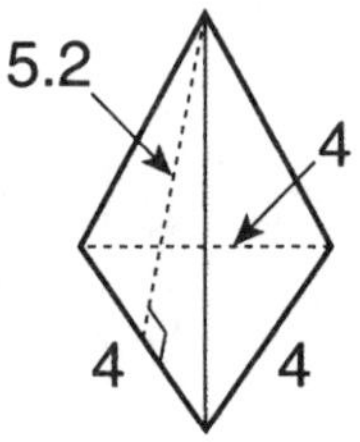
5.2
4
4 4

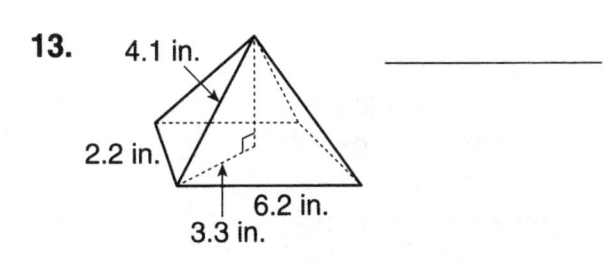

12. _____
12 in. 5 in.
3 in. 3 in.

18. the figure in Exercise 14 _____

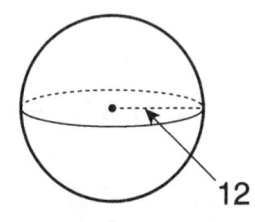

13. _____
4.1 in.
2.2 in.
6.2 in.
3.3 in.

12

19. Find the surface area of the sphere, both in terms of π and to the nearest tenth of a unit using 3.14 for π.

_____

20. Find the volume of the sphere, both in terms of π and to the nearest tenth of a unit using 3.14 for π.

_____

Copyright © by Holt, Rinehart and Winston.
All rights reserved.

Holt Pre-Algebra

Name _____ Date _____ Class _____

Chapter Test
CHAPTER 7 — Form A

Find two ratios that are equivalent to each given ratio.

1. $\dfrac{5}{10}$ _____

2. $\dfrac{4}{6}$ _____

3. $\dfrac{16}{4}$ _____

4. $\dfrac{21}{27}$ _____

Simplify to tell whether the ratios form a proportion.

5. $\dfrac{4}{12}$ and $\dfrac{2}{8}$ _____

6. $\dfrac{1}{2}$ and $\dfrac{4}{8}$ _____

7. $\dfrac{1}{3}$ and $\dfrac{2}{6}$ _____

Find the unit price for each offer and tell which is the better buy.

8. 20-oz box of cereal for $3.80; 15-oz box of cereal for $3.15

9. 10 blank CDs for $2.50; 15 blank CDs for $3.00

Find the appropriate factor for each conversion.

10. kilometers to meters

11. gallons to quarts

12. A car travels 5 miles in 6 minutes. What is its speed in miles per hour?

13. A home improvement store sells 2355 feet of wire per day. How many yards of wire does the store sell per day?

Solve each proportion.

14. $\dfrac{3}{9} = \dfrac{x}{3}$ _____

15. $\dfrac{32}{t} = \dfrac{4}{1}$ _____

16. $\dfrac{12}{22} = \dfrac{18}{p}$ _____

Holt Pre-Algebra

Name _____ Date _____ Class _____

CHAPTER 7 Chapter Test
Form A, continued

Dilate each figure by the given scale factor with the origin as the center of dilation.

17. Triangle with vertices A(1, 2), B(4, 1), C(4, 4), scale factor = 2

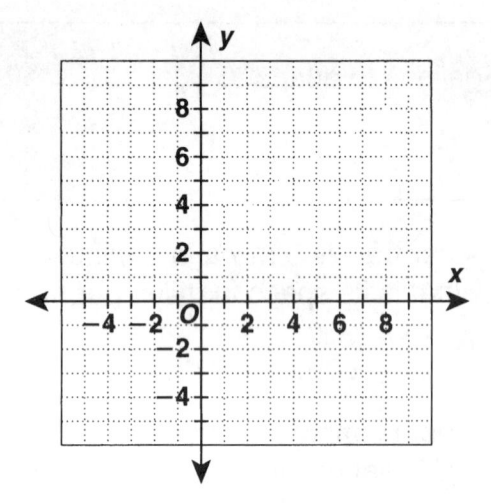

18. Triangle with vertices A(4, 2), B(8, 2), C(6, 6), scale factor $\frac{1}{2}$

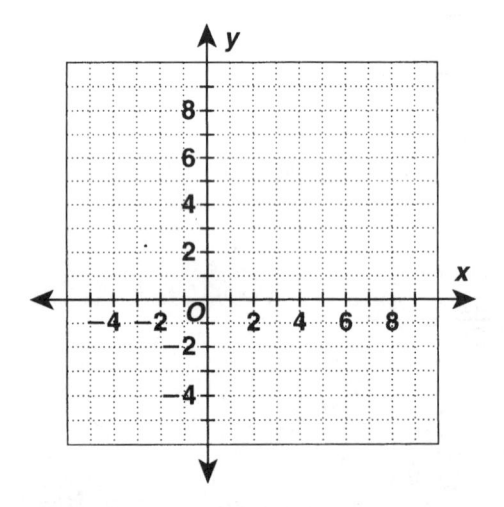

Use the properties of similar figures to answer each question.

19. Rectangle A has length 11 m and width 7 m. Rectangle B has length 33 m and width 21 m. Are rectangles A and B similar?

_____

20. A soccer field for 13-year olds measures 50 yards wide by 100 yards. 8-year olds play on a similar soccer field that is 20 yards wide. How long is the field for 8-year olds?

_____

21. Julia's room is 4 in. long on a scale drawing. If her room is actually 16 ft long, what is the scale?

_____

22. A drawing of a 78-foot long building was built using a scale of 1 in.:8 ft. What is the length of the drawing?

_____

Tell whether each scale reduces, enlarges, or preserves the size of the actual object.

23. 1 cm:12 m _____

24. 6 ft:10 in. _____

25. 1 km: 1000 m _____

A 3-in. cube is built from small cubes, each 1 in. on a side. Compare the following values.

26. the side lengths

_____

_____

27. the volumes

_____

_____

Holt Pre-Algebra

Name _____ Date _____ Class _____

Chapter Test
CHAPTER 7 Form B

Find two ratios that are equivalent to each given ratio.

1. $\dfrac{2}{9}$ _____

2. $\dfrac{50}{15}$ _____

3. $\dfrac{11}{17}$ _____

4. $\dfrac{18}{16}$ _____

Simplify to tell whether the ratios form a proportion.

5. $\dfrac{4}{26}$ and $\dfrac{2}{13}$ _____

6. $\dfrac{18}{60}$ and $\dfrac{3}{10}$ _____

7. $\dfrac{5}{25}$ and $\dfrac{15}{50}$ _____

Find the unit price for each offer and tell which is the better buy.

8. 20 blank CDs for $2.79; 12 blank CDs for $1.20

9. 6 paperback books for $19.00; 8 paperback books for $26.00

Find the appropriate factor for each conversion.

10. months to years

11. pounds to ounces

12. A woodworker can put together 2 wood toy trains per day. How many trains could the woodworker make in 8 weeks?

13. A store sells 8-ounce packages of mushrooms for $1.29. What is the cost of 3 pounds of mushrooms?

Solve each proportion.

14. $\dfrac{4}{10} = \dfrac{y}{20}$ _____

15. $\dfrac{r}{0.32} = \dfrac{3}{2}$ _____

16. $\dfrac{11}{q} = \dfrac{5}{2}$ _____

Name _____ Date _____ Class _____

Chapter Test
Chapter 7 Form B, continued

Dilate each figure by the given scale factor with the origin as the center of dilation.

17. Triangle with vertices A(2, 2), B(8, 4), C(4, 8), scale factor = $\frac{1}{4}$

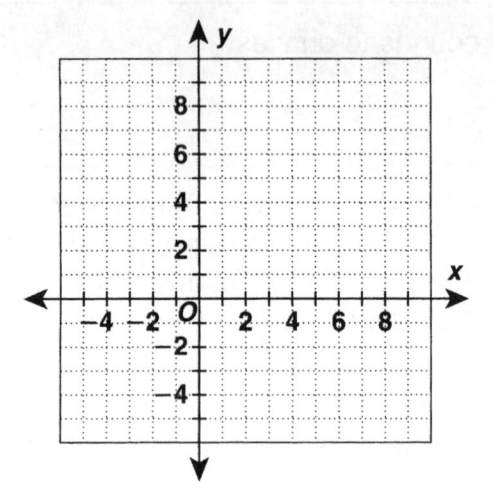

18. Quadrilateral with vertices A(2, 2), B(8, 2), C(8, 4), D(2, 6), scale factor 1.5

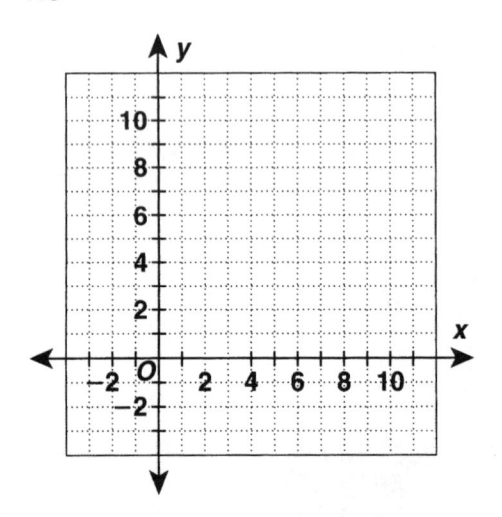

19. Jess made two picture frames. One frame is 8 inches by 14 inches. The other frame is 12 inches by 22 inches. Are the frames similar?

_____

20. Lisa is having a 5 in. by 7 in. photo made into a similar poster. If the poster is 2 ft wide, how long will it be?

_____

21. If the scale is 1 cm:8 m, how tall is a drawing of a 654-m skyscraper?

_____

22. A drawing of an airplane hangar was made using a scale of 1 in.:20 ft. If the hangar is actually 250 feet wide, how wide is the drawing?

_____

Tell whether each scale reduces, enlarges, or preserves the size of the actual object.

23. 15 ft:1 in. _____

24. 10 mm:1 cm _____

25. 2 m:10 km _____

An 8-cm cube is built from small cubes, each 1 cm on a side. Compare the following values.

26. the side lengths

_____
_____

27. the volumes

_____
_____
_____

Copyright © by Holt, Rinehart and Winston.
All rights reserved.

72

Holt Pre-Algebra

Name _____ Date _____ Class _____

Chapter 7 Chapter Test
Form C

Find two ratios that are equivalent to each given ratio.

1. $\frac{5}{9}$ _____

2. $\frac{2}{12}$ _____

3. $\frac{15}{6}$ _____

4. $\frac{14}{8}$ _____

Simplify to tell whether the ratios form a proportion.

5. $\frac{6}{16}$ and $\frac{9}{24}$ _____

6. $\frac{36}{28}$ and $\frac{10}{7}$ _____

7. $\frac{21}{27}$ and $\frac{7}{8}$ _____

Find the unit price for each offer and tell which is the better buy.

8. $7.98 for a 3-pound ham; $10.84 for a 5-pound ham

_____

9. A 12-ounce drink for $1.40; a 20-ounce drink for $2.70.

_____

Find the appropriate factor for each conversion.

10. weeks to hours

_____

11. miles to yards

_____

12. A car wash cleans automobiles at a rate of 35 per hour. How many cars do they clean in an 8-hour day?

_____

13. Zara biked for 2 hours at an average rate of 15 meters per second. How many kilometers did she bike?

_____

Solve each proportion.

14. $\frac{6}{4} = \frac{x}{5}$ _____

15. $\frac{33}{t} = \frac{4}{1}$ _____

16. $\frac{12}{1.5} = \frac{40}{p}$ _____

Holt Pre-Algebra

Name _____ Date _____ Class _____

Chapter 7 Chapter Test
Form C, continued

Identify the scale factor used in each dilation.

17.

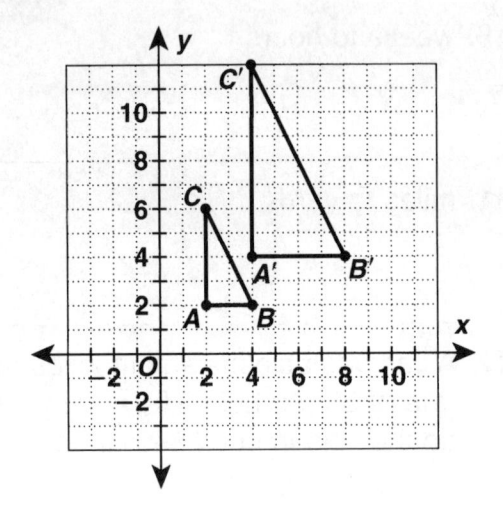

18.

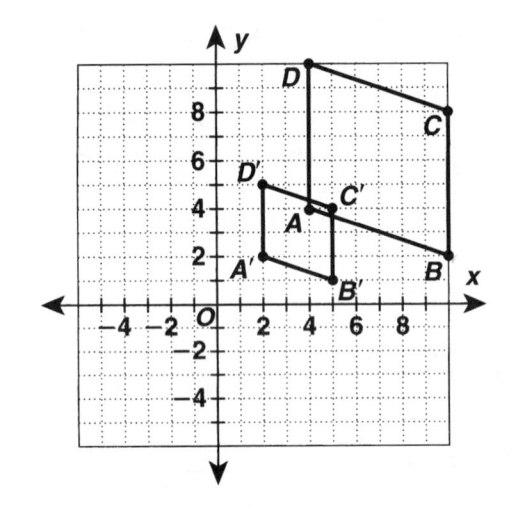

Use the properties of similar figures to answer each question.

19. Bart is using two different size triangles to make a tile design. One has sides of 4 in., 6 in., and 7 in. The other has sides of 12 in., 18 in., and 20 in. Are the two triangles similar? Explain.

20. The two triangles are similar. Use the scale factor to solve for x.

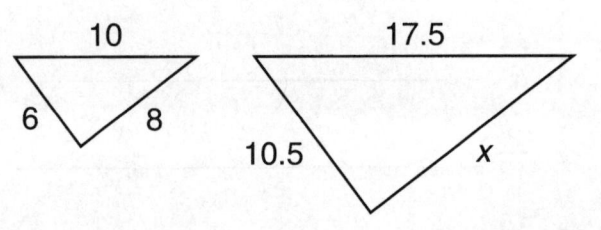

21. A scale drawing of a rectangular swimming pool is 6 in. by 10.5 in. If the scale is 0.25 in.:1 ft, what is the perimeter of the actual pool?

22. If the scale of a drawing is 2 in.:35 ft, how long would a 49-foot fence be in the drawing?

Find the scale factor and tell whether it reduces, enlarges, or preserves the size of the actual object.

23. 1 cm:12 m _____

24. 2 ft:10 in. _____

25. 8 in.:32 ft _____

For each cube, a reduced scale model is built using a scale factor of 0.25. Find the length of the model and the number of 1-cm cubes used to build it.

26. a 16-cm cube

27. a 48-cm cube

Name _____ Date _____ Class _____

CHAPTER 8 Chapter Test
Form A

Find the missing ratio or percent equivalent for each letter on the number line.

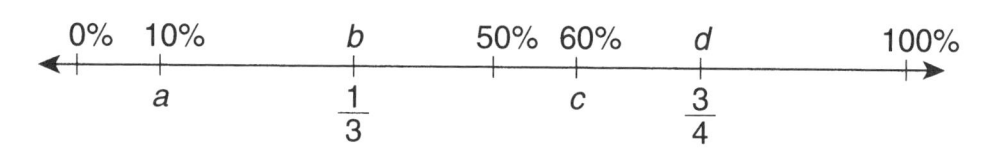

```
0%  10%       b      50% 60%    d         100%
          a       1/3        c      3/4
```

1. a = _____   2. b = _____   3. c = _____

4. Write $\frac{7}{10}$ as a percent.

5. Write 40% as a fraction.

6. What percent of 175 is 28?

7. What percent of 305 is 122?

8. 9.6 is 15% of what number?

9. 40% of what number is 114?

10. Estimate 25% of 203.

11. Estimate 12% of 80.

12. About 35 acres of a 125-acre farm are planted with corn. What percent of the farm's fields are planted with corn?

13. Jake runs the 100-meter dash in track. Erika runs a race 400% the distance of Jake's race. How long is Erika's race?

14. Derek can carry 65% of his weight in his backpack while camping. If his backpack weighs 88.4 pounds, how much does Derek weigh?

15. Kersten has 12 postcards from New York City. This is 30% of her total postcard collection. How many postcards does she have in her collection?

16. Find the percent increase or decrease from 25 to 20.

Copyright © by Holt, Rinehart and Winston.
All rights reserved.

Holt Pre-Algebra

Name _____ Date _____ Class _____

Chapter Test
CHAPTER 8 Form A, continued

Give answers to the nearest percent.

17. A toy store sold a toy train set for $49.95 last year. This year, the same train set costs $52.50. What was the percent increase of the cost of the train set?

_____

18. The president of a small company made $72,000 last year. She cut her salary this year to $60,000 because the company was not doing as well. What was the percent decrease in her salary to the nearest percent?

_____

19. A rain jacket costs $52. It is on sale for 20% off. Estimate the discount on the jacket.

_____

20. In a state with a sales tax rate of 6%, Alex bought paper for his printer for $17.99. How much was his sales tax to the nearest penny?

_____

21. Theo earned $3045 over the summer as a lifeguard. Of this, $669.90 was withheld for taxes. What percent of his income was withheld?

_____

22. Clarence earns a 10% commission on sales plus a $200 weekly salary. In one week, his sales totaled $2100. What was his total pay that week?

_____

23. Shannon invested $5000 in a bond. Her total simple interest on her investment after 3 years was $1200. What was the yearly interest rate on her investment?

_____

24. Tess borrowed $10,500 from the bank to buy a car. The length of her loan is 4 years with a simple interest rate of 7.5%. How much will she pay in interest if she pays off the loan at the end of the 4 years?

_____

25. Stephen deposited $2500 in a savings account that earned an annual simple interest rate of 4%. When he closed his account, he had $3000. How long did he have his account?

_____

Name _____ Date _____ Class _____

Chapter Test
Form B (Chapter 8)

Find the missing ratio or percent equivalent for each letter on the number line.

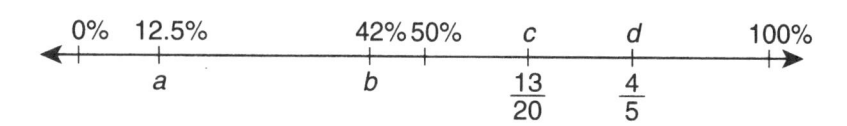

```
    0%  12.5%        42% 50%      c          d         100%
    |----+-------------+---+------+----------+-----------|
         a             b          13         4
                                  ──         ─
                                  20         5
```

1. a = _____   2. b = _____   3. c = _____

4. Write $\frac{17}{25}$ as a percent.

5. Write 29% as a decimal.

6. What percent of 67 is 134?

7. 48 is what percent of 192?

8. 12.5 is 20% of what number?

9. 120% of what number is 98.4?

10. Estimate 20% of 198.

11. Estimate 3015 out of 8999 as a percent.

12. Mille Lacs county covers 574 square miles of Minnesota. If Minnesota is 79,610 square miles, what percent of Minnesota is Mille Lacs county to the nearest tenth of a percent?

13. Elsa is 56 inches tall. Her brother's height is 65% of Elsa's height. How tall is her brother?

14. A store sold 156 winter coats during a sale. If this represented 60% of their total inventory, how many coats did they have before the sale?

15. Monica has 64 stamps commemorating Olympic games in her collection. This is 16% of her total collection. How many stamps does she have in her collection?

16. On sale, a sweater was reduced from $40 to $32. Find the percent of decrease.

Copyright © by Holt, Rinehart and Winston.
All rights reserved.

Holt Pre-Algebra

Chapter Test
Form B, continued

Give answers to the nearest percent.

17. A toy store sold a toy train set for $49.79 last year. This year, the same train set costs $51.29. What was the percent increase of the cost of the train set?

18. A small company had profits of $550,000 last year. This year, their profits were only $484,000. What was the percent decrease in their profits?

19. A pair of shoes are on sale for 25% off. They normally cost $42.95. Estimate the discount on the shoes.

20. In a state with a sales tax rate of 6.5%, Thomas bought a new DVD player for $256.99 and a DVD movie for $24.99. How much is the sales tax on his purchase to the nearest penny?

21. Kirk earned $4147 over the summer working as a waiter. $829.40 was taken out for taxes. What percent of his income was withheld for taxes?

22. Justin works as a car salesman where he earns 8% commission on his sales and no weekly salary. What will his weekly sales have to be to earn $3280 for the week?

23. Karen deposited $8500 in a college savings account for her grandson that earns an annual simple interest rate of 6.5%. What will be the total amount in the account in 10 years?

24. Sadie borrowed $3500 from a bank at an annual simple interest rate for a home remodeling project. After 4 years, she repaid the bank $4200. What was the interest rate of the loan?

25. If Aisha deposits $2250 in a savings account that earns 3.5% annual simple interest, how long must she keep the money in the account for its total to reach $2722.50?

Chapter Test
Form C

Gia's Budget

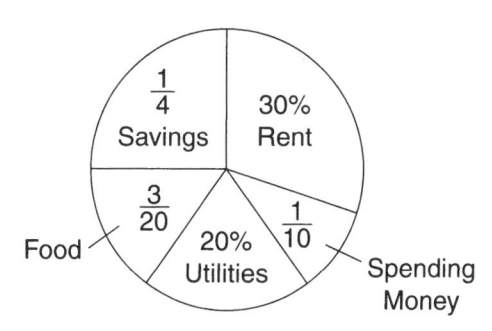

- 1/4 Savings
- 30% Rent
- 3/20 Food
- 20% Utilities
- 1/10 Spending Money

1. What percent of Gia's monthly income goes into savings?

2. What fraction of her income is used to pay for utilities?

3. What percent of her income is used to buy food?

4. Write $\frac{8}{25}$ as a percent.

5. Write 6% as a fraction.

6. What percent of 2950 is 531?

7. What percent of 460 is 621?

8. 51.45 is 35% of what number?

9. 23.5% of what number is 42.3?

10. Estimate 25% of 58.5.

11. Estimate 110% of 89.75.

12. Enrique has read 67.5% of a 354-page book. To the nearest page, how many pages has he read?

13. Sophia needs to earn at least 25% of her college tuition before her parents will help with the rest. Tuition for next year is $3410. How much does Sophia need to earn?

14. An artist is designing a statue of a man standing next to a horse. The man will be 8.5 feet tall and the horse will be 130% the height of the man. How tall will the horse be?

15. If 114 out of 355 students in the eighth grade play instruments, what percent of eighth grade students do not play instruments? Round your answer to the nearest percent.

Chapter 8 Test
Form C, continued

Give answers to the nearest hundredth of a percent.

16. The number of children taking swimming lessons at the community pool last summer was 274. The number registered for lessons this summer is 292. What is the percent increase?

17. A pair of Kelly's new pants shrunk in the wash. The inseam was 37.25 inches before they were washed and 36.375 inches after they were washed. What is the percent decrease in the length of the pants?

18. During a sale, the price of a DVD was decreased by 50%. By what percent must the sale price be increased to restore the original price?

19. The total area of the state of Florida is 58,560 square miles. The total land area of Florida is 54,252 square miles. Estimate what percent of Florida's total area is water.

20. Janine is paid $200 a week plus 12.5% commission on sales. What were her weekly sales if she earned $950?

21. Shaneece made $42,500 last year. Isak made $35,450 last year. If they are both taxed at 27.5% of the amount over $27,050, how much more tax did Shaneece pay than Isak?

22. 19.5% of Alisha's paycheck is withheld each week for taxes. If she earns $568 per week, how much money is her check written for?

23. A credit union advertises that if you put $3000 in a certificate of deposit, you can earn $1800 in ten years. What yearly simple interest rate does the certificate of deposit offer?

24. Salim borrowed $12,375 for 4 years at an annual simple interest rate of 7.5%. If he makes one payment at the end of the loan, how much interest will he have to repay?

25. Todd had to pay $1147.50 of interest on a loan with an annual simple interest rate of 8.5% that he had for 5 years. Assuming he makes one payment at the end of the loan, what was the principal of the loan?

Name _____ Date _____ Class _____

Chapter Test
Form A (Chapter 9)

An experiment consists of drawing 4 balls from a bag and counting the number of red balls. The table gives the probability of each outcome.

Number of Red Balls	0	1	2	3	4
Probability	0.025	0.36	0.27	0.282	0.063

1. What is the probability of drawing fewer than 2 red balls?

   _____

2. What is the probability of drawing more than 1 red ball?

   _____

3. Erin's soccer team has won 17 of their 20 games this season. What is the probability that they will win their next game?

   _____

4. A researcher conducted a survey of 324 high school students and found that 54 of them were enrolled in advanced chemistry. What is the probability that a randomly selected student is enrolled in advanced chemistry?

   _____

Use the table of random numbers to simulate each situation. Use at least 10 trials for each simulation.

33	35	71	65	22	33	04	35	56	99
63	41	51	27	76	48	30	84	63	20
57	62	81	29	54	61	35	22	35	44
62	61	22	24	35	12	73	42	64	46
33	20	52	77	88	15	73	82	19	97
31	96	04	29	74	36	44	42	38	26
53	14	76	41	98	83	53	64	15	91
24	83	42	19	61	12	52	62	28	32

5. Katy hits the ball 58% of the time she bats. Estimate the probability that she will hit the ball at least 3 times in her next 4 at bats.

   _____

6. Sandra hits a golf ball over 120 yards on her first drive 67% of the time. Estimate the probability that she will hit the ball over 120 yards at least 4 times in the next 7 times she drives.

   _____

An experiment consists of spinning a spinner with equal chances of landing on one of four colors – red, blue, green, or yellow.

7. What is the probability of landing on blue or yellow?

   _____

Copyright © by Holt, Rinehart and Winston.
All rights reserved.

Holt Pre-Algebra

Chapter Test
Chapter 9 Form A, continued

8. The PIN numbers for a cash card at a bank contain four digits 1–9. All codes are equally likely. Find the number of possible PIN numbers.

9. The flavors at an ice cream shop are chocolate, vanilla, mint, and strawberry. The cone choices are waffle or sugar. Describe all of the different ice cream cone options available.

10. 6 swimmers are competing in the 100-yard butterfly. In how many different orders can all of the swimmers finish the race?

11. Find the number of different 4-person teams that can be made from 14 people.

12. There are 3 apples, 5 oranges, and 2 tangerines in a bowl of fruit. Two pieces of fruit are chosen at random and not replaced. What is the probability of choosing an apple and then an orange?

13. 5 red dice and 5 blue dice are put into a bag. What is the probability that when two dice are taken out, one is blue and one is red?

14. There is a jar with 10 nickels and 5 dimes. If two coins are chosen at random, what is the probability of choosing first a nickel and then a dime?

15. The odds of winning a door prize at a birthday party are 1:12. What is the probability of winning a door prize?

16. Six people called a radio talk show: two lawyers, two doctors, a veterinarian, and an accountant. What is the probability that a randomly selected caller is not a doctor?

17. Jared has collected 30 contest game pieces. Of those, 4 were winning pieces. What are the odds against winning a prize?

Name _____ Date _____ Class _____

CHAPTER 9 Chapter Test
Form B

Trisha made an educated guess on four of the multiple choice questions on her drivers license examination. The table gives the probability of each possible result.

Number of Correct Answers	Probability
0	0.025
1	0.271
2	0.359
3	0.282
4	0.063

1. What is the probability of getting 3 or more correct?

2. What is Trisha's probability of failing this part of the exam (getting fewer than 2 correct)?

3. A ball was randomly drawn from a bag and then replaced. In 300 experiments, a green ball was chosen 58 times, a red ball was chosen 118 times, a yellow ball was chosen 99 times, and a blue ball was chosen 25 times. What is the probability of choosing a yellow ball?

4. A researcher conducted a survey of 425 high school students and found that 356 of them planned to attend college. Estimate to the nearest percent the probability that a randomly selected student plans to attend college.

Use the table of random numbers to simulate each situation. Use at least 10 trials for each simulation.

33	35	71	65	22	33	04	35	56	99
65	41	51	27	76	48	30	84	63	20
57	62	81	29	74	61	35	22	35	44
42	61	22	24	35	12	73	42	69	46
33	20	52	57	88	15	73	82	19	97
31	96	04	29	74	36	48	42	38	26
53	14	76	41	18	43	53	68	15	91
24	83	42	19	61	12	52	62	28	32

5. In a city in Alaska, snow falls on 64% of winter days. Estimate the probability that it will snow at least 6 out of 7 days during a week in January.

6. At a pizza place, about 45% of the customers order pepperoni on their pizza. Estimate the probability that at least 5 out of the next 6 customers will order pepperoni.

An experiment consists of rolling a fair eight-sided die.

7. What is the probability of rolling a 3, 4, or 5?

Copyright © by Holt, Rinehart and Winston.
All rights reserved.

83

Holt Pre-Algebra

Chapter 9 Chapter Test Form B, continued

8. Student ID codes at a university contain two letters followed by two digits 0–9 and then another two letters. All codes are equally likely. Find the number of possible student ID codes.

9. A store sells three different styles of fleece jackets in 8 different colors. Each style can also be purchased with or without a hood. How many different versions of fleece jackets does the store sell?

10. Maureen has 7 different plants that she wants to plant in a row in her flower bed. How many different ways can she arrange the plants?

11. If Maureen decides she only has room for 3 of the 7 plants in her flower bed, how many different selections of 3 plants does she have to choose from?

12. Karleen made a snack mix by mixing 30 raisins, 45 peanuts, and 15 marshmallows. If she randomly selects two pieces from her snack mix, what is the probability they will both be peanuts?

13. A fair die is rolled twice. What is the probability of getting 4 the first time but not the second?

14. There is a jar with 10 nickels, 5 dimes, and 6 quarters. If three coins are chosen at random, what is the probability of choosing all quarters?

15. If the odds of winning a free meal at a restaurant are 1:356, what is the probability of winning?

16. A class ordered 7 pizzas. 3 are pepperoni, 2 are mushroom, and 2 are cheese. The boxes are not labeled. What is the probability that the first box opened will not contain a pepperoni pizza?

17. If 5 of the first 175 customers that arrive at a boat sale will win free lifejackets, what are the odds against winning?

Name _____ Date _____ Class _____

Chapter Test
Chapter 9 Form C

Outcome	Probability
1	0.071
2	0.271
3	0.314
4	0.282
5	0.062

1. What is the probability of outcome 1 or 2 occurring?

   _____

2. What is the probability of neither outcome 1 nor outcome 4?

   _____

3. Among 692 patients who were tested for a certain disease, 295 tested negative, 104 tested positive, and 293 had inconclusive results. Estimate to the nearest thousandth the probability of a patient's testing negative.

   _____

Sport	Number of Students
Swimming	28
Basketball	31
Football	45
Tennis	24

4. 165 students were polled about the sports they participate in. Estimate to the nearest thousandth the probability that a randomly polled student does not participate in any sports.

   _____

Use the table of random numbers to simulate each situation. Use at least 10 trials for each simulation.

12	77	64	83	08	46	32	03	19	60
42	97	43	64	44	23	82	66	34	15
16	52	23	30	43	55	69	93	32	05
25	78	81	57	38	21	99	73	33	57
22	43	51	27	34	26	44	62	48	34
21	96	11	25	26	22	83	24	26	72
12	95	46	13	52	23	54	45	01	47
64	11	84	35	36	41	59	15	76	28

5. Mr. Guanara gives a pop quiz about 25% of the days his class meets. What is the probability that there will be a pop quiz at least 3 out of the next 6 times his class meets.

   _____

6. An inspector finds that about 3% of computer motherboards assembled are defective. What is the probability that 1 of the next 8 motherboards will be defective?

   _____

Three fair coins with sides marked A and B are tossed.

7. Find each probability. $P(ABA)$

   _____

Copyright © by Holt, Rinehart and Winston.
All rights reserved.

Holt Pre-Algebra

Chapter 9 Chapter Test
Form C, continued

8. Raffle tickets at a school have 6 letters followed by 2 digits 0–9. All possible raffle tickets are equally likely. Find the number of possible raffle tickets.

9. At a sandwich shop, you can choose from 5 different kinds of bread, 7 different meats, 10 different cheeses, and 6 different sandwich spreads. How many different kinds of sandwiches are offered?

10. 14 people are lined up to purchase movie tickets. How many different ways can they be lined up?

11. How many ways can a teacher choose the first, second, third, and fourth students to give a presentation from a class of 21?

12. A group of 11 professionals includes 3 lawyers, 1 accountant, 2 doctors, 4 teachers, and 1 salesperson. What is the probability to the nearest thousandth that a committee of 2 randomly chosen people will consist of 2 teachers?

13. Grace and her brothers are drawing straws to see who has to vacuum the living room. There are 8 long straws and 2 short straws. They will draw until someone gets a short straw. What is the probability that 3 long straws and then a short straw will be drawn?

14. There is a jar with 9 pennies, 10 nickels, 5 dimes, and 6 quarters. If two coins are chosen at random, what is the probability of choosing first a nickel and then a dime?

15. The probability of winning 3rd prize in a contest is $\frac{1}{500}$. What are the odds of winning 3rd prize?

16. A batch of 16 vials of medicine includes 5 vials of painkiller, 2 vials of arthritis medication, 3 vials of medication that inhibits blood clotting, and 6 vials of a medication for high blood pressure. What is the likelihood that a randomly selected vial contains a medication related to blood or blood circulation?

17. 8 students each have an equal chance of being selected for student council president. What are the odds against being selected?

Name _____ Date _____ Class _____

Chapter Test
Chapter 10 Form A

Solve.

1. $4t + 5 = 13$ _____

2. $3h - 2 = 1$ _____

3. $-2.1a + 1.3 = 5.5$ _____

4. $\frac{x}{2} + 1 = 4$ _____

5. $5p + 2p - 3 = 11$ _____

6. $-3b - 6 + 7b = 6$ _____

7. $\frac{s}{3} - \frac{2}{3} = \frac{1}{3}$ _____

8. $\frac{x}{2} + \frac{x}{6} = 2$ _____

9. $x - 1 = x + 7$ _____

10. $5t + 6 = 2t - 3$ _____

11. $4(g + 1) = 2g - 6$ _____

12. $\frac{a}{3} + \frac{a}{3} - 1 = a$ _____

Solve and graph.

13. $3c - 2 \geq 4$ _____

14. $-1 < 5z + 4$ _____

15. $t - 3t - 1 > 7$ _____

Solve each equation for the indicated variable.

16. Solve $a + b + 3 = 5$ for b.

   _____

17. Solve $P = 2(l + w)$ for l.

   _____

18. Solve $d = r \cdot t$ for t. _____

19. Solve $y = x^2$ for x. _____

Solve each equation for y and graph.

20. $y + 2x = 4$ _____

Name _____ Date _____ Class _____

Chapter Test
Chapter 10 Form A, continued

21. $y - 3x = -2$ _____

Determine if the ordered pair is a solution of the system of equations.

22. $(1, 3)$; $y = 2x - 3$, $y = x + 2$

_____

23. $(3, 6)$; $y = x + 3$, $y = 3x - 3$

_____

Solve each system of equations.

24. $y = x - 5$, $y = 2x + 1$

_____

25. $y = x + 4$, $y = -x + 2$

_____

26. $y = 3x + 4$, $y = 2x + 5$

_____

27. $x + y = 5$, $x - y = 7$

_____

Write and solve an equation, inequality, or system of equations to answer the question.

28. Jeff bought 5 loaves of bread that were each the same price. He used coupons worth $2.60. Angie bought 3 loaves of bread, without using coupons. They paid the same total amount. What was the price of each loaf?

_____

29. Maria has $50 to spend and would like to buy some music CDs. She also has a $10 gift certificate. What is the greatest amount that each CD can cost if Maria wants to buy 4 CDs? Assume all the CDs are the same price.

_____

30. Two numbers have a sum of 10, and a difference of 4. Find the two numbers.

_____

Name _____ Date _____ Class _____

Chapter Test
Chapter 10 Form B

Solve.

1. $8t + 5 = 37$ _____

2. $\dfrac{-b}{2} - 10 = 5$ _____

3. $-5.9a - 5.5 = 12.2$ _____

4. $\dfrac{g + 3}{5} = 9$ _____

5. $-5p - 3p + 4 = 36$ _____

6. $\dfrac{4c}{5} - \dfrac{3c}{5} + \dfrac{1}{5} = \dfrac{-2}{5}$ _____

7. $\dfrac{x}{2} + \dfrac{5x}{6} = 4$ _____

8. $\dfrac{2y}{7} + \dfrac{4y}{7} - \dfrac{3}{14} = \dfrac{1}{14}$ _____

9. $-5x - 9 = -2x + 6$ _____

10. $3t + 6 = 3t - 14$ _____

11. $5(g - 3) = 2g + 3$ _____

12. $\dfrac{8a}{3} + \dfrac{4a}{3} - 1 = a + 5$ _____

Solve and graph.

13. $4d - 5 > 11$ _____

14. $-1 < \dfrac{k}{4} + \dfrac{1}{2} + \dfrac{k}{2}$ _____

15. $-5s - 9 - 3s \geq 15$ _____

Solve each equation for the indicated variable.

16. Solve $P = a + b + c$ for b.

_____

17. Solve $A = \dfrac{1}{2} rp$ for p.

_____

18. Solve $y = 2x + 3$ for x.

_____

19. Solve $A = \pi r^2$ for r.

_____

Solve each equation for y and graph.

20. $2y - 3x = 4$ _____

Name _____ Date _____ Class _____

CHAPTER 10 Chapter Test
Form B, continued

21. $2y - 5x = 8$ _____

Determine if the ordered pair is a solution of the system of equations.

22. $(-1, 4)$; $y = -5x - 1$, $y = 2x + 6$

_____

23. $(2, 7)$; $y = 3x + 1$, $y = -x - 8$

_____

Solve each system of equations.

24. $y = x + 2$, $y = 3x + 6$ _____

25. $y = -3x + 7$, $y = -2x - 5$ _____

26. $x + y = 8$, $x - y = 12$ _____

27. $2x - 3y = 5$, $4x + y = 3$ _____

Write and solve an equation, inequality, or system of equations to answer the question.

28. Olivia bought 5 greeting cards that were each the same price. She used coupons worth $2.50. Alexandra bought 4 cards and used a $1.00 coupon. Each girl paid the same amount. What was the price of each card?

_____

29. Brandon is planning a surprise party for Ashley. He has $25 to spend for drinks and a cake. He knows that the cake will cost $18 and drinks will cost $0.50 per person. What is the greatest number of people he can invite to the party?

_____

30. Two numbers have a sum of 39. The larger number is 9 less than twice the smaller number. Find the two numbers.

_____

Name _____ Date _____ Class _____

Chapter Test
Chapter 10 Form C

Solve.

1. $\dfrac{3}{4} - \dfrac{5t}{6} = \dfrac{5}{12}$ _____

2. $-6.8h + 15.3 = -39.1$ _____

3. $-\dfrac{8-a}{12} = -6$ _____

4. $\dfrac{12+x}{5} + \dfrac{4}{5} = -20$ _____

5. $12p + 8 - 7p - 3 = -20$ _____

6. $\dfrac{3b}{4} - \dfrac{2}{3} + \dfrac{5b}{12} = -3$ _____

7. $\dfrac{2y}{27} - \dfrac{2}{9} + \dfrac{y}{3} = -\dfrac{5}{17}$ _____

8. $\dfrac{x+4}{5} - \dfrac{1}{3} + \dfrac{4x}{5} = \dfrac{1}{2}$ _____

9. $2(x-5) + \dfrac{3x}{4} = 4x$ _____

10. $-(t+5) - 4 = -2(t-6)$ _____

11. $3(4g+1) + 5 = 12g - 7$ _____

12. $-4\left(2a - \dfrac{1}{2}\right) + 5 = 2\left(4a - \dfrac{1}{2}\right)$ _____

Solve and graph.

13. $\dfrac{3n}{11} - 2 > 4$ _____

14. $-\dfrac{x}{2} < \dfrac{x}{4} + \dfrac{3}{8}$ _____

15. $2\left(\dfrac{v}{8} + \dfrac{1}{4}\right) \leq \dfrac{v+1}{6}$ _____

Solve each equation for the indicated variable.

16. Solve $C = \dfrac{5}{9}(F - 32)$ for F.

17. Solve $A = \dfrac{1}{2}h(b_1 + b_2)$ for b_1.

18. Solve $S = 6rs + 6sh$ for r.

19. Solve $V = \dfrac{1}{3}\pi r^2 h$ for r.

Holt Pre-Algebra

Name _____ Date _____ Class _____

Chapter Test
Chapter 10 Form C, continued

Solve each equation for y and graph.

20. $5y + 2x = 15$ _____

21. $y - 4 = -2(x + 3)$ _____

Determine if the ordered pair is a solution of the system of equations.

22. $(-1, 1)$; $2y - 3x = 5$, $x + 5y = 4$

_____

23. $(3, 5)$; $8x - 4y = 4$, $7y - 3x = -26$

_____

Solve each system of equations.

24. $y = 2x + 5$, $x = -2y - 5$

_____

25. $2y = -10x + 40$, $-6x + 4y = -24$

_____

26. $2x - 5y = -19$, $-3x + 4y = 18$

_____

27. $8x + 9y = -18$, $7x - y = 2$

_____

Write and solve an equation, inequality, or system of equations to answer the question.

28. Two sisters left at the same time to travel home. Susanna drove at 65 miles per hour and started 9 miles farther from home than Rachel. Rachel drove at 55 miles per hour. They reached home at the same time. How far was each girl from home?

_____

_____

29. Gerri received grades of 83, 94, 76, and 89 on four tests. What is the lowest average score she can afford to get on her next two tests to end up with an overall average of 90?

_____

30. The difference of two numbers is 18. Their sum is 13 more than 3 times the smaller number. Find the two numbers.

_____

Name _____ Date _____ Class _____

Chapter Test
Chapter 11 Form A

Graph each equation and tell whether it is linear.

1. $y = x^2$ _____

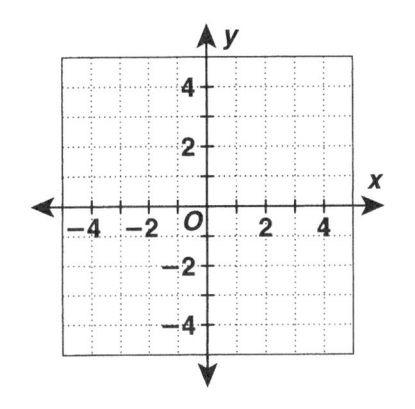

2. $y = x - 5$ _____

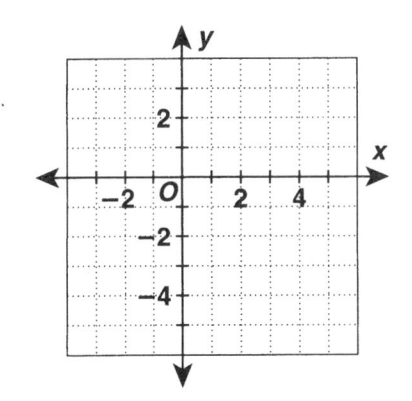

3. Find the slope of the line that passes through the points (5, 4) and (3, 1).

_____

Tell whether the lines passing through the given points are parallel or perpendicular.

4. line 1: (3, 0) and (2, 1); line 2: (4, 5) and (3, 6)

_____

Find the x-intercept and y-intercept of each line. Use the intercepts to graph the equation.

5. $y = -x + 5$ _____

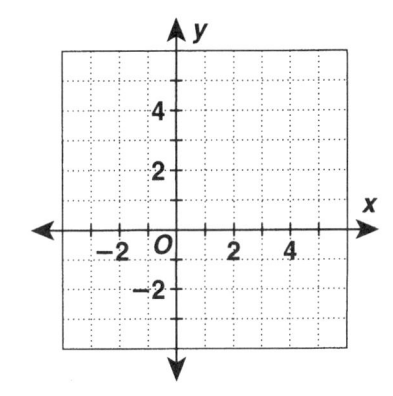

6. $x - y = -3$ _____

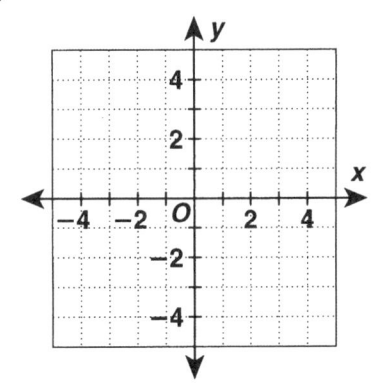

7. Write $x + y = 9$ in slope-intercept form and then find the slope and y-intercept.

_____

Copyright © by Holt, Rinehart and Winston.
All rights reserved.

Holt Pre-Algebra

Chapter 11 Chapter Test
Form A, continued

8. Write the equation of the line that passes through the points (1, 9) and (2, 6) in slope-intercept form.

9. Identify a point the line for $y - 4 = 5(x - 3)$ passes through and identify the slope of the line.

10. Write the point-slope form of the equation for a line with a slope -2 that passes through the point $(-1, 1)$.

Find each equation of direct variation, given that y varies directly with x.

11. y is 10 when x is 2

12. y is -3 when x is $\frac{1}{2}$

Graph each inequality.

13. $y > x - 1$

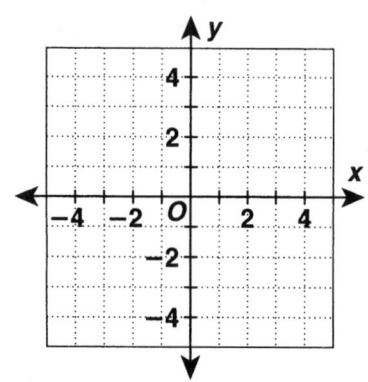

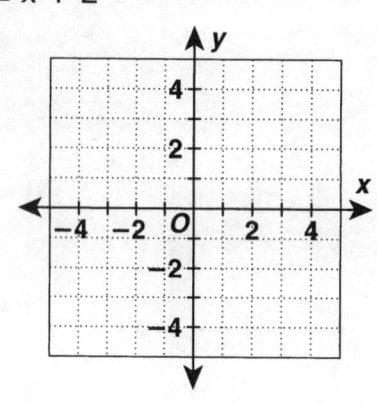

14. $y \leq x + 2$

Use a line of best fit to answer the question.

15. Find a line of best fit for the data. Use the equation of the line to predict the y-value for $x = 8$.

x	1	3	4	5	7
y	5	8	11	14	17

16. A mechanic charges $20 plus $40 for each hour he works on a car. The equation $y = 40x + 20$ represents the total amount he charges for working x hours. If he works 5 hours on a car, how much should he charge?

Name _____ Date _____ Class _____

Chapter Test
11 Form B

Graph each equation and tell whether it is linear.

1. $y = 2x + 1$ _____

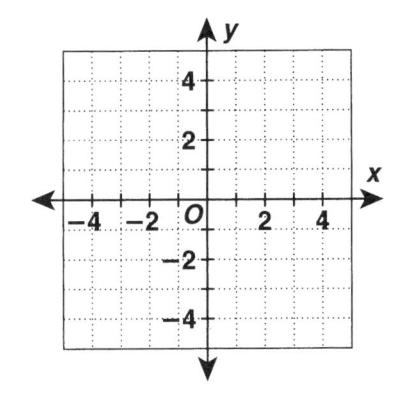

2. $y = x^2 - 5$ _____

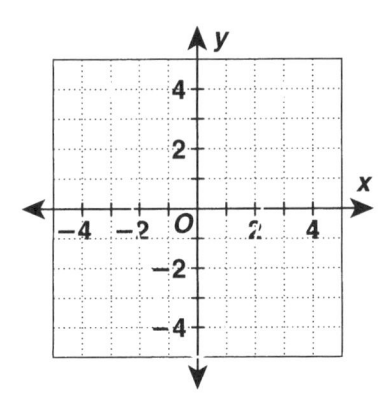

3. Find the slope of the line that passes through the points $(-3, 4)$ and $(5, -1)$.

_____

Tell whether the lines passing through the given points are parallel or perpendicular.

4. line 1: $(7, -2)$ and $(3, 1)$;
   line 2: $(-8, 2)$ and $(-5, 6)$

_____

Find the x-intercept and y-intercept of each line. Use the intercepts to graph the equation.

5. $3y = 2x - 3$ _____

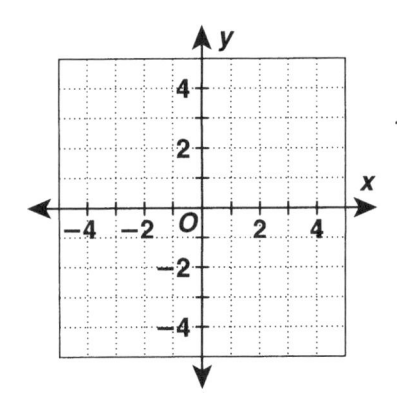

6. $x - 2y = -4$ _____

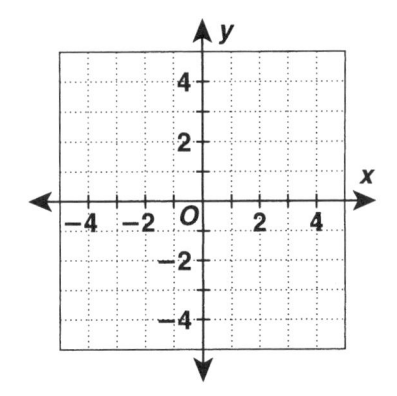

7. Write $5x - y = 8$ in slope-intercept form and then find the slope and y-intercept.

_____

Copyright © by Holt, Rinehart and Winston.
All rights reserved.

Holt Pre-Algebra

Name _____ Date _____ Class _____

Chapter Test
11 Form B, continued

8. Write the equation of the line that passes through the points (−2, 4) and (−4, 10) in slope-intercept form.

    _____

9. Identify a point the line for $y + 5 = -\frac{2}{3}(x - 3)$ passes through and identify the slope of the line.

    _____

10. Write the point-slope form of the equation for a line with a slope of −5 that passes through the point (−4, 3).

    _____

Find each equation of direct variation, given that y varies directly with x.

11. y is −3 when x is 12

    _____

12. y is 6 when x is 9

    _____

Graph each inequality.

13. $y > 2x - 1$

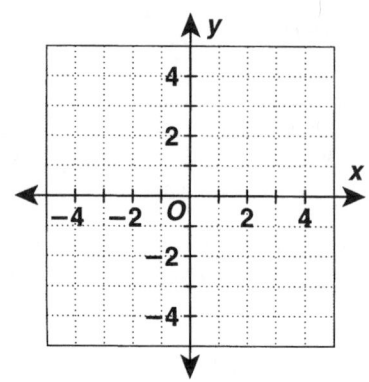

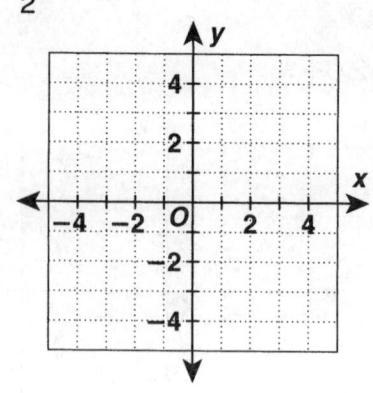

14. $y \leq \frac{1}{2}x + 2$

Use a line of best fit to answer the question.

15. The temperature one day in January dropped during a cold front. The data are shown below. Find a line of best fit for the data. Use the equation of the line to predict the temperature after 6 hours.

Hours	0	1	2	3	4
Temperature	40	35	32	29	27

    _____

16. A hiker walking down a mountain decreases his altitude by 20 feet each minute he hikes. The equation $y = -20x + 8000$ represents his altitude after hiking x minutes. If he hikes 30 minutes, what is his altitude?

    _____

Name _____ Date _____ Class _____

Chapter Test
Chapter 11 Form C

Graph each equation and tell whether it is linear.

1. $y = -\frac{1}{3}x - 2$ _____

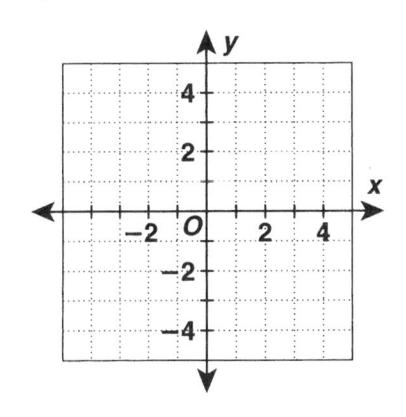

2. $y = 2x^2 - 3$ _____

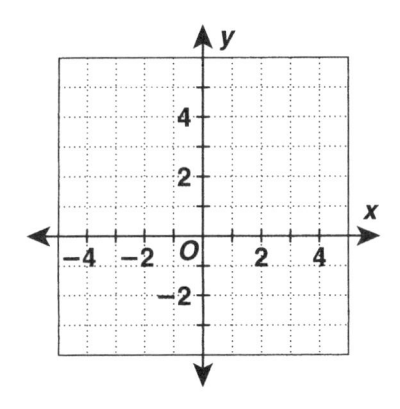

3. Find the slope of the line that passes through the points $(-7, -5)$ and $(4, -3)$.

_____

Tell whether the lines passing through the given points are parallel or perpendicular.

4. line 1: $(8, -4)$ and $(-2, 3.5)$;
   line 2: $(-4, -3)$ and $(6, -10.5)$

_____

Find the x-intercept and y-intercept of each line. Use the intercepts to graph the equation.

5. $-\frac{1}{3}y = -\frac{1}{2}x + 1$ _____

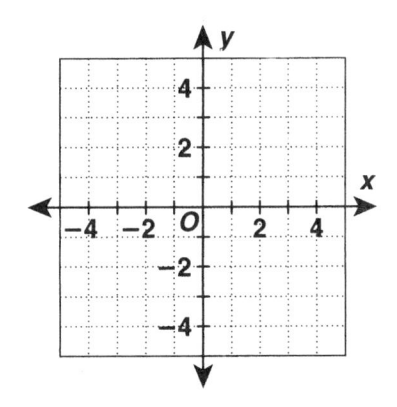

6. $\frac{3}{4}x - 2y = 3$ _____

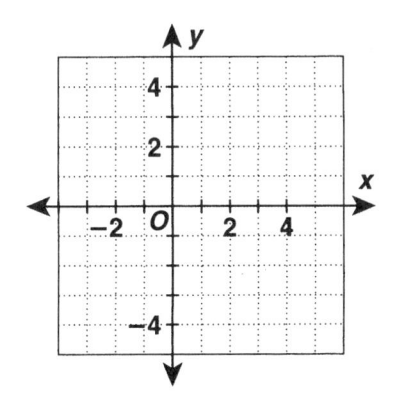

7. Write $5(4x - 2y) = 15$ in slope-intercept form and then find the slope and y-intercept.

_____

Name _____ Date _____ Class _____

CHAPTER 11 Chapter Test
Form C, continued

8. Write the equation of the line that passes through the points (−10, 6) and (15, 1) in slope-intercept form.

_____

9. Identify a point the line for $y + 7 = -\frac{3}{5}(x - 4)$ passes through and identify the slope of the line.

_____

10. Write the point-slope form of the equation for a line with a slope of $-\frac{1}{7}$ that passes through the point $\left(-21, \frac{1}{2}\right)$.

_____

Find each equation of direct variation, given that y varies directly with x.

11. y is −63 when x is 81

_____

12. y is $\frac{3}{5}$ when x is $\frac{1}{2}$

_____

Graph each inequality.

13. $\frac{1}{4}y > -\frac{1}{6}x + \frac{1}{2}$

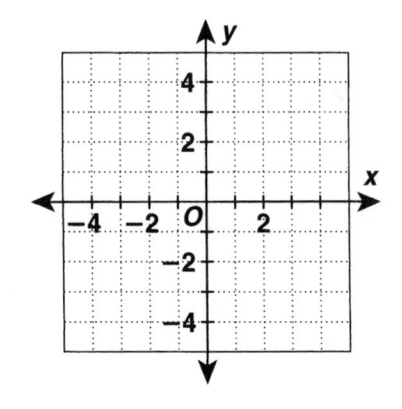

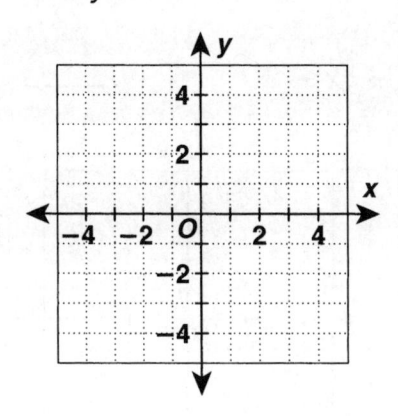

14. $-3x + 4y \leq 12$

Use a line of best fit to answer the question.

15. The value of a car decreases as it is driven more miles. The data are shown below. Find a line of best fit for the data. Use the equation to predict the value of the car at 24,000 miles.

Miles	Value of Car
10,000	$14,500
12,000	$14,150
14,000	$13,815
16,000	$13,650
18,000	$13,500

_____

16. In order to rent an apartment, Vik makes an initial payment of $1535, and each month he must pay $750. The equation $y = 750x + 1535$ represents the amount of money Vik has paid after x months. After 2 years, how much rent money has Vik paid?

_____

Copyright © by Holt, Rinehart and Winston.
All rights reserved.

Holt Pre-Algebra

Name _____ Date _____ Class _____

CHAPTER 12 Chapter Test
Form A

Determine if each sequence could be arithmetic. If so, give the common difference and find the specified term.

1. 1, 3, 5, 7, 9, ... ; 8th term.

   _____

2. 1, 3, 9, 27, 81, ... ; 7th term.

   _____

Determine if each sequence could be geometric. If so, give the common ratio and find the specified term.

3. 64, 32, 16, 8, ... ; 6th term.

   _____

4. 1, −1, 1, −1, ... ; 7th term.

   _____

Use the first and second differences to find the next three terms in the sequence.

5. 1, 2, 4, 7, 11, ...

   _____

Find the first five terms of the sequence specified by the rule.

6. $a_n = \dfrac{2}{n}$

   _____

Determine if each relationship represents a function.

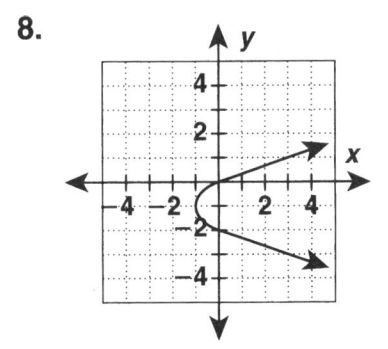

7.
x	1	2	3	4
y	6	−2	6	8

_____

8.

_____

Write the rule for each linear function.

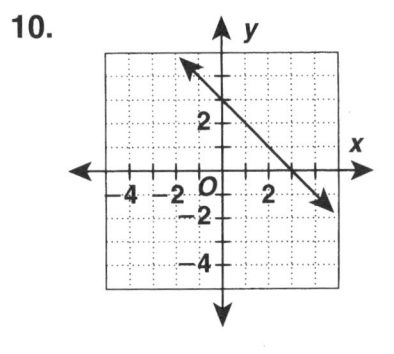

9.
x	−1	0	1	2
y	8	10	12	14

_____

10.

_____

Copyright © by Holt, Rinehart and Winston.
All rights reserved.

Holt Pre-Algebra

Chapter 12 Chapter Test
Form A, continued

Complete the table for each exponential function and use it to graph the function.

11. $f(x) = 3\left(\dfrac{1}{2}\right)^x$

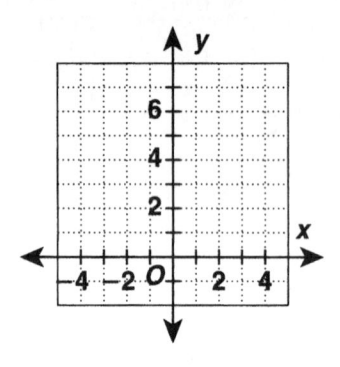

x	−1	0	1	2
y				

Complete the table for each quadratic function and use it to make a graph.

12. $f(x) = -x^2 + 2x$

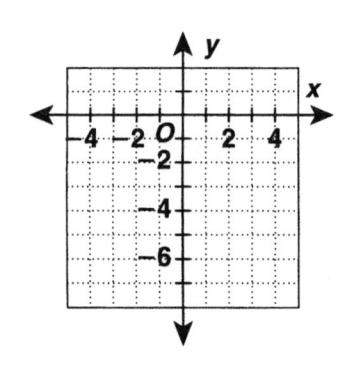

x	−2	−1	0	1	2	3
y						

13. $f(x) = (x - 2)(x + 2)$

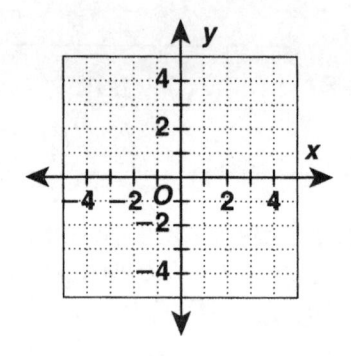

x	−2	−1	0	1	2
y					

Tell whether the relationship is an inverse variation.

14. The table below shows the distance driven in a given time.

Time (hours)	2	4	5	7
Miles Driven	120	240	300	420

_____

Graph the inverse variation.

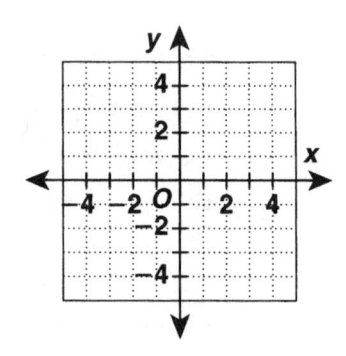

15. $f(x) = \dfrac{9}{x}$.

16. The height of a ball thrown horizontally from the top of a 75-meter tower is given by the function $f(t) = -5t^2 + 75$. What is the height after 2 seconds?

Name _____ Date _____ Class _____

CHAPTER 12 Chapter Test
Form B

Determine if each sequence could be arithmetic. If so, give the common difference and find the specified term.

1. $2, \frac{7}{3}, \frac{8}{3}, 3, \frac{10}{3}, \ldots$; 11th term.

_____

2. $-5, 6, 17, 28, 39, \ldots$; 20th term.

_____

Determine if each sequence could be geometric. If so, give the common ratio and find the specified term.

3. $625, 125, 25, 10, \ldots$; 9th term.

_____

4. $-\frac{1}{2}, 2, -8, 32, \ldots$; 7th term.

_____

Use the first and second differences to find the next three terms in the sequence.

5. $6, 11, 15, 18, 20, \ldots$

_____

Find the first five terms of the sequence specified by the rule.

6. $a_n = \dfrac{n^2 + 1}{n}$

_____

Determine if each relationship represents a function.

7.
x	0	4	0	9
y	6	9	5	-1

_____

8.

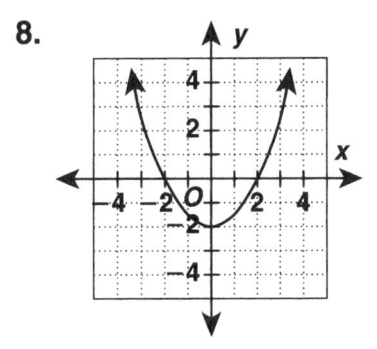

_____

Write the rule for each linear function.

9.
x	-2	1	4	6
y	15	3	-9	-17

_____

10.

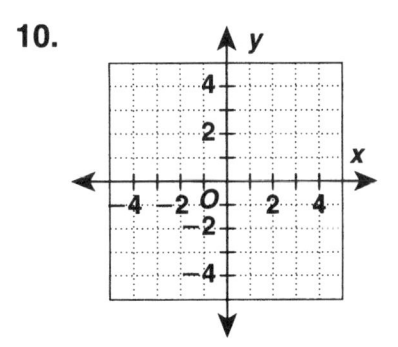

_____

101 Holt Pre-Algebra

Name _____ Date _____ Class _____

Chapter 12 Chapter Test
Form B, continued

Create a table for each function and use it to graph the function.

11. $f(x) = -2\left(\frac{1}{3}\right)^x$

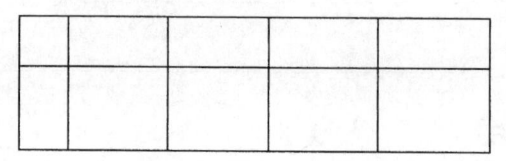

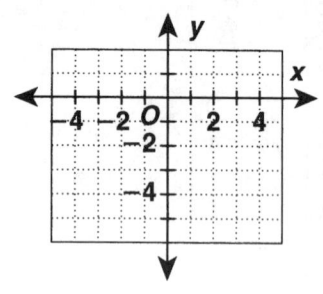

12. $f(x) = x^2 - 2x + 3$

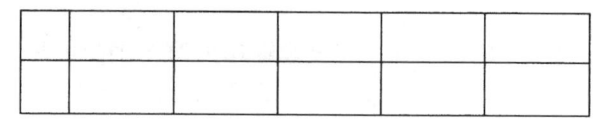

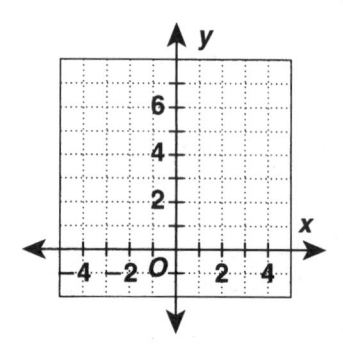

13. $f(x) = \frac{1}{2}(x - 2)(x + 1)$

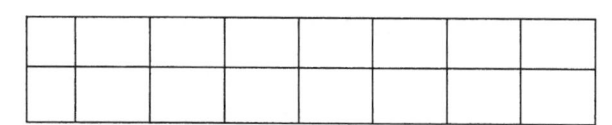

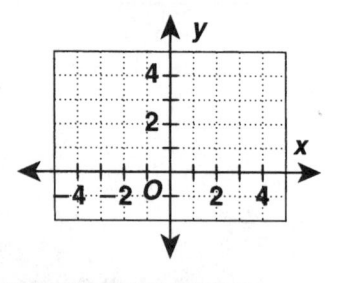

Tell whether the relationship is an inverse variation.

14. The table below shows the number of items purchased as a function of the cost per item.

Cost per Item ($)	7	10	12	15	20
Items Purchased	60	42	35	28	21

_____

Graph the inverse variation.

15. $f(x) = -\frac{4}{x}$

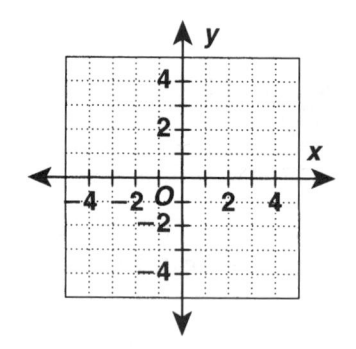

16. Ms. Suarez wants to earn $150 in interest over a 3-year period from a savings account. The principal she must deposit varies inversely with the interest rate of the account. If the interest rate is 0.08, she must deposit $625. If the interest rate is 0.064, how much must she deposit?

_____

Copyright © by Holt, Rinehart and Winston.
All rights reserved.

Holt Pre-Algebra

Name _____ Date _____ Class _____

CHAPTER 12 Chapter Test
Form C

Determine if each sequence could be arithmetic. If so, give the common difference and find the specified term.

1. $-\frac{2}{7}, \frac{6}{7}, 2, \frac{20}{7}, \frac{27}{7}, \ldots$; 21st term.

2. $-\frac{11}{3}, -\frac{2}{3}, \frac{7}{3}, \frac{16}{3}, \frac{25}{3}, \ldots$; 33rd term.

Determine if each sequence could be geometric. If so, give the common ratio and find the specified term.

3. $1, \frac{1}{2}, \frac{1}{3}, \frac{1}{4}, \ldots$; 11th term.

4. 2, 16, 128, 512, ... ; 9th term.

Use the first and second differences to find the next three terms in the sequence.

5. $-\frac{18}{5}, -\frac{11}{5}, -\frac{3}{5}, \frac{6}{5}, \frac{16}{5}, \ldots$

Find the first five terms of the sequence specified by the rule.

6. $a_n = \frac{n(n+1)(n+2)}{6}$

Determine if each relationship represents a function.

7.
x	−2	6	11	17
y	11	17	−2	6

8.

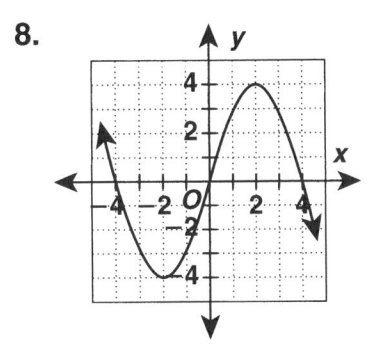

Write the rule for each linear function.

9.
x	6	8	2	−4
y	−19	−29	1	31

10.

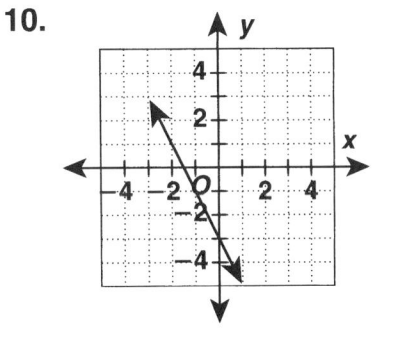

103 Holt Pre-Algebra

Name _____ Date _____ Class _____

CHAPTER 12 Chapter Test
Form C, continued

Create a table for each function and use it to graph the function.

11. $f(x) = -\frac{1}{2}\left(\frac{1}{3}\right)^x$

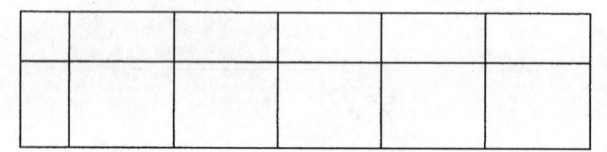

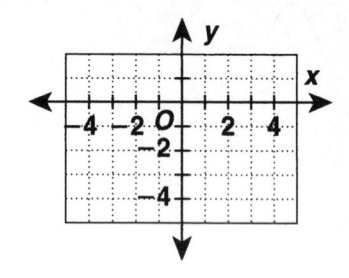

12. $f(x) = \frac{1}{2}x^2 + x - 2$

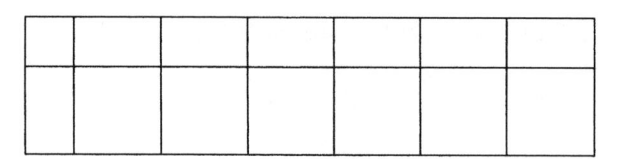

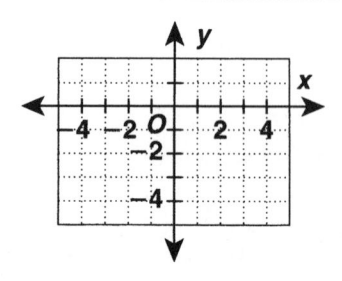

13. $f(x) = \frac{1}{2}(x-3)(x+2)$

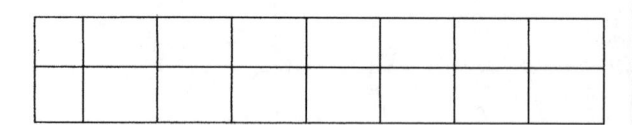

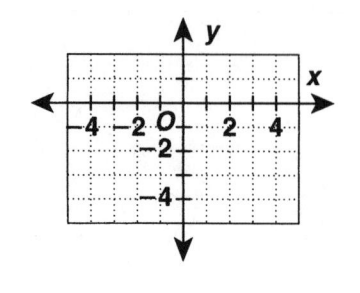

Tell whether the relationship is an inverse variation.

14. The table below shows the speed of a car in relation to time needed to cover a given distance.

Time (min)	10	15	24	30	50
Speed (kph)	60	40	25	20	12

Graph the inverse variation.

15. $f(x) = \frac{-8}{(3x)}$.

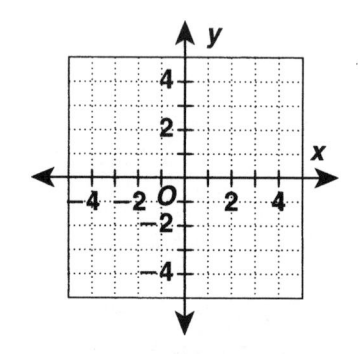

16. The resistance of a 30-m piece of wire varies inversely with the square of the diameter. If the diameter of the wire is 2 mm, it has a resistance of 0.2 ohms. What is the resistance of a wire with a diameter of 0.4 mm?

Holt Pre-Algebra

Name _____ Date _____ Class _____

CHAPTER 13 Chapter Test
Form A

Classify each expression as monomial, binomial, trinomial, or not a polynomial.

1. $5x^2$

2. $9y^2$

3. $3a^2 + b^2$

4. $7x^2 + x^3 - 8x$

Find the degree of each polynomial.

5. $3x^2 + 4x$

6. $4n^3 + 2n^8$

7. $9a^5 - 3a^3 - 5a^2$

8. $k^3 + k^4 + k^7 + k$

Identify the like terms in each polynomial.

9. $4n^2 + 3 + 5n^2$

10. $6x^3 - 5x^2 + 5x^3$

Simplify.

11. $7y^2 + 3y - 4y^2$

12. $4n^2 + 8 + 8n^2$

13. $5t^2 + t^2 - 3s^2$

Add.

14. $(6x + 2) + (5x + 2)$

15. $(8n + 8) + (6n - 5)$

16. $(4x + 2x^2) + (4x^2 + 2x)$

17. $(5n^2 + 3n) + (2n^2 - 2n)$

18. $(8a^3 - 6b) + (9b - 4a^3)$

19. $(2x - 2x^3) + (8x^3 + 4x)$

20. $(7xy + 4x) + (3xy - 2x)$

Copyright © by Holt, Rinehart and Winston.
All rights reserved.

Holt Pre-Algebra

Name _____ Date _____ Class _____

Chapter Test
Chapter 13 Form A, continued

Find the opposite of each polynomial.

23. $5n^3$

24. $6x^4y^4$

Subtract.

27. $5x - (2x - 4x^2)$

28. $(6n + 4n^2) - (2n^2 + 4n)$

29. $(4x^2 + 2y^2) - (3x^2 - 9y^2)$

30. $(8y^2 - 4) - (-5y^2 - 8)$

31. $(7n^2 - 5n) - (-7n + 5n^2)$

32. $(9x + 7 + 10x^4) - (2x + 3 - 5x^4)$

33. $(4n - 2n^2 + 6a^2) - (-8n^2 - 6a^2)$

Multiply.

34. $(3x^2)(5x^2)$

35. $(2n^2m^3)(3n^3m^2)$

36. $5x(4x^2 + 8)$

37. $9x^2(2x - 6x^3)$

38. $3ab^2(2a^3b^3 + 2ab + 5b^2)$

39. $(n + 3)(n + 4)$

40. $(x + 2)(y + 7)$

41. $(n + 6)(n - 6)$

42. $(y + 5)(y - 4)$

43. $(2a + b)(a - 4b)$

44. A rectangle has a length of $2w + 3$ and a width of $w - 5$. Write and simplify an expression for the area of the rectangle.

Name _____ Date _____ Class _____

CHAPTER 13 Chapter Test
Form B

Classify each expression as monomial, binomial, trinomial, or not a polynomial.

1. $6a^2 - 4a$

2. $\left(\frac{1}{7}\right)x^5y^3$

3. $8p^2 + p^{2.2}$

4. $4m^2n^2 + 5n^3 - 4n$

Find the degree of each polynomial.

5. $8x^5 + 6x^4$

6. $-6a^3 + 8a - 2a^5$

7. $5 + 3m^4 - 2m^7 + m^8$

8. $b^3 - b^9 - 3 - 3b^2$

Identify the like terms in each polynomial.

9. $t^2 + 9s^3 - 5t^4 + 8 + 4s^3$

10. $4a^3b^2 - 5a^2b^2 + a^3b^2$

Simplify.

11. $v^2 + 6v - 8v^2 + 4v^2$

12. $-4x^2 + 9 + 5x^2$

13. $5m^2n^2 + n^3 - 3m^2n^2 + 7n^3$

Add.

14. $(13x - 4) + (5x + 5)$

15. $(6a + 9) + (6a^2 - 5a - 3)$

16. $(3x + 2x^3 - 7) + (4x^3 - 2x + 7)$

17. $(6z^4 + 7y^3 + 4) + (2y^3 - z^4 - 3)$

18. $(7r^3 + 7s + 7r^2) + (4s - 4r^3)$

19. $(5x - 2x^3 - 9) + (8x^3 - 3x - 5)$

20. $(7ab + 6b^2 - 5a) + (ab - 3a)$

Holt Pre-Algebra

Chapter 13 Chapter Test Form B, continued

Find the opposite of each polynomial.

23. $9a^5b^5c^3$

24. $9y^3 - 6x$

Subtract.

27. $8b - (8a^2 + 7b - 5)$

28. $(4x + 5x^2) - (3x^2 + x)$

29. $(x^3 - 5x + 2x^2) - (3x - 3x^3 + 9)$

30. $(6y^2z^6 + 3yz - 4z^2) - (-5yz - 9)$

31. $(n^4 + 9n^3m^7 - 5n) - (-6n - n^3m^7)$

32. $(6x - 6 + x^3) - (4 - 2x - 5x^3)$

33. $(-5s + 7s^2 + 4r^2 - 4) - (-3s^2 - 6s)$

Multiply.

34. $(5x^2)(2y^7)(4x^3)(2y^3)$

35. $(9a^2b^5)(6ab^4)$

36. $6n(3n^2 - 7)$

37. $-4ab(5a^2 - 7b^3)$

38. $6st^2(7s^3t^3 + st - 5t^2)$

39. $(a - 3)(y + 5)$

40. $(x + 4)(x + 5)$

41. $(b + 6)^2$

42. $(y + 5)(y - 4)$

43. $(5y - 7)^2$

44. A square has a side length of $3e + 4$. Write and simplify an expression for the area of the square.

Name _____ Date _____ Class _____

CHAPTER 13 Chapter Test
Form C

Classify each expression as monomial, binomial, trinomial, or not a polynomial.

1. $9x^2 - \dfrac{7}{x^2}$

2. 88

3. $5n^{11} + m^2 - nm$

4. $6a^2b^2 + \left(\dfrac{1}{2}b\right)n^{3.1} + 8n$

Find the degree of each polynomial.

5. $-2x + 8x$

6. $-10n^8 + 10n^8 + n^8$

7. $4 + 5y^3 - 8y^2 - 4y^9$

8. $35j^4 + 28j^{12} - 15j^7$

Identify the like terms in each polynomial.

9. $11x^2 + 11y^2 - t^4 + 9y^2 + 4x^{22}$

10. $g^2h^2 - 8gh^2 + g^2h + 8g^2h^2$

Simplify.

11. $3(5x^2 + 3y) - 4x^2 + 4y^2$

12. $5n^2 - 10m^3 - 4n^2 + m^3 + n^2$

13. $3(3a^2b^2 + b^3) - 3a^2b^2 + 2(7a^3 + 4b^3)$

Add.

14. $(3x^3 - 4x) + (-2x + 5 - x^3)$

15. $(9a + 5b^2) + (3a^2 - 2a) + (6b^2 - 5a^2)$

16. $(4x + x^3) + (4x^3 - x + 7) + (3x + 9x^3)$

17. $(2a^3 + 6b^3 + 7) + (5b^3 - a^3b^4 - 3a^3)$

18. $(6m^3n^3 + 2m + 12m^3n^3) + (-8m - n^3)$

19. $(xy^3 - 2x^3 - 9xy^3) + (8xy^3 - 2x^3 - 2)$

20. $(7cd + 6d^2) + (cd - 3c) + (-8d - d^2)$

Copyright © by Holt, Rinehart and Winston.
All rights reserved.

109

Holt Pre-Algebra

Name _____ Date _____ Class _____

Chapter 13 Chapter Test
Form C, continued

Find the opposite of each polynomial.

23. $12x^8yz^3 - 7xy + 23$

24. $-11a^3 - 6ab^3 + 12a^4b^3$

Subtract.

27. $(5b + 2a^2) - (9a^2 + 12b - 8)$

28. $(4n^2 + 10n^5m^5 - 14n) - (3n - 8n^5m^5)$

29. $(6r^4s^4 + 11r^2 + 9r^2s^2) - (14r^2s^2 - 5r^2)$

30. $(12y^2z - 9yz - 4z^5) - (-5yz - 9y^2z)$

31. $(9s - 13s^2 - r^2 + 4) - (5r^2 + s - 18s^2)$

32. $(13rs^8 + 9s^4 - 3r^6s^2) - (-4r^6s^2 + 9s^4)$

33. $(6yz^5 + 16yz - 4z^2) - (5yz^5 - 9z^2 + z)$

Multiply.

34. $8x^2(5x^2y^2 + 9x^3y^3)$

35. $(-rs^5)(7r^5s)$

36. $6x^6y(6x^2y^3 - 7xy^4 + 12x^3)$

37. $13fg(-9g^2 - 5f^3 + f^5g^2)$

38. $10x^2y^2(8xy - 5xy^3 + 4x^3y)$

39. $(9x + 2)(x + 5)$

40. $(2x + 4)(4x - 6)$

41. $(5a - 4)^2$

42. $(6y + x)(y - 8x)$

43. $(8r - 6s)(4r - 10s)$

44. A parallelogram has a base length of $5p - 1$ and a height of $2p + 3$. Write and simplify an expression for the area of the parallelogram.

Name _____ Date _____ Class _____

Chapter Test
CHAPTER 14 Form A

Use the correct symbol to make each statement true. Choose ∈ or ∉.

1. 13 ☐ {prime numbers}

2. pentagon ☐ {parallelograms}

Tell whether the first set is a subset of the second set. Use the correct symbol.

3. N = {positive integers}; R = {whole numbers}

4. E = {circles}; G = {plane figures}

Find the intersection of the sets.

5. A = {0, 1, 2, 3}; B = {2, 3, 4, 5, 6}

6. R = {−2, −1, 0}; T = {0, 1, 2, 3, 4}

For 16–20, find the union of the sets.

7. M = {0, 1, 2, 3}; N = {4, 5, 6}

8. G = {multiples of 4}; H = {factors of 8}

For 9–11, use the Venn diagram.

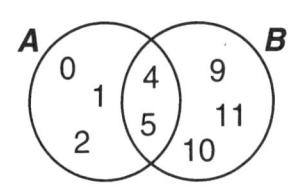

9. Identify the intersection of sets A and B.

10. Identify the union of sets A and B.

11. True or False: $A \subset B$.

12. Make a truth table for the conjunction P and Q.
    P: A number is a multiple of 5.
    Q: A number is a multiple of 2.

13. Complete the truth table.
    P: Today is Monday.
    Q: There is school today.

P	Q	P and Q	P or Q
T	T		T
T		F	T
F	T		
	F		

Chapter Test

14 Form A, continued

For 14–15, name the hypothesis and the conclusion in each statement.

14. If you sleep, you will feel rested.

   _____

   _____

15. In order to feel well, eat a good breakfast.

   _____

   _____

16. Make a conclusion, if possible, from the deductive argument.
   If a rectangle has 4 equal sides, it is a square. Rectangle ABCD has 4 equal sides.

   _____

For 17–19, use the graph.

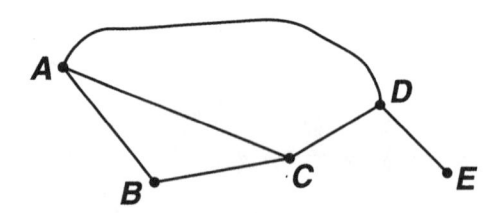

17. Find the degree of each vertex.

   _____

   _____

18. Is the graph connected? How do you know?

   _____

   _____

   _____

19. Determine whether the graph can be traversed through an Euler circuit. If your answer is yes, describe an Euler circuit in the graph.

   _____

   _____

For 20–21, use the graph.

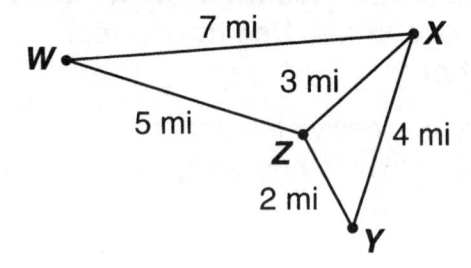

20. A Hamiltonian circuit must pass through every vertex. How is it different from an Euler circuit?

   _____

   _____

21. What is the length of the shortest Hamiltonian circuit that begins and ends at W?

   _____

Name _____ Date _____ Class _____

Chapter Test
Chapter 14 Form B

Insert the correct symbol to make each statement true. Choose $\in$ or $\notin$.

1. $x^2 - 2x = 0$ ☐ {quadratic equations}

   _____

2. Delaware ☐ {United Nations}

   _____

Determine whether the first set is a subset of the second set. Use the correct symbol.

3. N = {natural numbers};
   R = {real numbers}

   _____

4. P = {polygons}; T = {triangles}

   _____

Find the intersection of the sets.

5. $A = \{2, 4, 6, 8\}$; $B = \{6, 8, 10, 12\}$

   _____

6. J = {negative integers};
   $K = \{-4, -3, -2, -1, 0, 1\}$

   _____

For 16–20, find the union of the sets.

7. G = {multiples of 2};
   H = {multiples of 4}

   _____

8. X = {quadrilaterals};
   Y = {polygons}

   _____

For 9–11, use the Venn diagram.

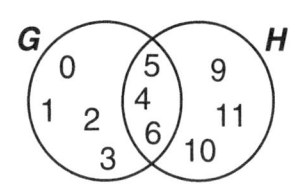

G: 0, 1, 2, 3   5, 4, 6   H: 9, 11, 10

9. Identify the intersection of sets G and H.

   _____

10. Identify the union of sets G and H.

    _____

11. Identify a subset in the diagram.

    _____

12. Make a truth table for the conjunction P and Q.
    P: A number is an even number.
    Q: A number is a factor of 90

13. Complete the truth table.
    P: Today is Saturday.
    Q: There is practice today.

P	Q	P and Q	P or Q
T	T		T
T		F	T
F	T		
	F		

Copyright © by Holt, Rinehart and Winston.
All rights reserved.

Holt Pre-Algebra

Name _____ Date _____ Class _____

Chapter Test
14 Form B, continued

For 14–15, identify the hypothesis and the conclusion in each conditional.

14. If you study, the test will be easy.

_____
_____

15. To be healthy, eat well.

_____
_____

16. Make a conclusion, if possible, from the deductive argument.
If a quadrilateral has 4 right angles, it is a rectangle. Quadrilateral *ABCD* has angle measures of 90°, 90°, 120°, and 60°.

_____

For 17–19, use the graph.

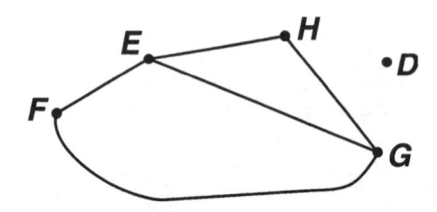

17. Find the degree of each vertex.

_____
_____

18. Is the graph connected? Explain.

_____
_____

19. Determine whether the graph can be traversed through an Euler circuit. If your answer is yes, describe an Euler circuit in the graph.

_____
_____

For 20–21, use the graph.

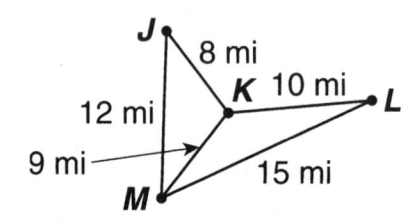

20. Describe how a Hamiltonian circuit differs from an Euler circuit.

_____
_____
_____

21. Determine the length of the shortest Hamiltonian circuit beginning at *L*.

_____

Name _____ Date _____ Class _____

Chapter Test
Chapter 14 Form C

Insert the correct symbol to make each statement true. Choose ∈ or ∉.

1. $y = x^3 - 3x$ ☐ {quadratic functions}

_____

2. Ohio River ☐ {rivers in the United States}

_____

Determine whether the first set is a subset of the second set. Use the correct symbol.

3. $P = $ {perimeter}; $A = $ {area}

_____

4. $V = \{x^2, 3x^3, 2x\}$; $M = $ {monomials}

_____

Find the intersection of the sets.

5. $A = \{x \mid x < 9\}$; $B = \{x \mid x > 12\}$

_____

6. $C = $ {first 5 prime numbers}; $D = $ {first 10 positive integers}

_____

Find the union of the sets.

7. $G = $ {multiples of 2}; $H = $ {factors of 12}

_____

8. $W = $ {square roots of 16, 36, 64}; $Z = $ {squares of 2, 4, 6, 8}

_____

For 9–11, use the Venn diagram.

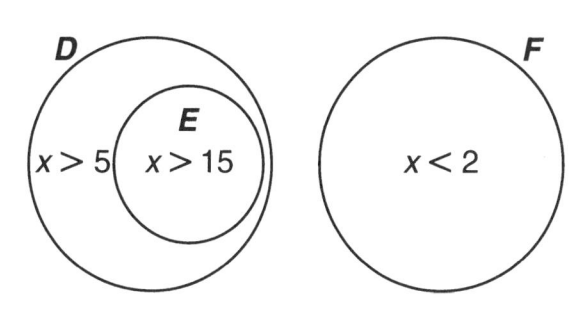

D — $x > 5$ — E: $x > 15$ F: $x < 2$

9. Identify the union of D and F.

_____

10. Identify the intersections.

_____

11. Identify any subsets.

_____

12. Make a truth table for the disjunction P or Q.
    P: A figure is a quadrilateral.
    Q: A figure is a regular polygon.

Chapter Test
14 Form C, continued

13. Complete the truth table.
P: Ann is on the basketball team.
Q: Ann plays the violin.

P	Q	P and Q	P or Q
T	T		T
T		F	T
F	T		
	F		

For 14–15, identify the hypothesis and the conclusion in each conditional.

14. Work hard, and you will be rewarded.

_____

_____

15. Plants need water to grow.

_____

_____

16. Make a conclusion, if possible, from the deductive argument.
If a rectangle has sides of 8 ft and 12 ft, then its area is 96 ft².
Rectangle WXYZ has an area of 96 ft².

_____

_____

For 17–19, use the graph.

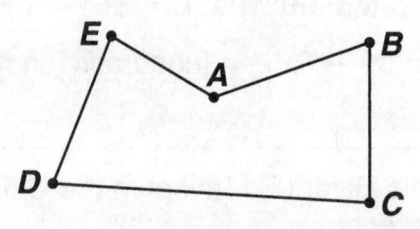

17. Find the degree of each vertex.

_____

18. Is the graph connected? Explain.

_____

_____

19. Determine whether the graph can be traversed through an Euler circuit. If your answer is yes, describe an Euler circuit in the graph.

_____

_____

For 20–21, use the graph.

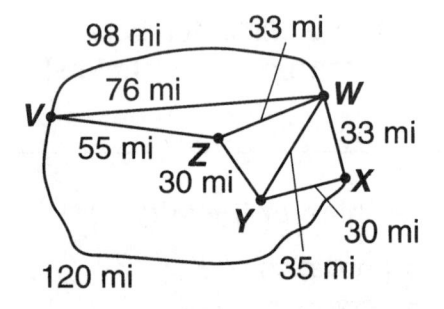

20. Determine the length of the shortest Hamiltonian circuit beginning at V.

_____

21. Determine the length of the longest Hamiltonian circuit beginning at X.

_____

Name _____ Date _____ Class _____

Performance Assessment Teacher Support
CHAPTER 1 *Algebra Toolbox*

Purpose:
To assess student understanding of relating events to graphs.

Time:
20–30 minutes

Grouping:
Individuals or partners

Preparation Hints:
Review labeling axes and discuss how certain events could be represented by a graph.

Introduce the Task:
Students are given two situations and asked to draw a representative graph. They are also given a graph and asked to think of a situation that could be modeled by the graph. They label the axes of the graphs and write several sentences describing the graph.

Performance Indicators:
____ Draw two representative graphs

____ Think of a situation that could be modeled by the graph.

____ Label the axes of the graph.

____ Write several sentences describing the graph.

Scoring Rubric:
Level 4: Student solves problems correctly and gives good explanations.
Level 3: Student solves problems but does not give satisfactory explanations.
Level 2: Student solves some problems but does not give satisfactory explanations.
Level 1: Student is not able to solve any of the problems.

Name _____ Date _____ Class _____

CHAPTER 1
Performance Assessment
Algebra Toolbox

Demonstrate your knowledge by giving a clear, concise solution to the problems presented. Be sure to include all-important information.

1. Draw a graph that is representative of the following statements.

   - A bus stops to pick up students for school.

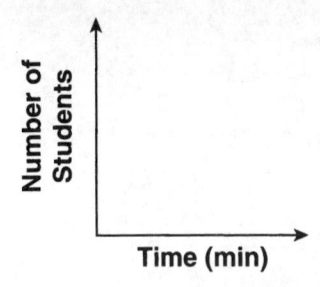

   [Graph with axes: Number of Students (y-axis), Time (min) (x-axis)]

   - On a very hot day in July you enter your hot house and turn on the air conditioner.

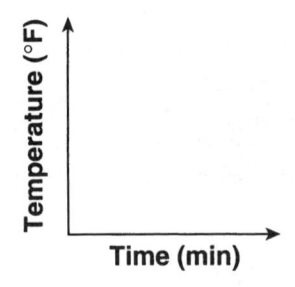

   [Graph with axes: Temperature (°F) (y-axis), Time (min) (x-axis)]

Think of a situation that could be modeled by the graph.

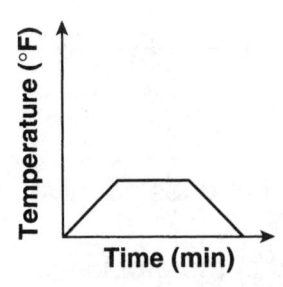

[Graph with axes: Temperature (°F) (y-axis), Time (min) (x-axis), showing a trapezoidal shape]

2. Label the axes of the graph and then write several sentences describing the graph.

_____
_____
_____

Name _____ Date _____ Class _____

Performance Assessment Teacher Support
CHAPTER 2 — Chapter 2

Purpose
To assess understanding of positive and negative numbers

Time
30–45 minutes

Grouping
Individuals or partners

Preparations Hints
Review what a time line is. Have students make timelines of important events in their lives. When they started to walk, when they went to kindergarten, lost a first tooth, etc.

Overview
Students are to create a time line of the birth dates for famous mathematicians and answer some questions pertaining to these dates.

Introduce the Task
Students are given the birth and death dates of 10 famous mathematicians.

Performance Indicators

_____ Completes the time line correctly.

_____ Explains how a time line and number line are similar.

_____ Is able to determine that Thales and Pythagoras were alive at the same time.

_____ Calculates the ages of the mathematicians.

_____ Lists the mathematicians in order from youngest to oldest correctly.

Scoring Rubric
Level 4: Student solves problems correctly and gives good explanations.
Level 3: Student solves problems but does not give satisfactory explanations.
Level 2: Student solves some problems but does not give satisfactory explanations.
Level 1: Student is not able to solve any of the problems.

Name _____ Date _____ Class _____

Performance Assessment Teacher Support
CHAPTER 2 — *Chapter 2*

Ten Famous Mathematicians	
Euclid (325–265 B.C.) Best known for his book *The Elements*	Al-Khwarizmi (A.D. 780–850) Considered one of the first users of Algebra.
Omar Khayyam (A.D. 1048–1131) The first mathematician to deal with every type of cubic equation that yields a positive square root	Archimedes (287–212 B.C) Best known for his practical applications to physics topics such as the lever and water displacement.
Plato (429–348 B.C.) Known to the world as a Greek philosopher	Mandelbrot (A.D. 1924–) The father of fractal geometry
Pythagoras (570–500 B.C.) Established a school in Croton, in Southern Italy and one of the most interesting figures in the history of mathematics.	Thales (624–547 B.C.) He is considered the "Father of Geometry" for his early discoveries and proofs of geometric relationships.
Hypatia (A.D. 370–415) Known as the first woman mathematician in history	Descartes (A.D. 1596–1650) Inventor of analytic geometry

1. Create a timeline showing the birthdates of the mathematicians.

←—+—+—+—+—+—+—+—+—+—+—→

2. Explain how a timeline is different than a number line. What do A.D. and B.C. represent?

_____

_____

3. List the mathematicians in order from youngest to oldest by age at death.

_____

_____

Copyright © by Holt, Rinehart and Winston.
All rights reserved.

Holt Pre-Algebra

Name _____ Date _____ Class _____

CHAPTER 3 Performance Assessment Teacher Support
Rational and Real Numbers

Purpose
Assess students' understanding of rational numbers and fractions.

Time
30–45 minutes

Grouping
Individuals

Preparation Hints
Review how to reduce fractions.

Overview
Students are given a square divided into several pieces. They must determine what fractional part each piece represents and give an explanation of how they determined the fractions. As an option, they are asked to design their own fraction squares.

Introduce the Task
Review with students the purpose of the assessment. Go through the tasks with the students and answer any questions they might have. As an option for the third question give the students a set of constraints in which to build their fraction squares.

Performance Indicators
_____ Able to determine each fractional section

_____ Gives a reasonable explanation for determining each section.

_____ Designs own fraction square (optional)

Scoring Rubric
Level 4: Student solves problems correctly and gives good explanations.
Level 3: Student solves problems but does not give satisfactory explanations.
Level 2: Student solves some problems but does not give satisfactory explanations.
Level 1: Student is not able to solve any of the problems.

Name _____ Date _____ Class _____

CHAPTER 3
Performance Assessment
Rational and Real Numbers

The large outer square represents one whole unit. It has been divided into smaller pieces labeled A to K.

1. Decide what each fraction of the whole unit each piece represents.

    A _____   E _____   I _____

    B _____   F _____   J _____

    C _____   G _____   K _____

    D _____   H _____

2. Write a brief explanation of how you determined the fraction for each of the following pieces:

    B _____   I _____

    _____   _____

    _____   _____

    F _____

    _____   J _____

    _____   _____

    _____   _____

3. Design your own fraction square and give the fractions for the pieces.

    _____

    _____

    _____

Copyright © by Holt, Rinehart and Winston.
All rights reserved.

Holt Pre-Algebra

Name _____ Date _____ Class _____

CHAPTER 4 — Performance Assessment Teacher Support
Collecting, Displaying, and Analyzing Data

Purpose
This performance task assesses the students understanding of their ability to draw a graph, calculate mean, median and mode, create a scatter plot, and draw a trend line.

Time
30–45 minutes

Grouping
Individuals

Preparation Hints
Review constructing a graph, including what information is needed and where labels are located.

Overview
Students are asked to represent different data sets graphically and make mean, median and mode calculations.

Introduce the Task
Review with students the purpose of the assessment. Go through the tasks with the students and answer any questions they might have.

Performance Indicators
____ Makes two representations of the data.
____ Calculates the mean, median, and mode of the ticket price data.
____ Calculates the mean, median, and mode of the roller coaster data.
____ Makes a scatterplot of the data.
____ Draws a trend line.

Scoring Rubric
Level 4: Student solves problems correctly and gives good explanations.
Level 3: Student solves problems but does not give satisfactory explanations.
Level 2: Student solves some problems but does not give satisfactory explanations.
Level 1: Student is not able to solve any of the problems.

Holt Pre-Algebra

Name _____ Date _____ Class _____

CHAPTER 4 Performance Assessment
Collecting, Displaying, and Analyzing Data

Park	One-Day Ticket Price	Number of Roller Coasters
Thrill Rides	$42.00	15
Flying Coasters Park	$44.99	15
Family Fun	$37.50	11
Gooseberry Farm	$48.00	10
Ride Monster	$39.99	9
Discovery Mountain	$34.95	8
Fun Town	$25.00	5

1. Make two bar graphs to represent the data. On one graph show the one-day ticket prices at each park. On the other, show the number of roller coasters.

2. Find the mean, median, and mode of the ticket price data.

   Mean _____   Median _____   Mode _____

3. Find the mean, median, and mode of the number of roller coaster data.

   Mean _____   Median _____   Mode _____

4. Make a scatter plot of ticket price vs the number of roller coasters and draw a trend line.

Name _____ Date _____ Class _____

Performance Assessment Teacher Support
CHAPTER 5 Plane Geometry

Purpose
This performance task assesses the students' understanding of parallel and perpendicular lines. It also assesses their understanding of rotating, reflecting, and translating a figure in the coordinate plane.

Time
15 minutes

Grouping
Individuals

Preparation Hints
Review plotting points on a coordinate plane.

Overview
Students are asked to draw lines and manipulate a given figure around the coordinate plane.

Introduce the Task
Review with students the purpose of the assessment. Go through the tasks with the students and answer any questions they might have.

Performance Indicators
____ Graphs two parallel lines.
____ Graphs two perpendicular lines and gives correct rules.
____ Plots the original points correctly on the coordinate grid.
____ Rotates the given figure appropriately.
____ Reflects the given figure appropriately.
____ Translates the given figure appropriately.

Scoring Rubric
Level 4: Student solves problems correctly and gives good explanations.
Level 3: Student solves problems but does not give satisfactory explanations.
Level 2: Student solves some problems but does not give satisfactory explanations.
Level 1: Student is not able to solve any of the problems.

Copyright © by Holt, Rinehart and Winston.
All rights reserved.

125

Holt Pre-Algebra

Name _____ Date _____ Class _____

CHAPTER 5 — Performance Assessment
Plane Geometry

Complete the tasks given below.

1. Graph two parallel lines and two perpendicular lines. What are the rules for determining whether two lines are parallel or perpendicular?

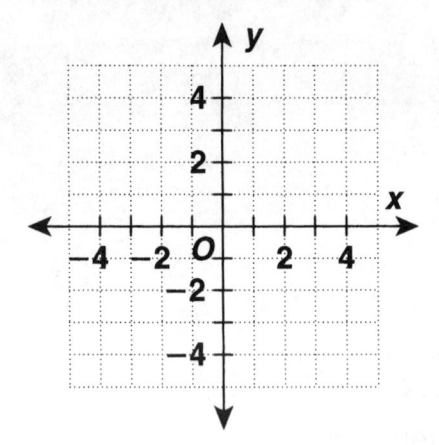

   _____
   _____
   _____
   _____
   _____
   _____
   _____

2. Plot the points (0, 0), (4, 0), (3, 3) and (1, 3) on the coordinate grid. Connect the points in the order given and connect the last point to the first point.

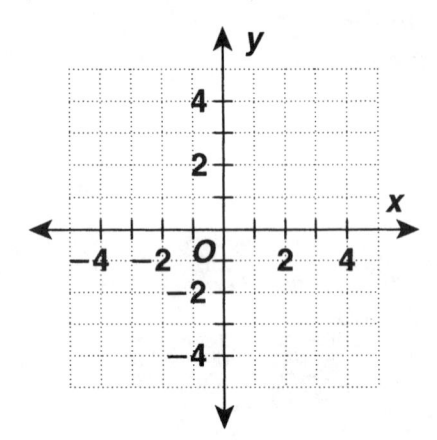

3. Rotate the original figure 180 degrees. Show the result of the transformation of the coordinate grid.

   _____

4. Reflect the original figure across the x-axis. Show the result of the transformation of the coordinate grid.

   _____

5. Translate the original figure 4 units to the left. Show the result of the transformation of the coordinate grid.

   _____

Copyright © by Holt, Rinehart and Winston.
All rights reserved.

Holt Pre-Algebra

Name _____ Date _____ Class _____

CHAPTER 6 Performance Assessment Teacher Support
Perimeter, Area, and Volume

Purpose
This performance task assesses students' understanding of volume and surface area of different shaped containers.

Time
30–45 minutes

Grouping
Individuals

Preparation Hints
Review formulas for finding surface area and volume.

Overview
Students are asked to make decisions regarding a best buy based on the volume or surface area of a container.

Introduce the Task
Review with students the purpose of the assessment. Go through the tasks with the students and answer any questions they might have.

Performance Indicators
_____ Determines the volume of a cone.
_____ Determine the volume of a cylinder.
_____ Determines the volume of a prism.
_____ Determines surface area of cone, cylinder and prism.
_____ Determines the size of a sphere.

Scoring Rubric
Level 4: Student solves problems correctly and gives good explanations.
Level 3: Student solves problems but does not give a satisfactory explanation.
Level 2: Student solves some problems but does not give a satisfactory explanation.
Level 1: Student is not able to solve any of the problems.

Holt Pre-Algebra

Name _____ Date _____ Class _____

CHAPTER 6 Performance Assessment
Perimeter, Area, and Volume

1. Your soccer club wants to sell frozen yogurt to raise money for new uniforms. The club has the choice of the two different size containers shown below. Each container costs the club the same amount. The club plans to charge customers $2.50. Which container should the club buy? Explain.

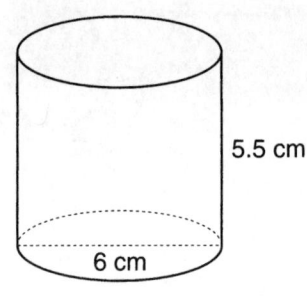

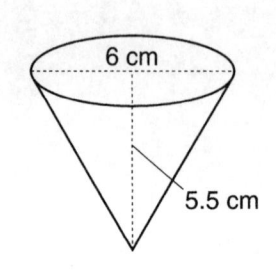

   [Cylinder: 6 cm diameter, 5.5 cm height]   [Cone: 6 cm diameter, 5.5 cm height]

   _____
   _____
   _____

2. A new movie theater is going to sell popcorn. The manager has the choice of the three different size containers shown below. The manager plans to charge $4.75 for a container of popcorn.

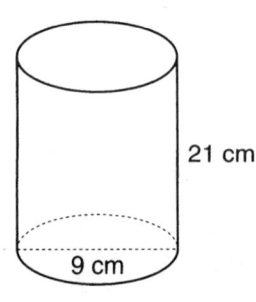

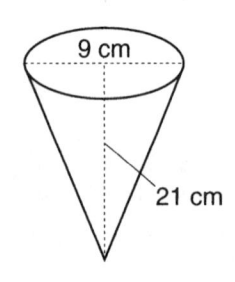

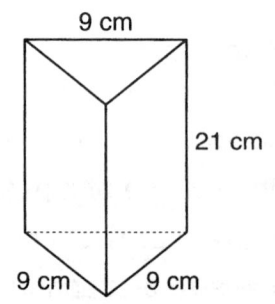

   [Cylinder: 9 cm diameter, 21 cm height]   [Cone: 9 cm diameter, 21 cm height]   [Triangular prism: 9 cm sides, 21 cm height]

   a. Which container would you choose as the manager of the movie theater? Explain.

   _____
   _____

   b. If all of the containers are made using the same material, which container would cost the least to make?

   _____
   _____

Copyright © by Holt, Rinehart and Winston.
All rights reserved.

Holt Pre-Algebra

Name _____ Date _____ Class _____

Performance Assessment Teacher Support
CHAPTER 7 *Ratios and Similarity*

Purpose
This performance task assesses the students' understanding of ratio and proportion using scale drawings.

Time
30–45 minutes

Grouping
Individuals

Preparation Hints
Review solving proportions.

Overview
Students are asked to measure lines and make conversions using a scale.

Introduce the Task
Review with students the purpose of the assessment. Go through the tasks with the students and answer any questions they might have.

Performance Indicators
___ Determines the width of the bedroom and closet doors

___ Determines the dimensions of the bedroom

___ Determines the number of feet of baseboard needed and cost

___ Determines the number of square yards and cost of carpeting

___ Accurately determines the width of the windows and dimensions of the closet

___ Calculates the area of the room and number of gallons of paint needed

Scoring Rubric
Level 4: Student solves problems correctly and gives good explanations.
Level 3: Student solves problems but does not give a satisfactory explanation.
Level 2: Student solves some problems but does not give a satisfactory explanation.
Level 1: Student is not able to solve any of the problems.

Name _____ Date _____ Class _____

CHAPTER 7 — Performance Assessment
Ratios and Similarity

Complete the tasks given below.

The diagram shown below is an architect's plan for a bedroom area. Using a ruler, make the necessary measurements to answer the questions that follow. Give the actual measurements.

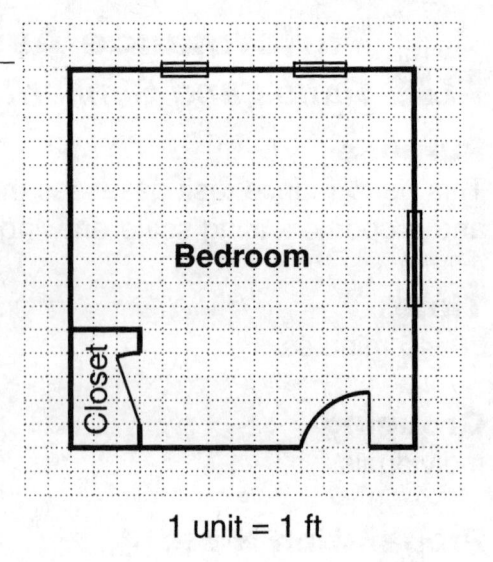

1 unit = 1 ft

1. How many feet wide is the door leading into the bedroom?

2. How many feet wide is the door leading into the closet?

3. How wide is the widest part of the bedroom?

4. What are the dimensions of the bedroom?

5. How many feet of baseboard will be needed to go around the bedroom, including the inside of the closet? (Hint: baseboard does not go in the doorways.)

6. If baseboard costs $2.35 per linear foot, what is the total cost for baseboard?

7. How many square yards of carpeting will be needed for the bedroom and closet? (round to the nearest yard)

8. If carpet costs $27.99 per square yard, what is the total cost for carpeting?

9. What are the widths of the windows?

10. What are the dimensions of the closet?

11. If the walls are 8 feet high, what is the total area of the walls and ceiling, including the inside and outside of the closet? Subtract 95 square feet for windows and doors.

12. If one gallon of paint covers 300 square feet, how many gallons of paint will be needed to paint the walls and ceiling excluding windows and doors? Round up to the nearest gallon.

Holt Pre-Algebra

Name _____ Date _____ Class _____

CHAPTER 8 Performance Assessment Teacher Support
Percents

Purpose
This performance task assesses the students' understanding of percents and organizing data.

Time
15–20 minutes

Grouping
Individuals

Preparation Hints
Review finding percents.

Overview
Students are asked to determine percents and construct a circle graph.

Introduce the Task
Review with students the purpose of the assessment. Go through the tasks with the students and answer any questions they might have.

Performance Indicators
____ Determines the correct percent for each restaurant.

____ Determines the correct sector size for each restaurant.

____ Accurately constructs the circle graph.

____ Gives a reasonable recommendation.

Scoring Rubric
Level 4: Student solves problems correctly and gives good explanations.
Level 3: Student solves problems but does not give satisfactory explanations.
Level 2: Student solves some problems but does not give satisfactory explanations.
Level 1: Student is not able to solve any of the problems.

Copyright © by Holt, Rinehart and Winston.
All rights reserved.

131

Holt Pre-Algebra

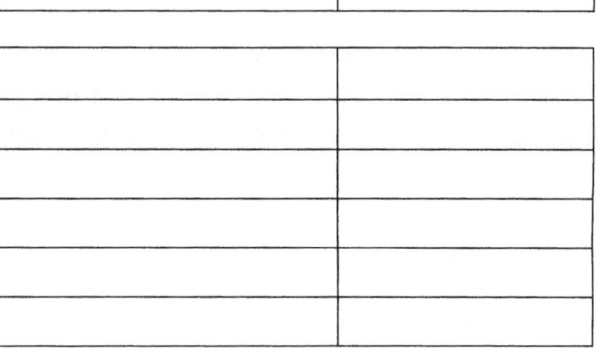

Name _____ Date _____ Class _____

CHAPTER 8 Performance Assessment
Percents

You have been hired by Pizza Palace to do a new marketing campaign. The campaign should show that Pizza Palace has the best food.

The table shows the results of a survey to find out which restaurant people liked best.

Restaurant	Response
Taco Town	128
Pizza Palace	248
Mr. Cluck's Chicken	28
Burger Barn	12
Salad Haven	184

1. What percent of the people surveyed chose each of the five restaurants? Round to the nearest hundredth of a percent.

2. In your presentation to Pizza Palace you want to give them a colorful representation of the results of the survey. You have chosen to make a circle graph. Calculate the size of each sector. Round to the nearest whole degree.

3. Construct the circle graph.

4. What is your recommendation to Pizza Palace?

_____

_____

Name _____ Date _____ Class _____

Performance Assessment Teacher Support
CHAPTER 9 Probability

Purpose
This performance task assesses the students' understanding of probability.

Time
30 minutes

Grouping
Individuals

Preparation Hints
Review how to use a random number table.

Overview
Students are asked to determine which spinner would produce the data set, use an organized method to find a combination of letters, and determine results with a random number table.

Introduce the Task
Review with students the purpose of the assessment. Go through the tasks with the students and answer any questions they might have.

Performance Indicators
____ Determines which spinner produced the desired results
____ Organizes a list
____ Uses a random number table
____ Makes predictions
____ Gives reasonable explanations

Scoring Rubric
Level 4: Student solves problems correctly and gives good explanations.
Level 3: Student solves problems but does not give satisfactory explanations.
Level 2: Student solves some problems but does not give satisfactory explanations.
Level 1: Student is not able to solve any of the problems.

Name _____ Date _____ Class _____

CHAPTER 9 — Performance Assessment
Probability

Complete the tasks given below.

1. Which spinner was most likely the spinner that produced the following data set? Explain your answer.
   Data Set: yellow, green, green, yellow, yellow, yellow, green, yellow, yellow, yellow

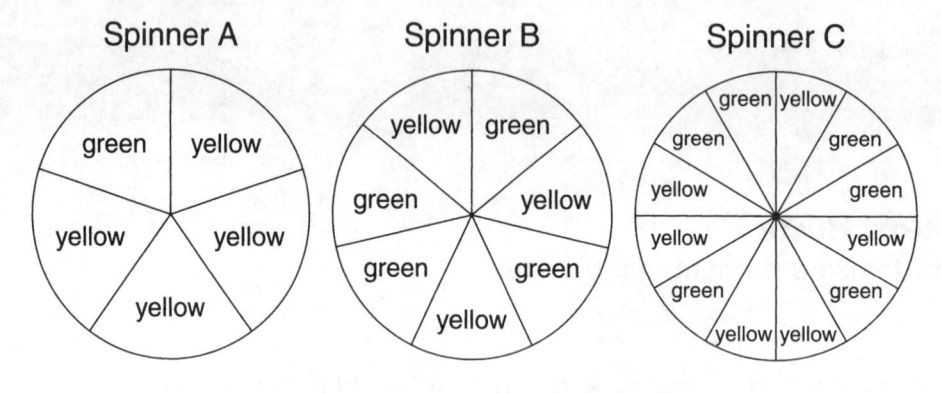

   Spinner A   Spinner B   Spinner C

   _____

   _____

   _____

2. Show several ways to find the number of 3-letter combinations in the word EQUAL. Write your answers in an organized list.

   _____

   _____

3. Use a random number table and write the first fifty numbers as single digits. Example: 43215 would be 4, 3, 2, 1, 5.

   A. How many 5's did you get? _____

   B. How many numbers larger than 6 did you get? _____

   C. How many odd numbers did you get? _____

   D. If you selected the next fifty random numbers, how many 5's would you expect to get? _____

   E. How many odd numbers would you expect to get? _____

Copyright © by Holt, Rinehart and Winston.
All rights reserved.

Holt Pre-Algebra

Name _____ Date _____ Class _____

Performance Assessment Teacher Support
CHAPTER 10 *More Equations and Inequalities*

Purpose
This performance task assesses the students' understanding of solving linear equations and inequalities and solving systems of equations.

Time
30 minutes

Grouping
Individuals

Preparation Hints
Review writing a linear inequality and a system of equations.

Overview
Students are asked to explain how to solve different types of linear equations, to write and solve a linear inequality, and to write and solve a system of equations.

Introduce the Task
Review with students the purpose of the assessment. Go through the tasks with the students and answer any questions they might have.

Performance Indicators
____ Explains how to solve each linear equation.

____ Writes and solves a linear inequality.

____ Determines the maximum amount of time an item could be rented for.

____ Determines which company has the best price for rental given certain restrictions.

Scoring Rubric
Level 4: Student solves problems correctly and gives good explanations.

Level 3: Student solves problems but does not give satisfactory explanations.

Level 2: Student solves some problems but does not give satisfactory explanations.

Level 1: Student is not able to solve any of the problems.

Name _____ Date _____ Class _____

CHAPTER 10 Performance Assessment
More Equations and Inequalities

Complete the tasks given below.

1. Solve each of the equations shown below. Describe the steps needed.

   A. $3x - 4 = 17$

   _____

   B. $\frac{z}{6} - 3 = 2$

   _____

   C. $6x + 3 = 2x - 13$

   _____

2. A1-Rental charges $35 plus $6 per hour to rent a mini-loader, while BilJax charges $15 plus $8 per hour. Russell Casius needs to use a mini-loader for about 7 hours.

   A. For what number of hours would BilJax cost no more than A1-Rental? Write and solve a linear inequality.

   _____

   B. If Russell can spend no more than $75 to excavate some land in his yard, what is the *maximum* amount of time he could use the rented mini-loader from A1? _____

   C. What is the *maximum* amount of time he could use the mini-loader from BilJax? _____

   D. A1 is 20 minutes from Russell's home, and BilJax is half an hour away. Taking drive time into consideration, where should Russell rent the mini-loader? Explain.

   _____
   _____
   _____
   _____

Name _____ Date _____ Class _____

Performance Assessment Teacher Support
CHAPTER 11 Graphing Lines

Purpose
This performance task assesses the students' understanding of inequalities, parallel and perpendicular lines, and writing equations of a line.

Time
30 minutes

Grouping
Individuals

Preparation Hints
Review plotting points on a coordinate plane.

Overview
Students are asked to draw a graph showing the possible dimensions of a garden, to determine the dimensions that maximize the area, and to interpret roads on a map as lines on a graph, having slopes represented by equations.

Introduce the Task
Review with students the purpose of the assessment. Go through the tasks with the students and answer any questions they might have.

Performance Indicators
____ Graphs the possible dimensions of a garden.
____ Determines the dimensions that give the maximum area.
____ Determines the slope of a line.
____ Determines the equation of a line.
____ Explains that parallel lines have the same slope.
____ Explains that perpendicular lines have slopes that are negative reciprocals.

Scoring Rubric
Level 4: Student solves problems correctly and gives good explanations.
Level 3: Student solves problems but does not gives satisfactory explanations.
Level 2: Student solves some problems but does not give satisfactory explanations.
Level 1: Student is not able to solve any of the problems.

Copyright © by Holt, Rinehart and Winston.
All rights reserved.

Holt Pre-Algebra

Name _____ Date _____ Class _____

CHAPTER 11 Performance Assessment
Graphing Lines

Complete the tasks given below.

1. Salina and Roberto are working on sectioning off a portion of the backyard for a rectangular garden. They have a total of 110 feet of fencing. Due to the shape of the yard, the garden has to be at least 25 feet long.

   a. Draw a graph that shows all the possible dimensions of the garden.

   b. Decide which dimensions would give the maximum gardening space. Explain your reasoning.

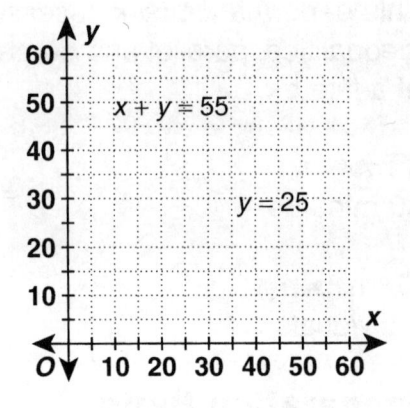

   _____

   _____

2. Martin Norris is a land developer who plans to expand a subdivision. He will need to add three additional roads, the plans for which are shown in the graph. Washington Drive will be parallel to Kennedy Drive, and Lincoln Drive will be perpendicular to the other two roads.

   a. Find the slope of each road.

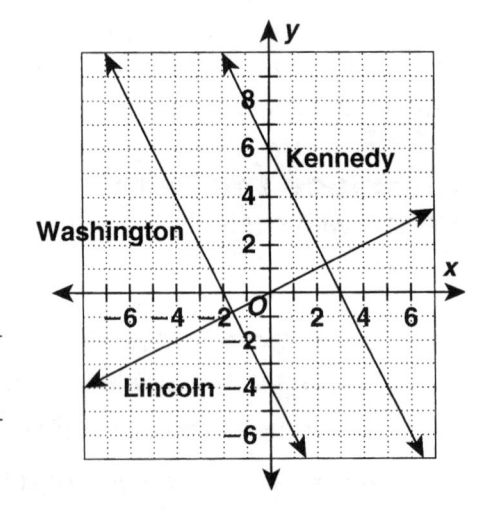

   _____

   _____

   b. Write an equation for each line.

   _____

   c. What has to be true for Kennedy Drive and Washington Drive to be parallel?

   _____

   d. What has to be true for Lincoln Drive to be perpendicular to Washington Drive and Kennedy Drive?

   _____

   _____

   _____

Copyright © by Holt, Rinehart and Winston.
All rights reserved.

Holt Pre-Algebra

Name _____ Date _____ Class _____

Performance Assessment Teacher Support
CHAPTER 12 Sequences and Functions

Purpose
This performance task assesses the students' understanding of functions, and of inverse and direct variation.

Time
30 minutes

Grouping
Individuals

Preparation Hints
Review direct and inverse variation.

Overview
Students are asked to explain the concept of a function, to describe different kinds of functions, to describe relations of direct and inverse variation, and to graph an inverse variation relation.

Introduce the Task
Review with students the purpose of the assessment. Go through the tasks with the students and answer any questions they might have.

Performance Indicators
___ Explains the concepts of function, domain, and range.
___ Gives an example of a function and a relation that is not a function.
___ Identifies the different types of functions—linear, quadratic, and exponential.
___ Explains the difference between direct and inverse variation.

Scoring Rubric
Level 4: Student solves problems correctly and gives good explanations.
Level 3: Student solves problems but does not give satisfactory explanations.
Level 2: Student solves some problems but does not give satisfactory explanations.
Level 1: Student is not able to solve any of the problems.

Name _____ Date _____ Class _____

CHAPTER 12 Performance Assessment
Sequences and Functions

Complete the tasks given below.

1. Write a paragraph explaining the meaning of the terms function, domain and range. Give examples to support your answer.

   _____
   _____
   _____

2. Identify the functions below as linear, exponential, or quadratic. Explain how you can classify each function.

   A. $f(x) = 9 + 8x$ _____

   B. $f(x) = x^2 + 3x - 2$ _____

   C. $f(x) = \left(\dfrac{1}{2}\right)^x$ _____
   _____

   D. $y = 4x - 9$ _____

   E. $y = x^2 - 3$ _____

3. Explain the difference between direct and inverse variation. Explain how the y-values change.

   _____
   _____
   _____
   _____

Copyright © by Holt, Rinehart and Winston.
All rights reserved.

Holt Pre-Algebra

Name _____ Date _____ Class _____

CHAPTER 13 Performance Assessment Teacher Support
Polynomials

Purpose
To assess students' ability to solve problems involving operations with polynomials and reasoning about polynomials.

Time
30–35 minutes

Grouping
Individuals or partners

Preparation Hints
Remind students that polynomials can be written in many ways but that descending order of the variable is preferred.

Introduce the Task
Students combine and simplify like terms to find the expressions for perimeter and multiply binomials to find expressions for the area. Students then use reasoning skills to order sections of the garden by size.

Performance Indicators

____ Uses knowledge of algebra and geometry to accurately complete measures of missing sides of figure.

____ Accurately combines like terms to determine an expression for the perimeter of the figure.

____ Accurately multiplies binomials to determine the area of the figure and correctly interprets the results of the multiplication.

Scoring Rubric
Level 4: Student solves problems correctly and gives good explanations.
Level 3: Student solves problems but does not give a satisfactory explanation.
Level 2: Student solves some problems but does not give a satisfactory explanation.
Level 1: Student is not able to solve any of the problems.

Copyright © by Holt, Rinehart and Winston.
All rights reserved.

141

Holt Pre-Algebra

Name _____ Date _____ Class _____

CHAPTER 13 Performance Assessment
Polynomials

You have 34 meters of fencing to use in constructing the perimeter of a rectangular garden. The garden will have 4 sections as shown below. After you have used all the fencing to build the outside walls, you will need to buy more fencing to separate the sections.

```
        7 m         x
     ┌────────┬────────┐
     │        │        │
  x  │        │        │
     ├────────┼────────┤
  2 m│        │        │
     └────────┴────────┘
```

- Write an expression in simplified form for the perimeter of the garden.

_____

- Write as many expressions as you can for the area of the entire garden, using the variable x.

_____

- What is the value of x in the drawing?

_____

- How much additional fencing will you need to build the inside walls of the garden?

_____

- From least to greatest in area, the sections of the garden will be used for radishes, peppers, squash, and tomatoes. Find the area that will be used to plant each crop. Explain.

_____

_____

_____

_____

Name _____ Date _____ Class _____

CHAPTER 14 Performance Assessment Teacher Support
Analyzing a Graph

Purpose
To assess students' ability to solve problems involving networks

Time
30–35 minutes

Grouping
Individuals or partners

Preparation Hints
Remind students of the difference between an Euler circuit and a Hamiltonian circuit. Student should understand that a graph could be a Hamiltonian circuit without being an Euler circuit.

Introduce the Task
Students examine a graph to determine whether it represents an Euler circuit, a Hamiltonian circuit, both, or neither.

Performance Indicators
____ Uses knowledge of networks and circuits to discriminate between definitions of types and to determine the applicability of each.

____ Accurately determines specific details about particular types of graphs.

____ Accurately extends reasoning beyond the given problem situation.

Scoring Rubric
Level 4: Student solves problems correctly and gives good explanations.
Level 3: Student solves problems but does not give a satisfactory explanation.
Level 2: Student solves some problems but does not give a satisfactory explanation.
Level 1: Student is not able to solve any of the problems.

Name _____ Date _____ Class _____

CHAPTER 14 Performance Assessment
Analyzing a Graph

The graph represents a highway system connecting 6 towns: *A, B, C, D, E,* and *F.* A salesperson lives in one of the towns and travels between the towns on his route.

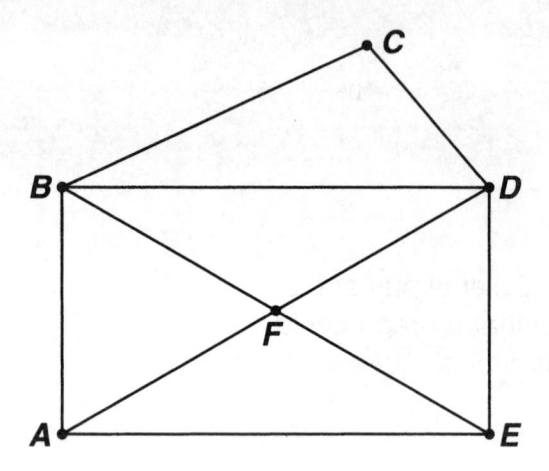

- Why would the salesperson be more interested in whether this graph is a Hamiltonian rather than an Euler circuit?

    _____

- Name at least two Hamiltonian circuits on the graph.

    _____

- Another worker, a highway inspector, also lives in one of the towns. How can you tell without tracing the graph whether it would be possible for her to visit each town and return home without driving on the same road twice?

    _____

- Does the graph represent an Euler circuit? Explain.

    _____

- How could you convert this graph into an Euler circuit?

    _____

    _____

Name _____ Date _____ Class _____

CHAPTER 1 Cumulative Test
Form A

1. What is the median of the following data?
   12, 2, 6, 10, 8, 4, 9
   - **A** 4
   - **B** 6
   - **C** 8
   - **D** 10

2. Find $1.2 \cdot 10^3$.
   - **A** 12
   - **B** 120
   - **C** 1200
   - **D** 12,000

3. What is the greatest common factor of 8 and 32?
   - **A** 2
   - **B** 4
   - **C** 8
   - **D** 16

4. Simplify $8 - (-2)$.
   - **A** 6
   - **B** −6
   - **C** 10
   - **D** −10

5. Simplify $-4 \cdot 3$.
   - **A** −12
   - **B** 12
   - **C** −7
   - **D** 7

6. Which of the following is the best estimate of $4.9 \cdot 5.1$?
   - **A** 20
   - **B** 25

7. Simplify $3\frac{3}{4} + 2\frac{1}{8}$.
   - **A** $5\frac{7}{8}$
   - **B** $5\frac{1}{3}$

8. Which graph shows no correlation?

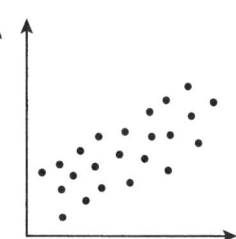

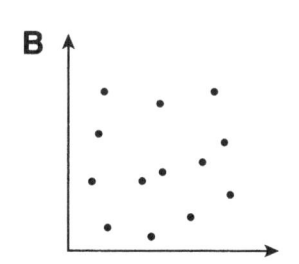

   A

   B

145

Holt Pre-Algebra

Name _____ Date _____ Class _____

CHAPTER 1 Cumulative Test
Form A, continued

9. Which of the following pairs of figures are NOT similar?

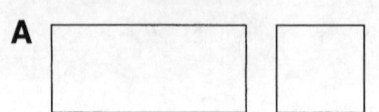

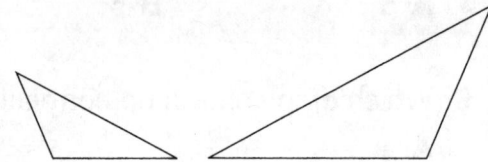

   A
   B

10. Solve the proportion $\frac{4}{3} = \frac{w}{21}$.
    A $w = 20$
    B $w = 28$
    C $w = 63$
    D $w = 82$

11. 25 is 10% of what number?
    A 2.5
    B 25
    C 250
    D 2500

12. Classify the triangle according to its sides and angles.

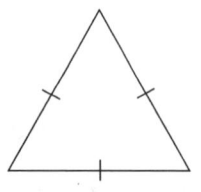

    A isosceles, acute
    B equilateral, equiangular
    C scalene, obtuse
    D isosceles, obtuse

13. Convert 25 m to cm.
    A 25 cm
    B 250 cm
    C 2500 cm
    D 25,000 cm

14. Find the volume of the prism.

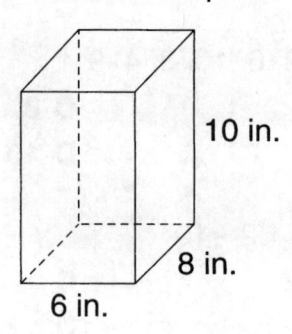

    10 in.
    8 in.
    6 in.

    A 24 in^3
    B 48 in^3
    C 480 in^3
    D 960 in^3

15. Use the Pythagorean Theorem to find the missing measure.

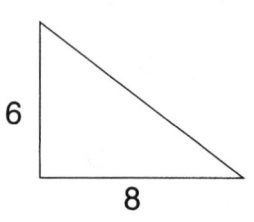

    6
    8

    A 2
    B 10
    C 14
    D 24

16. What is the slope and the y-intercept of $y = 2x + 1$?
    A 2, 1
    B 2, −1

17. Maria wants to arrange four books on her shelf. How many ways could she arrange them?
    A 24
    B 12
    C 6
    D 1

Name _____ Date _____ Class _____

Cumulative Test
Form A, continued
CHAPTER 1

18. What are the next two terms in this sequence?
3, 7, 11, 15 …
A 19, 23
B 20, 25

19. Evaluate 3w for w = 5.
A 2   **C** 15
B 8   **D** 18

20. Use the formula $A = lw$ to find the area of the rectangle.

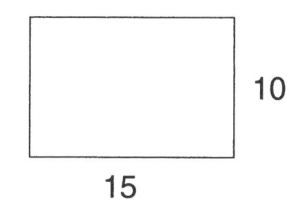

[rectangle with width 10 and length 15]

A 15   **C** 60
B 50   **D** 150

21. Which is an algebraic expression for 18 more than a number y.
A 18 − y   **C** 18 ÷ y
B y + 18   **D** 18y

22. Sam read x books today; he had read 15 books before that. How many total books did Sam read? Which expression represents this problem?
A x + 15   **C** x − 15
B 15x   **D** 15 − x

23. Solve x + 7 = 9.
A x = 1.27   **C** x = 16
B x = 2   **D** x = 63

24. Solve 3x = 30.
A x = 10   **C** x = 60
B x = 27   **D** x = 90

25. Which is the graph of x + 2 > 5?

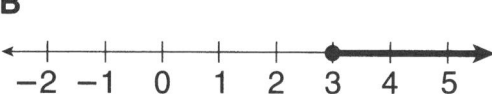

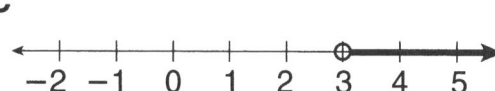

A [number line shaded left to 3, closed circle at 3]

B [number line, closed circle at 3]

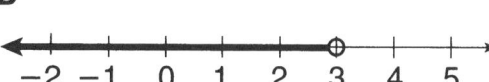

C [number line, open circle at 3, shaded right]

D [number line, open circle at 3, shaded left]

Cumulative Test

Form A, continued

26. Simplify $8w + 3w$.

A $5w$ **C** $11w$
B $11w$ **D** $24w$

27. Which ordered pair is a solution of $y = 6x$?

A (12, 2) **C** (3, 18)
B (4, 18) **D** (6, 1)

28. What are the coordinates of point B?

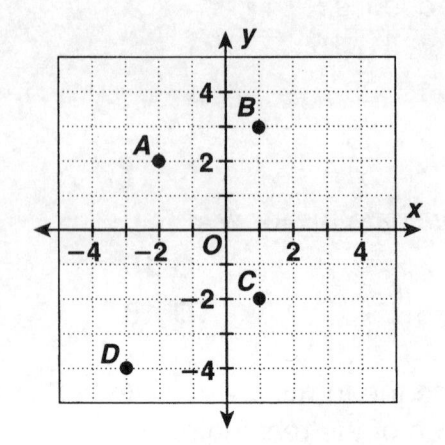

A (3, 1) **C** (1, −2)
B (1, 3) **D** (−3, −4)

29. Which graph matches the equation $y = 2x$?

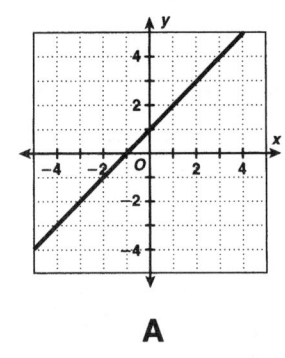

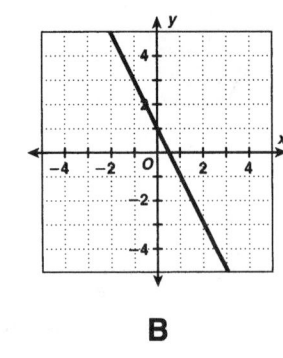

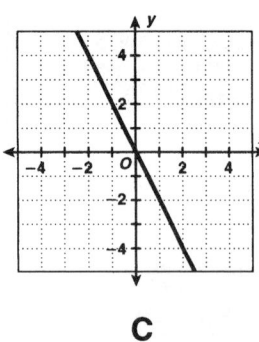

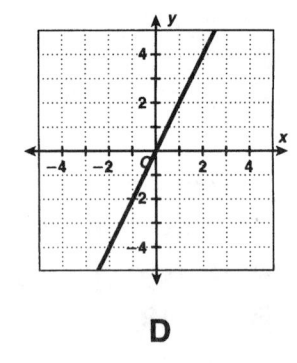

A B C D

30. On a highway, a driver sets a car's cruise control for a constant speed of 55 mi/h. Which graph shows the speed of the car?

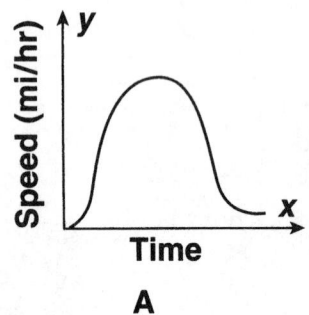

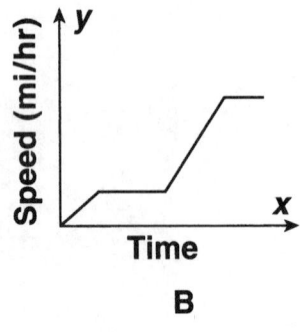

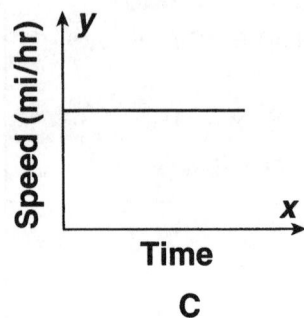

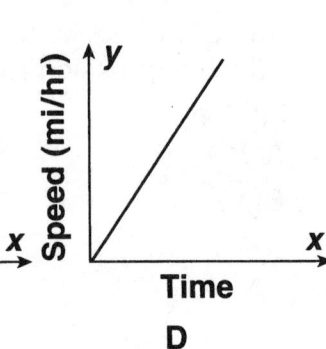

A B C D

Name _____ Date _____ Class _____

CHAPTER 1 Cumulative Test
Form B

1. What is the median of the following data?
   12, 26, 16, 14, 23, 28, 19, 20
   A 17.5 **C** 19.5
   B 19   **D** 20

2. Find $1.47 \cdot 10^4$.
   F 14,700 **H** 147
   G 1470   **J** 14.700

3. What is the greatest common factor of 18, 27, and 45?
   A 3 **C** 6
   B 5 **D** 9

4. Simplify $-7 - (-5)$.
   F −12 **H** 2
   G −2   **J** 12

5. Simplify $-3 \cdot 28$.
   A 84 **C** −78
   B 78 **D** −84

6. Which of the following is the best estimate of $4.12 \cdot 26.2$?
   F 98   **H** 75
   G 100 **J** 130

7. Simplify $5\frac{2}{5} + 3\frac{4}{15}$.
   A $8\frac{1}{3}$ **C** $9\frac{2}{5}$
   B $8\frac{2}{3}$ **D** $9\frac{1}{3}$

8. Which graph shows a negative correlation?

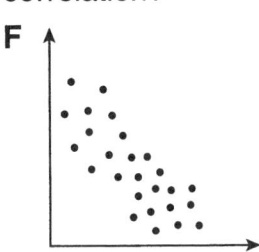

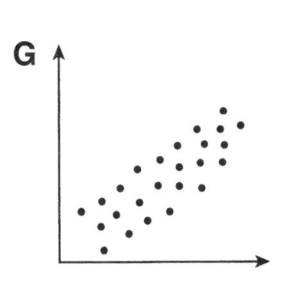

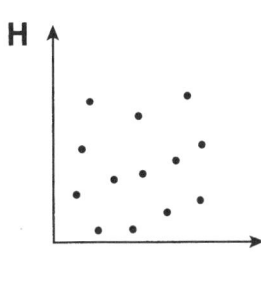

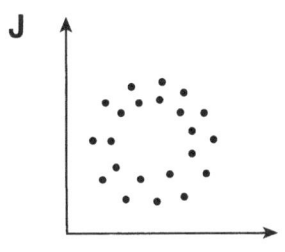

   F

   G

   H

   J

CHAPTER 1 Cumulative Test
Form B, continued

9. Which of the following pairs of figures are NOT similar?

   A 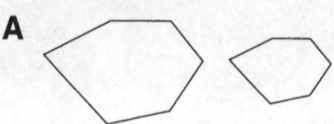
   B 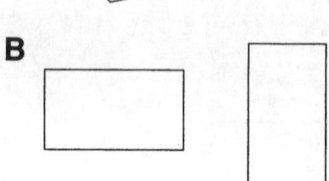
   C
   D

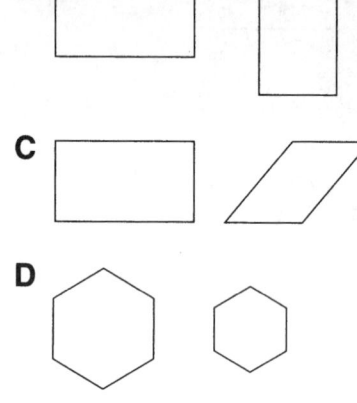

10. Solve the proportion $\frac{5}{w} = \frac{7.5}{15}$.
    - F $w = 2.5$
    - G $w = 6$
    - H $w = 9$
    - J $w = 10$

11. 35 is 7% of what number?
    - A 500
    - B 700
    - C 50
    - D 70

12. Classify the triangle according to its sides and angles. 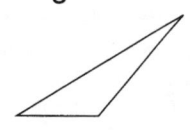
    - F isosceles, acute
    - G equilateral, equiangular
    - H scalene, obtuse
    - J isosceles, obtuse

13. Convert 350 m to km.
    - A 350,000 km
    - B 3500 km
    - C 0.35 km
    - D 0.035 km

14. Find the volume of the sphere to the nearest tenth. Use 3.14 for π. 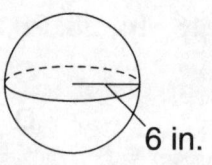

    6 in.
    - F 288.0 in^3
    - G 904.3 in^3
    - H 226.1 in^3
    - J 2713.0 in^3

15. Use the Pythagorean Theorem to find the missing measure. 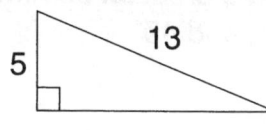
    - A 8
    - B 10
    - C 12
    - D 14

16. What is the slope and the y-intercept of $y = 3x - 2$?
    - F 2, −3
    - G 3, −2
    - H 2, 3
    - J −2, 3

17. Alana wants to arrange six trophies on her shelf. How many ways could she arrange them?
    - A 720
    - B 540
    - C 120
    - D 6

Cumulative Test
Form B, continued

18. What are the next three terms in this sequence?
1, 4, 9, 16, ...
- F 19, 22, 25
- G 23, 30, 37
- H 25, 36, 49
- J 24, 29, 36

19. Evaluate $3w - y$ for $w = 6$ and $y = 5$.
- A 13
- B 14
- C 4
- D 23

20. Use the formula $p = 2h + 2b$ to find the perimeter of the rectangle.

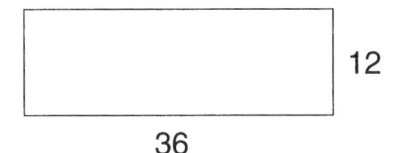

[rectangle with width 12 and length 36]

- F 48
- G 84
- H 96
- J 432

21. Which is an algebraic expression for 2 plus the product of 3 and w.
- A $3(2w)$
- B $2(3w)$
- C $2 + 3w$
- D $2w + 3$

22. To add to his 19-card collection, Kevin bought 2 baseball cards each week for x weeks. How many does he have now? Which expression represents this problem?
- F $19 + 2x$
- G $2(3 + x)$
- H $2x - 19$
- J $19 \div 2x$

23. Solve $w - 24 = 56$.
- A $w = 19$
- B $w = 2.3$
- C $w = 32$
- D $w = 80$

24. Solve $\frac{p}{5} = 450$.
- F $p = 9$
- G $p = 90$
- H $p = 445$
- J $p = 2250$

25. Which of the following is the graph of $4x + 9 > 21$?

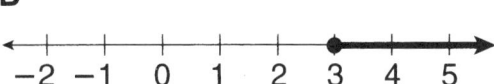

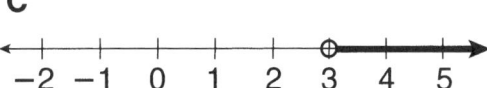

A [number line from −1 to 6, shaded left with closed circle at 3]

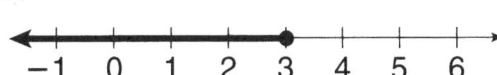

B [number line from −2 to 5, shaded right with closed circle at 3]

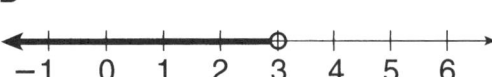

C [number line from −2 to 5, shaded right with open circle at 3]

D [number line from −1 to 6, shaded left with open circle at 3]

Name _____ Date _____ Class _____

CHAPTER 1 Cumulative Test
Form B, continued

26. Simplify $14t - 4t + 7w + 3t$.
- **F** $13 + 7w$
- **G** $14wt$
- **H** $13t + 7w$
- **J** $10t + 7w$

27. Which ordered pair is a solution of $y = 2x - 7$?
- **A** $(-5, -17)$
- **B** $(-3, 1)$
- **C** $(0, 7)$
- **D** $(2, 11)$

28. What are the coordinates of point A? 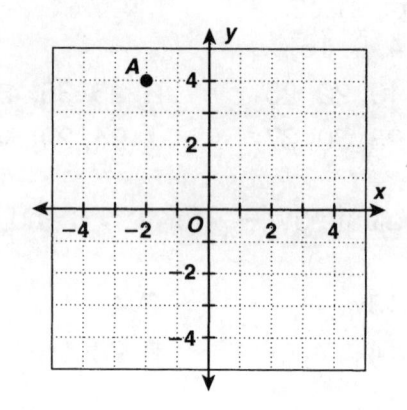
- **F** $(-2, -4)$
- **G** $(-2, 4)$
- **H** $(2, -4)$
- **J** $(2, 4)$

29. Which graph matches the equation $y = 2x + 1$?

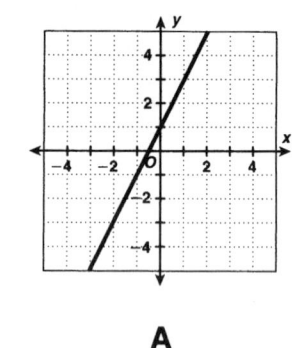

A 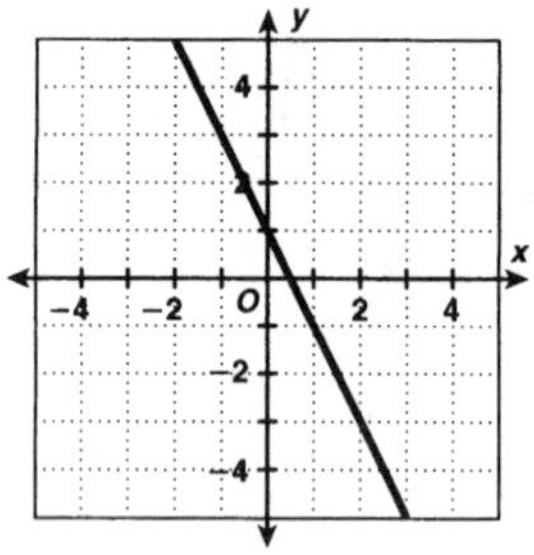   B 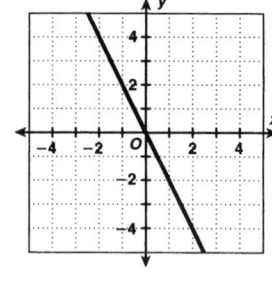   C 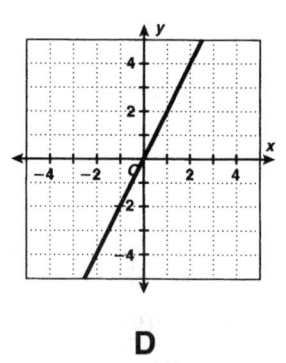   D

30. Which graph shows the speed of a scooter as you push for 2 minutes and then coast until coming to a rest?

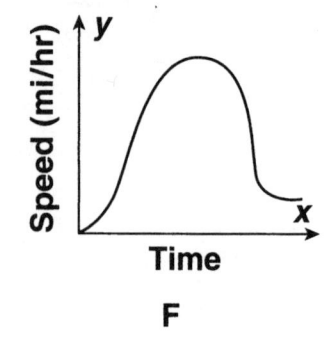

F 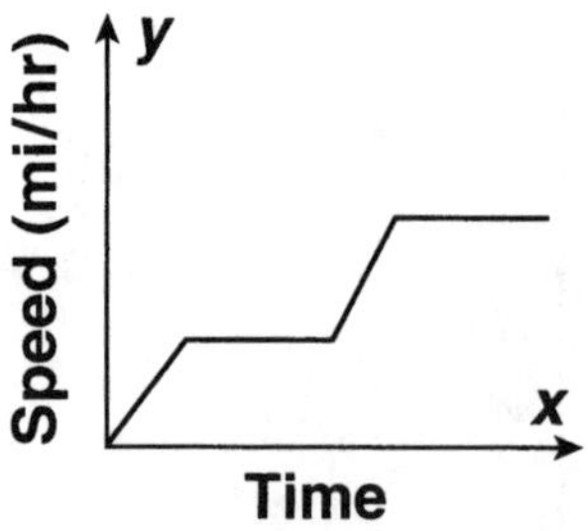   G 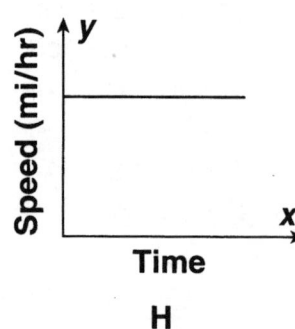   H 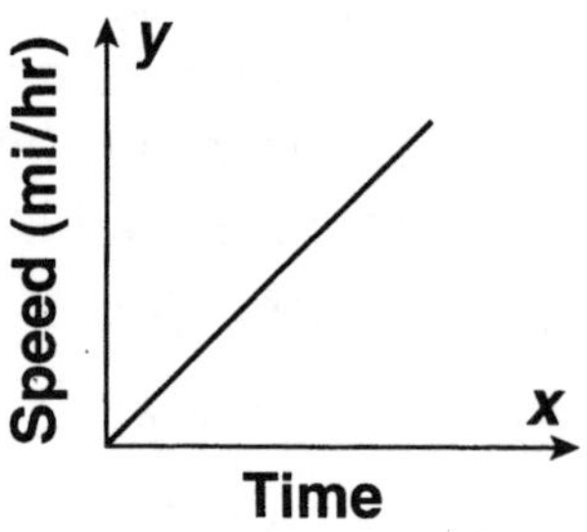   J

Copyright © by Holt, Rinehart and Winston.
All rights reserved.

Holt Pre-Algebra

Name _____ Date _____ Class _____

CHAPTER 1 Cumulative Test
Form C

1. What is the median of the following data?
   212, 226, 216, 214, 223, 228, 219, 220

   A 217.5 **C** 219.5
   B 219 **D** 220

2. Find $2.482 \cdot 10^6$.

   F 24,820 **H** 2,482,000
   G 248,200 **J** 24,820,000

3. What is the greatest common factor of 16, 40, and 72?

   A 2 **C** 8
   B 4 **D** 16

4. Simplify $-32 - (-18)$.

   F 14 **H** 50
   G -14 **J** -50

5. Simplify $-9 \cdot -28$.

   A -37 **C** -252
   B 37 **D** 252

6. Which of the following is the best estimate of $8.725 \cdot 59.41$?

   F 72 **H** 480
   G 284 **J** 540

7. Simplify $18\frac{3}{16} + 122\frac{9}{40}$.

   A $140\frac{33}{80}$ **C** $140\frac{3}{14}$
   B $140\frac{12}{56}$ **D** 140

8. Which graph shows a positive correlation?

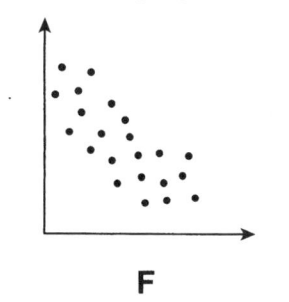

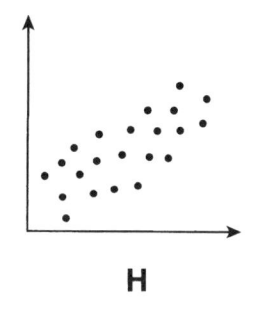

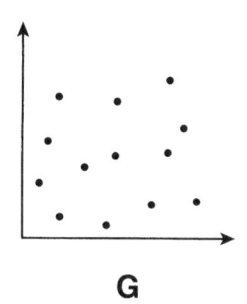

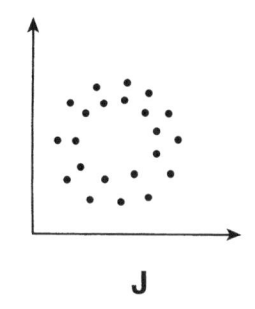

   F   H   G   J

Name _____ Date _____ Class _____

CHAPTER 1 **Cumulative Test**
Form C, continued

9. Which of the following pairs of figures are NOT similar?

   A 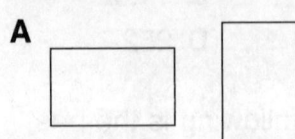

   B 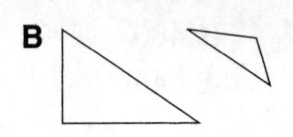

   C 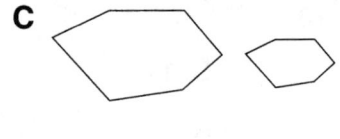

   D

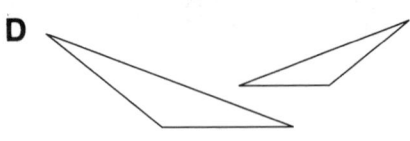

10. Solve the proportion $\frac{26}{w} = \frac{19.5}{18}$.

    F 9   H 28.2
    G 15   J 24

11. 120 is 0.5% of what number?

    A 240   C 24,000
    B 2400   D 240,000

12. Classify the triangle according to its sides and angles. 

    F isosceles, acute
    G equilateral, equiangular
    H scalene, obtuse
    J isosceles, obtuse

13. Convert 3725 cm to km.

    A 0.003725 km   C 0.3725 km
    B 3.725 km   D 0.03725 km

14. Find the volume of the sphere. 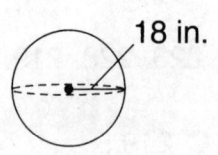

    18 in.

    F 24,416.64 in³   H 6104.16 in³
    G 18,321.48 in³   J 1356.48 in³

15. Use the Pythagorean Theorem to find the missing measure. 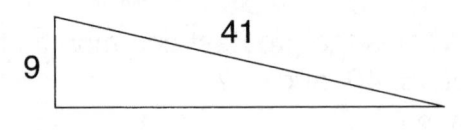

    9, 41

    A 50   C 40
    B 42   D 24

16. What is the slope and the *y*-intercept of $8x + 4y = 4$?

    F 8, 4   H 2, −1
    G 4, 8   J −2, 1

17. Marta wants to arrange eight bottles of perfume on her dresser. How many ways could she arrange them?

    A 52,430   C 1024
    B 40,320   D 36

Cumulative Test
Form C, continued

18. What are the next three terms in this sequence?
1000, 500, 250, 125, …
- F 62.5, 31.25, 15.625
- G 100, 75, 50
- H 25, 36, 49
- J 70, 50, 20

19. Evaluate $3(w - 2y + z)$ for $w = 14$, $y = 3$, and $z = 2$.
- A 20
- B 26
- C 30
- D 114

20. Use the formula $p = 2h + 2b$ to find the perimeter of the rectangle.

(rectangle: 10.1 by 3.4)

- F 13
- G 27
- H 34.34
- J 40

21. Which is an algebraic expression for 12 more than the product of 6 and a number.
- A $12 + w$
- B $12(6 - w)$
- C $12(6 + w)$
- D $6w + 12$

22. Renee bought a pounds of apples at $2 per pound and p pounds of peaches at $3 per pound. Which expression represents this problem?
- F $2a - 3p$
- G $2a + 3p$
- H $2 + p$
- J $2(a + p)$

23. Solve $142 = w - 85$.
- A $w = 57$
- B $w = 147$
- C $w = 227$
- D $w = 320$

24. Solve $5w - 4 = 21$.
- F $w = 5$
- G $w = 7$
- H $w = 6$
- J $w = 8$

25. Which of the following is the graph of $8y - 17 \leq 47$?

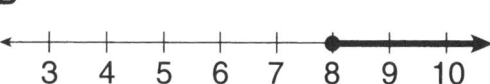

A (number line 3–10, shaded from left through closed circle at 8)

B (number line 3–10, shaded from closed circle at 8 to right)

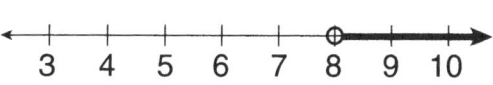

C (number line 3–10, shaded from left through open circle at 8)

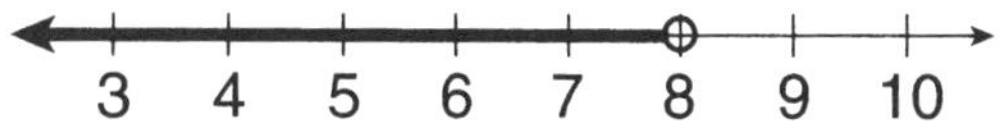

D (number line 3–10, shaded from open circle at 8 to right)

Name _____ Date _____ Class _____

CHAPTER 1 Cumulative Test
Form C, continued

26. Simplify $4(2x + 3) - 3y + 5 + 2x$.
  F $10x - 3y + 17$ H $10x + 17$
  G $8x - 3y + 8$ J $17xy$

27. Which ordered pair is a solution of the equation $7x - 4y = 5$?
  A (3, 4)   C (0, −4)
  B (2, 3)   D (3, 24)

28. What are the coordinates of point A?

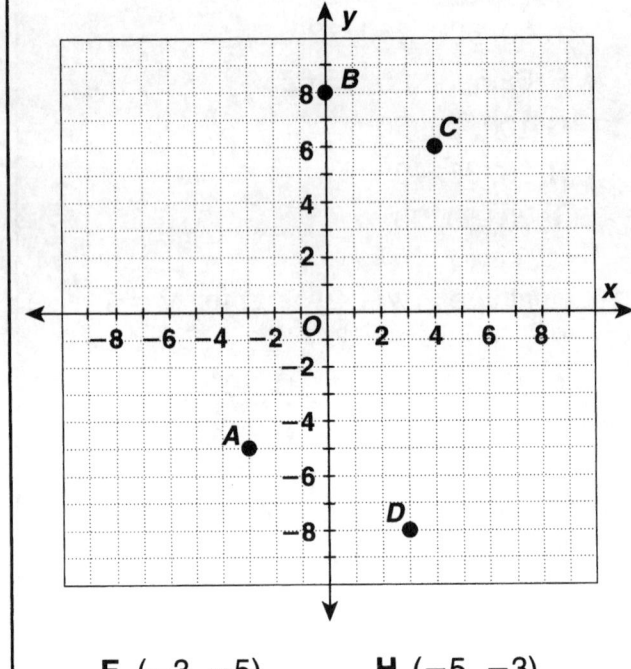

  F (−3, −5)   H (−5, −3)
  G (3, −5)   J (5, −3)

29. Which graph matches the equation $y = 2x - 1$?

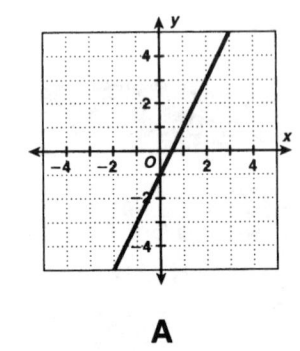

  A 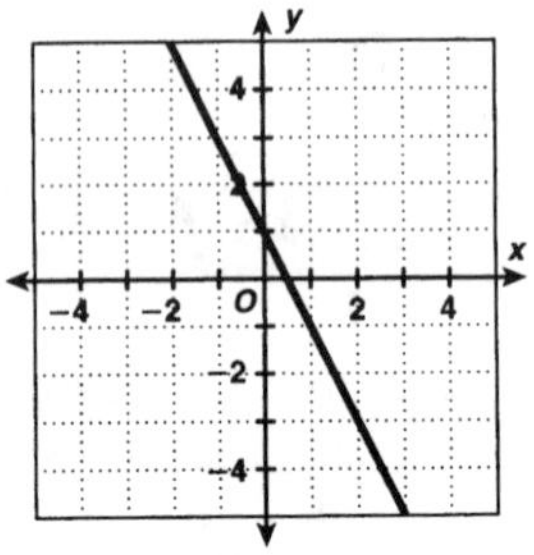   B 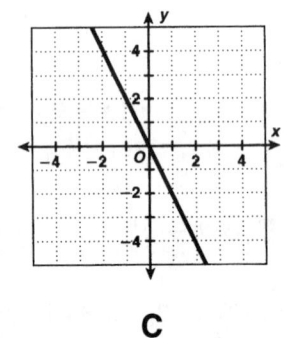   C 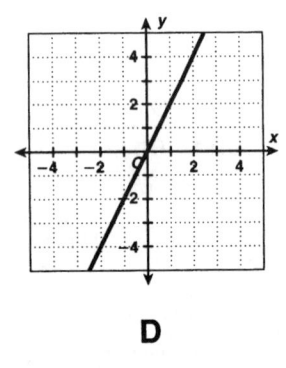   D

30. Which graph displays the speed of a car as it accelerates from a stopped position?

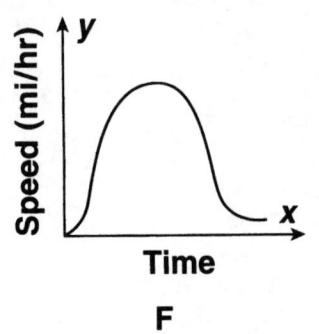

  F 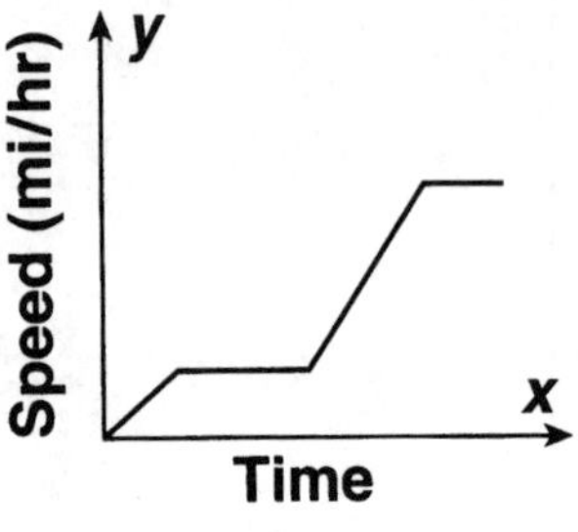   G 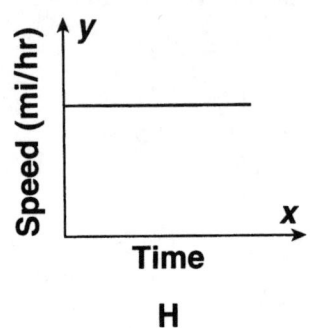   H 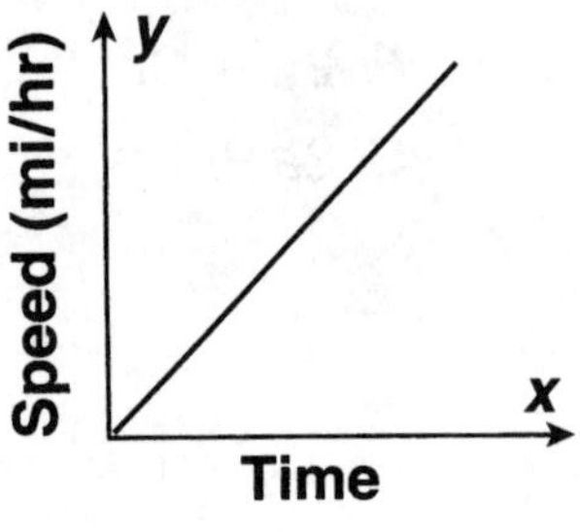   J

Name _____ Date _____ Class _____

Cumulative Test
CHAPTER 2 Form A

1. Evaluate $a + 3$ for $a = 8$.
   A 3
   B 11

2. Evaluate $5m + 9n + 1$ for $m = 0$ and $n = 2$.
   A 24
   B 19

3. Evaluate $-8v - 2$ for $v = -2$.
   A -18
   B 14

4. Subtract $-11 - (-6)$.
   A -17 C -5
   B -12 D 5

5. What is the best estimate for the weight of an adult?
   A 70 mg   C 70 kg
   B 70 g   D 95 mL

6. Simplify $-5a + (-3a)$.
   A $-2a$
   B $-8a$

7. Simplify $-9m - 3n - 6n$.
   A $-9m - 3n$
   B $-9m - 9n$

8. Solve $8 - r = 5$.
   A $r = -3$ C $r = 2$
   B $r = -2$ D $r = 3$

9. Solve $s + 7 = 22$.
   A $s = 29$
   B $s = 15$

10. A straight path that extends without end in opposite directions is a _____.
    A ray   C plane
    B point   D line

11. Simplify $(-11) \cdot (-11)$.
    A -121
    B 121

12. Simplify $(-37) \cdot (0)$.
    A 37   C 0
    B 1   D -37

13. Evan drove 300 miles at the average rate of 60 miles per hour. How long did his trip take?
    A 6 hr   C 5 hr
    B $\frac{1}{5}$ hr   D 4 hr

14. Express 1.7×10^5 using standard notation.
    A 170,000
    B 1,700,000

15. Express 14,000,000 using scientific notation.
    A 14×10^6
    B 1.4×10^7

Cumulative Test
Form A, continued

16. Find the median of the data in the stem and leaf plot.

A 64
B 65

```
5 | 0 0 4 7
6 | 1 3 4 5
7 | 1 4 6 8 9
```

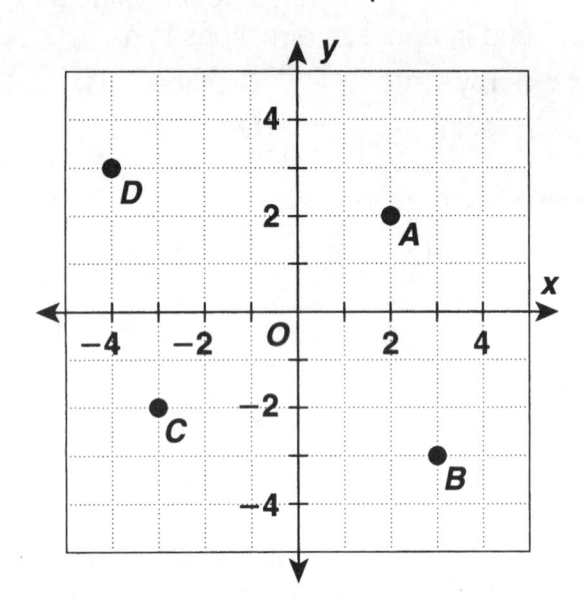

Day	1	2	3	4	5	6	7
High	73°	78°	80°	84°	77°	69°	70°

20. Using the table above, determine the average high temperature to the nearest degree.

A 66° B 76°

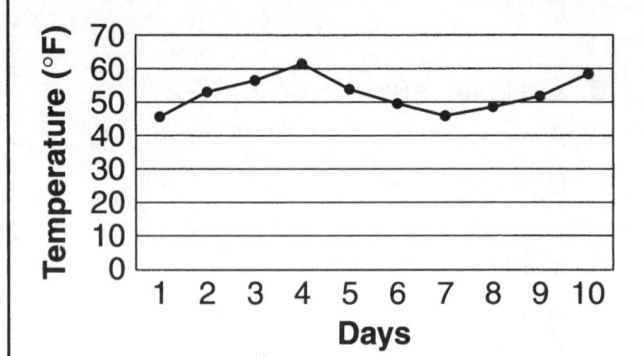

17. Give the coordinates of point A as shown on the graph above.

A (2, 2)   B (3, 3)

18. Give the coordinates of point B as shown on the graph above.

A (−3, −2)   C (3, −3)
B (−3, 3)   D (−4, 3)

19. Express 4 • 4 • 4 • 4 • 4 • 4 using exponents.

A 4^6 B 4^5

21. Use the graph above to determine the approximate temperature on day 7.

A 50 degrees   C 44 degrees
B 52 degrees   D 47 degrees

22. Simplify −3 + (−16).

A −19   C 13
B −13   D 19

23. Express $\dfrac{m^9}{m^5}$ as one power.

A m^4 B $\dfrac{1}{m^4}$

Name _____ Date _____ Class _____

CHAPTER 2 Cumulative Test
Form A, continued

24. Which ordered pair is a solution of $3x = y$?
   A (3, 1)
   B (1, 3)

25. Which ordered pair is a solution of $y = \frac{1}{2}x$?
   A (1, 2)   C (6, 3)
   B (4, 1)   D (3, 6)

26. Natasha has $77 in her bank account when she writes a check for $48. She makes a deposit of $31. How much is now in the account?
   A $156
   B $60

27. Classes of 30 math students each met in the cafeteria to take achievement tests. If exactly 5 students sat at each table and 24 tables were used, how many classes took the tests?
   A 6 classes   C 4 classes
   B 17 classes   D 7 classes

28. Express the phrase, "11 is greater than a number n," as an algebraic expression.
   A $11 > n$
   B $n > 11$

29. A boy who is 5 feet tall casts a shadow 3 feet long. He is standing near a tree that casts a shadow of 24 feet. How tall is the tree?
   A 36 feet   C 44 feet
   B 40 feet   D 50 feet

30. Simplify 3^{-4}.
   A 81
   B $\frac{1}{81}$

31. Simplify $\frac{x^2}{x^4}$.
   A x C $2x$
   B x^2 D $\frac{1}{x^2}$

32. An acute angle measures _____.
   A less than 90°   C more than 90°
   B exactly 90°   D exactly 180°

33. Solve $\frac{x}{3} = 4$.
   A $x = 6$ C $x = 12$
   B $x = 7$ D $x = 1$

34. Express the phrase, "three times a number g," as an algebraic expression.
   A $3 + g$ C $3g$
   B $3 - g$ D $\frac{3}{g}$

Cumulative Test
Form A, continued

35. Find the area of a parallelogram with base length 15 m and height 8 m.
- **A** 60 m²
- **B** 44 m²
- **C** 48 m²
- **D** 120 m²

36. Express the phrase, "the quotient of a number m and six," as an algebraic expression.
- **A** $6m$
- **B** $\dfrac{m}{6}$

37. Which expression could be used to find the approximate area of the base of a cylinder which has a height of 4 inches and a radius of 5 inches?
- **A** $5 \cdot \pi \cdot 3.14$
- **B** $5^2 \cdot 3.14$

38. Li has $1\dfrac{1}{2}$ pounds of tuna. She wants to serve each person a 6-oz serving. How many people can she serve?
- **A** 2 people
- **B** 4 people

39. Select the graph that is a solution to the inequality, $9x \leq 54$.

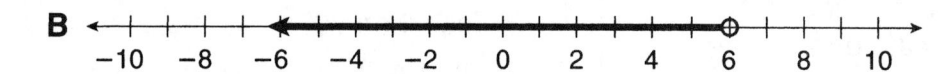

40. Select the graph that is a solution to the inequality, $x + 2 \leq -3$.

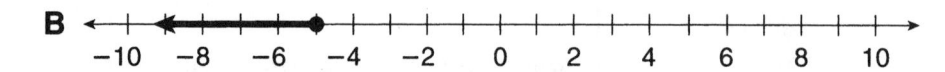

Name _____ Date _____ Class _____

CHAPTER 2 Cumulative Test
Form B

1. Evaluate $\frac{x+3}{3}$ for $x = 6$.
   - A 9
   - B 3
   - C 1
   - D 6

2. Evaluate $2m + 3n + 1$ for $m = -1$ and $n = 2$.
   - F 9
   - G 10
   - H 5
   - J 4

3. Evaluate $63 - (-4z)$ for $z = 12$.
   - A 111
   - B 59
   - C 15
   - D −15

4. Simplify $12x - (-17y) - 32y + (-43x)$.
   - F $55x + 49y$
   - G $25x - 75y$
   - H $-31x - 60y$
   - J $-31x - 15y$

5. What is the best estimate for the length of a baseball bat?
   - A 1 m
   - B 10 m
   - C 3 cm
   - D 30 km

6. Simplify $-6y - 4x - 2x$.
   - F $-12y$
   - G $-12x$
   - H $-6y - 6x$
   - J $-6y - 6x^2$

7. Simplify $-2 - (3 - 5t)$.
   - A $5 + 5t$
   - B $5t - 5$
   - C $1 + 5t$
   - D $-1 + 5t$

8. Solve $u + 47 = 238$.
   - F 164
   - G 191
   - H 236
   - J 285

9. Solve $v + 3 = 4$.
   - A −7
   - B −1
   - C 7
   - D 1

10. When two angles have the same measure, they are said to be _____.
    - F complementary
    - G supplementary
    - H right
    - J congruent

11. Simplify $(-4) \cdot (-4) \cdot (-4) \cdot (1)$.
    - A −64
    - B −65
    - C 64
    - D 65

12. Simplify $-\frac{143}{13}$.
    - F 13
    - G 11
    - H −11
    - J −13

13. A house worth $124,000 was assessed taxes based on $\frac{3}{4}$ of its value. What is the assessed value of the house?
    - A $93,000
    - B $124,000
    - C $165,333
    - D $930,000

14. Express 0.000086 using scientific notation.
    - F 8.6×10^{-5}
    - G 86×10^{-4}
    - H 8.6×10^{-4}
    - J 8.6×10^{4}

15. Express 3.5×10^{-6} using standard notation.
    - A 0.000035
    - B 0.0035
    - C 0.0000035
    - D 0.00035

Name _____ Date _____ Class _____

CHAPTER 2 Cumulative Test
Form B, continued

16. Find the mode of the data in the stem and leaf plot.
    F 50   H 64.8
    G 64   J 65

    ```
    5 | 0 0 4 7
    6 | 1 3 4 5
    7 | 1 4 6 8 9
    ```

Day	1	2	3	4	5	6	7
High	73°	85°	81°	85°	77°	69°	70°

20. Using the table above, determine the temperature difference between the warmest day and coolest day.
    F 3 degrees   H 16 degrees
    G 8 degrees   J 7.5 degrees

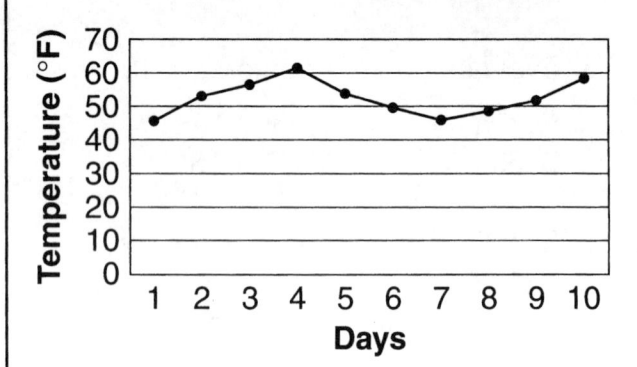

21. Use the graph above to determine the approximate change in temperature from day 1 to day 10.
    A 8 degrees   C 15 degrees
    B 12 degrees   D 10 degrees

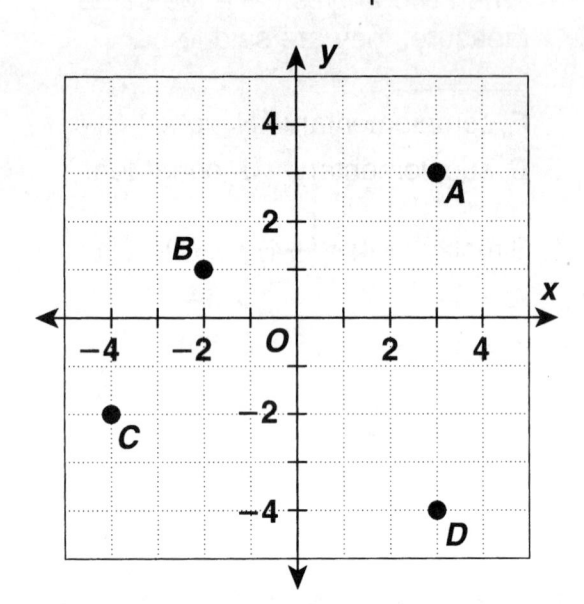

17. Give the coordinates of point D as shown on the graph above.
    A (3, 4)   C (−2, 1)
    B (3, −4)   D (−4, −1)

18. Give the coordinates of point C as shown on the graph above.
    F (4, 2)   H (2, 4)
    G (−2, 4)   J (−4, −2)

19. Simplify $(12 - 4^3)$.
    A 76   C −52
    B 64   D −64

22. Simplify $15 - (-4) + 0$.
    F 11   H 19
    G −11   J −19

23. Express the product $t^3 \cdot t^3 \cdot t^4$ as one power.
    A t^{36} C $3t^{10}$
    B t^{10} D t^{18}

Name _____ Date _____ Class _____

CHAPTER 2 Cumulative Test
Form B, continued

24. Which ordered pair is a solution of $3x = 4y$?
F (8, 6)   H (3, 4)
G (6, 8)   J (1, 2)

25. Which ordered pair is a solution of $\frac{3x}{2} = y$?
A $(-1, -2)$ C $(-2, 3)$
B $(-1, -\frac{3}{2})$ D $(-\frac{3}{2}, -1)$

26. Damon purchased CDs on sale for $2.45 each. If the total bill before sales tax was $34.30, how many CDs did he buy?
F 13   H 31
G 14   J 140

27. Winnie earned $2700 last summer. She spent $1200 to pay off her car debt, $540 for clothes, $360 for gifts, and put $600 in the bank. What fractional part of her income did she not spend?
A $\frac{2}{9}$ C $\frac{4}{9}$
B $\frac{2}{15}$ D $\frac{3}{15}$

28. Express the phrase, "8 more than three times a number is less than or equal to 34," as an algebraic expression.
F $3x + 8 > 34$ H $3x + 8 \leq 34$
G $3x - 8 \leq 34$ J $3x + 8 < 34$

29. How tall is a giraffe that casts a shadow 320 cm long, if a man standing nearby who is 180 cm tall casts a shadow 100 cm long?
A 177.78 cm   C 576 cm
B 232 cm   D 626 cm

30. Simplify 5^{-4}.
F -20 H $\frac{1}{20}$
G 625   J $\frac{1}{625}$

31. Simplify $\frac{b^{45}}{b^{62}}$.
A $\frac{1}{b^{17}}$ C $\frac{1}{b^{107}}$
B b^{17} D $\frac{1}{b^{-17}}$

32. A right angle measures _____.
F less than 90°   H more than 90°
G exactly 90°   J exactly 180°

33. Solve $4r + 3 = 19$.
A $r = 12$ C $r = 4$
B $r = 1$ D $r = 16$

34. Express the phrase, "12 is less than twice a number x," as an algebraic expression.
F $12 < 2x$ H $2x < 12$
G $12 - 2x$ J $2x - 12$

Cumulative Test
Form B, continued

35. Find the area of a parallelogram with base length 18 m and height 12 m.
- A 30 m²
- B 60 m²
- C 216 m²
- D 108 m²

36. Express the phrase, "2 times the quotient of a number y and nine," as an algebraic expression.
- F $\dfrac{9}{2y}$
- G $2y - 9$
- H $18y$
- J $2\left(\dfrac{y}{9}\right)$

37. The diameter of a bicycle wheel is 20 inches. How far does it travel in one complete revolution? Use 3.14 for π.
- A 31.4 in.
- B 62.8 in.
- C 314 in.
- D 1256 in.

38. How many one-cup servings of milk can be poured from 2 gallons of milk?
- F 16
- G 32
- H 64
- J 128

39. Select the graph that is the solution to the inequality $x \geq 4$.

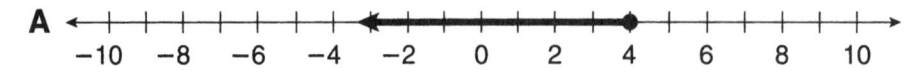

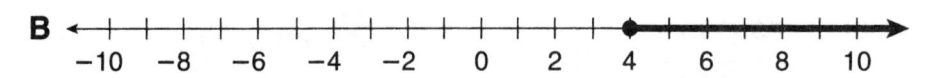

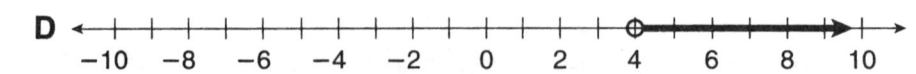

40. Select the graph that is the solution to the inequality $x - 4 \geq -10$.

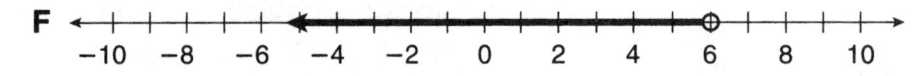

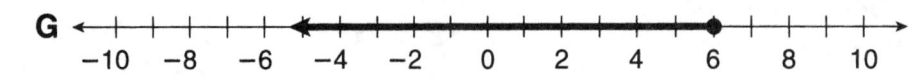

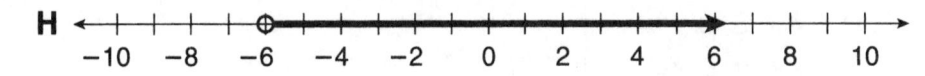

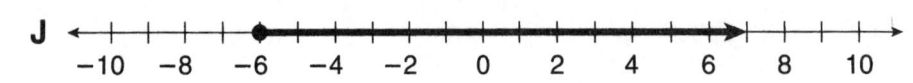

Name _____ Date _____ Class _____

CHAPTER 2 Cumulative Test
Form C

1. Evaluate $\frac{2x + 3}{10}$ for $x = 1$.
   A 5   C 4
   B $\frac{2}{5}$ D $\frac{1}{2}$

2. Evaluate $-4m + 2n + 1$ for $m = 3$ and $n = -2$.
   F -15 H -7
   G 16   J 9

3. Evaluate $22 - 6s$ for $s = -7$.
   A 924   C -112
   B 64   D -20

4. Simplify $157 - (-235)$.
   F 392   H -78
   G 78   J -392

5. What is the best estimate for the capacity of a large juice bottle?
   A 1 mL   C 1 kL
   B 1 L   D 1 kg

6. Simplify $5(v + 4) + 8v$.
   F $20v$ H $13v + 20$
   G $5v + 5$ J $13v + 4$

7. Simplify $-9m - 3n - 5n$.
   A $-9m - 2n$ C $-9m + 2n$
   B $-9m - 8n$ D $-17mn$

8. Together, Manuel and Fatima collected $342 in donations for new band uniforms. Their collections were $6539 less than the total collected amount. What was the total?
   F $6881   H $3119
   G $6197   J $342

9. Joshua has 248 more baseball cards than Sarah, who has 63 more than Calvin. If Joshua has 752 cards, how many cards does Calvin have?
   A 689   C 441
   B 504   D 248

10. Movement of a figure along a straight line is called _____.
    F translation   H reflection
    G rotation   J protraction

11. Divide $\frac{18(-5)}{3(15)}$.
    A -90 C -2
    B -45 D 45

12. Simplify $(-3) \cdot (-3) \cdot (3)$.
    F -27 H -9
    G 27   J 9

13. Elijah wishes to buy a car for $16,550 before tax. If the sales tax is 7%, what will be the total cost of the car?
    A $1158.50   C $17,708.50
    B $15,391.50   D $28,135.00

14. Express 4.6×10^6 in standard notation.
    F 4,600,000   H 460,000
    G 46,000,000   J 46,000

15. Express 0.00163 in scientific notation.
    A 1.63×10^{-3} C 1.63×10^{-4}
    B 163×10^{-4} D 1.63×10^4

Name _____ Date _____ Class _____

CHAPTER 2 Cumulative Test
Form C, continued

16. Find the mean of the data in the stem and leaf plot to the nearest tenth.

15	0 0 4 7
16	1 3 4 5
17	1 4 6 8 9

F 150.0 H 164.8
G 164.0 J 165.0

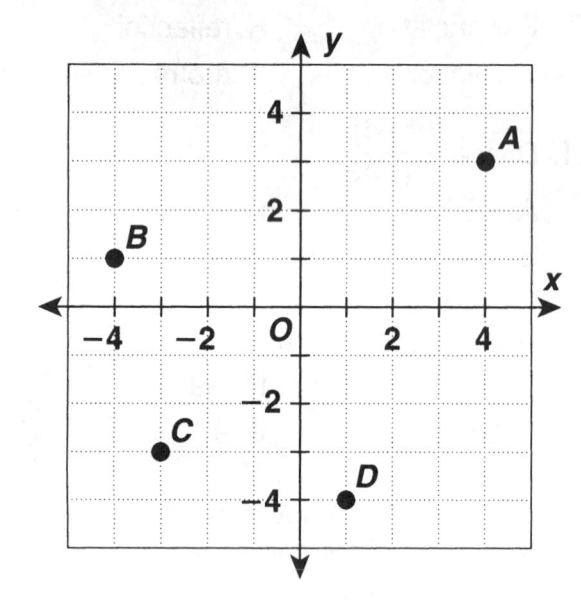

17. Give the coordinates of point B as shown on the graph above.
A (4, 1)   C (1, −4)
B (−4, 1) D (−4, −1)

18. Give the coordinates of point C as shown on the graph above.
F (3, −4) H (3, 4)
G (3, −3) J (−3, −3)

19. Simplify 4^4.
A 256 C 3
B 44   D 16

20. What is the difference between the average temperatures for the first three days and the last three days?

Day	1	2	3	4	5	6	7
High	73°	78°	81°	85°	77°	69°	70°

F 77.3° H 5.3°
G 72°   J 0.3°

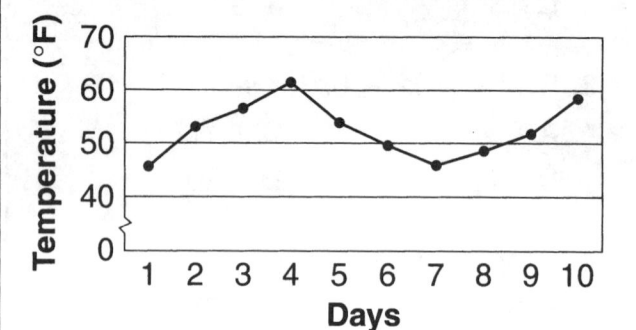

21. Which statement is NOT true.
A On day 4, the temperature reached above 60°.
B The change in temperature from day 9 to day 10 was greater than from day 3 to day 4.
C On most days, the temperature was below 50°.
D The highest temperature reached on any day was approximately 61°.

22. Simplify 76 + (−15) + (−32).
F 123 H 44
G 59   J 29

23. Express the product $\dfrac{m^{12}}{m^0} \cdot m^2$ as one power.
A m^{24} C m^{14}
B 1   D $\dfrac{1}{m^{12}}$

Name _____ Date _____ Class _____

CHAPTER 2 Cumulative Test
Form C, continued

24. Which ordered pair is a solution of $2x = -y$?
    - **F** (0, −6)
    - **G** (−3, −6)
    - **H** (3, −6)
    - **J** (3, 6)

25. Which ordered pair is a solution of $5x + 1 = y$?
    - **A** (−1, −4)
    - **B** (−5, −4)
    - **C** (−5, −5)
    - **D** (1, −4)

26. Your bank account has $79 in it right now. You write checks for $41, $22, and $11. Then you make deposits of $58 and $24. How much is in the account after these transactions?
    - **F** $71
    - **G** $98
    - **H** $87
    - **J** $8

27. For Charmaine to get a B in math she needs to average 80 on four tests. Her scores on the first three tests were 78, 81, and 75. What is the lowest score she can receive on the next test and still get a B?
    - **A** 78
    - **B** 96
    - **C** 80
    - **D** 86

28. Express the phrase, "the quotient of 4 times a number, and the number times itself is greater than or equal to 12," as an algebraic expression.
    - **F** $\frac{x^2}{4x} \geq 12$
    - **G** $\frac{4x}{x^2} \leq 12$
    - **H** $\frac{4x}{x^2} \geq 12$
    - **J** $\frac{4x}{x^2} > 12$

29. On a particular scale drawing a measurement of 25 feet is represented by 2 inches. If two apartment buildings are actually 62.5 feet apart, what is the distance between them on the drawing?
    - **A** 2.5 in.
    - **B** 5 in.
    - **C** 11.4 in.
    - **D** 312.5 in.

30. What is the sum of 13 and −7, raised to the negative 4 power?
    - **F** −1296
    - **G** $\frac{1}{24}$
    - **H** $\frac{1}{1296}$
    - **J** 1296

31. Simplify $(16 - 11)^{-5}$.
    - **A** 3125
    - **B** $\frac{1}{3125}$
    - **C** $\frac{1}{25}$
    - **D** −3125

32. If two angles of a triangle measure 40° and 60°, how many degrees does the third angle measure?
    - **F** 70°
    - **G** 80°
    - **H** 90°
    - **J** 100°

33. Solve $\frac{x}{15} = 21$.
    - **A** $x = 6$
    - **B** $x = 36$
    - **C** $x = 315$
    - **D** $x = 1201$

34. Express the phrase, "the product of 5 and 9 more than a number y," as an algebraic expression.
    - **F** $5(y + 9)$
    - **G** $5 + 9 \cdot y$
    - **H** $(5 + 9)y$
    - **J** $5 \cdot 9 + y$

Cumulative Test
Chapter 2 Form C, continued

35. What is the area of a trapezoid with bases 15 m and 12 m and a height of 9 m.
 A 36 m²
 B 121.5 m²
 C 54 m²
 D 1620 m²

36. Express the phrase, "nine is equal to twelve minus three times a number n cubed," as an algebraic expression.
 F $12 = 9 - 3n^2$
 G $12 = 9 - 3n^3$
 H $9 = 12 - 3n^3$
 J $9 = 12 - 3n^2$

37. One bicycle wheel has a diameter of 25 inches, another wheel has a diameter of 27 inches. How much farther does the 27-inch wheel go than the 25-inch wheel in one rotation? Use 3.14 for π.
 A 2 inches
 B 6.28 inches
 C 78.5 inches
 D 84.78 inches

38. Tasha has ordered 2 pairs of jeans that weigh 12 oz each and 1 pair of running shoes that weigh 8 oz. If the cost to ship the items is $4.99 per pound, how much will it cost to ship the items to Tasha?
 F $4.99
 G $5.21
 H $9.98
 J $12.04

39. Select the graph that is a solution to $2x - 1 < 3$.

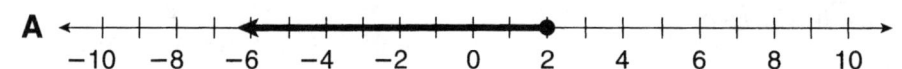

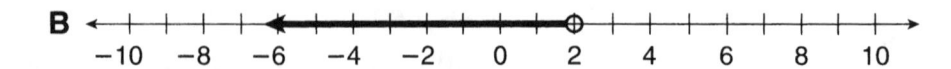

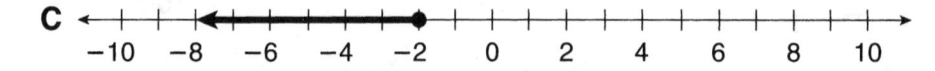

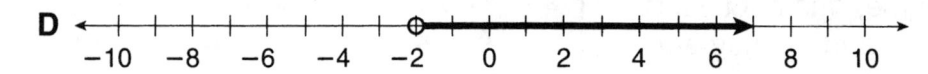

40. Select the graph that is a solution to $x \leq -3$.

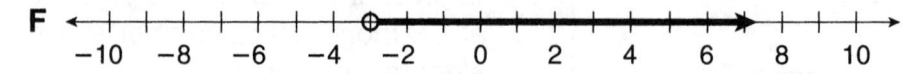

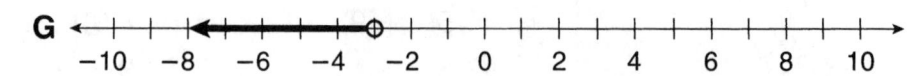

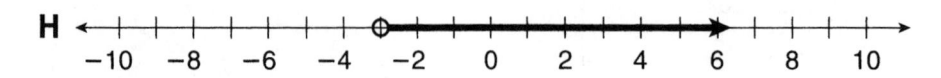

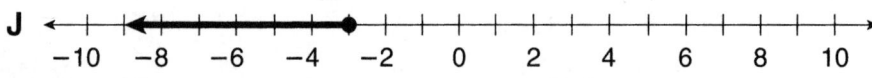

Name _____ Date _____ Class _____

CHAPTER 3 Cumulative Test
Form A

1. Simplify $9 + (-5)$.
   A -4 B 4

2. Simplify $-7 + 21$.
   A 14 B -28

3. Simplify $3(10x + 7) + 2x$.
   A $30x + 21$ B $32x + 21$

4. Solve $11b - 3b = 32$.
   A $b = 4$ C $b = 8$
   B $b = 14$ D $b = 32$

5. How many days are there in 3 weeks?
   A 21 B 15

6. Simplify $\left(-\frac{1}{2}\right)\left(\frac{4}{19}\right)$.
   A $\frac{1}{19}$ B $-\frac{2}{19}$

7. Simplify $-\frac{18}{9}$.
   A 2 B -2

8. Which does *not* apply to the quadrilateral below?

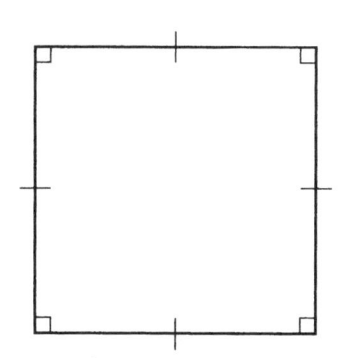

   A rhombus   C rectangle
   B parallelogram   D trapezoid

9. Evaluate $r + 3s$ for $r = 5$ and $s = 0$.
   A 5 B 15

10. Evaluate $\frac{c-2}{2}$ for $c = -3$.
    A $-\frac{1}{2}$ B $-\frac{5}{2}$

11. Simplify $\frac{4}{11} - \left(-\frac{6}{11}\right)$.
    A $\frac{10}{11}$ B $-\frac{2}{11}$

12. Multiply $7\left(-\frac{6}{7}\right)$.
    A 6 C $\frac{13}{7}$
    B $-\frac{13}{7}$ D -6

13. Express 8.87×10^3 in standard notation.
    A 8870 B 887

14. Find the perimeter of the rectangle below.

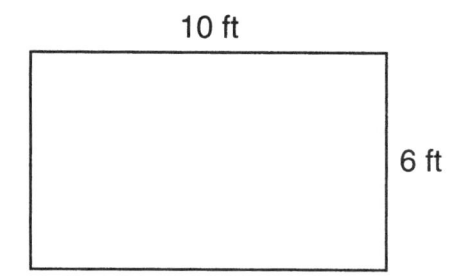

    10 ft
    6 ft

    A 60 ft   B 32 ft

Name _____ Date _____ Class _____

CHAPTER 3 Cumulative Test
Form A, continued

15. Are the lines shown below parallel or perpendicular?

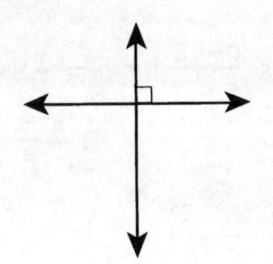

A parallel   **B** perpendicular

16. Solve $4n = -8$.
- **A** $n = -2$
- **B** $n = 2$
- **C** $n = 0.5$
- **D** $n = -0.5$

17. Solve $-7 + t = 2$.
- **A** $t = -9$
- **B** $t = 9$

18. Solve $3.2a - 4 = 4.2a - 13$.
- **A** $a = -6$
- **B** $a = 9$

19. Express $x \cdot x \cdot x \cdot x \cdot x \cdot x \cdot x$ using exponents.
- **A** x^7
- **B** $7x$

20. Solve $\frac{1}{2}m = 5$.
- **A** $m = 10$
- **B** $m = 2.5$
- **C** $m = 0.25$
- **D** $m = 1$

21. Simplify 10^3.
- **A** 30
- **B** 1000

22. Choose the correct classification for the triangle below.

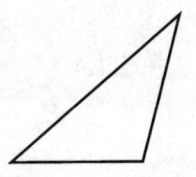

A acute   **B** obtuse

23. Simplify 7^{-3}.
- **A** $\frac{1}{343}$
- **B** $\frac{1}{49}$

24. Simplify $\frac{x^4}{x^3}$.
- **A** $\frac{1}{x}$
- **B** x^7
- **C** x
- **D** $7x$

25. Express 0.3 as a fraction.
- **A** $\frac{3}{10}$
- **B** $\frac{3}{100}$

26. Simplify $\frac{8}{12}$.
- **A** $\frac{3}{4}$
- **B** $\frac{2}{3}$

27. What is the phrase, "the product of 7 and b, minus the product of 3 and b," as an algebraic expression?
- **A** $3b - 7b$
- **B** $7b - 3b$
- **C** $3 \cdot b \cdot 7 \cdot b$
- **D** $(3 + 7)b$

28. Find the area of the triangle below.

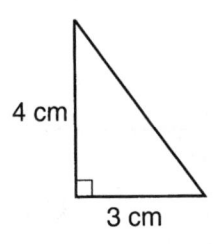

4 cm, 3 cm

- **A** 12 cm^2
- **B** 10 cm^2
- **C** 7 cm
- **D** 6 cm^2

Name _____ Date _____ Class _____

CHAPTER 3 Cumulative Test
Form A, continued

29. Divide $\frac{5}{12} \div 4$.

   A $\frac{20}{12}$ **B** $\frac{5}{48}$

30. Find the volume of the prism below.

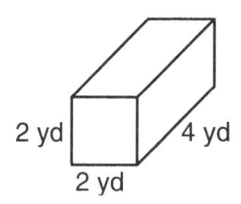

   (2 yd × 2 yd × 4 yd)

   A 8 yd³   **B** 16 yd³

31. What is the quotient $\frac{b^3}{b^2}$ as one power?

   A b^5 **B** b

32. An advertising company pays $275 a week plus a bonus of $23 for each new service contract. What is the total pay if 5 service contracts were sold in one week?

   A $390   **C** $115
   B $275   **D** $490

33. A sawmill trims a 2-inch board to be $1\frac{9}{16}$ inches wide. How much is trimmed off?

   A $\frac{11}{16}$ in.   **B** $\frac{7}{16}$ in.

34. A rectangular scarf measures $2\frac{1}{2}$ feet by $3\frac{1}{3}$ feet. What is the distance around the scarf?

   A $5\frac{2}{5}$ ft   **B** $11\frac{2}{3}$ ft

35. Venecia has $750 in her checking account. This month the bank charged $15 for checks, and she wrote checks for $95 and $55. What is the balance in the account?

   A $585   **B** $615

36. To the nearest tenth, find the volume of a sphere with a radius of 1 in. Use 3.14 for π.

   A 2.4 in³   **B** 4.2 in³

37. Simplify $\sqrt{\frac{100}{25}}$.

   A 2   **B** $\frac{5}{2}$

38. Find the surface area of the sphere below to the nearest tenth. Use 3.14 for π.

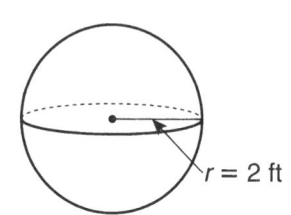

   ($r = 2$ ft)

   A 50.2 ft²   **B** 16.7 ft²

39. What kind of number is $\sqrt{-25}$?

   A rational   **C** not real
   B irrational   **D** negative

40. Translate $\frac{v}{2} + 1$ into words.

   A the quotient of v and 2, plus 1
   B the sum of two times v and 1
   C the sum of v and 1 divided by 2
   D the product of v and 2 and one

Name _____ Date _____ Class _____

CHAPTER 3 Cumulative Test
Form A, continued

41. Find the length of a side of a square with an area of 121 ft².
    A 44 ft. B 11 ft

42. To the nearest tenth, find the approximate distance around a square with an area of 110 cm².
    A 10.5 cm B 42.0 cm

43. Select the graph that is the solution of $x > 3$.

    A
    [number line from −10 to 10 with open circle at 3, shaded right]

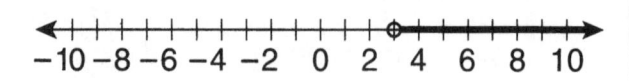

    B
    [number line from −10 to 10 with closed circle at −3, shaded right]

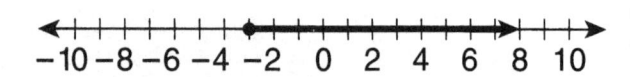

44. Select the graph that is the solution of $\frac{1}{2}x > 1$.

    A
    [number line from −10 to 10 with open circle at 2, shaded right]

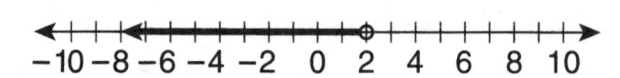

    B
    [number line from −10 to 10 with open circle at 2, shaded right]

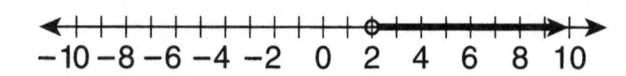

45. Solve $2.1x \leq -2.1$.
    A $x \geq -1$ C $x \leq 1$
    B $x \leq -1$ D $x \geq 1$

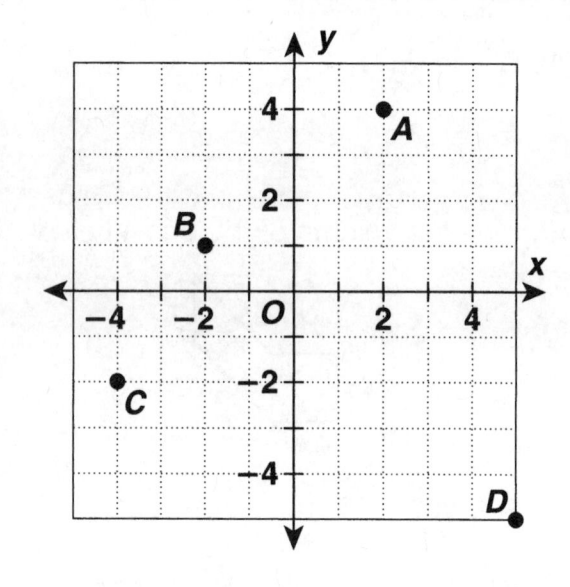

[Coordinate graph showing points A, B, C, D]

46. Use the graph above to give the coordinates of point B.
    A (−2, 1) B (−2, −1)

47. Use the graph above to give the coordinates of point A.
    A (2, 4) B (2, −4)

48. Which ordered pair is a solution of $2x + y = 5$?
    A (4, 1)   C (8, 1)
    B (16, 1)   D (1, 3)

49. Which ordered pair is a solution of $x + 3y = 18$?
    A (3, 5) B (2, 3)

Name _____ Date _____ Class _____

Cumulative Test
Chapter 3 Form B

1. Simplify −23 + (−14).
   - A −9
   - B 37
   - C −37
   - D 9

2. Simplify 10 + (−7).
   - F 17
   - G 3
   - H −17
   - J −3

3. Simplify 18v(4 − 2) − 25v + 72.
   - A 72 − 4v
   - B 18 + 2v
   - C 72 + 11v
   - D 18 + 11v

4. Simplify 32x − (7x − 9)5.
   - F 68x − 45
   - G 25x − 45
   - H −45 + 3x
   - J 45 − 3x

5. How many minutes are there in 7.6 hours?
   - A 402
   - B 420
   - C 456
   - D 465

6. Simplify $\left(-\frac{4}{15}\right)\left(\frac{3}{7}\right)$.
   - F $-\frac{4}{35}$
   - G $\frac{2}{29}$
   - H $-\frac{24}{29}$
   - J $-\frac{2}{29}$

7. Simplify $\frac{42}{70}$.
   - A $\frac{6}{12}$
   - B $\frac{6}{7}$
   - C $\frac{3}{5}$
   - D $\frac{14}{15}$

8. Give all the names that apply to the quadrilateral below.

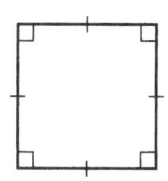

   - F square, rhombus, kite, parallelogram
   - G square, rhombus, rectangle, parallelogram
   - H square, rhombus, trapezoid
   - J square, rhombus, parallelogram

9. Evaluate 5r + 9s + 6 for r = 0 and s = 8.
   - A 78
   - B 46
   - C 83
   - D 15

10. Evaluate $\frac{c + 4}{9}$ for c = 4.
    - F $\frac{22}{9}$
    - G $\frac{16}{9}$
    - H $\frac{9}{8}$
    - J $\frac{8}{9}$

11. Divide $\frac{3}{8} \div \frac{2}{5}$.
    - A $\frac{15}{16}$
    - B $\frac{3}{20}$
    - C $\frac{31}{40}$
    - D $\frac{1}{40}$

12. Multiply $\frac{5}{8}\left(-\frac{7}{12}\right)$.
    - F $-\frac{35}{96}$
    - G $-\frac{60}{56}$
    - H $\frac{12}{20}$
    - J $-\frac{2}{4}$

13. What is 5.6 × 10^6 in standard notation?
    - A 56,000,000
    - B 56,000
    - C 56,000,000
    - D 5,600,000

173

Holt Pre-Algebra

Name _____ Date _____ Class _____

CHAPTER 3 Cumulative Test
Form B, continued

14. Find the circumference of the circle to the nearest tenth. Use 3.14 for π. 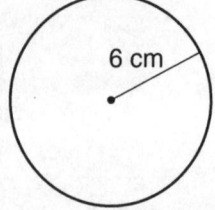

    F 37.7 cm
    G 37.6 cm
    H 18.8 cm
    J 113.0 cm

    (6 cm radius)

15. State whether the lines shown are parallel, perpendicular, skew, or none of these. 

    A parallel
    B perpendicular
    C skew
    D none of these

16. Solve $\frac{1}{2}m = -4$.

    F $m = -2$
    G $m = -8$
    H $m = 1$
    J $m = 2$

17. Solve $-b - 2 = 18$.

    A $b = 16$
    B $b = 20$
    C $b = -16$
    D $b = -20$

18. Solve $3.7a + 4 = 2.7a - 8$.

    F $a = -11$
    G $a = -2$
    H $a = -13$
    J $a = -12$

19. Express $(-g) \cdot (-g) \cdot (-g) \cdot (-g) \cdot (-g)$ using exponents.

    A $-g^4$
    B $\frac{1}{g^5}$
    C $(-g)^5$
    D $-\frac{1}{g^5}$

20. Solve $-55.2 = -6.9c$.

    F $c = 8.0$
    G $c = 2.0$
    H $c = 48.3$
    J $c = -48.3$

21. Simplify $(-5)^4$.

    A -625
    B -125
    C 625
    D 3125

22. Classify the triangle according to its sides and angles. 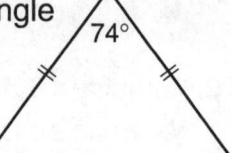

    F scalene acute triangle
    G right triangle
    H isosceles acute triangle
    J isosceles obtuse triangle

23. Simplify $\frac{4^7}{4^9}$.

    A $\frac{1}{16}$
    B $\frac{1}{8}$
    C 16
    D 8

24. Simplify $\frac{x^{17}}{x^5}$.

    F $\frac{1}{x^{12}}$
    G x^{12}
    H x^{22}
    J x^{85}

25. Express 1.25 as a fraction.

    A $\frac{5}{4}$
    B $1\frac{2}{5}$
    C $\frac{1}{4}$
    D $1\frac{3}{4}$

26. Simplify $\frac{17}{51}$.

    F 3
    G $\frac{1}{4}$
    H $\frac{1}{3}$
    J 4

27. What is the phrase, "the difference of 8 times a number r and 12," as an algebraic expression?

    A $12 - 8r$
    B $8 - r + 12$
    C $8r - 12$
    D $12 + 8r$

Name _____ Date _____ Class _____

Cumulative Test
Chapter 3 Form B, continued

28. Find the area of the triangle.
    - F 56 ft²
    - G 28 ft²
    - H 15 ft
    - J 112 ft²

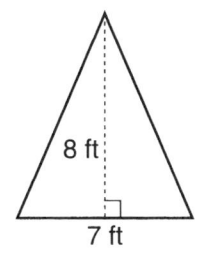

    (8 ft, 7 ft)

29. Divide $2 \div \frac{2}{15}$.
    - A $\frac{4}{15}$
    - B 15
    - C $\frac{1}{15}$
    - D $\frac{2}{15}$

30. To the nearest tenth, find the volume of the cylinder. Use 3.14 for π.
    - F 63.6 in³
    - G 763.0 in³
    - H 254.3 in³
    - J 84.8 in³

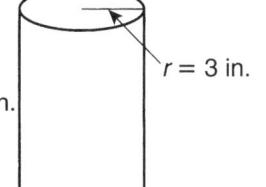

    $r = 3$ in., $h = 9$ in.

31. Express the quotient $\frac{m^5 m^3}{m^2}$ as one power.
    - A m^{17}
    - B m^{13}
    - C m^6
    - D m^4

32. Keisha earns $325 a week plus a bonus of $23 for each service contract she sells. What is her pay if she sells 3 service contracts in one week?
    - F $325
    - G $394
    - H $494
    - J $69

33. Martin has a 4-inch board that he wants to be $3\frac{11}{16}$ inches wide. How much does he need to trim off?
    - A $7\frac{11}{16}$ in.
    - B $\frac{5}{16}$ in.
    - C $\frac{11}{16}$ in.
    - D $4\frac{5}{16}$ in.

34. A rectangular garden measures $2\frac{2}{3}$ feet by $50\frac{1}{2}$ feet. What is the distance around the garden?
    - F $106\frac{1}{3}$ ft
    - G $53\frac{1}{6}$ ft
    - H $134\frac{2}{3}$ ft
    - J 109 ft

35. Raul has $506 in his checking account. This month the bank charged $6 for checks, and he wrote checks for $91 and $72. What is the balance in the account?
    - A $531
    - B $675
    - C $337
    - D $349

36. To the nearest tenth, find the volume of a sphere with a radius of 2 m. Use 3.14 for π.
    - F 25.1 m³
    - G 10.7 m³
    - H 33.5 m³
    - J 16.7 m³

37. Simplify $\sqrt{144}$.
    - A 72
    - B 13
    - C 12
    - D irrational

38. To the nearest tenth, find the surface area of the cylinder formed by the net. Use 3.14 for π.
    - F 75.4 cm²
    - G 150.7 cm²
    - H 87.9 cm²
    - J 138.2 cm²

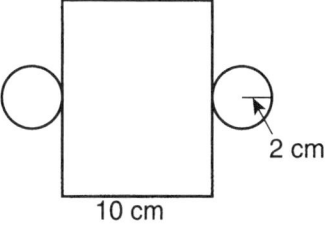

    (10 cm, 2 cm)

39. What kind of number is $-\sqrt{17}$?
    - A rational
    - B irrational
    - C not real
    - D positive

Name _____ Date _____ Class _____

CHAPTER 3 Cumulative Test
Form B, continued

40. Translate $\frac{v}{2} + 3v$ into words.

   F v divided by two, plus three times v
   G half a number plus 3
   H 2 times the quotient of v and 3
   J the product of half of v and 3 times v

41. To the nearest tenth, find the approximate length of a side of a square with an area of 652 cm².

   A 32.6 cm   C 26.2 cm
   B 25.5 cm   D 17.3 cm

42. To the nearest tenth, find the distance around a square with an area of 165 cm².

   F 51.4 cm   H 52.1 cm
   G 12.8 cm   J 50 cm

43. Which is the solution of $2x \leq 6$?

   A –10–8–6–4–2 0 2 4 6 8 10
   B –10–8–6–4–2 0 2 4 6 8 10
   C –10–8–6–4–2 0 2 4 6 8 10
   D –10–8–6–4–2 0 2 4 6 8 10

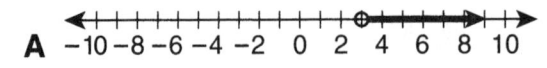

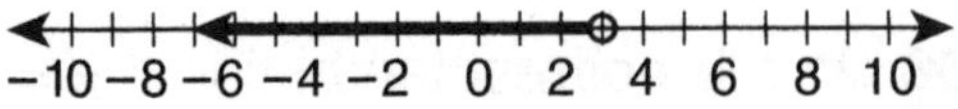

44. Which is the solution of $2x > -5$?

   F –10–8–6–4–2 0 2 4 6 8 10
   G –10–8–6–4–2 0 2 4 6 8 10
   H –10–8–6–4–2 0 2 4 6 8 10
   J –10–8–6–4–2 0 2 4 6 8 10

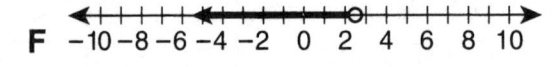

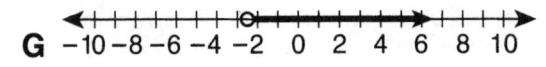

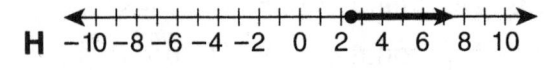

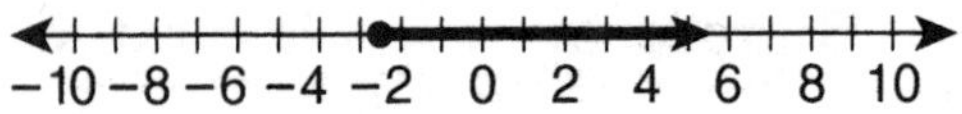

45. Solve $-\frac{1}{3}x + 1 \geq 3$.

   A $x \geq 12$ C $x \leq 6$
   B $x \leq 12$ D $x \leq -6$

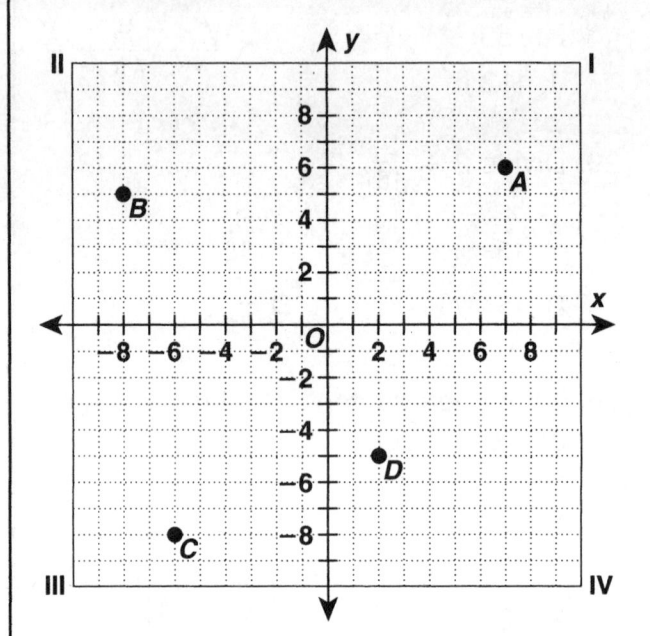

46. Use the graph above to give the coordinates of point B.

   F (2, –5)   H (–8, 5)
   G (5, 8)   J (–5, 8)

47. Use the graph above to give the coordinates of point C.

   A (8, 6)   C (6, –8)
   B (–8, 6)   D (–6, –8)

48. Which ordered pair is a solution of $5x - y = 6$?

   F (3, 9)   H (3, 15)
   G (5, 15)   J (45, 5)

49. Which ordered pair is a solution of $3y = 6x + 3$?

   A (7, 3)   C (3, 7)
   B (–6, –3)   D (–7, –12)

Name _____ Date _____ Class _____

CHAPTER 3 Cumulative Test
Form C

1. Simplify −9 + (4 − 9).
   A 4
   B 14
   C −4
   D −14

2. Simplify 26 + (−127).
   F 153
   G 127
   H −101
   J −153

3. Simplify 5(11 + 8y) − 2y.
   A 55 − 42y
   B 55 + 38y
   C 55 − 38y
   D 55 + 42y

4. Simplify 7(21 − 9m) − 13m.
   F 147 − 22m
   G 147 − 76m
   H 28 − 22m
   J 28 − 76m

5. How many minutes are there in 1 year?
   A 43,200
   B 525,600
   C 8760
   D 31,536,000

6. Simplify $\left(-\frac{7}{25}\right)\left(-\frac{1}{5}\right)$.
   F $\frac{7}{5}$
   G $\frac{7}{125}$
   H $\frac{7}{30}$
   J $\frac{7}{20}$

7. Simplify $-\frac{75}{15}$.
   A 15
   B $\frac{-75}{8+7}$
   C −5
   D 5

8. Give all the names that apply to the quadrilateral.
   F trapezoid
   G kite, trapezoid
   H trapezoid, parallelogram
   J trapezoid, rhombus

9. Evaluate 6(t − 6) for t = −3.
   A −18
   B −54
   C −24
   D −9

10. Evaluate −5st for s = −2 and t = −4.
    F −20
    G 40
    H −80
    J −40

11. Simplify $-\frac{5}{13} - \left(-\frac{9}{13}\right)$.
    A $-\frac{14}{13}$
    B $\frac{4}{13}$
    C $\frac{14}{13}$
    D $-\frac{4}{13}$

12. Multiply $2\frac{8}{9}\left(\frac{3}{7}\right)$.
    F $\frac{21}{26}$
    G $\frac{26}{21}$
    H $\frac{16}{21}$
    J $\frac{21}{16}$

13. What is 2.321×10^{-6} in standard notation?
    A 0.0000002321
    B 2,321,000
    C 0.000002321
    D 0.002321

14. Find the perimeter of the polygon.
    F 29 cm
    G 28 cm
    H 30 cm
    J 31 cm

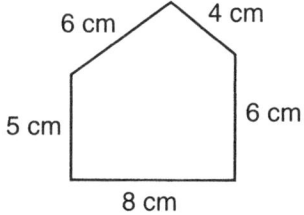

15. State whether the lines shown are parallel, perpendicular, skew, or none of these.
    A parallel
    B perpendicular
    C skew
    D none of these

Cumulative Test
Chapter 3 Form C, continued

16. Solve $-2(n - 6) = -4$.
    F $n = 8$
    G $n = 10$
    H $n = 4$
    J $n = 5$

17. Solve $-t + 20 = -54$.
    A $t = -74$
    B $t = 74$
    C $t = 34$
    D $t = -34$

18. Solve $-0.111c + 1 = 0.889c$.
    F $c = 10.0$
    G $c = 0.778$
    H $c = 1$
    J $c = -0.778$

19. Express $\frac{5}{6} \cdot \frac{5}{6} \cdot \frac{5}{6}$ using exponents.
    A $\frac{5^3}{6}$
    B $\left(\frac{5}{6}\right)^2$
    C $\left(\frac{5}{6}\right)^3$
    D $\frac{5}{6^3}$

20. Solve $\frac{1}{2}m = -\frac{5}{2}$.
    F $m = -2$
    G $m = -5$
    H $m = 3$
    J $m = 2$

21. Simplify $(-2)^9$.
    A $-\frac{5}{2}$
    B $\frac{5}{2}$
    C -256
    D -512

22. Classify the triangle according to its sides and angles.
    F scalene acute triangle
    H isosceles right triangle
    G isosceles acute triangle
    J scalene obtuse triangle

23. Simplify $(25 - 9)^{-2}$.
    A $\frac{1}{16}$
    B $\frac{1}{256}$
    C -16
    D -64

24. Simplify $y^8 \cdot y^3$.
    F y^{11}
    G y^{-24}
    H y^{-5}
    J y^{24}

25. What is 0.00625 as a fraction?
    A $\frac{625}{1000}$
    B $\frac{1}{625}$
    C $\frac{1}{160}$
    D $\frac{62}{100,000}$

26. Simplify $\frac{80}{104}$.
    F $\frac{1}{3}$
    G $\frac{40}{52}$
    H $\frac{10}{13}$
    J $\frac{8}{10}$

27. What is the phrase, "the product of two and six less than twice b," as an algebraic expression?
    A $2 - 6 \cdot b$
    B $2(2b - 6)$
    C $2 \cdot 6 - b$
    D $(2 - 6)b$

28. Find the area of the trapezoid below.
    F 13.5 m^2
    G 12 m^2
    H 27 m^2
    J 18 m^2

Name _____ Date _____ Class _____

CHAPTER 3 Cumulative Test
Form C, continued

29. Divide $4\frac{4}{5} \div 12\frac{1}{5}$.
    - A 17
    - B $\frac{3}{8}$
    - C $\frac{24}{61}$
    - D $1\frac{8}{9}$

30. Find the volume of the prism.
    - F 19 ft³
    - G 50 ft³
    - H 200 ft³
    - J 100 ft³

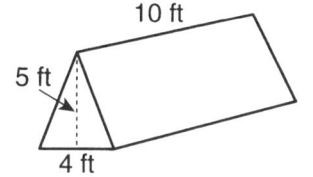

31. Express the quotient $\frac{m^5 m^3 n^2}{m^4 n^2}$ as one power.
    - A m^{22}
    - B m^{11}
    - C m^4
    - D $m^4 n$

32. Nela borrowed $430 from her sister. If it is paid back in 6 monthly payments of $95, how much is the sister charging for the loan?
    - F $140
    - G $570
    - H $45
    - J $670

33. Marla put $7\frac{1}{2}$ lb, $3\frac{1}{3}$ lb, and 2 lb of meat in the freezer. What is the total amount of meat?
    - A $\frac{2}{7}$ lb
    - B $12\frac{5}{6}$ lb
    - C $12\frac{2}{5}$ lb
    - D $3\frac{1}{2}$ lb

34. A rectangular scarf measures $5\frac{1}{3}$ inches by $49\frac{1}{2}$ inches. What is the distance around the scarf?
    - F 264 in.
    - G $54\frac{5}{6}$ in.
    - H $109\frac{2}{3}$ in.
    - J 115 in.

35. A football team gained 20 yards on the 1st play, lost 3 yards on the 2nd play, and then gained another 14 yards. What was the net gain or loss of yardage?
    - A 31 yards
    - B −3 yards
    - C 37 yards
    - D 3 yards

36. Find the volume of a cone that is 10 cm high and has a base with a radius of 3 cm. Use 3.14 for π.
    - F 282.6 cm³
    - G 94.2 cm³
    - H 31.4 cm³
    - J 141.3 cm³

37. Simplify $\sqrt{\frac{529}{49}}$.
    - A 11
    - B $\frac{23}{8}$
    - C $\frac{23}{7}$
    - D $\frac{24}{7}$

38. Find the surface area of the prism formed by the net below. 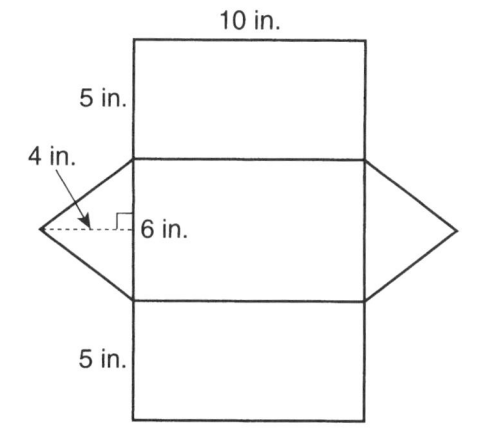
    - F 172 in²
    - G 174 in²
    - H 294 in²
    - J 184 in²

39. What kind of number is $\sqrt{\frac{25}{4}}$?
    - A rational
    - B irrational
    - C not real
    - D negative

Name _____ Date _____ Class _____

Cumulative Test
CHAPTER 3 Form C, continued

40. Translate $\left(\dfrac{v}{2}\right)^2$ into words.
    F 2 times the quotient of v and 2
    G the square of the product of v and 2
    H the square of the quotient of v and 2
    J the product of half v and 2

41. A square has an area of 37 mm². To the nearest tenth, what is the length of one side?
    A 19.0 mm   C 5.8 mm
    B 6.1 mm   D 1396.0 mm

42. What is the distance around a square that has an area of 1500 cm²?
    F 38.7 cm   H 154.9 cm
    G 50.0 cm   J 375.0 cm

43. Which is the solution of $5x < -28 + x$?

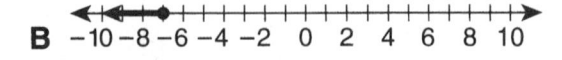

    A [number line from −10 to 10]
    B [number line from −10 to 10]
    C [number line from −10 to 10]
    D [number line from −10 to 10]

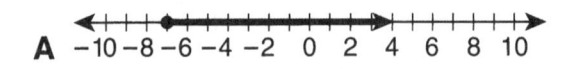

44. Which is the solution of $-2 < \dfrac{2n}{7}$?

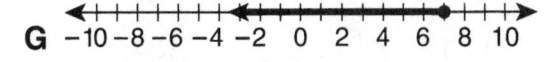

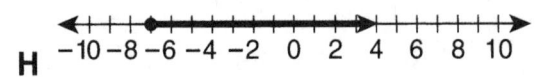

    F [number line from −10 to 10]
    G [number line from −10 to 10]
    H [number line from −10 to 10]
    J [number line from −10 to 10]

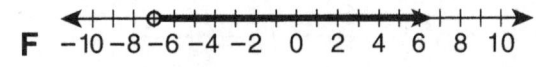

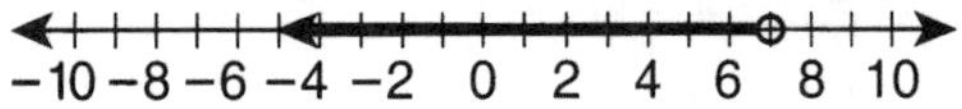

45. Solve $6.8x \geq 8.5$.
    A $x \geq 0.8$ C $x \leq 1.25$
    B $x \leq 0.8$ D $x \geq 1.25$

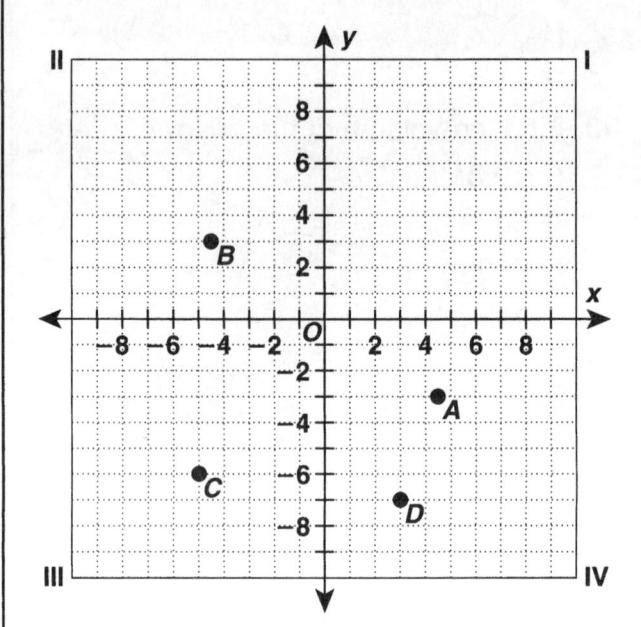

46. Use the graph above to give the coordinates of point D.
    F (3, −7)   H (−3, 7)
    G (−7, 3)   J (−3, −7)

47. Use the graph above to give the coordinates of point A.
    A (3, 4)   C (4.5, −3)
    B (4, −7)   D (−4.5, −3)

48. Which ordered pair is a solution of $0.25x + 1.5y = 4$?
    F (2, 4)   H (2, 1.5)
    G (0.5, 1.5)   J (4, 2)

49. Which ordered pair is a solution of $\dfrac{1}{3}x + \dfrac{2}{5}y = \dfrac{11}{15}$?
    A $\left(\dfrac{6}{5}, \dfrac{5}{3}\right)$ C (1, 1)
    B $\left(\dfrac{2}{5}, \dfrac{1}{3}\right)$ D $\left(\dfrac{1}{3}, \dfrac{2}{5}\right)$

Name _____ Date _____ Class _____

CHAPTER 4 Cumulative Test
Form A

1. Simplify $\sqrt{-4}$.
   A −2
   B not a real number

2. Which is equivalent to 2^4?
   A 2 • 2 • 2 • 2   B 2 • 4

3. Describe and give the value of $-\sqrt{9}$.
   A Irrational, −3   B Rational, −3

4. Express 1900 in scientific notation.
   A 1.9×10^3 B 10×1.9^5

5. Express 4.5×10^4 in standard notation.
   A 45   C 4500
   B 450,000   D 45,000

6. Find the two square roots of $\frac{9}{4}$.
   A $\frac{3}{2}, -\frac{3}{2}$ B $\sqrt{\frac{3}{2}}, \sqrt{-\frac{3}{2}}$

7. Which of the following is the same as the phrase "the sum of three times a number and 1 is 13"?
   A $n + 3 = 13$ B $3n + 1 = 13$

8. Solve $2x = 16$.
   A $x = -4$ C $x = 4$
   B $x = 8$ D $x = 32$

9. Simplify $5^8 \div 5^6$.
   A 5^{14} B 5^2

10. Which is equal to $(n)(n)(n)(n)$?
    A $4n$ B n^4

11. Solve $5b = -10$.
    A $b = -2$ B $b = 2$

12. This frequency table represents which data set?

Age When Learned to Catch	2	3	4	5
Frequency	4	3	2	1

    A 3 2 4 2 5 2 3 5 2 3
    B 4 2 4 2 5 2 3 4 2 3
    C 3 2 4 2 5 2 3 4 2 3
    D 2 2 4 2 5 2 3 4 2 3

13. Evaluate 3^{-2}.
    A −9   B $\frac{1}{9}$

14. Simplify $5 - (-3)$.
    A 8   C −8
    B 2   D −2

15. Find the area of a rectangle that measures $2\frac{2}{3}$ in. by 3 in.
    A $6\frac{1}{3}$ in^2 B 8 in^2

16. A 4-H club has 64 members. There are 20 more boys than girls. How many girls are there?
    A 22 girls   C 20 girls
    B 44 girls   D 64 girls

Name _____ Date _____ Class _____

CHAPTER 4 Cumulative Test
Form A, continued

17. The area of the top of a square picnic table is 9 ft². What is the measurement of the top?

   A 3 feet × 3 feet

   B 2 feet × 7 feet

18. Luke biked 45 miles at an average rate of 15 miles per hour. How long did the trip take?

   A 3 hr   B $\frac{1}{3}$ hr

19. Convert the following to an algebraic expression "Anna earned $20 a day at her job cleaning up at the fairgrounds. If she works one day a week, how much did she earn in *n* weeks?"

   A $20 + n$ B $20n$

20. What is possibly misleading about the phrase, "The average school lunch price is $2.00"?

   A the average may be skewed by one high-priced item

   B school lunches are always the same price

   C the sample size may be too large

   D most students bring their lunches

21. Simplify $\frac{9}{-3}$.

   A 3   B -3

22. Ten eighth graders said that hip-hop was their favorite kind of music, so Sarah states in her report that all eighth graders like hip-hop. What is possibly wrong with this statement?

   A the sample size is too small

   B too many people were asked

23. To make a fruit salad, Marla bought $2\frac{1}{4}$ pounds of apples, $1\frac{1}{3}$ pounds of peaches, and 1 pound of bananas. How much fruit salad did this make?

   A $4\frac{7}{12}$ lb   B $4\frac{2}{7}$ lb

24. Simplify $\frac{2}{5} - \frac{3}{5}$.

   A $\frac{1}{5}$ B $-\frac{1}{5}$

25. Which choice represents the data from the stem-and-leaf plot?

Stem	Leaves
2	7 8
3	5 8
4	7 8

   A 27, 21, 28, 38, 47, 48

   B 27, 28, 28, 35, 47, 48

   C 27, 28, 35, 38, 45, 48

   D 27, 28, 35, 38, 47, 48

26. Solve $\frac{1}{2} b > \frac{3}{4}$.

   A $b > 1\frac{1}{2}$ B $b > \frac{3}{8}$

27. Find the range for the data set. 11, 12, 10, 14, 13.

   A 4   C 12

   B 1   D 2

Name _____ Date _____ Class _____

CHAPTER 4 Cumulative Test
Form A, continued

28. Use the data to find the median. May's normal monthly rainfall (in inches) for 5 different U.S. cities: 3.1, 1.2, 2.4, 3.2, 4.1.

  A 2.8 **B** 3.1

29. Find the first quartile of the data set. 18, 14, 14, 23, 29, 10, 19

  A 12 **B** 14

30. What type of correlation is shown in the scatter plot?

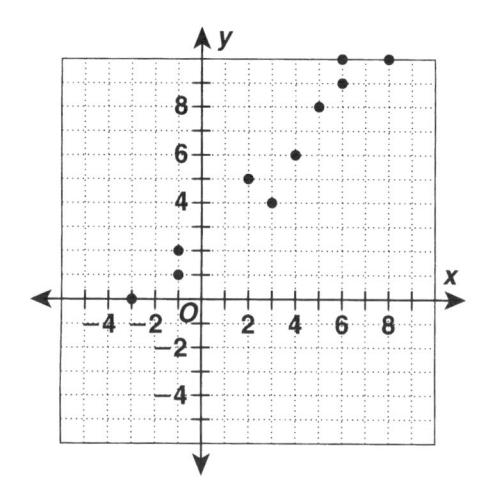

  A strong positive correlation
  B strong negative correlation
  C weak positive correlation
  D weak negative correlation

31. Divide $\frac{2}{3} \div \frac{1}{5}$.

  A $3\frac{1}{3}$ **B** $\frac{3}{10}$

32. Kay's scores for her past six quizzes were 15, 10, 14, 12, 9, and 13. Find the average of her quiz scores to the nearest tenth.

  A 12.2 **C** 12.0
  B 14.6 **D** 11.5

33. If a waiter at Rudy's restaurant received the following tips for the last five nights, what is his range of tips? $30, $30, $10, $50, and $25

  A $20 **B** $40

34. Solve $2x \geq -6$.

  A $x \leq -3$ **B** $x \geq -3$

Cumulative Test
Chapter 4 Form A, continued

35. Evaluate $\frac{3a - 2b}{9}$ when $a = 4$ and $b = 1$.
    A $\frac{10}{9}$ B $\frac{1}{9}$

36. Simplify $6g - 8 + 6g$.
    A $4g$ C $20g$
    B $12g + 8$ D $12g - 8$

37. Express 1.2 as a fraction.
    A $1\frac{1}{5}$ B $1\frac{2}{5}$

38. Solve $0.25a + 5 = 1.25a$.
    A $a = 7.5$ C $a = -5$
    B $a = 5$ D $a = 10$

39. Solve $-3 < \frac{b}{2}$.
    A $b > -6$ B $b < -6$

40. Determine which ordered pair is a solution of the equation $3x + y = 8$.
    A $(5, 0)$ B $(3, -1)$

41. On the graph, Point A represents which ordered pair?

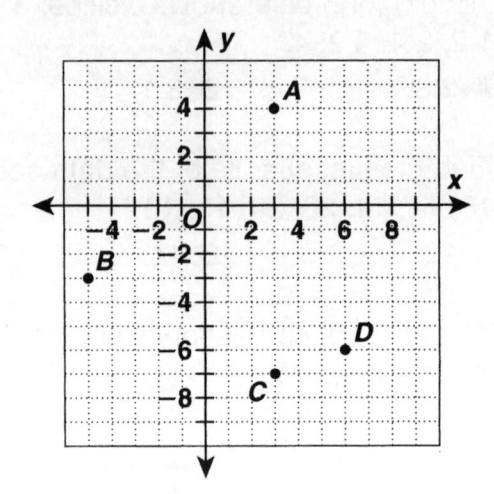

    A $(3, -4)$ B $(3, 4)$

42. Simplify $-1 + (-5)$.
    A 4   C -6
    B 6   D -4

43. Simplify $2(5 - 3m)$.
    A $10 - 3m$ B $10 - 6m$

44. Simplify $(2 + 7)^2$.
    A 51   C 81
    B -32 D 12

Name _____ Date _____ Class _____

Cumulative Test
Chapter 4 Form B

1. Simplify $\sqrt{-16}$.
   A −4
   B 8
   C 4
   D not real

2. Which is equivalent to 12^4?
   F $12 \cdot 12 \cdot 12 \cdot 12$
   G $12 \cdot 4$
   H $12 \cdot 12 \cdot 12 \cdot 12 \cdot 12$
   J $12 \div 4$

3. Describe and give the value of $-\sqrt{100}$.
   A Irrational, −50
   B Not a real number
   C Irrational, −10
   D Rational, −10

4. Express 1,900,000 in scientific notation.
   F 1.9×10^5
   G 10×1.9^5
   H 1.9×10^6
   J 10×1.9^6

5. Express 7.3×10^5 in standard notation.
   A 73
   B 7,300,000
   C 73,000
   D 730,000

6. Find the two square roots of $\frac{81}{25}$.
   F $\frac{9}{5}, -\frac{9}{5}$
   G 9, −9
   H $\sqrt{\frac{9}{5}}, \sqrt{-\frac{9}{5}}$
   J 9, 5

7. Which of the following is the same as the phrase, "The product of twice a number and 3 is 84"?
   A $6a = 84a$
   B $6a = 84$
   C $2a + 3 = 84$
   D $2a = 3(84)$

8. Solve $10x - 4 = 36$.
   F $x = -4$
   G $x = 6$
   H $x = 4$
   J $x = 32$

9. Simplify $6^3 \cdot 6^7$.
   A 6^{21}
   B 6^{10}
   C 6^4
   D 36^{21}

10. Which expression is equal to $(-3n)(-3n)(-3n)(-3n)$?
    F $-12n$
    G $-(3n)^4$
    H $-3n^4$
    J $(-3n)^4$

11. Solve $-3b = -18$.
    A $b = 6$
    B $b = -6$
    C $b < -6$
    D $b \leq -6$

12. This frequency table represents which data set?

Age When Learned to Ride a Bike	4	5	6	7
Frequency	2	9	5	2

    F 5 4 5 5 5 7 6 5 5 5 5 6 5 6 4 6 5 6
    G 3 4 5 5 5 7 6 5 5 5 5 6 6 7 4 6 5 6
    H 5 4 5 5 5 7 6 5 5 5 5 6 6 7 4 6 5 6
    J 5 4 5 5 5 7 6 5 4 4 5 6 6 7 4 6 5 6

13. Evaluate 4^{-3}.
    A −12
    B $\frac{1}{64}$
    C −81
    D $-\frac{3}{4}$

14. Simplify $12 - (-13)$.
    F −25
    G 1
    H −1
    J 25

Cumulative Test

CHAPTER 4 Form B, continued

15. Find the area of a rectangle that measures $2\frac{2}{3}$ in. by $36\frac{1}{2}$ in.

    A $78\frac{1}{3}$ in² C $97\frac{1}{3}$ in²

    B 81 in² D $39\frac{1}{6}$ in²

16. The eighth grade at Byrnedale Junior High School has 464 students. There are 212 more boys than girls. How many boys are there?

    F 212 boys H 126 boys
    G 338 boys J 464 boys

17. The area of the top of a square table is 240 in². What are the dimensions of the top?

    A 12 inches × 20 inches
    B 15.49 inches × 15.49 inches
    C 24 inches × 10 inches
    D 61.97 inches × 4 inches

18. Jacob drove 450 miles at an average rate of 65 miles per hour. Rounded to the nearest tenth of an hour, how long did the trip take?

    F $6\frac{9}{10}$ hr H $6\frac{93}{100}$ hr
    G $\frac{1}{10}$ hr J $\frac{14}{100}$ hr

19. Convert the following to an algebraic expression "Leana earned $25 a day at her job. If she works five days a week, how much did she earn in n weeks?"

    A $25 + n$ C $25n$
    B $125n$ D $25 + 5n$

20. What is possibly misleading about the statistic, "Three out of four students prefer to ride the bus to school"?

    F The average is skewed by one outlier.
    G The measurements are not the same.
    H The sample size may be too small.
    J Students don't like to ride buses.

21. Simplify $\frac{10}{-5}$.

    A 5 C −5
    B 2 D −2

22. Fifty families who live near a freeway were asked whether the freeway should be expanded. Why might this sample be biased?

    F The average is skewed.
    G These people don't drive.
    H The people near the freeway are not likely to want it expanded.
    J Too many people were asked.

23. To obtain a party mix, Marla mixed $2\frac{1}{4}$ pounds of peanuts, $5\frac{1}{3}$ pounds of chocolate candies, and 5 pounds of raisins. What was the total weight?

    A $\frac{5}{18}$ lb C $12\frac{7}{12}$ lb
    B $3\frac{3}{5}$ lb D $\frac{12}{151}$ lb

24. Simplify $-\frac{2}{7} - \left(-\frac{4}{7}\right)$.

    F $-\frac{6}{7}$ H $\frac{2}{7}$
    G $-\frac{2}{7}$ J $\frac{6}{7}$

Cumulative Test
Chapter 4, Form B, continued

25. Which choice represents the data from the stem-and-leaf plot?

Stem	Leaves
9	0 2 2
10	2 9
11	3 3 9

A 90, 92, 99, 102, 109, 113, 113, 119
B 90, 92, 92, 102, 109, 113, 113, 119
C 92, 92, 102, 102, 109, 113, 113, 119
D 90, 92, 92, 102, 109, 112, 113, 119

26. Solve $4b \leq -\frac{2}{3}$.

F $b \leq -\frac{1}{6}$
G $b \leq -\frac{8}{3}$
H $b \geq -\frac{8}{3}$
J $b \geq -\frac{1}{6}$

27. Find the range for the data set. 326, 467, 588, 401, 326, 515

A 189
B 262
C 121
D 48

28. The average August monthly precipitation (in inches) for 10 different U.S. cities is given below. Find the median.
3.5, 1.6, 2.4, 3.7, 4.1, 3.9, 1.0, 3.6, 4.2, 3.4

F 3.5 in.
G 3.7 in.
H 3.6 in.
J 3.55 in.

29. Find the first quartile for the data set.
30, 38, 35, 38, 49, 38, 49, 39, 45

A 46.5
B 38
C 36.5
D 35

30. Which type of correlation is shown in the scatter plot?

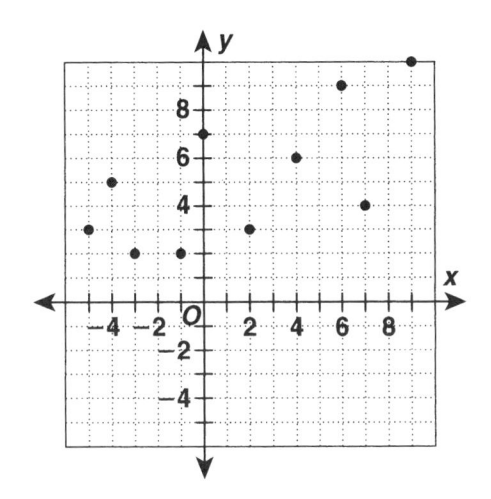

F strong positive correlation
G strong negative correlation
H weak positive correlation
J weak negative correlation

31. Divide $\frac{7}{8} \div \frac{5}{6}$.

A $\frac{35}{42}$
B $1\frac{1}{20}$
C $\frac{20}{21}$
D $\frac{6}{7}$

32. Jill's last grocery bills were $65.72, $55.82, $68.70, $78.19, $64.80, and $40.66. Find the average bill and round your answer to the nearest cent.

F $93.47
G $74.78
H $62.78
J $62.32

33. If you received the scores of 30, 37, 11, 50, and 53 on five math quizzes, what is the range of your scores?

A 42
B 20
C 37
D 36.2

Name _____ Date _____ Class _____

CHAPTER 4 Cumulative Test
Form B, continued

34. Solve $10x \geq -70$.
  F $x > -7$
  G $x \geq -7$
  H $x = -7$
  J $x \leq -7$

35. Evaluate $\dfrac{7a - 5b}{6}$ when $a = 6$ and $b = 7$.
  A $\dfrac{19}{6}$
  B $\dfrac{20}{3}$
  C $\dfrac{77}{6}$
  D $\dfrac{7}{6}$

36. Simplify $6v - 8 - 6v + 16$.
  F 24
  G $v + 24$
  H 8
  J $12v + 8$

37. Express 2.125 as a fraction.
  A $2\dfrac{1}{8}$
  B $21\dfrac{1}{4}$
  C $2\dfrac{2}{5}$
  D $2\dfrac{1}{3}$

38. Solve $0.667a + 5 = 1.667a$.
  F $a = 1.667$
  G $a = -5$
  H $a = 5$
  J $a = 10$

39. Solve $-7 < \dfrac{b}{3}$.
  A $b < -21$
  B $b > -21$
  C $b < 21$
  D $b > 21$

40. Determine which ordered pair is a solution of the equation $y + 3x = 15$.
  F (5, 0)
  G (6, 3)
  H (5, 1)
  J (−6, 3)

41. On the graph, Point C represents which ordered pair?

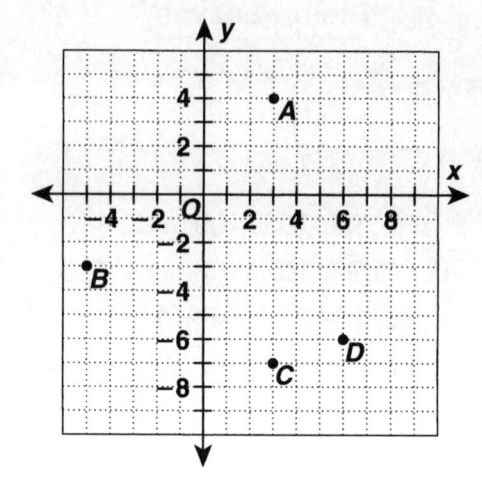

  A (6, 6)
  B (3, 7)
  C (3, −7)
  D (−3, 7)

42. Simplify $-4 + (-12)$.
  F −8
  G −16
  H 16
  J 8

43. Simplify $-1 + 9(5 - 3m)$.
  A $44 + 27m$
  B $45 - 27m$
  C $44 - 3m$
  D $44 - 27m$

44. Simplify $(2 - 6)^{-3}$.
  F $\dfrac{1}{64}$
  G 64
  H $-\dfrac{1}{64}$
  J 128

Name _____ Date _____ Class _____

CHAPTER 4 **Cumulative Test**
Form C

1. Simplify $-\sqrt{-10}$.
   - A −5
   - B −3.16
   - C 3.16
   - D not real

2. Which is equivalent to 9^5?
   - F $9 \cdot 9 \cdot 9 \cdot 9$
   - G $9 \cdot 5$
   - H $9 \cdot 9 \cdot 9 \cdot 9 \cdot 9$
   - J $9 \div 5$

3. Describe and give the value of $-\sqrt{200}$.
   - A Irrational, −20
   - B Not a real number
   - C Irrational, −14.14
   - D Rational, −20

4. Express 4,600,000,000 in scientific notation.
   - F 4.6×10^9
   - G 10×4.6^8
   - H 4.6×10^8
   - J 10×4.6^9

5. Express 9.5×10^{-4} in standard notation.
   - A 0.0095
   - B 0.00095
   - C 9500
   - D 95,000

6. Find the two square roots of $\frac{169}{196}$.
   - F $\frac{13}{14}, -\frac{13}{14}$
   - G 13, 14
   - H $\sqrt{\frac{13}{14}}, \sqrt{-\frac{13}{14}}$
   - J 14, 13

7. Which of the following is the same as the phrase, "The product of a number and 5 is less than the number cubed minus 47"?
   - A $5a > a^3 - 47$
   - B $5a < a^3 - 47$
   - C $5a = a^3 - 47$
   - D $5a < a^3 + 47$

8. Solve $9x - 27 - 6x + 3 = 0$.
   - F $x = -8$
   - G $x = 3$
   - H $x = 6$
   - J $x = 8$

9. Simplify $16^4 \cdot 16^7$.
   - A 36^{21}
   - B 16^{-3}
   - C 16^3
   - D 16^{11}

10. Which is equal to $(-12q)^3$?
    - F $(-12q)(-12q)(-12q)$
    - G $-[(-12q)(-12q)(-12q)]$
    - H $12q^{-3}$
    - J $\frac{1}{12q^3}$

11. Solve $-7d = 105$.
    - A $d = -15$
    - B $d = 15$
    - C $d = 135$
    - D $d = -135$

12. This frequency table represents which data set?

Age of students in karate class	6	7	8	9
Frequency	2	2	11	3

    - F 7 9 8 8 7 8 6 8 6 9 7 6 8 8 9 8 8 8
    - G 8 9 8 8 7 8 5 8 6 9 7 6 8 8 9 8 8 8
    - H 8 9 8 8 7 8 8 8 6 9 7 6 8 8 9 8 8 8
    - J 6 9 8 8 7 8 8 8 6 9 7 6 8 8 9 8 8 8

13. Evaluate 7^{-4}.
    - A −2401
    - B $\frac{1}{2401}$
    - C −28
    - D $-\frac{1}{28}$

14. Simplify $-12 - (-13) + 21 - 3^2$.
    - F −10
    - G 6
    - H 13
    - J 24

Copyright © by Holt, Rinehart and Winston.
All rights reserved.

189

Holt Pre-Algebra

Cumulative Test
Form C, continued

15. Find the area of a table top that measures $2\frac{7}{9}$ in. by $4\frac{3}{4}$ in.

    A $7\frac{2}{5}$ in²   C $13\frac{7}{36}$ in²

    B 665 in²   D $29\frac{1}{6}$ in²

16. A store has an inventory of 1500 paint brushes. There are 372 more 4-inch paint brushes than 2-inch paint brushes. How many 2-inch paint brushes are there?

    F 225   H 1128
    G 564   J 1872

17. The area of a square game board is 245 in². What are the dimensions of the game board?

    A 15 × 5 inches
    B 16 × 16 inches
    C 2.45 × 100 inches
    D 15.65 × 15.65 inches

18. Jennifer rode a plane for 865 miles. The plane averaged 450 miles per hour. Rounded to the nearest tenth of an hour, how long did the trip take?

    F 1.9 hr   H 1.92 hr
    G 0.9 hr   J 0.92 hr

19. Convert the following to an algebraic expression "Lila earned $37.50 a day in tips. If she worked six days a week, how much did she earn in n weeks?"

    A $37.50 + n$ C $37.50n$
    B $225n$ D $37.50 + 6n$

20. What is possibly misleading about the statistic, "Maggie owns a surfing shop that made $12,000 from June to August and $500 from October to December."?

    F The average is skewed.
    G The sample size is too small.
    H The incomes are measured at different times.
    J The income amounts are too different.

21. Simplify $\frac{-125}{-5}$.

    A 25   C −25
    B 20   D −20

22. Twenty students were asked whether the school day should be lengthened. Why might this sample be biased?

    F The average is skewed.
    G Students can't vote.
    H Students won't likely want a longer school day.
    J Too many people were asked.

23. To obtain a trail mix, Marla mixed $3\frac{3}{4}$ pounds of candy pieces, $5\frac{1}{3}$ pounds of cashews, and $5\frac{1}{3}$ pounds of dried cherries. What was the total weight?

    A $12\frac{7}{12}$ lb   C $14\frac{5}{12}$ lb
    B $12\frac{5}{9}$ lb   D $13\frac{1}{9}$ lb

24. Simplify $-\frac{15}{13} - \left(-\frac{9}{13}\right)$.

    F $-\frac{6}{13}$ H $\frac{6}{13}$
    G $-\frac{24}{13}$ J $\frac{24}{13}$

Cumulative Test
Chapter 4 Form C, continued

25. Find the original data from the stem-and-leaf plot.

Stem	Leaves
8	0 3 7
9	4 6 6
10	1 1 4

A 80, 83, 86, 94, 96, 96, 101, 104, 101

B 84, 87, 93, 94, 96, 96, 101, 101, 104

C 80, 83, 87, 49, 96, 96, 101, 104, 101

D 80, 83, 87, 94, 96, 96, 101, 104, 101

26. Solve $-6c > -\frac{4}{3}$.

F $c < -\frac{2}{9}$ **H** $c > \frac{2}{9}$

G $c < \frac{2}{9}$ **J** $c > -\frac{2}{9}$

27. Find the range for the data set.
120, 116, 112, 117, 120

A 0   **C** 117
B 8   **D** 120

28. The average January monthly snowfall (in inches) for 10 different U.S. cities is given below. Find the median. 3.23, 1.67, 2.45, 3.45, 4.12, 3.61, 5.56, 3.68, 4.24, 3.40

F 3.53 in.   **H** 3.89 in.
G 3.45 in.   **J** 3.565 in.

29. Find the third quartile for the data set. 120, 138, 146, 138, 149, 138, 149, 123, 137

A 125   **C** 147.5
B 149   **D** 143.5

30. Which type of correlation is shown in the data set illustrated in the scatter plot?

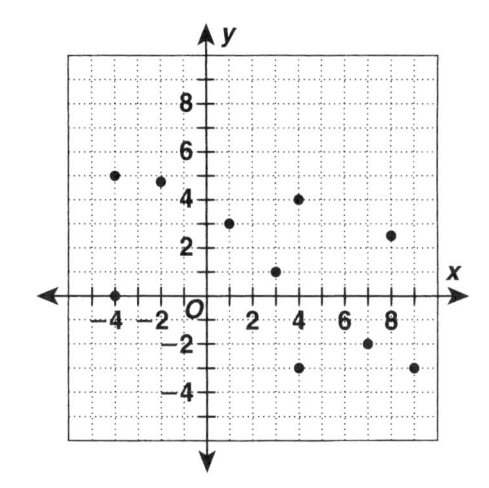

F strong positive correlation
G strong negative correlation
H weak positive correlation
J weak negative correlation

31. Divide $\frac{5}{11} \div \frac{2}{15}$.

A $\frac{75}{22}$ **C** $\frac{22}{75}$

B $\frac{2}{33}$ **D** $\frac{7}{15}$

32. Janelle's long distance bills for the last semester of college were $174.94, $154.72, $181.11, $183.12, $166.61 and $150.41. Find the average bill and round your answer to the nearest cent.

F $182.18   **H** $170.18
G $102.73   **J** $168.49

Cumulative Test
Form C, continued

33. If the high temperatures for five days were 89.6°F, 87.3°F, 90.4°F, 88.9°F, and 86.1°F, what is the range of high temperatures for these five days?
A 88.46°F   C 3.5°F
B 4.3°F   D 88.9°F

34. Solve $18x + 30 < -60$.
F $x > -5$ H $x = -5$
G $x < -5$ J $x < -7$

35. Evaluate $\frac{13x - 19y}{6}$ when $x = -3$ and $y = -2$.
A $\frac{22}{6}$ C $-\frac{77}{6}$
B $\frac{4}{6}$ D $-\frac{1}{6}$

36. Simplify $14d - 18 - 12d + (-22)$.
F $22d + 4$ H $22d - 40$
G $2d - 40$ J $2d + 4$

37. Express 12.375 as a fraction.
A $12\frac{3}{8}$ C $12\frac{2}{5}$
B $12\frac{3}{4}$ D $12\frac{1}{3}$

38. Solve $-0.257b + 12 = 1.343b$.
F $b = 7.5$ H $b = 5$
G $b = -7.5$ J $b = 12$

39. Solve $-5 < \frac{b}{3} + 2$.
A $b < -21$ C $b < 21$
B $b > 21$ D $b > -21$

40. Determine which ordered pair is a solution of the equation $15 - x = 5y$.
F $(-5, 4)$ H $(5, 3)$
G $(20, 1)$ J $(2, 5)$

41. The equation $y = x + 1$, shown on the graph below, contains which ordered pair?

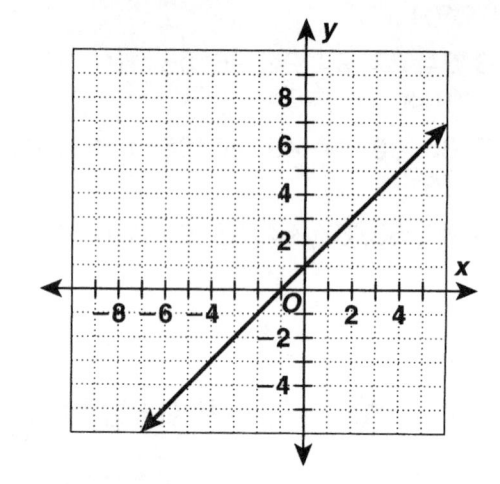

A $(-3, -2)$ C $(2, 2)$
B $(0, 0)$ D $(1, 4)$

42. Simplify $-3 + (-21) - 17$.
F -35 H 35
G 1 J -41

43. Simplify $-1 + 9(5 - 3m) + m$.
A $44 + 27m$ C $44 - 2m$
B $44 - 27m$ D $44 - 26m$

44. Simplify: $3(2 - 6)^3 + 3^0$.
F -191 H -193
G -1728 J 192

Name _____ Date _____ Class _____

Cumulative Test
Chapter 5 Form A

Select the best answer.

1. Simplify $2 - (-1) \cdot (-4)$.
   A 4   B −2

2. Solve $\frac{1}{2}v \leq 30$.
   A $v \leq 60$ B $v \leq 15$

3. What type of correlation probably exists between the amount of time you watch television and the amount of time you spend reading?
   A negative   C positive
   B neutral   D no correlation

4. Multiply $\frac{2}{3} \cdot \frac{5}{6}$.
   A $\frac{7}{9}$ B $\frac{5}{9}$

5. Subtract $1\frac{1}{4} - \frac{1}{2}$.
   A $\frac{5}{8}$ B $\frac{3}{4}$

6. Solve $c - 3 = 4$.
   A $c = 7$ B $c = 1$

7. Solve $2n > 4.4$.
   A $n > 8.8$ C $n > 2.2$
   B $n < 2.2$ D $n < 8$

8. Simplify $-z - 3z$.
   A $4z$ B $-4z$

9. Determine if the following number is rational, irrational, or not real and give the value if rational or irrational. $\sqrt{-9}$
   A not real   B rational, −3

10. Find the algebraic expression for, "4 is less than 7 plus x."
    A $4 < 7 + x$ B $7 < x + 4$

11. Find the slope of $\overleftrightarrow{AB}$.

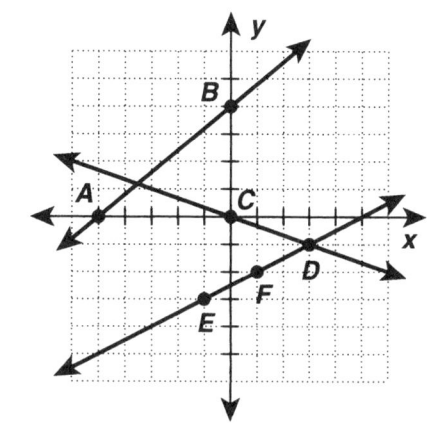

    A $\frac{4}{5}$ C $\frac{1}{2}$
    B $-\frac{1}{3}$ D $-\frac{5}{4}$

12. Express 1.3×10^{-4} in standard notation.
    A 0.00013   B 0.0013

13. Find both square roots of $\sqrt{100}$.
    A 5, 20   B 10, −10

Name _____ Date _____ Class _____

Cumulative Test
CHAPTER 5 Form A, continued

14. Find the sum of the angle measures in a regular pentagon.
  A 180°
  B 540°
  C 900°
  D 720°

15. Find the original data from the stem-and-leaf plot.

Stem	Leaves
1	1 8
2	1 1 2
3	1 8

  A 11, 18, 21, 21, 22, 31, 38
  B 118, 211, 212, 318

16. Find the mean amount spent for snacks over the last four months. Round your answer to the nearest cent.
  $58.39, $89.53, $82.24, $70.94
  A $75.28
  B $76.59

17. Find the range for the data set.
  11, 8, 11, 8, 8, 15
  A 9.5
  B 7

18. Solve the equation $\frac{1}{3}x = 4$.
  A $x = \frac{4}{3}$
  B $x = 12$

19. Simplify $6^3 \cdot 6^2$ using positive exponents.
  A 6^5
  B 6^6
  C 36^5
  D 36^6

20. What is a quadrilateral with 4 congruent sides?
  A rhombus
  B rectangle

21. In the graph, how has figure DEF been transposed from figure ABC?

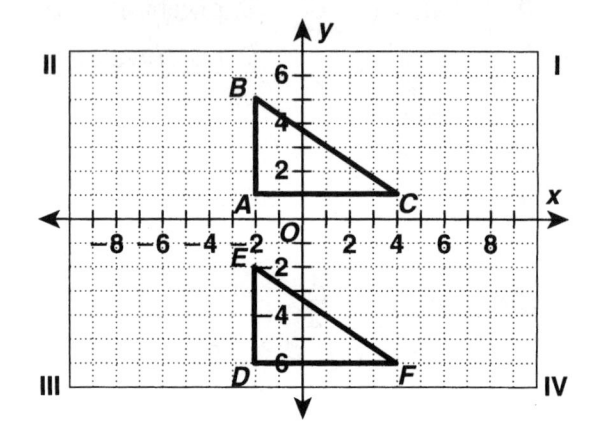

  A reflection across the y-axis
  B translation 7 units down

22. A piece of fabric is 18 inches wide. If $7\frac{1}{4}$ inches are trimmed off, how wide is the fabric?
  A $10\frac{3}{4}$ in.
  B $10\frac{1}{4}$ in.

Cumulative Test
Chapter 5, Form A, continued

23. Which ordered pair is in the third quadrant?
  A (3, 2) C (2, 3)
  B (−2, 3) D (−2, −3)

24. Which graph represents the inequality $x \leq 7$?

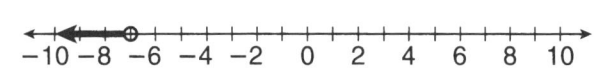

  A

−10 −8 −6 −4 −2 0 2 4 6 8 10

  B

−10 −8 −6 −4 −2 0 2 4 6 8 10

  C

−10 −8 −6 −4 −2 0 2 4 6 8 10

  D

−10 −8 −6 −4 −2 0 2 4 6 8 10

25. Simplify -2^2.
  A 4 B −4

26. Find the mode in the data set.
  11, 8, 11, 15, 8, 15, 8
  A 11 B 8

27. Which is $\frac{7}{8}$ as a decimal?
  A 0.875 B 7.8

28. Determine which ordered pair is a solution of the equation $y = 1 + 3x$.
  A (1, 4) B (4, 1)

29. Identify the type of triangle.

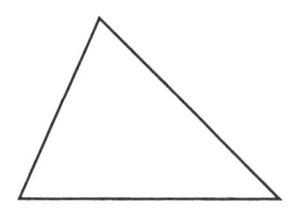

  A right C obtuse
  B acute D isosceles

Refer to the following figure for problems 30–32.

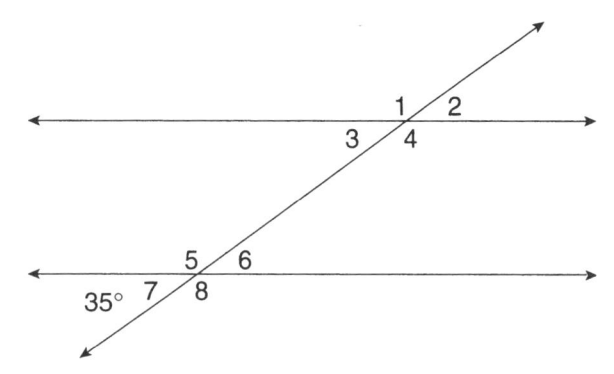

30. Find $m\angle 1$ in the figure above.
  A 35° B 145°

31. In the figure above, which angle is vertical to $\angle 8$?
  A $\angle 5$ C $\angle 4$
  B $\angle 6$ D $\angle 7$

32. In the figure above, name an angle that is a supplementary to $\angle 2$.
  A $\angle 3$ B $\angle 4$

Cumulative Test
Chapter 5 Form A, continued

33. Express 42,000 in scientific notation.
  A 4.2×10^5 B 4.2×10^4

34. Polygon $ABCD \cong LMNO$. Find s.

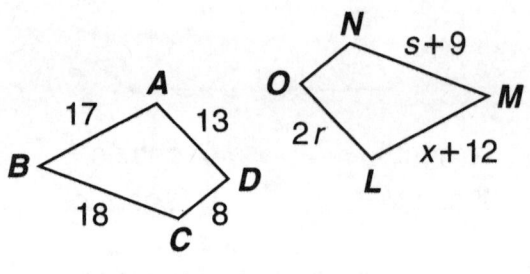

  A 9   B 28

35. Express 2.75 as a fraction.
  A $2\frac{7}{5}$ B $2\frac{3}{4}$

36. Line $a \parallel$ line b.
  If $m\angle 1 = 70°$, find $m\angle 4$.
  A 110°
  B 70°
  C 20°
  D 250°

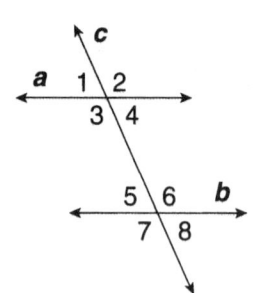

37. At Peterson Junior High School an average of 100 students check out books in the school library every day. On Wednesday, the first 25 students who came into the library were asked what types of books they like to read. Identify the sample of the survey.
  A all students at the school
  B the 25 that were surveyed

38. Which number is equivalent to 2^{-2}?
  A -4 C $\frac{1}{4}$
  B $-\frac{1}{4}$ D 4

39. A recipe for cookies calls for 2 cups of sugar. How much sugar would be needed if the recipe is tripled?
  A $1\frac{1}{2}$ C $\frac{1}{3}$ cup
  B 3 cups   D 6 cups

40. The surface area of the top of a square table is 81 cm². What are the dimensions of the top of the table?
  A 9 cm × 9 cm   B 8 cm × 11 cm

Name _____ Date _____ Class _____

CHAPTER 5 Cumulative Test
Form B

Select the best answer.

1. Simplify $10 - (-7) \cdot (-4) + 9$.
   A 15   C 12
   B −59   D −9

2. Solve $\frac{5}{6}m > 12$.
   F $m > 14\frac{2}{5}$ H $m > 10$
   G $m < 14\frac{2}{5}$ J $m < 10$

3. What type of correlation exists between the amount of time you exercise and the number of calories you use?
   A negative   C positive
   B neutral   D no correlation

4. Multiply $\frac{20}{23} \cdot \frac{15}{16}$.
   F $\frac{5}{7}$ H $\frac{75}{92}$
   G $\frac{35}{39}$ J $\frac{301}{368}$

5. Subtract $\frac{11}{15} - 1\frac{1}{5}$.
   A $\frac{23}{25}$ C $\frac{11}{18}$
   B $-\frac{7}{15}$ D $\frac{11}{180}$

6. Solve $m - 13 = -16$.
   F $m = 3$ H $m = -29$
   G $m = -3$ J $m = 29$

7. Solve $0.25n > 2$.
   A $n > 8$ C $n > 0.5$
   B $n < 0.125$ D $n < 4$

8. Simplify $-7z - 3z - 7z$.
   F $17z$ H $-17z$
   G $-17z^2$ J $17z^2$

9. Determine if the following number is rational, irrational, or not real and give the value if rational or irrational. $\sqrt{64}$
   A rational, 8   C rational, 32
   B not real   D irrational, 32

10. Find the algebraic expression for "fourteen is less than or equal to seven subtracted from x."
    F $14 < 7 - x$ H $14 \leq x - 7$
    G $14 \leq 7 - x$ J $x - 7 < 14$

11. Find the slope of CD.

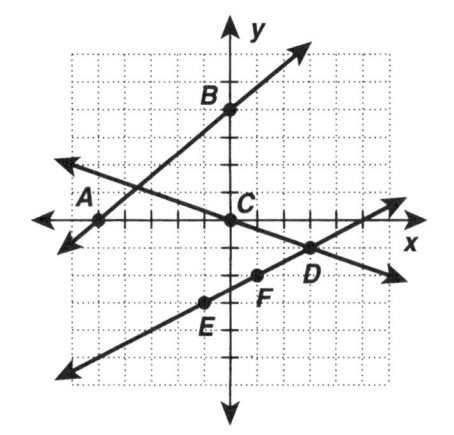

    A $\frac{4}{3}$ C $\frac{1}{2}$
    B $-\frac{1}{3}$ D 3

12. Express 4.45×10^{-5} in standard notation.
    F 0.0000445   H 0.000445
    G 0.00445   J 44,500

13. Find both square roots of $\sqrt{169}$.
    A 13, −13   C 15, −15
    B 16, −16   D 26, −26

Name _____ Date _____ Class _____

CHAPTER 5 Cumulative Test
Form B, continued

14. Find the sum of the angle measures in a regular hexagon.
  F 540° H 900°
  G 720° J 1080°

15. Find the original data from the stem-and-leaf plot.

Stem	Leaves
4	1 8
5	1 1 2 7
6	1 8 8

  A 418, 51127, 6188
  B 41, 48, 51, 51, 52, 57, 61, 68, 68
  C 41, 48, 51, 52, 57, 61, 68
  D 418, 511, 527, 618, 618

16. Find the mean amount spent for groceries over the last eight weeks. Round your answer to the nearest cent.
  $58.39, $89.53, $82.24, $70.94, $58.39, $87.74, $58.17, $61.42
  F $72.63 H $70.86
  G $66.18 J $70.85

17. Find the range for the data set.
  11, 25, 11, 28, 28, 15
  A 17 C 4
  B 19.5 D 11

18. Solve the equation $\frac{3}{5}x + 2 = \frac{2}{5}x - 4$.
  F $x = \frac{26}{5}$ H $x = \frac{1}{5}$
  G $x = -\frac{7}{5}$ J $x = -30$

19. Simplify $3^5 \cdot 3^3 \cdot 3^2$ using positive exponents.
  A 3^{10} C 3^{30}
  B 27^{10} D 27^{30}

20. What is a quadrilateral with 4 congruent angles and 4 congruent sides?
  F trapezoid H square
  G rhombus J rectangle

21. In the graph, how has figure ADE been transposed from figure ABC?

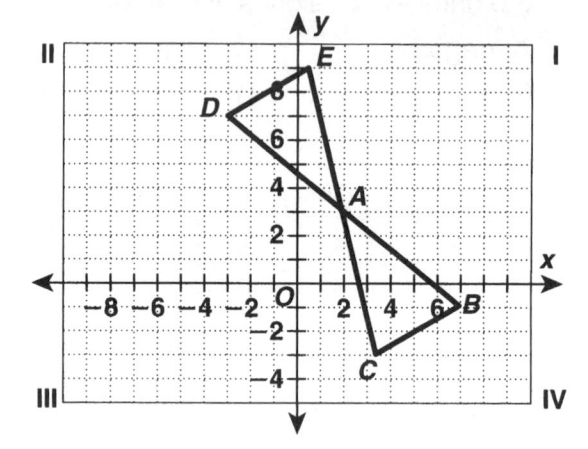

  A reflection across the x-axis
  B a 180° rotation about (2, 3)
  C reflection across the y-axis
  D translation 2 units down

22. A piece of fabric is 2 yards long and 8 inches wide. If the width is trimmed to $7\frac{11}{16}$ inches, how much was trimmed off the width?
  F $\frac{123}{16}$ in. H $\frac{5}{16}$ in.
  G $3\frac{3}{5}$ in. J $\frac{11}{16}$ in.

Name _____ Date _____ Class _____

CHAPTER 5 Cumulative Test
Form B, continued

23. Which ordered pair is in the fourth quadrant?
- **A** (3, −5)
- **B** (5, 3)
- **C** (3, 5)
- **D** (−5, 3)

24. Which graph represents the inequality $4x < -28$?

F 
G 
H 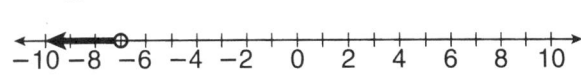
J

25. Simplify -5^2.
- **A** −25
- **B** 25
- **C** 10
- **D** −10

26. Find the mode in the data set.
11, 28, 11, 15, 28, 15, 28
- **F** 11
- **G** 28
- **H** 17
- **J** 15

27. Which is $\frac{7}{25}$ as a decimal?
- **A** 0.28
- **B** 1.4
- **C** 3.5
- **D** 7.25

28. Determine which ordered pair is a solution of the equation $2y = 6 + 3x$.
- **F** $\left(-1, 1\frac{1}{2}\right)$
- **G** $(-1, 3)$
- **H** $\left(-1, \frac{9}{2}\right)$
- **J** $(3, 2)$

29. Identify the type of triangle.

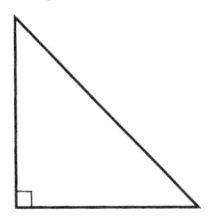

- **A** right
- **B** acute
- **C** obtuse
- **D** equilateral

Refer to the following figure for problems 30–32.

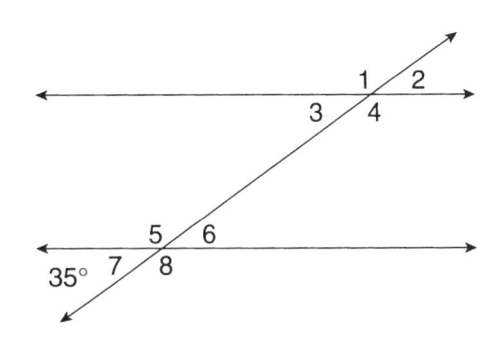

30. Find $m\angle 1$ in the figure above.
- **F** 35°
- **G** 145°
- **H** 90°
- **J** 180°

31. In the figure above, which angle is vertical to $\angle 3$?
- **A** $\angle 2$
- **B** $\angle 5$
- **C** $\angle 4$
- **D** $\angle 1$

32. In the figure above, name the angles that are supplementary to $\angle 8$.
- **F** $\angle 5$ and $\angle 6$
- **G** $\angle 3$ and $\angle 4$
- **H** $\angle 6$ and $\angle 7$
- **J** $\angle 1$ and $\angle 4$

Holt Pre-Algebra

Name _____ Date _____ Class _____

CHAPTER 5 Cumulative Test
Form B, continued

33. Express 2,230,000 in scientific notation.
    A 2.23×10^6
    B 2.23×10^7
    C 0.223×10^{-7}
    D 22.3×10^6

34. Polygon $ABCD \cong LMNO$. Find x.

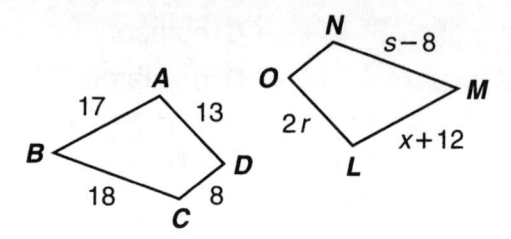

    F 1.667
    G −5
    H 5
    J 10

35. Express 0.225 as a fraction.
    A $\frac{9}{40}$
    B $22\frac{1}{2}$
    C $2\frac{1}{4}$
    D $2\frac{2}{5}$

36. Line $a \parallel$ line b.
    If $m\angle 2 = 128°$, find $m\angle 8$.
    F 128°
    G 52°
    H 38°
    J 218°

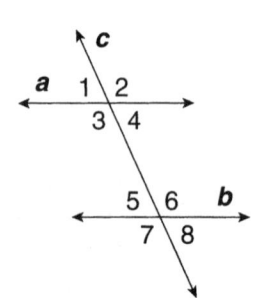

37. Of the 350 students at Byrnedale Junior High School, 250 buy their lunch in the school cafeteria. On Wednesday, the first 100 students eating in the cafeteria were asked their favorite food that was served in the cafeteria that day. Identify the population of the survey.
    A all students at the school
    B all students in school on Wednesday
    C 100 that were surveyed
    D 250 that buy lunch in cafeteria

38. Which number is equivalent to 6^{-2}?
    F −36
    G $-\frac{1}{36}$
    H $\frac{1}{36}$
    J $\frac{1}{12}$

39. A recipe for cookies calls for $1\frac{1}{2}$ cups of sugar. How much sugar would be needed if the recipe is tripled?
    A $4\frac{1}{2}$ cups
    B 3 cups
    C $\frac{1}{2}$ cup
    D 6 cups

40. The surface area of the top of a square table is 110.25 in.^{2}. What are the dimensions of the top of the table?
    F $10\frac{1}{2}" \times 1"$
    G $10\frac{1}{4}" \times 11\frac{1}{4}"$
    H $10\frac{1}{2}" \times 10\frac{1}{2}"$
    J $11\frac{1}{2}" \times 10"$

Cumulative Test
Chapter 5 Form C

Select the best answer.

1. Simplify $15 - (-5) \div (-5) \cdot 2$.
   - A -2
   - B -8
   - C 13
   - D -4

2. Solve $\frac{7}{12}m + 2 \le 16$.
   - F $m \le 24$
   - G $m \ge 24$
   - H $m \le \frac{18}{7}$
   - J $m \ge \frac{18}{7}$

3. What type of correlation likely exists between the amount of time you spend playing video games and the amount of time you spend studying?
   - A negative
   - B neutral
   - C positive
   - D no correlation

4. Simplify $2\frac{11}{15} \cdot 1\frac{4}{5}$.
   - F $5\frac{7}{15}$
   - G $3\frac{23}{15}$
   - H $4\frac{4}{5}$
   - J $4\frac{23}{25}$

5. Simplify $1\frac{23}{24} - 1\frac{17}{20}$.
   - A $3\frac{97}{120}$
   - B $2\frac{1}{2}$
   - C $\frac{13}{120}$
   - D $-\frac{13}{120}$

6. Solve $3m - 10 = -14 - m$.
   - F $m = -1$
   - G $m = -6$
   - H $m = -12$
   - J $m = -2$

7. Solve $1.25n \ge -2$.
   - A $n > -3.25$
   - B $n \ge -1.6$
   - C $n \ge 3.25$
   - D $n > -2.5$

8. Simplify $-6z - 3z - 6z^2$.
   - F $-15z$
   - G $-15z^2$
   - H $-9z - 6z^2$
   - J $9z^2$

9. Determine if the following number is rational, irrational, or not real and give the value if rational or irrational.
   $-\sqrt{30}$
   - A rational, -5.48
   - B not real
   - C irrational, 5.48
   - D irrational, -5.48

10. Find the algebraic expression for the phrase "twice x is less than or equal to three less than three times x."
    - F $2x \le 3 - 3x$
    - G $2x \le 3x - 3$
    - H $2x \ge 3 - 3x$
    - J $2x < 3x - 3$

11. Find the slope of EF.
    - A $\frac{4}{3}$
    - B $-\frac{1}{3}$
    - C $\frac{1}{2}$
    - D 2

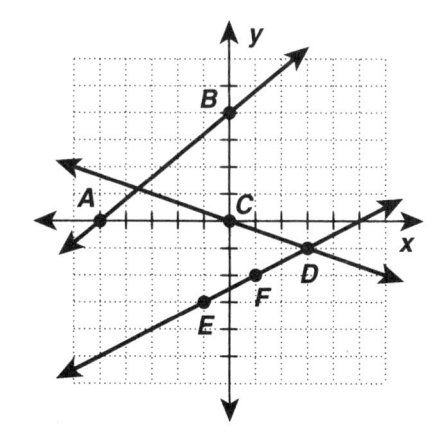

12. Express 1.05×10^{-6} in standard notation.
    - F 0.0000105
    - G 0.00105
    - H 0.000105
    - J 0.00000105

13. Find both square roots of $\sqrt{484}$.
    - A $24, -24$
    - B $22, -22$
    - C $12, -12$
    - D $44, -44$

Cumulative Test
Form C, continued

14. Find the sum of the angle measures in a nonagon.
- F 540°
- G 720°
- H 1260°
- J 1620°

15. Find the original data from the stem-and-leaf plot.

Stem	Leaves
4.	1 8
5.	1 1 2 7
6.	1 8

- A 4.18, 5.1127, 6.188
- B 41, 48, 51, 51, 52, 57, 61, 68, 68
- C 41, 48, 51, 52, 57, 61, 68
- D 4.1, 4.8, 5.1, 5.1, 5.2, 5.7, 6.1, 6.8

16. Find the mean amount that Jean spent on long distance telephone calls over the past 8 months. Round your answer to the nearest cent.
$161.42, $158.39, $158.39, $170.94, $189.53, $182.24, $187.74, $158.17,
- F $455.61
- G $116.18
- H $170.85
- J $172.63

17. Find the range for the data set.
185, 246, 372, 801, 675, 212
- A 415
- B 616
- C 309
- D 110

18. Solve $\frac{3}{5}x + 2 = \frac{2}{5}\left(x + \frac{9}{4}\right)$.
- F $-1\frac{1}{10}$
- G $1\frac{1}{4}$
- H $-5\frac{1}{2}$
- J $14\frac{1}{2}$

19. Simplify using positive exponents.
$(3a)^5 \cdot (3a)^3 \cdot (3a)^2 \cdot (3a)^0$
- A $3a^{10}$
- B $(3a)^{10}$
- C $(3a)^{30}$
- D $27a^{30}$

20. What is a quadrilateral with 4 congruent sides where, $\angle A \cong \angle C$ and $\angle B \cong \angle D$, but $\angle A \not\cong \angle B$?
- F rhombus
- G square
- H trapezoid
- J rectangle

21. In the graph, how has figure DEF been transposed from figure ABC?

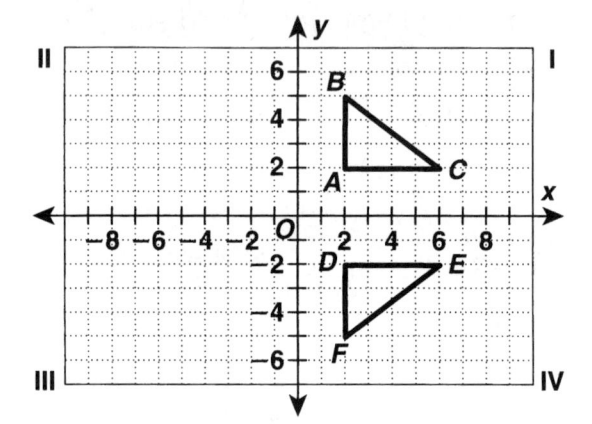

- A reflection across the x-axis
- B a 180° rotation about (6, 2)
- C reflection across the y-axis
- D translation 2 units down

22. A piece of fabric is 2 feet long and 18 inches wide. If both width and length are trimmed by $7\frac{11}{16}$ in., what are the new dimensions?
- F 2 ft × $10\frac{5}{6}$ in.
- G 2 ft × $7\frac{11}{16}$ in.
- H $16\frac{5}{16}$ in. × $10\frac{5}{16}$ in.
- J $16\frac{5}{16}$ in. × 18 in.

Cumulative Test
Chapter 5 Form C, continued

23. Which ordered pair is not in the fourth quadrant?

A $\left(2\frac{1}{2}, -6\right)$ C $\left(3\frac{1}{2}, 6\right)$

B $\left(6, -3\frac{1}{2}\right)$ D $\left(6, -2\frac{1}{2}\right)$

24. Which graph represents the inequality $3x \geq -28 - x$?

F [number line with closed circle at -7, shaded right]

G [number line with closed circle at 7, shaded left]

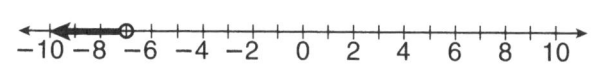

H [number line with open circle at -7, shaded left]

J [number line with open circle at 7, shaded left]

25. Simplify $-5^2 \cdot 6^2$.

A 1   C 900
B −900   D 810,000

26. Find the mode in the data set.
110, 280, 110, 150, 280, 150, 280

F 110   H 170
G 150   J 280

27. Which is $\frac{7}{11}$ as a decimal?

A 0.63   C 1.3
B $0.\overline{63}$ D $1.\overline{3}$

28. Determine which ordered pair is a solution of the equation $3y + 1 = 2x$.

F $\left(1, 1\frac{1}{2}\right)$ H $\left(\frac{3}{2}, \frac{2}{3}\right)$

G $(-1, 3)$ J $(2, 3)$

29. Identify the type of triangle.

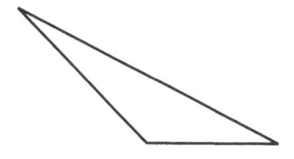

A right   C obtuse
B acute   D isosceles

Refer to the following figure for problems 30–32.

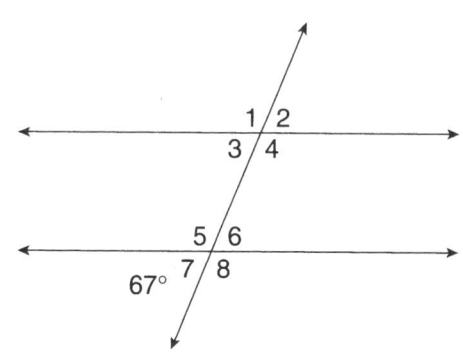

30. Find $m\angle 3$ in the figure above.

F 67°   H 77°
G 113°   J 180°

31. In the figure above, which angle is vertical to $\angle 4$?

A $\angle 2$ C $\angle 7$
B $\angle 6$ D $\angle 1$

32. In the figure above, name the angles that are supplementary to $\angle 3$.

F $\angle 5$ and $\angle 6$ H $\angle 1$ and $\angle 2$
G $\angle 1$ and $\angle 4$ J $\angle 2$ and $\angle 4$

CHAPTER 5 Cumulative Test
Form C, continued

33. Express 343,000,000 in scientific notation.
 A 3.43×10^8 C 343×10^8
 B 3.43×10^6 D 34.3×10^8

34. Polygon $ABCD \cong LMNO$. Find r.

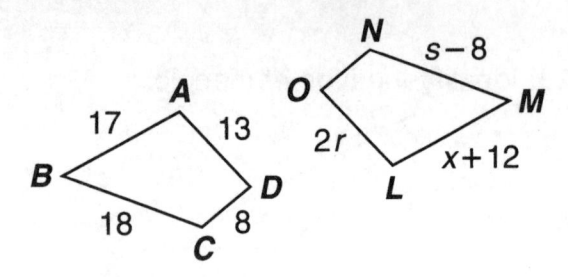

 F 8.5   H 4
 G 9.5   J 6.5

35. Express 2.325 as a fraction.
 A $2\frac{13}{40}$ C $2\frac{1}{4}$
 B $23\frac{1}{2}$ D $2\frac{3}{5}$

36. Line $a \parallel$ line b. If $m\angle 5 = 79°$, find $m\angle 2$.
 F 79°
 G 101°
 H 11°
 J 259°

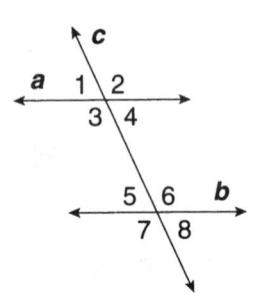

37. Of the 350 students at Burnet Junior High School, an average of 20 visit the nurse's office every day. On Wednesday, the first 5 students who visited the nurse's office were asked to describe their symptoms. Identify the sample in the survey.
 A the 5 who were surveyed
 B all students at the school
 C all students in school on that day
 D the 20 who visit the nurse's office

38. Which number is equivalent to 5^{-3}?
 F -125 H $\frac{1}{125}$
 G $-\frac{1}{15}$ J $\frac{1}{15}$

39. A recipe for cookies calls for $1\frac{1}{2}$ cups of sugar. How much sugar would be needed if the recipe is increased $1\frac{1}{2}$ times?
 A $\frac{1}{2}$ cup   C 3 cups
 B $2\frac{1}{4}$ cups   D 6 cups

40. The surface area of the top of a square table is 150.0625 in^2. What are the dimensions of the top of the table?
 F 10.1" × 15"   H 10.5" × 10.6"
 G $10\frac{1}{4}$" × $15\frac{1}{4}$"   J $12\frac{1}{4}$" × $12\frac{1}{4}$"

Name _____ Date _____ Class _____

CHAPTER 6 Cumulative Test Form A

1. Simplify $-3 + 7 - 1 - 7$.
   - A -4
   - B 17
   - C 11
   - D -8

2. Simplify $-2 - (-3 + 3)$.
   - A 0
   - B -4
   - C -8
   - D -2

3. A cone has a radius of 2.3 cm and a height of 4.0 cm. Find its volume to nearest tenth. Use 3.14 for π.
   - A 22.1 cm^3
   - B 66.4 cm^3

4. Find the square root of 8 to the nearest hundredth.
   - A 4.00
   - B 2.83
   - C 1.14
   - D 64.00

5. Simplify $-6 + (-3)$.
   - A -9
   - B -3

6. Simplify $mn - 2mn$.
   - A $-mn$
   - B -1

7. Simplify $-2(d + 1)$.
   - A $-2d - 1$
   - B $-2d - 2$

8. Evaluate $\dfrac{c + 4}{4}$ for $c = -1$.
   - A $\dfrac{3}{4}$
   - B -1

9. Solve $p - 4 = 4$.
   - A $p = 0$
   - B $p = 4$
   - C $p = -8$
   - D $p = 8$

10. Clark received the following scores on biology lab assignments: 12, 30, 15, 45, 5. Find the range of his scores.
    - A 21.4
    - B 7
    - C 40
    - D 15

11. Express $m \cdot m \cdot m$ using exponents.
    - A $3m$
    - B m^3
    - C $3m^6$
    - D 3^m

12. Find the measure of an angle whose supplement is 2 times the angle.
    - A $60°$
    - B $45°$

13. Express 1.2×10^{-3} in standard notation.
    - A 0.00012
    - B 0.012
    - C 0.0012
    - D 1.2000

14. Express 360,000 in scientific notation.
    - A 3.6×10^3
    - B 36×10^3
    - C 36×10^5
    - D 3.6×10^5

15. A cabinetmaker installs $\dfrac{1}{4}$-in. thick laminate material on top of $\dfrac{3}{8}$-in. plywood. What is the thickness of the completed countertop?
    - A $\dfrac{5}{8}$ in.
    - B $\dfrac{6}{12}$ in.

Copyright © by Holt, Rinehart and Winston.
All rights reserved.

205

Holt Pre-Algebra

Name _____ Date _____ Class _____

CHAPTER 6 Cumulative Test
Form A, continued

16. Find the perimeter of the polygon.

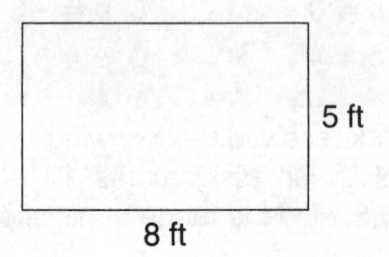

5 ft
8 ft

- **A** 13 ft
- **B** 40 ft
- **C** 26 ft
- **D** 80 ft

17. Find the average number of e-mails that Joshua received each day if the number of e-mails he received each day for the past week were 20, 5, 10, 30, 4, 4, and 5. Round to the nearest tenth.

- **A** 10
- **B** 11.1
- **C** 10.6
- **D** 5

18. Find the missing measurement in the triangle to the nearest tenth.

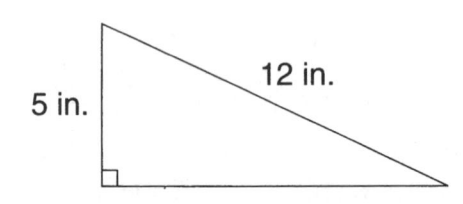

5 in.
12 in.

- **A** 8.5 in.
- **B** 10.9 in.
- **C** 12.5 in.
- **D** 13.0 in.

19. Solve the equation $1.5 = s + 2.8$.

- **A** $s = -1.3$
- **B** $s = 1.3$
- **C** $s = 4.3$
- **D** $s = -4.3$

20. Find the circumference of a circle with $d = 10$ mi. Use 3.14 for π.

- **A** 78.5 mi
- **B** 31.4 mi
- **C** 62.8 mi
- **D** 314 mi

21. Anissa writes one page in $\frac{1}{2}$ of an hour. At this rate, how many pages can she write in 3 hours?

- **A** 6 pages
- **B** $1\frac{1}{2}$ pages

22. Find the sum of the angles of a regular quadrilateral.

- **A** 360°
- **B** 720°
- **C** 1080°
- **D** 1620°

23. Find the volume of the figure to the nearest hundredth. Use 3.14 for π.

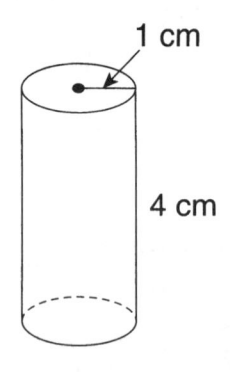

1 cm
4 cm

- **A** 25.12 cm^2
- **B** 12.56 cm^2

24. Find the area of the triangle.

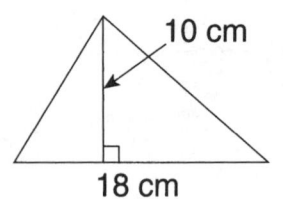

10 cm
18 cm

- **A** 90 cm^2
- **B** 180 cm^2

Cumulative Test
Chapter 6 Form A, continued

25. Simplify $-\frac{16}{8}$.
    A 2
    B −2

26. Find the value of 10^2.
    A 20
    B 200
    C 100
    D 10

27. Which equation means "five minus a number d is −4"?
    A $d - 5 = -4$
    B $5 - d = -4$

28. Solve for the hypotenuse in the right triangle ABC with leg $a = 3$ and leg $b = 4$.
    A $c = 5$
    B $c = 7$

29. Express $2^4 \cdot 2^3$ using positive exponents.
    A 2^7
    B 2^{12}
    C 4^7
    D 4^{12}

30. Solve $\frac{3}{5}y = \frac{5}{8} + \frac{2}{5}y + \frac{5}{8}$.
    A $y = \frac{5}{4}$
    B $y = \frac{25}{4}$

31. Which of the following inequalities means "two is less than or equal to x plus 2"?
    A $2 < x + 2$
    B $2 \leq x + 2$
    C $x - 2 \leq 2$
    D $2 = x - 2$

32. In $\triangle ABC$, $\angle A$ and $\angle B$ have the same measure. $\angle C$ is 30° larger than each of the other angles. Find m$\angle C$.
    A 50°
    B 30°
    C 80°
    D 150°

33. To the nearest tenth, find the volume of a sphere with a radius of 10 cm. Use 3.14 for π.
    A 3140 cm^3
    B 4186.7 cm^3
    C 1046.7 cm^3
    D 2355 cm^3

34. What is the name for a quadrilateral with all sides having equal length and angles having equal measurement?
    A parallelogram
    B square

35. The perimeter of a square is 100 inches. Find the length of one side.
    A 10 in.
    B 25 in.

36. Find the area of a circle with $d = 10$ mi. Use 3.14 for π.
    A 31.4 mi^2
    B 785 mi^2
    C 314 mi^2
    D 78.5 mi^2

37. Using the stem-and-leaf plot, find the mode.

Stem	Leaves
3	1 8
4	1 1 1 5
5	1 2 2 8 9

    A 46.27
    B 41
    C 50
    D no mode

Cumulative Test
Form A, continued

38. How much will it cost to carpet a 15 ft by 12 ft room if carpeting costs $10 per square yard?
  A $1800   **B** $200

39. Find the volume of the figure.

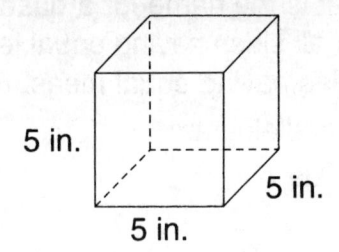

5 in., 5 in., 5 in.

  A 25 in³   **B** 125 in³

40. Express 5.25 as a fraction.
  A $7\frac{1}{8}$ **B** $5\frac{1}{4}$

41. Find the surface area of the figure.

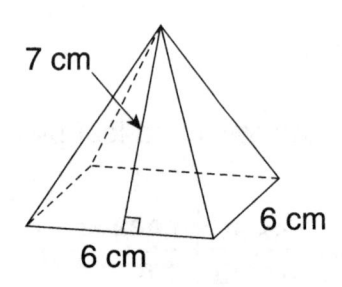

7 cm, 6 cm, 6 cm

  A 120 cm²   **C** 84 cm²
  B 252 cm²   **D** 168 cm²

42. Name the figure.

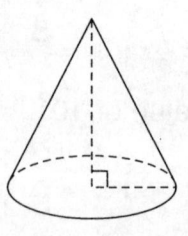

  A circle   **C** cone
  B pyramid   **D** prism

43. Find the surface area of the figure. Use 3.14 for π.

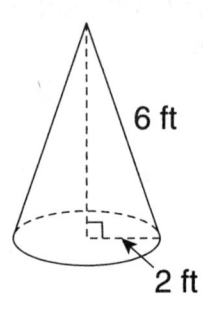

6 ft, 2 ft

  A 50.2 ft²   **B** 25.1 ft²

44. Find the area of the figure.

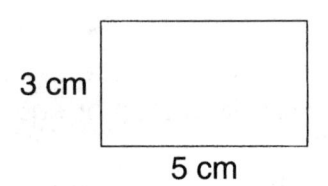

3 cm, 5 cm

  A 16 cm²   **B** 15 cm²

45. Find all roots of $\sqrt{\frac{1}{16}}$.
  A 4, −4   **B** $\frac{1}{4}, -\frac{1}{4}$

Name _____ Date _____ Class _____

CHAPTER 6 Cumulative Test
Form B

1. Simplify −3 + 17 − 1 −7.
   - A 6
   - B 28
   - C −22
   - D −3

2. Simplify −1 − (−4 + 3) + (−1).
   - F 1
   - G −5
   - H 5
   - J −1

3. An ice cream cone has a diameter of 7.7 cm and a height of 11.0 cm. Use 3.14 for π to find the volume to the nearest tenth.
   - A 170.7 cm^3
   - B 657.0 cm^3
   - C 682.6 cm^3
   - D 1971.0 cm^3

4. Find $\sqrt{808}$ to the nearest tenth.
   - F 808.0
   - G 28.4
   - H 28.3
   - J 28.422

5. Simplify −36 ÷ (−3) + (−4).
   - A −43
   - B $\frac{36}{7}$
   - C 8
   - D 16

6. Simplify 13mn − 13mn.
   - F 26mn
   - G 0
   - H −mn
   - J mn

7. Simplify −2(d + 3).
   - A −2d − 3
   - B −2d − 6
   - C −2d + 6
   - D −2d + 3

8. Evaluate $\frac{c+4}{9}$ for c = −9.
   - F $\frac{13}{9}$
   - G 4
   - H $-\frac{13}{9}$
   - J $-\frac{5}{9}$

9. Solve p − 8 = −4 + 12.
   - A p = 0
   - B p = 24
   - C p = −16
   - D p = 16

10. The prices for five meal choices in a school cafeteria are the following: $5.20, $4.49, $4.11, $3.80, $5.05. Find the range of these prices.
    - F $4.53
    - G $1.40
    - H $1.25
    - J $0.15

11. Express m•m•m•m•m•m using exponents.
    - A 6m
    - B 6m^6
    - C m^6
    - D 6^m

12. Find the measure of an angle whose supplement is 3 times the angle.
    - F 60°
    - G 45°
    - H 135°
    - J 90°

13. Express 4.23 × 10^{-5} in standard notation.
    - A 0.000423
    - B 0.00423
    - C 0.0000423
    - D 4.230000

14. Express 5,360,000 in scientific notation.
    - F 53.6 × 10^3
    - G 5.36 × 10^3
    - H 5.36 × 10^5
    - J 5.36 × 10^6

15. A cabinetmaker installs $\frac{3}{16}$-in. thick laminate material on top of $\frac{3}{4}$-in. plywood. What is the thickness of the completed countertop?
    - A $\frac{15}{16}$ in.
    - B $\frac{3}{10}$ in.
    - C $1\frac{1}{2}$ in.
    - D $\frac{1}{16}$ in.

Name _____ Date _____ Class _____

CHAPTER 6 Cumulative Test
Form B, continued

16. Find the perimeter of the polygon.

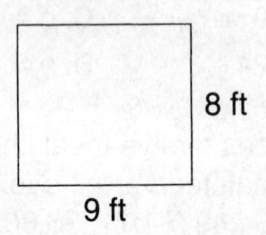

8 ft
9 ft

F 72 ft   H 32 ft
G 34 ft   J 17 ft

17. Find the average number of e-mails that Joshua received during the past week: 22, 4, 11, 38, 43, 47, 5. Round to the nearest tenth.
A 22.5   C 42
B 24.3   D 22

18. Find the missing measurement in the triangle to the nearest tenth.

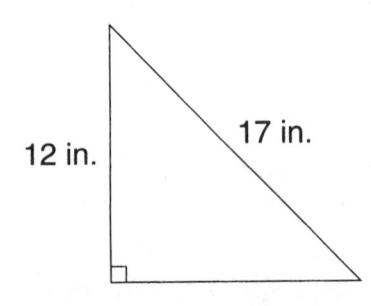

12 in.   17 in.

F 14.5 in.   H 72.5 in.
G 20.8 in.   J 12.0 in.

19. Solve the equation $4.3 = s + 7.2$.
A $s = -2.9$ C $s = 0.60$
B $s = 11.5$ D $s = 2.9$

20. Find the circumference of a circle with $d = 19$ mi to the nearest tenth. Use 3.14 for π.
F 29.8 mi   H 119.3 mi
G 59.7 mi   J 1133.5 mi

21. Ann reads $\frac{3}{4}$ of a book in 2 days. At this rate, how many books can she read in $4\frac{1}{3}$ days?
A $3\frac{1}{4}$ books   C $1\frac{1}{2}$ books
B 2 books   D $1\frac{5}{8}$ books

22. Find the sum of the angles of a regular 9-gon.
F 900°   H 1080°
G 1260°   J 1620°

23. Find the volume of the figure to the nearest tenth. Use 3.14 for π.
2 cm
6 cm
A 301.4 cm^3
B 75.4 cm^3
C 37.7 cm^3
D 25.1 cm^3

24. Find the area of the triangle.

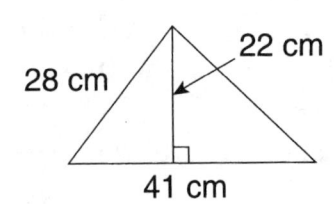

28 cm   22 cm
41 cm

F 242 cm^2 H 308 cm^2
G 902 cm^2 J 451 cm^2

Cumulative Test
Chapter 6 Form B, continued

25. Simplify $\frac{56}{-8}$.
- A 7
- B 8
- C −8
- D −7

26. Find the value of 10^4.
- F 40
- G 10,000
- H 1000
- J 100,000

27. Which of the following is the same as "five less than a number d is −40"?
- A $5 - d = -40$
- B $-5d = -40$
- C $d - 5 = -40$
- D $5 + d = -40$

28. Solve for the unknown side in the right triangle ABC: $b = 4$, $c = 5$.
- F $a = 3$
- G $a = 9$
- H $a = 6.4$
- J $a = 4.5$

29. Express $9^4 \cdot 9^9$ using positive exponents.
- A 9^{36}
- B 9^{13}
- C 81^{36}
- D 81^{13}

30. Solve $\frac{3}{5}y + \frac{2}{9} = \frac{5}{8} - \frac{2}{5}y + \frac{5}{8}$.
- F $y = \frac{53}{36}$
- G $y = \frac{29}{72}$
- H $y = \frac{37}{36}$
- J $y = \frac{53}{36}$

31. Choose an inequality that means "14 is less than or equal to x minus 2."
- A $14 < x - 2$
- B $14 \leq x - 2$
- C $x - 2 \leq 14$
- D $14 = x - 2$

32. In $\triangle ABC$, $\angle A$ and $\angle B$ have the same measure. $\angle C$ is 42° larger than each of the other angles. Find m$\angle C$.
- F 92°
- G 88°
- H 46°
- J 134°

33. To the nearest tenth, find the volume of a sphere with a diameter of 16 cm. Use 3.14 for π.
- A 539.9 cm^3
- B 2143.6 cm^3
- C 1607.7 cm^3
- D 17,148.6 cm^3

34. What is the name for a quadrilateral that is a parallelogram with all sides having equal length?
- F equilateral
- G rectangle
- H rhombus
- J trapezoid

35. The perimeter of a square is 88 inches. Find the length of one side.
- A 22 in.
- B 9.4 in.
- C 44 in.
- D 88 in.

36. To the nearest tenth find the area of a circle with $d = 19$ mi. Use 3.14 for π.
- F 1133.5 mi^2
- G 119.3 mi^2
- H 283.4 mi^2
- J 59.7 mi^2

37. Using the stem-and-leaf plot, find the mode.

Stem	Leaves
7	1 8
8	1 1 1 3 5
9	1 3 3 8 9 9
10	3 5

- A 81
- B 1
- C 91
- D no mode

Name _____ Date _____ Class _____

CHAPTER 6 Cumulative Test
Form B, continued

38. How much will it cost to paint a 15.5 ft by 12.5 ft ceiling if the painter charges $23 per square yard?
 F $495.14 H $4456.25
 G $193.75 J $1485.42

39. Find the volume of the figure.

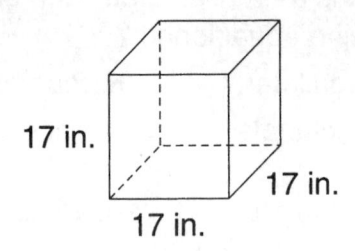

17 in.
17 in.
17 in.

 A 51 in³ C 289 in³
 B 578 in³ D 4913 in³

40. Express 7.125 as a fraction.
 F $7\frac{1}{4}$ H $7\frac{1}{8}$
 G $7\frac{1}{12}$ J $7\frac{1}{16}$

41. Find the surface area of the figure.

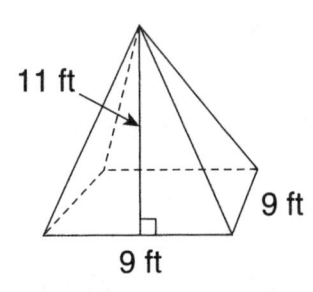

11 ft
9 ft
9 ft

 A 279 ft² C 234 ft²
 B 180 ft² D 477 ft²

42. Name the figure.

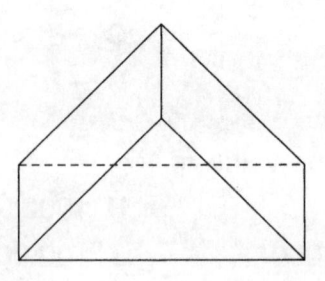

 F triangle   H prism
 G pyramid   J quadrilateral

43. Find the surface area of the figure to the nearest tenth. Use 3.14 for π.

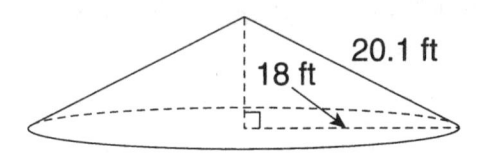

20.1 ft
18 ft

 A 2153.4 ft² C 2289.1 ft²
 B 3052.1 ft² D 1526 ft²

44. Find the area of the figure.

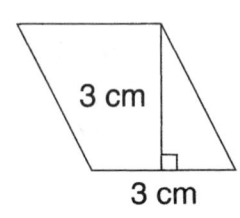

3 cm
3 cm

 F 6 cm²   H 9 cm²
 G 4.5 cm² J 12 cm²

45. Find all roots of $\sqrt{\frac{121}{16}}$.
 A 12, 11, 4   C 11, 4
 B $\frac{11}{4}, -\frac{11}{4}$ D $\frac{13}{4}, -\frac{13}{4}$

Name _____ Date _____ Class _____

CHAPTER 6 Cumulative Test
Form C

1. Simplify $-3 + 17 - (-2)$.
   - A 18
   - B 12
   - C -22
   - D 16

2. Simplify $-1 - 2(-4 + 3)$.
   - F 1
   - G -4
   - H -4
   - J -1

3. An ice cream cone has a diameter of 7.2 cm and a height of 11.2 cm. Use 3.14 for π to find the volume to the nearest tenth.
   - A 151.9 cm³
   - B 455.8 cm³
   - C 506.4 cm³
   - D 607.7 cm³

4. Find $\sqrt{1808}$ to the nearest tenth.
   - F 1808
   - G 32,400
   - H 904
   - J 42.5

5. Simplify $-36 \div (-12) \cdot (-4)$.
   - A $-\frac{3}{4}$
   - B -12
   - C $\frac{3}{4}$
   - D 12

6. Simplify $13mn - 13mn^2$.
   - F $26mn$
   - G 0
   - H $-mn^2$
   - J $13mn - 13mn^2$

7. Simplify $-2(-d^2 - 3)$.
   - A $2d^2 + 6$
   - B $2d - 6$
   - C $-2d^2 + 6$
   - D $-2d + 3$

8. Evaluate $\frac{2c + 5}{18}$ for $c = -9$.
   - F $\frac{5}{8}$
   - G 5
   - H $-\frac{5}{9}$
   - J $-\frac{13}{18}$

9. Solve $2p - 8 = -3p + 12$.
   - A $p = 4$
   - B $p = -4$
   - C $p = -20$
   - D $p = 20$

10. Jack records the average temperature for five days during the winter and obtains the following: 23.8°F, 10°F, 0.7°F, −5.3°F, 15.5°F. Find the range of these temperatures.
    - F 8.3°F
    - G 29.1°F
    - H 8.94°F
    - J 18.5°F

11. Express $m \cdot m \cdot m \cdot n \cdot n \cdot n$ using exponents.
    - A $6m$
    - B $3mn$
    - C $m^3 n^3$
    - D 3^{mn}

12. Find the measure of an angle whose complement is 3 times the angle.
    - F 22.5°
    - G 67.5°
    - H 45°
    - J 135°

13. Express 1.413×10^{-5} in standard notation.
    - A 0.0001413
    - B 0.001413
    - C 0.00001413
    - D 14.130000

14. Express 50,600,000 in scientific notation.
    - F 50.6×10^6
    - G 5.06×10^3
    - H 5.06×10^5
    - J 5.06×10^7

15. A chef spreads a $\frac{1}{4}$-in. thick layer of icing on top of a cake that is $2\frac{2}{3}$-in. in height. What is the height of the completed cake?
    - A $2\frac{11}{12}$ in.
    - B $3\frac{1}{4}$ in.
    - C $\frac{11}{12}$ in.
    - D $2\frac{7}{8}$ in.

Cumulative Test
Chapter 6, Form C, continued

16. Find the perimeter of the polygon.

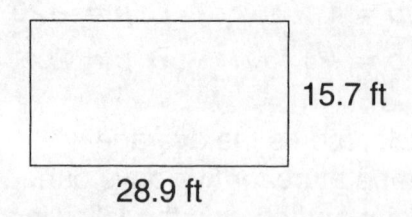

15.7 ft
28.9 ft

- F 44.6 ft
- G 89.2 ft
- H 112.3 ft
- J 453.7 ft

17. To the nearest tenth find the average number of e-mails that Joshua received during the past two weeks: 22, 4, 0, 11, 38, 43, 47, 0, 5, 10, 10, 0, 10, 11. Round to the nearest tenth.
- A 19.2
- B 15.1
- C 10
- D 105.5

18. Find the missing measurement in the triangle to nearest tenth.

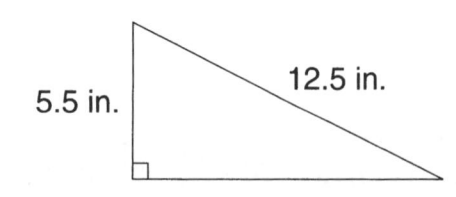

5.5 in.   12.5 in.

- F 9.0 in.
- G 34.4 in.
- H 11.2 in.
- J 13.7 in.

19. Solve the equation $8.62 = s + 14.5$.
- A $s = -5.88$
- B $s = 23.12$
- C $s = 0.60$
- D $s = 5.88$

20. Find the circumference of a circle with $d = 20.2$ mi to the nearest tenth. Use 3.14 for π.
- F 126.9 mi
- G 63.4 mi
- H 320.3 mi
- J 31.7 mi

21. Mark runs $3\frac{1}{4}$ miles in $\frac{1}{3}$ of an hour. At this rate, how long would it take him to run 6 miles?
- A $\frac{8}{13}$ hr
- B $6\frac{1}{2}$ hr
- C $1\frac{1}{12}$ hr
- D $\frac{2}{3}$ hr

22. Find the sum of the angles of a regular 11-gon.
- F 1980°
- G 1260°
- H 1080°
- J 1620°

23. Find the volume of the figure to the nearest tenth. Use 3.14 for π.

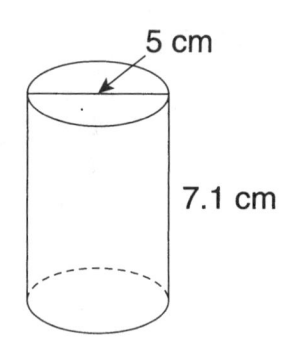

5 cm
7.1 cm

- A 557.4 cm³
- B 139.3 cm³
- C 791.4 cm³
- D 137.4 cm³

24. Find the area of the triangle to the nearest tenth.

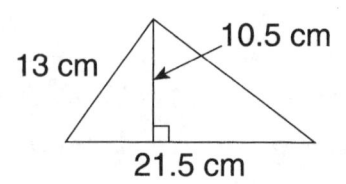

13 cm   10.5 cm
21.5 cm

- F 64.5 cm²
- G 225.8 cm²
- H 112.9 cm²
- J 451.9 cm²

CHAPTER 6 Cumulative Test
Form C, continued

25. Simplify $\frac{-56}{-16}$.
   - A 5
   - B $\frac{56}{16}$
   - C -5
   - D $3\frac{1}{2}$

26. Find the value of 10^6.
   - F 60
   - G 100,000
   - H 1000
   - J 1,000,000

27. Which of the following is the same as "five less than twice a number d is negative 4"?
   - A $5 - 2d = -4$
   - B $2d - 5 = -4$
   - C $2d - 5 = 4$
   - D $5 + 2d = -4$

28. To the nearest tenth, solve for the unknown side in the right triangle ABC: $a = 9$, $c = 11.4$.
   - F $b = 7.0$
   - G $b = 9$
   - H $b = 14.5$
   - J $b = 10.2$

29. Express $9^4 \cdot 9^9 \cdot 9^0$ using positive exponents.
   - A 9^{36}
   - B 9^{13}
   - C 0
   - D 81^{13}

30. Solve $\frac{3}{5}y + \frac{2}{9} = \frac{5}{8} - \frac{2}{5}y + \frac{5}{9}$.
   - F $y = \frac{115}{24}$
   - G $y = -\frac{11}{72}$
   - H $y = \frac{23}{24}$
   - J $y = -\frac{55}{72}$

31. Choose the inequality for "four is greater than two times x minus 2."
   - A $4 > 2x - 2$
   - B $4 \leq 2x - 2$
   - C $2x - 2 \leq 4$
   - D $4 = 2x - 2$

32. In $\triangle ABC$, $\angle A$, $\angle B$, and $\angle C$ have a 1:2:3 ratio to each other. Find m$\angle C$.
   - F 30°
   - G 90°
   - H 60°
   - J 135°

33. To the nearest tenth, find the volume of a sphere with a diameter of 21 cm. Use 3.14 for π.
   - A 3247.2 cm^3
   - B 4846.6 cm^3
   - C 3634.9 cm^3
   - D 38,772.7 cm^3

34. What is the name for a quadrilateral that has exactly one pair of parallel sides?
   - F equilateral
   - G rectangle
   - H rhombus
   - J trapezoid

35. The perimeter of a square is 188 inches. Find the length of one side.
   - A 13.7 in.
   - B 94 in.
   - C 47 in.
   - D 3.7 in.

36. To the nearest tenth, find the area of a circle with $d = 20.2$ mi. Use 3.14 for π.
   - F 63.4 mi^2
   - G 126.9 mi^2
   - H 320.3 mi^2
   - J 1281.2 mi^2

37. Using the stem-and-leaf plot, find the mode.

Stem	Leaves
7	1 8
8	1 2 3 5 6
9	1 3 8 9
10	3 5

   - A 86
   - B 1
   - C 90
   - D no mode

Name _____ Date _____ Class _____

CHAPTER 6 Cumulative Test
Form C, continued

38. How much will it cost to build a 57.5 ft by 72.25-ft skating rink if the builders charge $27.10 per square yard?

F $37,527.85   H $112,583.56
G $12,509.28   J $4154.38

39. Find the volume of the figure.

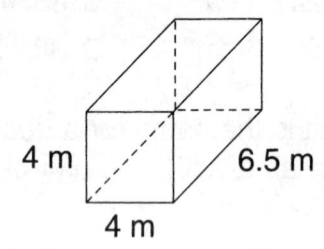

4 m   6.5 m
4 m

A 22.5 m³   C 26 m³
B 136 m²   D 104 m³

40. Express 17.125 as a fraction.

F $7\frac{1}{8}$ H $17\frac{1}{8}$
G $17\frac{1}{12}$ J $17\frac{1}{16}$

41. Find the surface area of the figure.

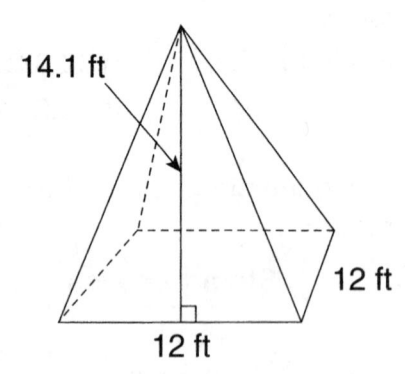

14.1 ft
12 ft
12 ft

A 482.4 ft²   C 314.4 ft²
B 388.8 ft²   D 338.4 ft²

42. Name the figure.

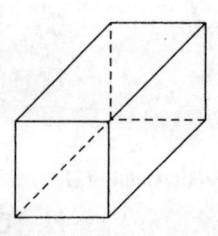

F triangle   H prism
G pyramid   J quadrilateral

43. Find the surface area of the figure. Use 3.14 for π.

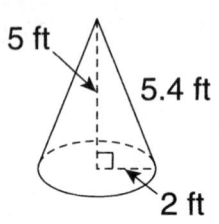

5 ft
5.4 ft
2 ft

A 46.5 ft²   C 20.9 ft³
B 54 ft²   D 102.4 ft³

44. Find the area of the figure to the nearest tenth.

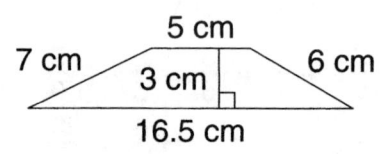

7 cm   5 cm   6 cm
3 cm
16.5 cm

F 32.3 cm²   H 46.5 cm²
G 32.5 cm²   J 93.0 cm²

45. Find all roots of $\sqrt{\dfrac{169}{81}}$.

A 13, 16, 9   C 16, 9
B $\dfrac{13}{9}, -\dfrac{13}{9}$ D $\dfrac{16}{9}, -\dfrac{16}{9}$

Name _____ Date _____ Class _____

CHAPTER 7 Cumulative Test
Form A

1. Evaluate $2ab$ for $a = 3$ and $b = -1$.
   - **A** 7
   - **B** 5
   - **C** −6
   - **D** 6

2. Solve: $m + 4 = -4$.
   - **A** $m = 0$
   - **B** $m = -16$
   - **C** $m = 8$
   - **D** $m = -8$

3. A length on a map is 4 in. The scale is 1 in:1.5 mi. What is the actual distance?
   - **A** 5.5 mi
   - **B** 6 mi

4. Solve $2(s - 3) = 6$.
   - **A** $s = 6$
   - **B** $s = 0$

5. A photograph 5 in. wide by 7 in. high is to be enlarged to 12.5 in. wide. How high will the enlarged photo be?
   - **A** 14.5 in.
   - **B** 17.5 in.
   - **C** 19.5 in.
   - **D** 8.9 in.

6. Which algebraic expression means "73 times a number"?
   - **A** $73n$
   - **B** $\frac{73}{n}$

7. Find the area of a rectangular poster measuring $\frac{1}{2}$ ft by $1\frac{1}{4}$ ft.
   - **A** $\frac{3}{4}$ ft^2
   - **B** $1\frac{3}{4}$ ft^2
   - **C** $3\frac{1}{2}$ ft^2
   - **D** $\frac{5}{8}$ ft^2

8. You received science test scores of 70, 40, 70, 90, and 40. What was your mean score?
   - **A** 62
   - **B** 70
   - **C** 66.7
   - **D** 40

9. A block of wood has dimensions of 4 in. × 2 in. × 3 in. What is its volume?
   - **A** 9 in^3
   - **B** 24 in^3

10. A poultry farm hatched the following number of chicks during the past week: 50, 10, 10, 40, 30, 25, 20. What was the median number of chicks hatched?
    - **A** 25
    - **B** 26.4

11. Find $\sqrt{16}$.
    - **A** 4
    - **B** 8

12. Solve for the unknown side in the right triangle.

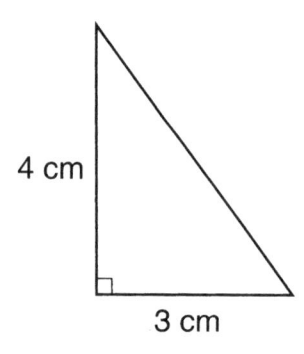

    4 cm
    3 cm

    - **A** 5 cm
    - **B** 12 cm
    - **C** 3 cm
    - **D** 4 cm

13. Solve $2.5 = n + 6.5$.
    - **A** $n = 4$
    - **B** $n = -4$

217

Holt Pre-Algebra

Cumulative Test
Form A, continued

14. Marla's pay for a week was $84.12. If she worked 12 hr, what was her rate of pay?
    A $7.01 per hr   B $8.59 per hr

15. Find the unknown number in the proportion: $\frac{4}{f} = \frac{10}{8}$.
    A $f = 3.2$ B $f = 8$

16. A 9-in. cube is built from small cubes, each 3 in. on a side. The volume of the large cube is how many times that of a small cube?
    A 3   B 27

17. Convert 12 cups to pints.
    A 6 pt   C 4 pt
    B 24 pt   D 2 pt

18. Express 3 • 3 • 3 • 3 in simplest form using exponents.
    A 3^{27} C 24
    B 9^4 D 3^4

19. Simplify $(2)^{-3}$.
    A $-\frac{1}{16}$ B $\frac{1}{8}$

20. The population of a small town is 45,000. Express this number using scientific notation.
    A 4.5×10^4 B 4.5×10^3

21. You received scores of 70, 40, 70, 90, and 40. Find the first quartile.
    A 40   B 70

22. In the figure, which angle is a supplement to $\angle ABE$?

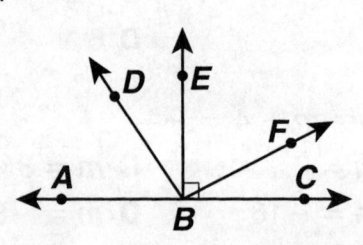

    A $\angle DBE$ B $\angle EBC$

23. Find the unknown length in the pair of similar triangles.

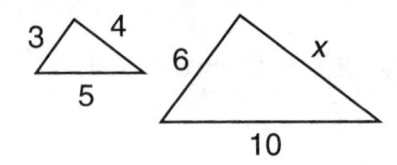

    A 8   C 9
    B 4   D 12

24. Find the volume of the figure.

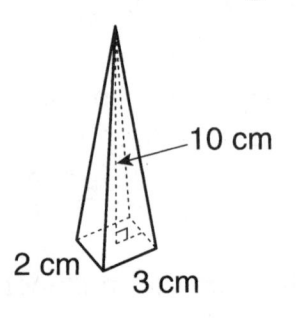

    A 60 cm³   B 20 cm³

25. Which angle is congruent to $\angle AOD$?

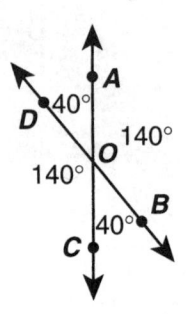

    A $\angle BOC$ B $\angle AOB$

Cumulative Test

Chapter 7 Form A, continued

26. Simplify $\dfrac{d^5}{d^5}$.

　A d^{10} 　　　　**B** 1

27. What are the coordinates of point A on the graph?

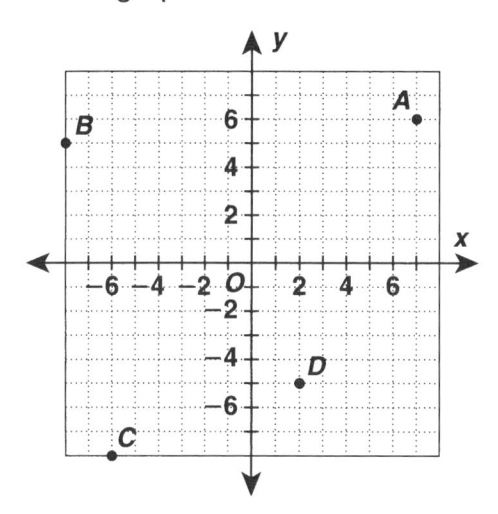

　A (5, −8) 　　　**C** (7, 6)
　B (−7, 6) 　　　**D** (6, 7)

28. Simplify $1\dfrac{1}{3} - \dfrac{5}{12}$.

　A $\dfrac{11}{12}$ 　　　　**B** $\dfrac{5}{12}$

29. Triangle ABC is similar to triangle DEF. Find x.

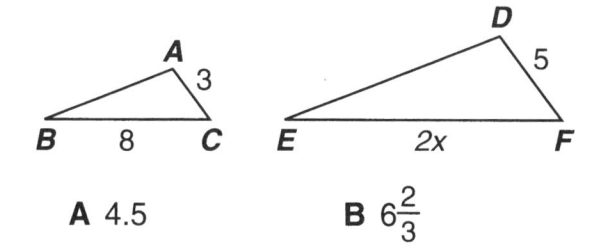

　A 4.5 　　　　**B** $6\dfrac{2}{3}$

30. The graph represents which inequality?

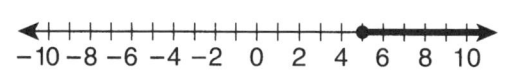

　A $x \geq -5$ 　　**B** $x \geq 5$

31. Simplify $2\dfrac{3}{8} + \dfrac{1}{4}$.

　A $2\dfrac{1}{3}$ 　　　**C** $2\dfrac{5}{8}$
　B $2\dfrac{2}{3}$ 　　　**D** $2\dfrac{3}{32}$

32. In △ABC, ∠A and ∠B have the same measure. m∠C is 90°. Find m∠A.

　A 45° 　　　　**B** 90°

33. What is the term for part of a line that is between two points?

　A point 　　　**C** line
　B ray 　　　　**D** line segment

34. Which of the following graphs represents $\dfrac{x}{3} \leq 3$?

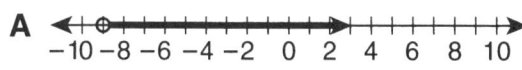

　A, **B**, **C**, **D**

35. In a town of 4500 voters, a pollster asks 400 people if they will support the next school levy. Identify the population.

　A 4500 voters
　B All people in the town
　C All school children
　D 400 people surveyed

219　Holt Pre-Algebra

Name _____ Date _____ Class _____

CHAPTER 7 Cumulative Test
Form A, continued

36. Name the figure.

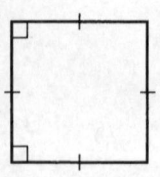

    A pentagon **C** hexagon
    B octagon   **D** square

37. Which of the following pairs of ratios forms a proportion?

    A $\frac{3}{4}$ and $\frac{6}{8}$

    B $\frac{4}{6}$ and $\frac{5}{8}$

38. Classify $-\sqrt{11}$ and give the approximate value.

    A rational; −3.3   **B** irrational; −3.3

39. If a drawing has a scale 1 cm:1 mi, is the drawing a reduction or an enlargement of the actual object?

    A enlargement   **B** reduction

40. Express 1.5 as a fraction.

    A $1\frac{1}{5}$ **C** $1\frac{1}{2}$

    B $1\frac{1}{4}$ **D** $1\frac{1}{10}$

41. Find $\sqrt{20}$ to the nearest hundredth.

    A 4.47   **B** 4.52

42. Solve $3c = -9$.

    A $c = 3$ **C** $c = 27$
    B $c = -3$ **D** $c = -27$

43. Solve $2(s - 2) + 2s$.

    A $4s - 10$ **C** $4s - 4$
    B $8s$ **D** $4s - 2$

44. Determine which of the following is a solution to $y = x - 3$.

    A (−1, −1)   **C** (2, 3)
    B (4, 1)   **D** (1, 2)

45. Find $(2 + 1)^2$.

    A 5   **B** 9

46. When dilating a triangle, using a scale factor greater than 1 will:

    A enlarge the triangle.
    B shrink the triangle.

Name _____ Date _____ Class _____

CHAPTER 7 Cumulative Test
Form B

1. Evaluate $6xy + 3$ for $x = -5$ and $y = 3$.
   - A -90
   - B 24
   - C 87
   - D -87

2. Solve $m + 5 = 11$.
   - F $m = 16$
   - G $m = -16$
   - H $m = 6$
   - J $m = -6$

3. The scale on a map is 1 in.:2.5 mi. What is the length of 42 miles on the map?
   - A 105 in.
   - B 16.8 in.
   - C 44.5 in.
   - D 19.3 in.

4. Solve $8(s - 3) = 24$.
   - F $s = 6$
   - G $s = 0$
   - H $s = 3$
   - J $s = -3$

5. A picture 6 in. wide by 9 in. high is enlarged to 15 in. wide. What is the scale factor for the enlargement?
   - A 6
   - B 2
   - C 2.5
   - D 9

6. Which algebraic expression means "73 minus a number"?
   - F $n - 73$
   - G $-n + 73$
   - H $-73 - n$
   - J $73 - n$

7. Find the area of a table top measuring $\frac{1}{3}$ ft by $\frac{2}{7}$ ft.
   - A $\frac{3}{7}$ ft^2
   - B $\frac{3}{10}$ ft^2
   - C $\frac{3}{21}$ ft^2
   - D $\frac{2}{21}$ ft^2

8. Bill received tips of $73, $48, $73, $92, and $48 over the past 5 nights. What were his mean nightly tips?
   - F $66.80
   - G $73
   - H $70
   - J $71

9. A small box has dimensions of 5 in. × 2 in. × 9 in. What is its volume?
   - A 225 in^3
   - B 32 in^3
   - C 20 in^3
   - D 90 in^3

10. A restaurant served the following number of steaks during the past week: 50, 9, 7, 42, 33, 25, 20. What was the median number of steaks served?
    - F 33
    - G 26.6
    - H 42
    - J 25

11. Find $\sqrt{43}$ and round your answer to the nearest hundredth.
    - A 43.0
    - B 6.5
    - C 6.56
    - D 6.55

12. Solve for the unknown side in the right triangle.
    - F 20 cm
    - G 19 cm
    - H 14 cm
    - J 16 cm

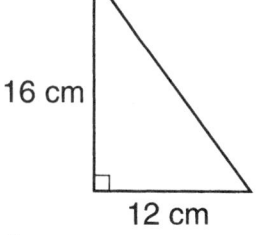

    (16 cm, 12 cm)

13. Solve $7.8 = n + 6.6$.
    - A $n = 14.4$
    - B $n = 1.18$
    - C $n = 1.2$
    - D $n = -1.2$

14. Jen's pay for a week was $82.20. If she worked 12 hr, what was her rate of pay?
    - F $6.85 per hr
    - G $6.73 per hr
    - H $6.98 per hr
    - J $7.29 per hr

Name _____ Date _____ Class _____

CHAPTER 7 Cumulative Test
Form B, continued

15. Find the unknown number in the proportion: $\frac{5}{f} = \frac{20}{8}$.
    A $f = 0.08$
    B $f = 20$
    C $f = 2$
    D $f = 12.5$

16. A 15 cm cube is built from small cubes, each 3 cm on a side. The surface area of the large cube is how many times that of a small cube?
    F 5
    G 12
    H 9
    J 25

17. Convert 24 pints to cups.
    A 6 c
    B 48 c
    C 3 c
    D 12 c

18. Express 4 • 4 • 4 • 4 • 4 • 4 using exponents.
    F 4^6
    G 6^4
    H 24
    J 4^5

19. Find $(-2)^{-4}$.
    A 16
    B $-\frac{1}{16}$
    C -16
    D $\frac{1}{16}$

20. The population of a small country is 5,740,000. Express this number using scientific notation.
    F 5.74×10^6
    G 5.74×10^5
    H 5.74×10^4
    J 57.4×10^4

21. You received scores of 73, 48, 73, 92, and 48. Find the first quartile.
    A 44
    B 48
    C 60.5
    D 73

22. In the figure, which angle is a complement to $\angle EBF$?

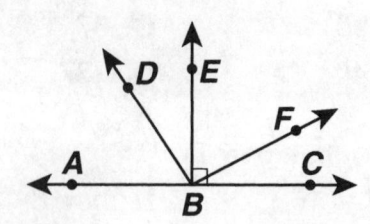

    F $\angle DBE$
    G $\angle EBA$
    H $\angle CBF$
    J $\angle ABD$

23. Find the unknown length in the pair of similar triangles.

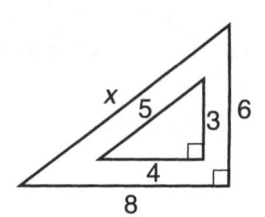

    A 11
    B 5
    C 15
    D 10

24. Find the volume of the figure.

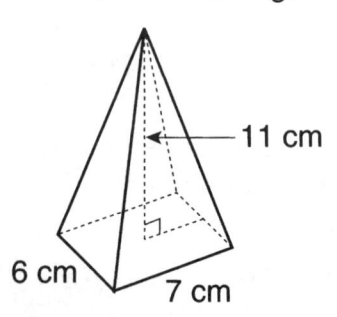

    F 185 cm^3
    G 154 cm^3
    H 90 cm^3
    J 462 cm^3

Name _____ Date _____ Class _____

CHAPTER 7 Cumulative Test
Form B, continued

25. Identify a pair of congruent angles.

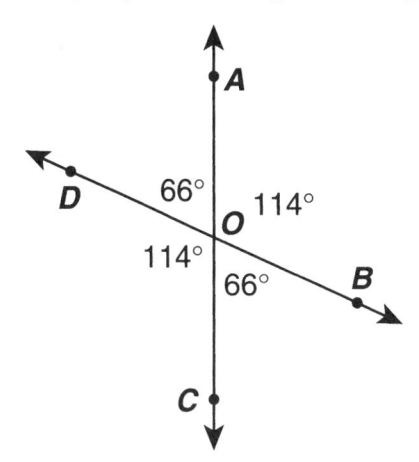

A ∠AOD and ∠BOC
B ∠AOD and ∠AOB
C ∠BOC and ∠COD
D ∠AOB and ∠BOC

26. Simplify $\dfrac{d^3}{d^5}$.

F d^2
G $\dfrac{1}{d^2}$
H d^{-5}
J $\dfrac{d^3}{d^2}$

27. What are the coordinates of point B on the graph?

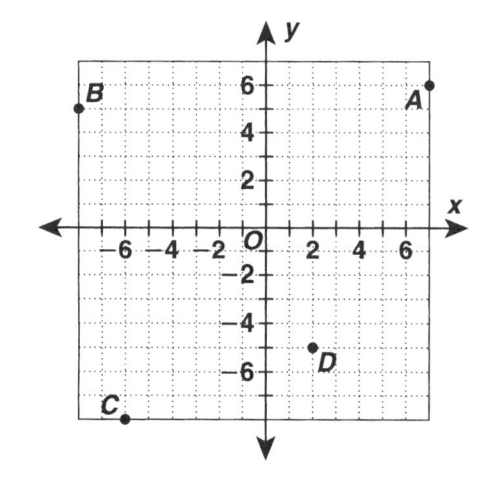

A (5, −8)
B (5, 8)
C (−8, 5)
D (8, 5)

28. Simplify $1\dfrac{2}{3} - \dfrac{11}{18}$.

F $\dfrac{19}{18}$
G $\dfrac{3}{7}$
H $\dfrac{10}{21}$
J $1\dfrac{4}{21}$

29. Triangle ABC ≅ triangle DEF. Find z.

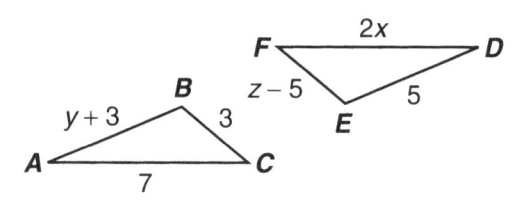

A 2
B 8
C 3
D 5

30. The graph represents which inequality?

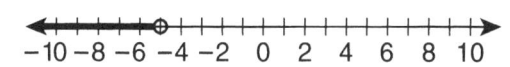

F $x \geq -5$
G $x > -5$
H $x < 5$
J $x < -5$

31. Simplify $2\dfrac{5}{6} + \dfrac{1}{3}$.

A $3\dfrac{7}{12}$
B $3\dfrac{1}{6}$
C $3\dfrac{1}{3}$
D $2\dfrac{1}{3}$

32. In △ABC, ∠A and ∠B have the same measure. m∠C is 92°. Find m∠A.

F 44°
G 88°
H 46°
J 134°

33. What is the term for part of a line that starts at one point and extends infinitely in one direction?

A point
B ray
C line
D line segment

CHAPTER 7 Cumulative Test
Form B, continued

34. Which of the following graphs represents: $\frac{x}{3} \geq -3$?

F. (number line -10 to 10, open circle at -8, arrow right)
G. (number line -10 to 10, closed circle at -8, arrow right)
H. (number line -10 to 10, closed circle at 4, arrow left)
J. (number line -10 to 10, closed circle at 8, arrow right)

35. In a town of 5500 voters, a pollster asks 500 people if they will support the challenger in the mayoral race. Identify the population.
A 500 people surveyed
B All people in the town
C All supporters
D 5500 voters

36. Name the figure.
F pentagon
G octagon
H hexagon
J heptagon

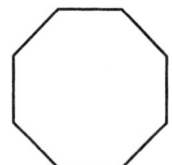

37. Which of the following pairs of ratios forms a proportion?
A $\frac{3}{8}$ and $\frac{6}{16}$ C $\frac{1}{2}$ and $\frac{2}{3}$
B $\frac{4}{9}$ and $\frac{5}{10}$ D $\frac{11}{20}$ and $\frac{1}{2}$

38. Classify $-\sqrt{25}$ as rational, irrational, or not real. If rational or irrational, give the approximate equivalent value.
F rational; −5   H irrational; 5
G irrational; −5   J not real

39. Choose which scale preserves the size of the actual object.
A 12 in.:6 ft   C 1 m:100 mm
B 1 cm:1 mi   D 36 in.:1 yd

40. Express 2.25 as a fraction.
F $2\frac{2}{5}$ H $2\frac{1}{4}$
G $2\frac{1}{25}$ J $2\frac{1}{5}$

41. Find $-\sqrt{21}$ to the nearest tenth.
A 2.1   C 10.5
B −4.6   D −441.0

42. Solve $3(3n) = -9$.
F $n = 1$ H $n = 9$
G $n = -1$ J $n = -9$

43. Solve $5(s - 2) - s$.
A $5s - 10$ C $4s - 10$
B $4s - 2$ D $6s - 2$

44. Determine which of the following is a solution to $y = 2x - 3$.
F (−1, −1)   H (2, 3)
G (1, −1)   J (1, 2)

45. Find $(-2)^{-2}$.
A 8   C 4
B $-\frac{1}{4}$ D $\frac{1}{4}$

46. When dilating a triangle, to enlarge the triangle you must use a scale factor that is:
F > 1   H = 1
G < 1   J = 0

Name _____ Date _____ Class _____

CHAPTER 7 Cumulative Test
Form C

1. Evaluate $12m + 15n$, for $m = -2$ and $n = -3$.
   - A −21
   - B −66
   - C 54
   - D −69

2. Solve $p + 5 = 11 - p$.
   - F $p = 8$
   - G $p = 3$
   - H $p = 6$
   - J $p = -16$

3. If a scale of 1 cm:3.5 m is used, how long is an object 4.8 cm long in a drawing?
   - A 8.3 m
   - B 16.8 m
   - C 1.37 m
   - D 20.3 m

4. Solve: $5(3s - 3) = 25$.
   - F $s = 1.9$
   - G $s = 2\frac{2}{3}$
   - H $s = 5$
   - J $s = 25$

5. A scale factor of 3.2 is used to enlarge a 15 cm by 30 cm picture. What are the dimensions of the enlargement?
   - A 18.2 cm × 33.2 cm
   - B 45 cm × 95 cm
   - C 48 cm × 96 cm
   - D 4.7 cm × 9.4 cm

6. Which algebraic expression means "44 less than a number"?
   - F $44 - n$
   - G $-n + 44$
   - H $-44 - n$
   - J $n - 44$

7. Find the area of a mirror measuring $1\frac{1}{3}$ ft by $\frac{7}{12}$ ft.
   - A $\frac{7}{9}$ ft^2
   - B $1\frac{3}{4}$ ft^2
   - C $1\frac{11}{12}$ ft^2
   - D $2\frac{2}{7}$ ft^2

8. You received quiz scores of 7.3, 4.8, 7.3, 9.2, and 4.8. What was your mean score rounded to the nearest tenth?
   - F 6.7
   - G 7.3
   - H 7.1
   - J 33.4

9. A locker has dimensions of 5.2 in. × 2.5 in. × 9 in. What is its volume?
   - A 16.7 in^3
   - B 22.5 in^3
   - C 117 in^3
   - D 243.4 in^3

10. A dentist treated the following number of patients during the past week: 50, 19, 17, 32, 33, 25, 20. What was the median number of patients treated?
    - F 25
    - G 28
    - H 32
    - J 33

11. Find $-\sqrt{142}$ and round your answer to the nearest tenth.
    - A −3.5
    - B −11.8
    - C −11.9
    - D 12.0

12. Solve for the unknown side in the right triangle. Round to the nearest tenth.
    - F 14.0 cm
    - G 9.2 cm
    - H 19.9 cm
    - J 28.0 cm

    15.5 cm, 12.5 cm

13. Solve $17.8 - n = n + 6.6$.
    - A $n = -12.2$
    - B $n = 5.6$
    - C $n = 11.2$
    - D $n = -11.2$

Name _____ Date _____ Class _____

CHAPTER 7 Cumulative Test
Form C, continued

14. Sarah's pay for two weeks was $158.16. If she worked 8 hours the first week and 4 hours the second, what was her rate of pay?
    F $3.23 per hr   H $22.60 per hr
    G $13.18 per hr   J $31.64 per hr

15. Find the unknown number in the proportion $\frac{5}{f} = \frac{20}{4.4}$.
    A $f = 1.1$ C $f = 11$
    B $f = 8$ D $f = 22.7$

16. A 2-m cube is built from small cubes, each 5 cm on a side. The volume of the large cube is how many times that of a small cube?
    F 40,000   H 64,000
    G 1600   J 16,000

17. Convert 24 cups to quarts.
    A 3 qt   C 8 qt
    B 6 qt   D 24 qt

18. Express $(5 \cdot 5 \cdot 5 \cdot 5) \cdot (2 \cdot 2 \cdot 2)$ using exponents.
    F 24^7 H $5^6 \cdot 2^6$
    G $(2 \cdot 5)^7$ J $5^4 \cdot 2^3$

19. Find $(-3)^{-3}$.
    A $\frac{1}{27}$ C $-\frac{1}{27}$
    B 27   D -27

20. The population of a large city is 3,040,000. Express this number using scientific notation.
    F 3.04×10^6 H 3.04×10^4
    G 3.04×10^5 J 30.4×10^7

21. What is the third quartile for 730, 480, 730, 920, and 480?
    A 440   C 668
    B 480   D 825

22. In the figure, which angle is a complement to $\angle EBD$?

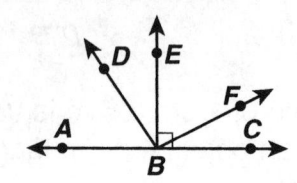

    F $\angle DBE$ H $\angle CBF$
    G $\angle EBC$ J $\angle ABD$

23. Find the unknown length in the pair of similar triangles.

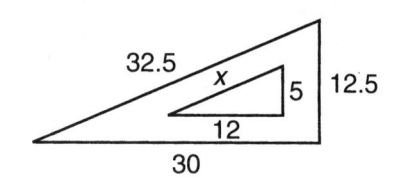

    A 15   C 13
    B 20   D 11

24. Find the height of the figure, if the volume is 154 cm³.
    F 33 cm
    G 11 cm
    H 3.7 cm
    J 9 cm

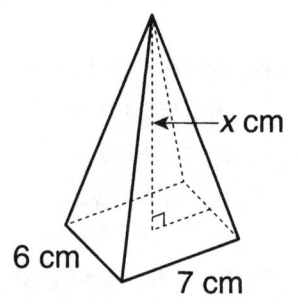

Cumulative Test
Form C, continued

25. Identify a pair of congruent angles.

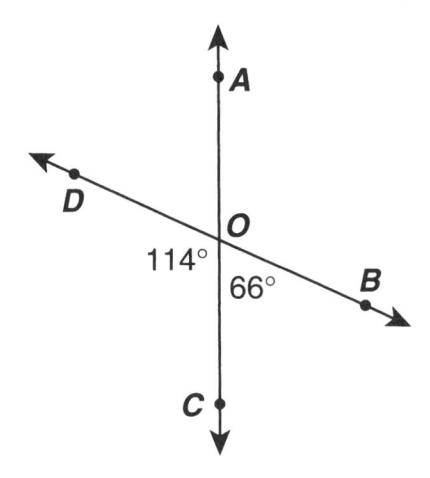

A $\angle AOD$ and $\angle COD$
B $\angle AOD$ and $\angle AOB$
C $\angle BOC$ and $\angle COD$
D $\angle AOB$ and $\angle COD$

26. Simplify $\dfrac{x^2 d}{d^6 x}$.

F 0
G $\dfrac{x}{d^5}$
H $d^5 x$
J 1

27. What are the coordinates of point C on the graph?

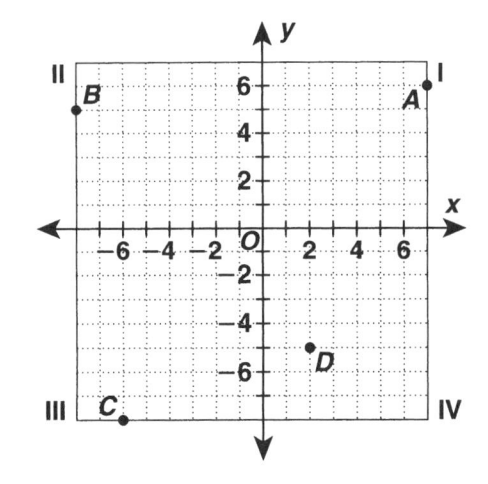

A (6, −8)
B (6, 8)
C (−6, −8)
D (8, 6)

28. Simplify $2\dfrac{2}{3} - 1\dfrac{5}{8}$.

F $\dfrac{8}{17}$
G $\dfrac{25}{24}$
H $1\dfrac{8}{9}$
J $4\dfrac{1}{2}$

29. Triangle ABC ≅ triangle DEF. Find y.

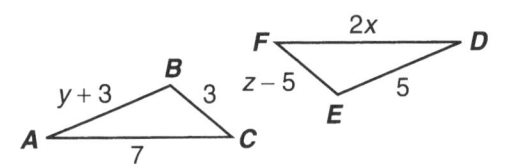

A 2
B 3
C 8
D 15

30. The graph represents which inequality?

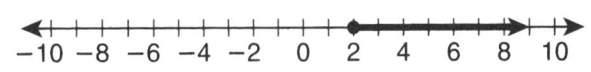

F $-x \geq 2$
G $x \leq 2$
H $-x + 4 \leq x$
J $x < 2$

31. Simplify $2\dfrac{5}{7} + 1\dfrac{1}{3}$.

A $2\dfrac{5}{21}$
B $4\dfrac{1}{21}$
C $3\dfrac{7}{10}$
D $3\dfrac{9}{10}$

32. In △ABC, ∠A and ∠B have the same measure. m∠C is 15° more than the other angles. Find m∠A.

F 55°
G 82.5°
H 60°
J 90°

33. Which is the term for a flat surface that extends infinitely in all directions?

A ray
B plane
C line
D line segment

Name _____ Date _____ Class _____

CHAPTER 7 Cumulative Test
Form C, continued

34. Which of the following graphs represents: $\dfrac{-x}{9} \geq -1$

F ← −10−8−6−4−2 0 2 4 6 8 10 →
G ← −10−8−6−4−2 0 2 4 6 8 10 →
H ← −10−8−6−4−2 0 2 4 6 8 10 →
J ← −10−8−6−4−2 0 2 4 6 8 10 →

35. In a town of 15,500 voters, a pollster asks 1500 people if they will support construction of a new landfill. Identify the sample.
 A 15,500 voters
 B All trash producers
 C All people in the town
 D 1500 people surveyed

36. Name the figure.
 F trapezoid
 G octagon
 H hexagon
 J rectangle

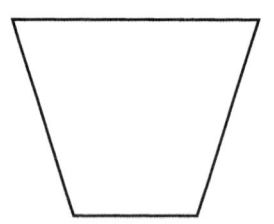

37. Which of the following pairs of ratios forms a proportion?
 A $\dfrac{3.5}{5.5}$ and $\dfrac{7}{11}$
 B $\dfrac{4}{9}$ and $\dfrac{2.5}{5}$
 C $\dfrac{2}{5}$ and $\dfrac{2}{3}$
 D $\dfrac{11}{20}$ and $\dfrac{1.1}{20}$

38. Classify $-\sqrt{-25}$ as rational, irrational, or not real. If rational or irrational, give the approximate equivalent value.
 F rational; −5
 G irrational; −5
 H irrational; 5
 J Not real

39. Choose the one which enlarges the most.
 A 1 cm:1 m
 B 2 m:15.5 km
 C 1.3 in.:0.8 mi
 D 4 in.:2 mi

40. Express 12.125 as a fraction.
 F $12\dfrac{2}{5}$
 G $12\dfrac{1}{25}$
 H $12\dfrac{1}{8}$
 J $12\dfrac{1}{5}$

41. Find $-\sqrt{110}$ to the nearest tenth.
 A −11
 B −10.5
 C not real
 D −12

42. Solve $3.5(3g) = -19.95$.
 F $g = -1.9$
 G $g = 5.7$
 H $g = 7.7$
 J $g = 6.65$

43. Simplify $2.5(s - 2) - 2.5s$.
 A −5
 B −2
 C $5s - 5$
 D $5s - 2$

44. Determine which of the following is a solution to $2y = 2x - 3$.
 F $(-1, -1)$
 G $(1, -\dfrac{1}{2})$
 H $(2, 3)$
 J $(1, 2)$

45. Find $(-3)^{-2} + (3)^2$.
 A 0
 B 1
 C −1
 D $9\dfrac{1}{9}$

46. When dilating a triangle, which of the following scale factors will reduce the image of the initial triangle?
 F $s > 1$
 G $s < 0$
 H $s = 1$
 J $0 < s < 1$

Name _____ Date _____ Class _____

CHAPTER 8 Cumulative Test
Form A

1. What is 9.8% as a decimal?
   A 0.098   **C** 0.98
   B 9.8   **D** 98

2. Allison needs 15 feet of ribbon for a craft project. How many yards is this?
   A 5 yd   **B** 45 yd

3. Simplify using exponents: $5^5 \cdot 5^4$.
   A 5^9 **B** 25^{20}

4. Determine which of the following is a solution to $y - 2 = 2x$.
   A (1, 4)   **B** (0, 1)

5. In the figure, $\triangle ABC$ is similar to $\triangle GHJ$. Find length x.
   A 5
   B 15

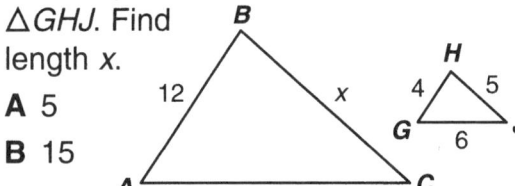

6. Solve $f - 10 = 40 - f$.
   A $f = 15$ **C** $f = 25$
   B $f = 30$ **D** $f = 50$

7. Find the mode of these numbers.
   7, 7, 2, 7, 13, 5, 9, 6
   A 7   **B** 13

8. Simplify the expression $\frac{v^2}{v^3}$.
   A $\frac{1}{v}$ **B** $\frac{1}{v^5}$

9. Construct a box-and-whisker plot for the data set:

15	24	34	18	20	48
26	18	24	19	39	32
20	28	25	24	35	23

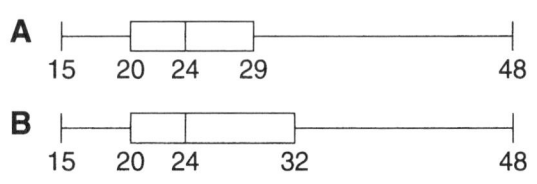

   A (box plot: 15, 20, 24, 29, 48)

   B (box plot: 15, 20, 24, 32, 48)

10. In the figure, which two angles are supplementary?

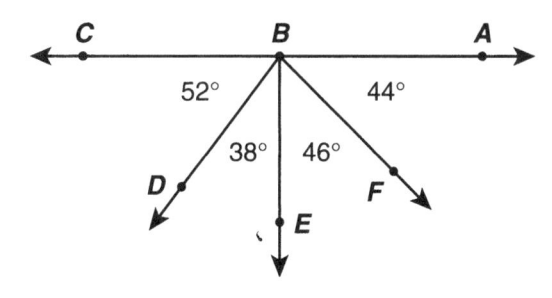

    A $\angle CBE$ and $\angle EBA$
    B $\angle CBD$ and $\angle DBE$

11. Find all square roots of $\sqrt{36}$.
    A 6.6   **C** 6, −6
    B 6, 6   **D** 36, −36

12. Evaluate $-3vw$ for $v = 3$ and $w = -1$.
    A −9   **C** −10
    B 9   **D** 10

229 Holt Pre-Algebra

Cumulative Test
Chapter 8 Form A, continued

13. Sarah is enclosing a square flower garden with timbers. The perimeter of the flower garden is 36 feet. Find the length of one side of the garden.
- A 4 ft
- B 6 ft
- C 9 ft
- D 12 ft

14. Find an algebraic expression for "2 plus the quotient of 13 and z."
- A $2 + \dfrac{13}{z}$
- B $\dfrac{2 + 13}{z}$

15. Find the original data from the stem-and-leaf plot.

Stem	Leaves
8	1 7
9	1 1 7
10	1 4 9

- A 8, 1, 7, 9, 11, 17, 10, 14, 19
- B 81, 87, 91, 91, 97, 101, 104, 109

16. Find the area of the figure.
- A 7.5 in²
- B 12 in²
- C 15 in²
- D 20 in²

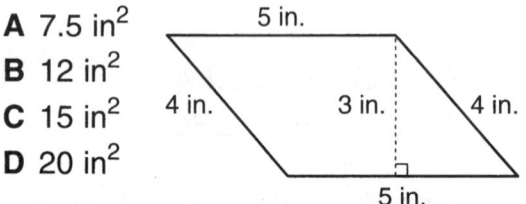

17. What is the value of x for the equivalent fraction? $\dfrac{12}{9} = \dfrac{x}{18}$
- A x = 2
- B x = 9
- C x = 12
- D x = 24

18. A circle has a radius of 2 inches. If the radius is doubled, what is the circumference to the nearest tenth? Use 3.14 for π.
- A 6.3 in.
- B 12.6 in.
- C 25.1 in.
- D 50.2 in.

19. A bin has 250 lag and toggle bolts. There are 80 more toggle bolts than lag bolts. How many lag bolts are in the bin?
- A 80
- B 85
- C 165
- D 170

20. A reception hall holds 192 people. If at a wedding reception the hall is filled to $\dfrac{3}{4}$ of the maximum capacity, how many people are at the reception?
- A 48
- B 144

21. Describe the radical as "rational," "irrational," or "not a real number." If a real number, give the value: $-\sqrt{81}$.
- A rational, 9
- B rational, −9
- C irrational, −9
- D not a real number

22. Find the best buy for a box of cereal and give the cost per ounce.
Fruity Pops: 12 oz for $2.34
Raisin Nut Crunch: 16 oz for $3.06
- A Fruity Pops, $0.19 per oz
- B Raisin Nut Crunch, $0.19 per oz
- C Fruity Pops, $0.20 per oz
- D Raisin Nut Crunch, $0.21 per oz

23. Find the area of the figure.
- A 60 ft²
- B 187.5 ft²
- C 375 ft²
- D 7500 ft²

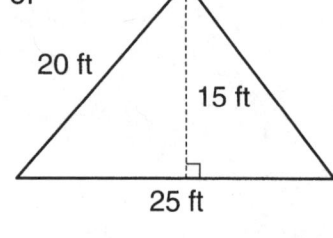

Cumulative Test
Form A, continued

24. Find the volume of the figure.

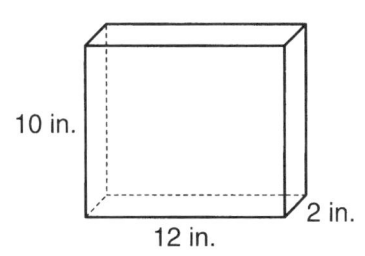

10 in.
12 in.
2 in.

- **A** 20 in³
- **B** 24 in³
- **C** 120 in³
- **D** 240 in³

25. What is the mean of the following temperatures?

Temperature Highs			
Mon.	55°	Fri.	55°
Tues.	65°	Sat.	60°
Wed.	55°	Sun.	70°
Thurs.	60°		

- **A** 55°
- **B** 60°
- **C** 65.5°
- **D** 70°

26. Find the surface area of the figure.

7 ft
4 ft
4 ft

- **A** 13 ft²
- **B** 27 ft²
- **C** 42 ft²
- **D** 72 ft²

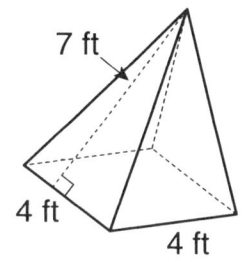

27. Below is a diagram of how a set of bracing leans against a wall of a house under construction. About how far from the house is the base of the bracing to the nearest tenth?

house
20 ft
30 ft

- **A** 22.4 ft
- **B** 36.1 ft

28. Simplify $1\frac{2}{3} \cdot \frac{5}{6}$.

- **A** $1\frac{7}{18}$
- **B** $1\frac{7}{9}$
- **C** $1\frac{5}{9}$
- **D** 3

29. Find the sum of the angle measures in a regular pentagon.

- **A** 360°
- **B** 540°
- **C** 720°
- **D** 900°

30. Solve $3x \le -30$.

- **A** $x \le -10$
- **B** $x < 10$
- **C** $x \ge -10$
- **D** $x > -10$

31. A recipe calls for $2\frac{1}{6}$ cups of flour and $1\frac{1}{3}$ cups of sugar. How many cups of dry ingredients are in the recipe?

- **A** $2\frac{1}{18}$ cups
- **B** $3\frac{2}{9}$ cups
- **C** $3\frac{1}{9}$ cups
- **D** $3\frac{1}{2}$ cups

32. In $\triangle ABC$, $m\angle A = m\angle B$ and $m\angle C$ is 30°. What is $m\angle B$ to the nearest tenth of a degree?

- **A** 75°
- **B** 150°

33. Seventy-five out of 100 students are going to the school dance. What is this number as a reduced fraction and as a percent?

- **A** $\frac{3}{4}$, 7.5%
- **B** $\frac{3}{4}$, 750%
- **C** $\frac{3}{4}$, 75%
- **D** $\frac{7.5}{10}$, 7.5%

Cumulative Test
Chapter 8 Form A, continued

34. An investment broker invests $60,000 in bonds. If he earns 5% per year on the investment, how much money is earned in just one year?
A $3000
B $6000
C $63,000
D $66,000

35. On a science test, a student got 25 questions correct. On a second test the student got 30 questions correct. What was the percent of increase in the score?
A 0.2%
B 20%

36. Find 5^{-2}.
A $\frac{1}{10}$
B $\frac{1}{25}$

37. Express 1.6×10^5 miles per hour in standard notation.
A 0.000016 mi/h
B 16,000 mi/h
C 160,000 mi/h
D 1,600,000 mi/h

38. Simplify $-10 - (-4) + 10$.
A 4
B -4
C 24
D 30

39. The Deerfield cross-country team ran 5.25 miles to prepare for a meet. What is this distance as a fraction?
A $5\frac{1}{5}$ mi
B $5\frac{2}{5}$ mi
C $5\frac{1}{4}$ mi
D $52\frac{1}{2}$ mi

40. The ratio of boys to girls on a baseball team is 5 to 2. If there are 4 girls on the team, how many boys are on the team?
A 12
B 10
C 9
D 8

Name _____ Date _____ Class _____

CHAPTER 8 Cumulative Test Form B

1. Express 79.8% as a decimal.
   A 0.0798
   B 7.98
   C 0.798
   D 79.8

2. Brad's new kite has 66 ft of string. How many yards is this?
   F 7.33 yd
   G 22 yd
   H 198 yd
   J 2376 yd

3. Simplify using exponents: $9^5 \cdot 9^4 \cdot 9^8$.
   A 9^{17}
   B 9^{160}
   C 729^{17}
   D 729^{160}

4. Determine which of the following is a solution to $2y - 2 = 2x$.
   F $(-1, -1)$
   G $(1, -1)$
   H $(0, 1)$
   J $(1, 3)$

5. In the figure, $\triangle ABC$ is similar to $\triangle GHJ$. Find length x.
   A 6
   B 20
   C 18
   D 24

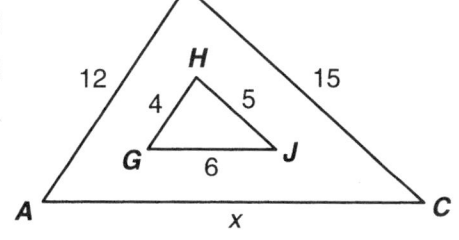

6. Solve $f - 18 = 48 - 10f$.
   F $f = 6$
   G $f = 12$
   H $f = -6$
   J $f = -12$

7. The following were temperatures of seven different cities in the United States on March 1, 2002: 77°, 25°, 77°, 13°, 25°, 29°, 56°, 77°. What was the mode of the temperatures?
   A 25°
   B 42.5°
   C 47.4°
   D 77°

8. Simplify the expression: $\dfrac{v^2}{v^6}$.
   F $\dfrac{v^2}{v^4}$
   G $\dfrac{v}{v^{12}}$
   H $\dfrac{1}{v^4}$
   J v^4

9. Construct a box-and-whisker plot for the data set:

   15 20 18 34 18 24 48
   29 39 23 24 26 18 29
   33 35 24 25 19 28 23

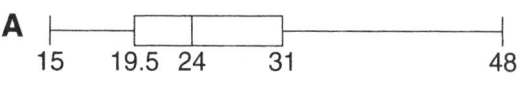

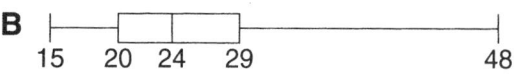

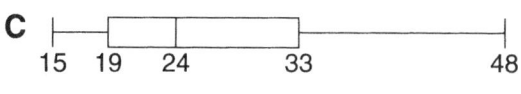

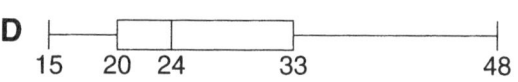

   A 15, 19.5, 24, 31, 48
   B 15, 20, 24, 29, 48
   C 15, 19, 24, 33, 48
   D 15, 20, 24, 33, 48

10. In the figure, which two angles are complementary?

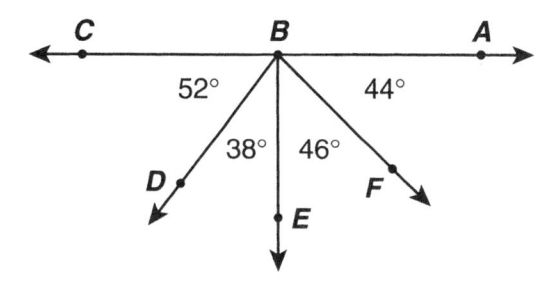

    F $\angle ABF$ and $\angle FBE$
    G $\angle CBE$ and $\angle EBA$
    H $\angle CBD$ and $\angle FBE$
    J $\angle FBE$ and $\angle EBD$

11. Find all square roots of $\sqrt{25}$.
    A 5.5
    B 5, 5
    C 5, −5
    D 25, −25

233 Holt Pre-Algebra

Cumulative Test
Form B, continued

12. Evaluate $-5v^2w$ for $v = 4$ and $w = -1$.
F 20
G 80
H −20
J −80

13. The perimeter of a square pavilion at a park is 128 feet. Find the length of one side of the pavilion.
A 5.7 ft
B 11.3 ft
C 32 ft
D 256 ft

14. Find an algebraic expression for "the quotient of 3 and the sum of z and 5."
F $\dfrac{3}{z+5}$
G $\dfrac{3}{z} \div 5$
H $\dfrac{z+5}{3}$
J $\dfrac{3}{5z}$

15. Find the original data from the stem-and-leaf plot.

Stem	Leaves
8	1 7
9	1 1 7
10	1 4 9
11	4

A 81, 17, 19, 19, 11, 14, 17, 19, 14
B 8, 81, 84, 91, 91, 97, 94, 101, 101, 114
C 81, 87, 91, 91, 97, 101, 104, 109, 114
D 81, 87, 91, 94, 97, 104, 107, 109, 114

16. Find the area of the figure.
F 14 in²
G 28 in²
H 24 in²
J 35 in²

17. What is the value of x for the equivalent fraction $\dfrac{12}{9} = \dfrac{x}{45}$?
A $x = 3$
B $x = 5$
C $x = 30$
D $x = 60$

18. A circle has a radius of 6 inches. If the radius is doubled, what is the circumference to the nearest tenth? Use 3.14 for π.
F 37.7 in.
G 75.4 in.
H 113.0 in.
J 452.2 in.

19. A jar has 415 red and blue balls. There are 87 more blue balls than red balls. How many blue balls are there in the jar?
A 77
B 164
C 251
D 338

20. A restaurant has a seating capacity of 187 customers. If the restaurant is $\dfrac{5}{11}$ full, how many customers are in the restaurant?
F 17
G 80
H 85
J 90

21. Describe the radical as "rational," "irrational," or "not a real number." If a real number, give the approximate value: $-\sqrt{191}$.
A irrational, −13.8
B irrational, 19.1
C rational, 19.1
D not a real number

Cumulative Test
Chapter 8 Form B, continued

22. Find the best buy and give the cost per ounce for the can of juice concentrate.
Libby: 12 oz for $1.32
Motts: 16 oz for $2.08
F Libby, $0.11 per oz
G Motts, $0.11 per oz
H Libby, $0.13 per oz
J Motts, $0.13 per oz

23. Find the area of the figure.
A 84 ft^2
B 399 ft^2
C 494 ft^2
D 798 ft^2

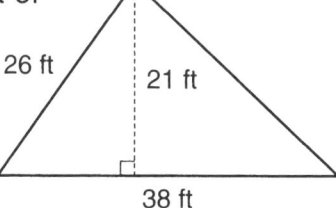

26 ft, 21 ft, 38 ft

24. Find the volume of the figure.
F 26 in^3
G 115 in^3
H 550 in^2
J 550 in^3

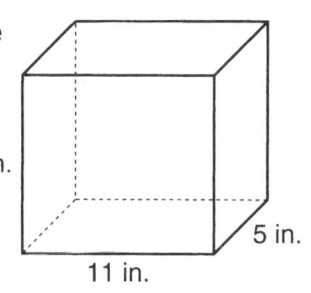

10 in., 5 in., 11 in.

25. Matthew has a birthday party at a local bowling alley. The following are the scores of the children's first game. What is the mean score?
55, 68, 69, 55, 69, 70, 55?
A 55
B 63
C 65.5
D 68

26. Find the surface area of the figure.
F 29 ft^2
G 43 ft^2
H 95 ft^2
J 105 ft^2

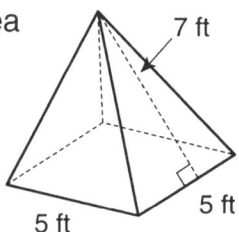

7 ft, 5 ft, 5 ft

27. Below is a diagram of how a ladder leans against one wall of a house. How far from the house is the base of the ladder to the nearest tenth?
A 8 ft
B 37.2 ft
C 20.4 ft
D 416 ft

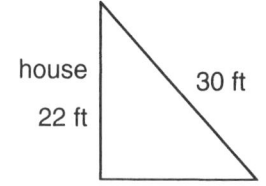

house, 30 ft, 22 ft

28. Simplify $2\frac{2}{3} \cdot \frac{15}{18}$.
F $2\frac{2}{9}$
G $2\frac{9}{21}$
H $3\frac{1}{5}$
J 10

29. Find the sum of the angle measures in a regular heptagon.
A 720°
B 900°
C 1080°
D 1260°

30. Solve $5x \leq -40$.
F $x \leq -8$
G $x < -8$
H $x \geq -8$
J $x > -8$

31. Kendra makes $2\frac{5}{6}$ quarts of fruit salad. She then decides to add $\frac{1}{3}$ quart of blueberries. How many quarts of fruit does she now have?
A $3\frac{7}{12}$ qt
B $3\frac{1}{6}$ qt
C $3\frac{1}{3}$ qt
D $2\frac{1}{3}$ qt

32. In $\triangle ABC$, $\angle A$ is 10° more than $\angle B$ and $\angle C$ is 30° more than $\angle B$. What is $m\angle A$ to the nearest tenth of a degree?
F 46.7°
G 56.7°
H 60.0°
J 76.7°

Name _____ Date _____ Class _____

CHAPTER 8 Cumulative Test
Form B, continued

33. For a fishing derby a pond is stocked with 100 trout. If 80 of the trout are caught, what is this ratio as a reduced fraction and a percent?

    A $\frac{4}{5}$, 8%   C $\frac{4}{5}$, 80%

    B $\frac{4}{5}$, 800% D $\frac{8}{10}$, 8%

34. Marty receives $74,000 from her grandmother's estate. If she invests the money in an IRA and earns 14% per year on the investment, how much money is earned in one year?

    F $10,360   H $52,857
    G $103,600   J $528,571

35. On a two-part history project, Joel got 25 points on the first part and 34 points on the second part. What was the percent of increase in the score?

    A 9%   C 26.5%
    B 36%   D 64%

36. Find 3^{-5}.

    F $\frac{1}{15}$ H $\frac{1}{81}$

    G $\frac{1}{243}$ J -15

37. The speed of light is 1.86×10^5 miles per second. What is this number in standard notation?

    A 0.0000186 mi/s
    B 186,000 mi/s
    C 1,860,000 mi/s
    D 18,600,000 mi/s

38. Simplify $7 - 15 - (-4) + 15$.

    F 3   H 30
    G 11   J -30

39. A carpenter cuts 15.3 inches off of a board of wood. What is this length as a fraction?

    A $15\frac{1}{3}$ in.   C $15\frac{3}{10}$ in.

    B $15\frac{2}{3}$ in.   D $15\frac{3}{5}$ in.

40. The ratio of juniors to seniors in the Lakewood High School Honor Society is 2:7. If there are 56 seniors, how many juniors are there in the Honor Society?

    F 14   H 18
    G 16   J 19

Name _____ Date _____ Class _____

CHAPTER 8 Cumulative Test
Form C

1. Express 179.8% as a decimal.
   - A 0.1798
   - B 17.98
   - C 1.798
   - D 179.8

2. A house in the country has a driveway that is 600 feet long. How many yards is this?
   - F 1800 yd
   - G 90 yd
   - H 100 yd
   - J 200 yd

3. Simplify using exponents:
   $9^5 \cdot 9^4 \cdot 9^4 \cdot 9^2$.
   - A 9^{11}
   - B 9^{15}
   - C 9^{160}
   - D 160^9

4. Determine which of the following is a solution to $-3y = 2x + 3$.
   - F $(1, -1)$
   - G $(0, -1)$
   - H $(2, 3)$
   - J $(1, 2)$

5. In the figure, $\triangle ABC$ is similar to $\triangle GHJ$. Find length y.
   - A 12
   - B 20
   - C 16
   - D 24

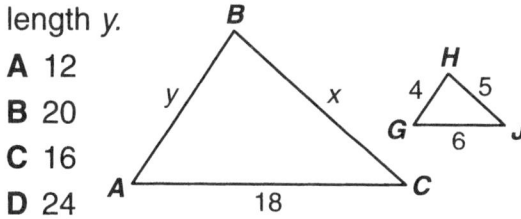

6. Solve $2f - 10 = 38 - 10f$.
   - F $f = 3.5$
   - G $f = 4$
   - H $f = 6$
   - J $f = -6$

7. A plumber has a bunch of scrap pipe in the following lengths. What is the mode for these lengths?
   7.7 ft, 2.5 ft, 7.7 ft, 1.3 ft, 2.5 ft, 2.9 ft, 5.6 ft, 7.7 ft
   - A 2.5 ft
   - B 4.3 ft
   - C 4.7 ft
   - D 7.7 ft

8. Simplify the expression $\dfrac{v^2 w^3}{v^6 w}$.
   - F $\dfrac{1}{v^2}$
   - G $\dfrac{w}{v^2}$
   - H $\dfrac{w^2}{v^4}$
   - J $\dfrac{(vw)^5}{(vw)^7}$

9. Construct a box-and-whisker plot for the data set:
   150 190 180 240 470 180 480
   290 390 260 180 240 230 290
   330 390 190 280 250 240 230
   - A 150 | 190 240 310 | 480
   - B 150 | 190 240 300 | 480
   - C 150 | 200 240 310 | 480
   - D 150 | 190 250 310 | 480

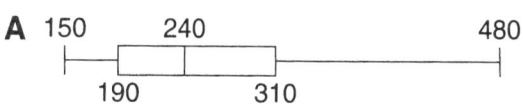

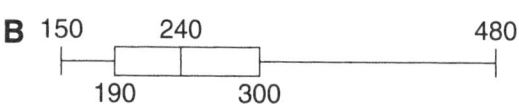

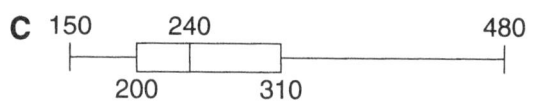

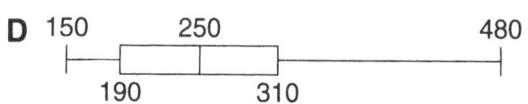

10. In the figure, give all combinations of angles that are complementary.
    - F $\angle ABF$ and $\angle FBE$ 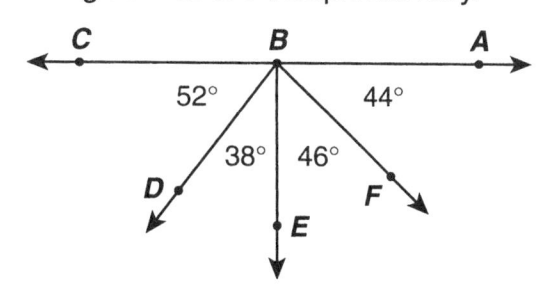
    - G $\angle CBE$ and $\angle EBA$
    - H $\angle CBD$ and $\angle DBE$; $\angle FBE$ and $\angle ABF$
    - J $\angle FBE$ and $\angle EBD$; $\angle ABF$ and $\angle CBD$

Copyright © by Holt, Rinehart and Winston.
All rights reserved.

Holt Pre-Algebra

Name _____ Date _____ Class _____

CHAPTER 8 Cumulative Test
Form C, continued

11. Find all square roots of $\sqrt{225}$.
    A 5, −5
    B 5
    C 15, −15
    D 25, −25

12. Evaluate $-6v^2w^2$ for $v = 4$ and $w = -1$.
    F 24
    G −24
    H 96
    J −96

13. The perimeter of a floor in a square shed is 222 feet. Find the length of one side of the floor.
    A 14.9 ft
    B 22 ft
    C 55.5 ft
    D 111 ft

14. Find an algebraic expression for "5 times the quotient of 2 and the sum of z and 5."
    F $5\left(\dfrac{2}{z+5}\right)$
    G $2\left(\dfrac{3}{z} \div 5\right)$
    H $\dfrac{2z+5}{3}$
    J $2\left(\dfrac{3}{5z}\right)$

15. Find the original data from the stem-and-leaf plot.

Stem	Leaves
8.	1 7
9.	1 1 4 7
10.	1 4 4 7 9
11.	4 5

    A 9, 15, 9, 9, 12, 15, 11, 11, 14, 17, 19, 15
    B 8.1, 8.7, 9.1, 9.1, 9.4, 9.7, 10.1, 10.4, 10.4, 10.7, 10.9, 11.4, 11.5
    C 81, 87, 91, 91, 94, 97, 101, 104, 104, 107, 109, 114, 115
    D 8.1, 8.7, 9.4, 9.4, 9.7, 10.4, 10.5, 10.7, 10.9, 11.4, 11.5

16. Find the area of the figure.

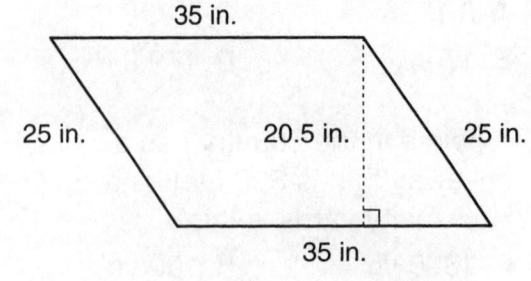

    35 in.
    25 in.   20.5 in.   25 in.
    35 in.

    F 61.5 in²
    G 80.5 in²
    H 358.8 in²
    J 717.5 in²

17. Solve $\dfrac{12}{5} = \dfrac{x}{18}$.
    A $x = 3.6$
    B $x = 12$
    C $x = 15$
    D $x = 43.2$

18. A circle has a diameter of 4.8 inches. If the diameter is doubled, what is the circumference to the nearest tenth? Use 3.14 for π.
    F 15.1 in.
    G 30.1 in.
    H 72.3 in.
    J 289.4 in.

19. A jar has 413 pink, purple, and yellow beads. There are 30 more purple beads than pink beads, and 20 more yellow than pink, and 10 more purple than yellow. How many purple beads are there in the jar?
    A 121
    B 141
    C 151
    D 292

20. A water dispenser holds 312 ounces of fluid. If the water dispenser is $\dfrac{7}{12}$ full, how many ounces of water are in the dispenser?
    F 26 oz
    G 130 oz
    H 182 oz
    J 312 oz

Name _____ Date _____ Class _____

CHAPTER 8 Cumulative Test
Form C, continued

21. Describe the radical as "rational," "irrational," or "not a real number." If a real number, give the approximate value: $-\sqrt{189}$.
    A irrational, 13.7
    B irrational, −13.7
    C rational, 13.7
    D not a real number

22. Find the best buy for laundry soap and give the cost per ounce.
    Brand A: 112 oz for $31.13
    Brand B: 116 oz for $37.58
    F Brand A, $0.278 per oz
    G Brand B, $0.278 per oz
    H Brand A, $0.324 per oz
    J Brand B, $0.324 per oz

23. Find the area of the figure.
    A 91 ft²
    B 372.8 ft²
    C 452.6 ft²
    D 745.5 ft²
    (25.5 ft, 21 ft, 35.5 ft)

24. Find the volume of the figure.
    F 76 ft³
    G 456 ft³
    H 988 ft³
    J 11,856 ft³
    (26 ft, 12 ft, 38 ft)

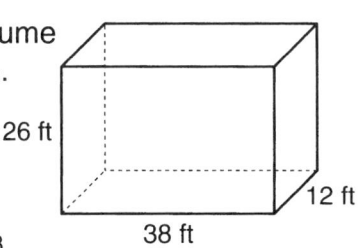

25. Find the mean of these numbers.
    55, 72, 69, 55, 79, 70, 55
    A 49.3 bags
    B 65 bags
    C 69 bags
    D 69.1 bags

26. Find the surface area of the figure.
    F 16 ft²
    G 44 ft²
    H 107.25 ft²
    J 134.8 ft²
    (7 ft, 5.5 ft, 5.5 ft)

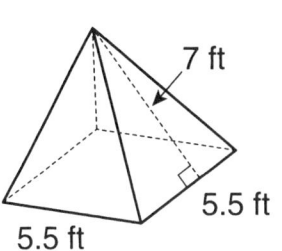

27. Below is a diagram of how a rope comes off the top of one tent pole. To the nearest tenth, how far from the tent is the rope staked?
    A 3.1 ft
    B 9.5 ft
    C 21.9 ft
    D 36.3 ft
    (20.5 ft, 30 ft)

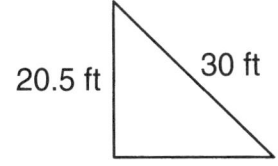

28. Simplify $12\frac{2}{3} \cdot 1\frac{5}{8}$.
    F $7\frac{31}{39}$
    G $13\frac{5}{12}$
    H $20\frac{7}{12}$
    J $47\frac{1}{2}$

29. Find the sum of the angle measures in a regular 10-gon.
    A 900°
    B 1080°
    C 1440°
    D 1800°

30. Solve for x in the inequality $-4x \leq -36$.
    F $x \geq 9$
    G $x \leq -9$
    H $x \geq -9$
    J $x \leq 9$

31. Peggy poured $1\frac{2}{5}$ liters of pop and $2\frac{1}{3}$ liters of juice into a punch bowl for a luncheon. How many liters of punch did she make?
    A $3\frac{2}{15}$ L
    B $3\frac{1}{4}$ L
    C $3\frac{3}{8}$ L
    D $3\frac{11}{15}$ L

Name _____ Date _____ Class _____

CHAPTER 8 Cumulative Test
Form C, continued

32. In △ABC, m∠A is 15° more than m∠B which is 10° more than m∠C. What is m∠A to the nearest tenth of a degree?
F 48.3°   H 63.3°
G 58.3°   J 73.3°

33. Carla has completed 18 out of the 27 questions on her science homework. What fraction, and what percent, of her science homework has she completed?
A $\frac{2}{3}$, 0.667%   C $\frac{2}{3}$, $66\frac{2}{3}$%
B $\frac{2}{3}$, 667%   D $\frac{18}{27}$, $6\frac{2}{3}$%

34. Mr. Harrington invests $174,500 in an IRA. If he earns 12.5% per year on the investment, how much money is earned in one year?
F $218.13   H $21,812.50
G $2181.25   J $152,687.50

35. The Bulldogs fifth-grade basketball team scored 23 points at their first game of the season. At the second game they scored 38 points. What was the percent of increase in the score to the nearest tenth of a percent?
A 15.0%   C 65.2%
B 39.5%   D 60.5%

36. Find $(-4)^{-4}$.
F $\frac{1}{16}$ H $-\frac{1}{16}$
G $\frac{1}{256}$ J $-\frac{1}{256}$

37. Express 2.06×10^{-6} in standard notation.
A 0.00000206   C 0.000000206
B 2,060,000   D 0.000206

38. Simplify $7 - 15 - (-4) \div (-2)$.
F −2   H 8
G −6   J −10

39. A cement truck carries 20.125 yards of concrete to a job. What is this amount as a fraction?
A $20\frac{1}{8}$ yd   C $20\frac{1}{4}$ yd
B $20\frac{1}{5}$ yd   D $20\frac{2}{5}$ yd

40. The ratio of cars to trucks in a parking lot is 12 to 5. If there are 90 trucks in the parking lot, how many cars are there?
F 38   H 216
G 142   J 324

Name _____ Date _____ Class _____

CHAPTER 9 Cumulative Test
Form A

1. A multiple-choice history test has questions with 4 possible choices. If you were to randomly guess on the first question, what is the probability that you would guess correctly?
   - **A** 0.10
   - **B** 0.25
   - **C** 0.798
   - **D** 79.8

2. Simplify 1^3.
   - **A** 0
   - **B** $\frac{1}{}$
   - **C** 1
   - **D** 3

3. Which is $\frac{9}{4}$ as a percent?
   - **A** $2\frac{1}{4}\%$
   - **B** $20\frac{1}{4}\%$
   - **C** 22.5%
   - **D** 225%

4. Tasha rode her bike for 300 minutes. How many hours did she bike?
   - **A** 5 hr
   - **B** 60 hr

5. Simplify the 3^{-2}.
   - **A** $\frac{1}{9}$
   - **B** -9

6. Find the perimeter of the figure.

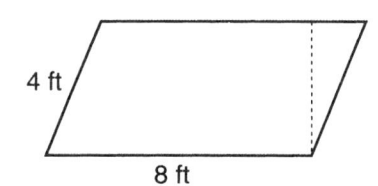

   - **A** 12 ft
   - **B** 24 ft

7. Ryan's chess club had a meeting every day for 5 days to practice for an upcoming tournament. What was the mean attendance at the meetings?

Chess Club	
Monday	5
Tuesday	12
Wednesday	8
Thursday	10
Friday	12

   - **A** 9.4
   - **B** 10

8. Roger has a garden in the shape of a triangle. What is the area?

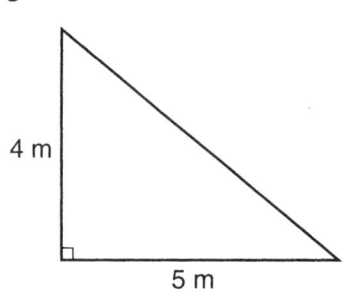

   - **A** 9 m^2
   - **B** 10 m^2
   - **C** 18 m^2
   - **D** 20 m^2

9. Find the square root to the nearest hundredth. $\sqrt{5}$
   - **A** 2.24
   - **B** 2.5

10. Solve the equation $1.2z = 2.4$.
    - **A** $z = 2$
    - **B** $z = 3.6$

11. A rectangular package is 6 in. high, 3 in. wide, and 2 in. deep. What is the volume of the package if the width is doubled?
    - **A** 36 in^3
    - **B** 72 in^3

Cumulative Test
Chapter 9 Form A, continued

12. Simplify $-2(a + 3) + 3a$.
 A $a - 6$ B $a + 6$

13. Evaluate ab^3 for $a = 2$ and $b = -1$.
 A -2 B 2

14. What are the coordinates of point B?

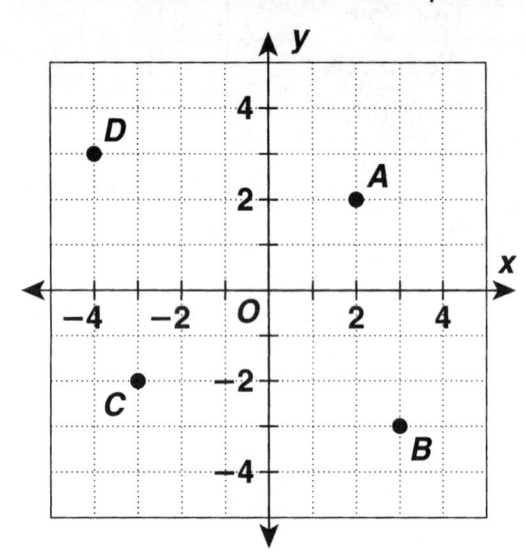

 A $(3, 3)$ C $(3, -3)$
 B $(-3, -3)$ D $(-3, 3)$

15. Tom can type one letter in $\frac{1}{2}$ an hour. How long do 7 letters take?
 A 1 hr   C 2 hr
 B 14 hr   D $3\frac{1}{2}$ hr

16. Deena has a circular flower garden with a diameter of 10 feet. Deena would like to edge the garden. How many feet of edging material would she need? Use 3.14 for π.
 A 31.4 ft   B 62.8 ft

17. Solve the proportion $\frac{10}{9} = \frac{5}{x}$.
 A $x = 4.5$ B $x = 18$

18. Sam took a survey of the number of hours of sleep that several family members received over the weekend. What is the mode?

Sleep Survey	
Member	Hours
Mom	8
Dad	8
Kim	13
Roger	14
Grandma	8
Grandpa	12

 A 10   B 8

19. Simplify using positive exponents: $\left(\frac{1}{6}\right)\left(\frac{1}{6}\right)\left(\frac{1}{6}\right)$.

 A $\left(\frac{1}{6}\right)^3$ C $6^{\frac{1}{6}}$
 B $\left(\frac{3}{6}\right)^3$ D $6\frac{1}{6}$

20. An auditorium currently has a seating capacity of 200 people. If the capacity is increased to 300 people, what is the percent of increase?
 A $33\frac{1}{3}\%$ B 50%

21. Which statement says that 3 minus a number equals twice the number?
 A $3 - n = 2n$ B $n - 3 = 2 + n$

22. Find all square roots of the number 81.
 A 1, 8   C -81
 B 9, -9 D 6561

Name _____ Date _____ Class _____

CHAPTER 9 Cumulative Test
Form A, continued

23. Express 2.9×10^3 in standard notation.
    A 0.0029 C 290
    B 0.029   D 2900

24. Solve $3c = -15$.
    A $c = -5$ B $c = -\frac{1}{5}$

25. Simplify $4 - (8) \cdot (2) + 5$.
    A -3 C -28
    B -7 D -92

26. Find the surface area of a sphere that has a radius of 3 m. Use 3.14 for π. Recall that $S = 4\pi r^2$.
    A 113.04 m^2
    B 37.68 m^2

27. Find the measure of an angle whose supplement is three times the measure of the angle.
    A 22.5°   B 45°

28. A dressmaker purchased $2\frac{1}{2}$ yd of denim, $3\frac{1}{2}$ yd of corduroy, and $1\frac{1}{4}$ yd of cotton. How much total fabric did the dressmaker purchase?
    A $7\frac{1}{4}$ yd   B $6\frac{1}{2}$ yd

29. Which inequality is represented by the graph?

    A $x \leq 6$ B $x \geq 6$

30. How many sides does a kite have?
    A 4   B 6

31. Two angles whose sum is 90° are known as what type of angles?
    A complementary angles
    B supplementary angles
    C congruent angles
    D obtuse angles

32. Carla had $400 in a checking account on Monday. On Tuesday she wrote a check for $200. What was the percent decrease in her account balance?
    A $33\frac{1}{3}\%$ B 50%

33. Identify the figure.

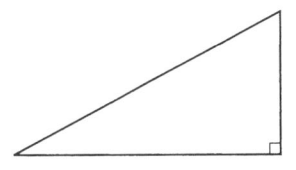

    A right triangle
    B obtuse triangle
    C isosceles triangle
    D equilateral triangle

34. A radio with an original price of $90 was marked down to $60. What is the percent decrease?
    A 50%   B $33\frac{1}{3}\%$

35. Larissa has read 12 out of 60 books that are required reading for her English class. What fraction and percent of the books has she read?
    A $\frac{1}{5}$, 20%   B $\frac{1}{5}$, 2.0%

CHAPTER 9 Cumulative Test
Form A, continued

36. What is the measure of $\angle EBD$?

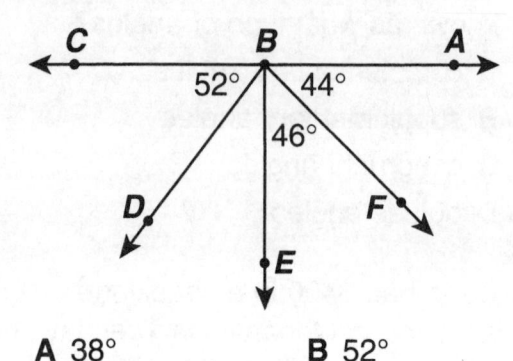

(diagram: ray BC at 52°, ray BA, with 44° between BA and BF, 46° between BF and BE)

A 38° **B** 52°

37. Find the volume of the figure. Use 3.14 for π.

(cylinder: radius 10 m, height 15 m)

A 1177.5 m³
B 4710 m³

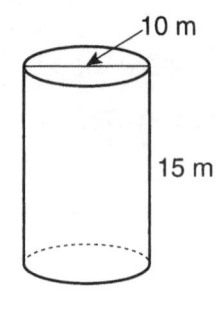

38. 300 people at a local mall were asked about their favorite cookie. Chocolate chip was the favorite of 125 people. Identify the sample.

A 125 preferring chocolate chip
B 300 asked their favorite cookie
C all people at the mall
D cannot be determined

39. If $P(A) = \frac{2}{3}$, find the odds in favor of A happening.

A 2:1 **B** 2:3

40. Lucinda has purchased $9\frac{3}{4}$ pounds of peaches. Which represents the decimal amount?

A 9.7   **C** 97
B 9.75 **D** 97.5

41. Simplify $\frac{5!}{3!}$.

A 20   **C** $\frac{5}{3}$
B $8\frac{1}{3}$ **D** 2

42. Kelly has 200 red and green beads in a bag. If there are 70 red beads, what is the probability of randomly selecting a red bead?

A 70% **B** 35%

43. You have 10 problems on a math test. The teacher is allowing you to select 4 problems to complete. How many different possible combinations are there?

A 210 **B** 5040

44. Find $_4P_3$.

A 12 **B** 24

Name _____ Date _____ Class _____

CHAPTER 9 Cumulative Test
Form B

1. Margaret is playing a carnival game. There are 5 possible prizes she could win by spinning a spinner (tiger, giraffe, bear, koala, or bird) divided into 5 equal sections. Every contestant wins a prize. What is the probability she will win the tiger?
   - **A** 0.15
   - **B** 0.20
   - **C** 0.50
   - **D** 0.80

2. Simplify -1^6.
   - **F** $\frac{1}{6}$
   - **G** -1
   - **H** 1
   - **J** -6

3. Express $\frac{10}{3}$ as a percent.
   - **A** $3\frac{1}{3}\%$
   - **B** $33\frac{1}{3}\%$
   - **C** $333\frac{1}{3}\%$
   - **D** 3000%

4. Becky drove 480 minutes. How many hours did she drive?
   - **F** 4 hr
   - **G** 8 hr
   - **H** 20 hr
   - **J** 192 hr

5. Simplify $(-4)^{-2}$.
   - **A** $\frac{1}{16}$
   - **B** 16
   - **C** $-\frac{1}{16}$
   - **D** -16

6. Find the perimeter.

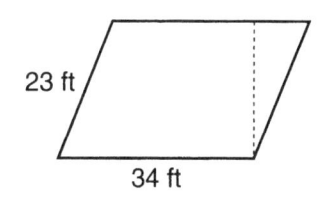

   23 ft / 34 ft

   - **F** 782 ft
   - **G** 114 ft
   - **H** 104 ft
   - **J** 57 ft

7. For the past 5 years Tony has had the following bonuses. What was Tony's mean bonus?

Year	Bonus
2002	$7830
2001	$4320
2000	$1570
1999	$2370
1998	$3600

   - **A** $1570
   - **B** $3600
   - **C** $3938
   - **D** $4320

8. Find the area.

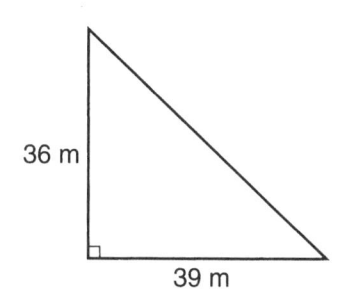

   36 m, 39 m

   - **F** 53 m²
   - **G** 75 m²
   - **H** 702 m²
   - **J** 1404 m²

9. What is $\sqrt{46}$?
   - **A** 6.77
   - **B** 6.78
   - **C** 6.79
   - **D** 46.00

10. Solve the equation $4.5z = 31.5$.
    - **F** $z = \frac{1}{7}$
    - **G** $z = 7$
    - **H** $z = 24.5$
    - **J** $z = 27$

11. A small planter measures 12 in. high by 6 in. wide by 3 in. deep. What is the volume of the box if the width is doubled?
    - **A** 1728 in³
    - **B** 432 in³
    - **C** 216 in³
    - **D** 42 in³

Name _____ Date _____ Class _____

Cumulative Test
CHAPTER 9 Form B, continued

12. Simplify $-4(a + 7) + 10a$.
    F $6a - 28$
    G $-14a - 28$
    H $6a + 28$
    J $-6a + 28$

13. Evaluate ab^3 for $a = 5$ and $b = -3$.
    A -22
    B 45
    C -45
    D -135

14. What are the coordinates of point D?

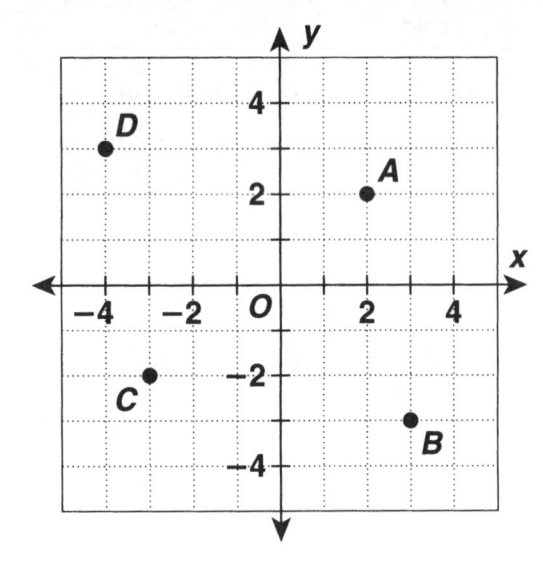

    F $(3, -4)$
    G $(-4, -3)$
    H $(4, 3)$
    J $(-4, 3)$

15. If you can paint one box in $\frac{2}{3}$ of an hour, how long do 5 boxes take?
    A 2 hr
    B 4 hr
    C $3\frac{1}{3}$ hr
    D $4\frac{2}{3}$ hr

16. A circular fountain has a diameter of 15 ft. What is the distance around the fountain? Use 3.14 for π.
    F 47.1 ft
    G 60 ft
    H 94.2 ft
    J 176.6 ft

17. Solve the proportion $\frac{25}{9} = \frac{15}{x}$.
    A $x = 1\frac{2}{3}$
    B $x = 5$
    C $x = -5$
    D $x = 5.4$

18. A group of fishermen on a charter boat have the following ages: 89, 49, 52, 49, 79, 39. What is the mode?
    F 59.5
    G 49
    H 60
    J 99

19. Simplify using positive exponents $\left(\frac{5}{6}\right)\left(\frac{5}{6}\right)\left(\frac{5}{6}\right)\left(\frac{5}{6}\right)\left(\frac{5}{6}\right)\left(\frac{5}{6}\right)$.
    A $\left(\frac{5}{6}\right)^6$
    B $\left(\frac{5}{6}\right)^5$
    C $6^{\frac{5}{6}}$
    D $6\frac{5}{6}$

20. Lindsay currently deposits $250 each month into her savings account. If she increases the amount to $350, what is the percent increase in her monthly deposit?
    F 140%
    G 100%
    H 40%
    J 28.6%

21. Which statement says that 56 minus twice a number equals the number?
    A $56 - 2n = 2n$
    B $2n - 56 = n$
    C $56 - 2 = n$
    D $56 - 2n = n$

22. Find all square roots of the number 225.
    F 15, −15
    G 16, −16
    H 20, −25
    J 25, −25

Name _____ Date _____ Class _____

CHAPTER 9 Cumulative Test
Form B, continued

23. Express 6.6×10^5 in standard notation.
    A 330
    B 66,000
    C 660,000
    D 6,600,000

24. Solve the equation $9c = -135$.
    F $c = 1$
    G $c = -15$
    H $c = -144$
    J $c = 144$

25. Simplify $9 - (-12) \cdot (-6) + 10$.
    A -116
    B -53
    C -29
    D -92

26. Find the surface area to the nearest tenth, of a sphere that has a radius of 4.5 m. Use 3.14 for π.
    F 56.5 m^2
    G 28.26 m^2
    H 254.3 m^2
    J 317.9 m^2

27. Find the measure of an angle whose complement is two times the measure of the angle.
    A 22.5°
    B 30°
    C 60°
    D 67.5°

28. The camp cook purchased $2\frac{1}{6}$ lb of hamburger, $6\frac{1}{2}$ lb of chicken, and $1\frac{3}{5}$ lb of pork. How many total pounds of meat did he purchase?
    F $9\frac{1}{5}$ lb
    G $9\frac{5}{13}$ lb
    H $9\frac{2}{3}$ lb
    J $10\frac{4}{15}$ lb

29. Which inequality is represented by the graph?

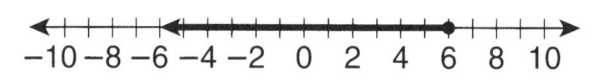

    ←++++++|←++++++++++●++++→
    −10 −8 −6 −4 −2 0 2 4 6 8 10

    A $2x - 3 \leq 6$
    B $-x + 3 \geq -3$
    C $x \leq 6 - x$
    D $x < 6$

30. How many sides does a rhombus have?
    F 4
    G 6
    H 8
    J 12

31. Two angles whose sum is 180° are known as what type of angles?
    A complementary angles
    B supplementary angles
    C congruent angles
    D acute angles

32. A submarine on maneuvers starts at a depth of 300 feet, rises 50 feet, then dives an additional 100 feet. What is the overall percent change from the original depth?
    F 10%
    G $16\frac{2}{3}$%
    H 50%
    J 110%

33. Identify the figure.
    A right triangle
    B obtuse triangle
    C isosceles triangle
    D equilateral triangle

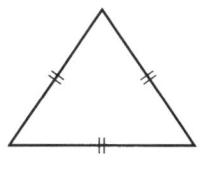

34. Selena was looking at a coat originally priced at $89.00. It was marked down to $56.99. What is the percent decrease in price?
    F 32.01%
    G 35.90%
    H 35.97%
    J 56.20%

Cumulative Test

Form B, continued

35. Ayna sent out 80 resumes. As of Friday, she had received 15 replies. What fraction and percent of replies did she receive?

A $\frac{3}{16}, 18\frac{3}{4}\%$ C $\frac{15}{18}, 18\%$

B $\frac{3}{16}, 5\frac{1}{3}\%$ D $\frac{3}{16}, 533\frac{1}{3}\%$

36. Find the measure of ∠ABD.

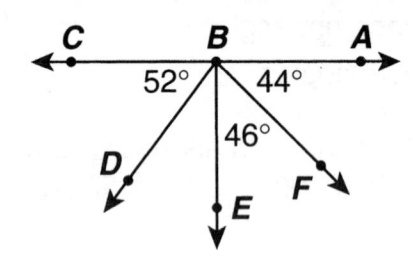

F 38°
G 90°
H 128°
J not enough information

37. Find the volume of the cylinder. Use 3.14 for π.

A 480 m³
B 753.6 m³
C 6028.8 m³
D 24,115.2 m³

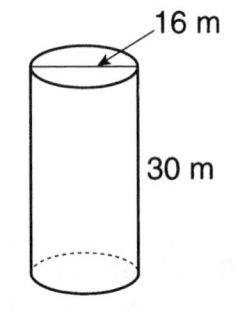

38. 350 people at a local carnival were asked about their favorite flavor of ice cream. Vanilla was the favorite of 115 people. Identify the sample.

F 115 preferring vanilla
G 350 asked their favorite ice cream
H all people at the carnival
J cannot be determined

39. If $P(A) = \frac{5}{6}$, what are the odds in favor of A happening?

A 1:5   C 5:1
B 5:6   D 6:5

40. A tentmaker uses $15\frac{3}{5}$ yards of fabric. What is $15\frac{3}{5}$ as a decimal?

F 15.3   H 15.8
G 15.6   J 15.9

41. Simplify $\frac{9!}{5!}$.

A 2!   C 3024
B $\frac{4}{5}$ D 45,000

42. A decorated tree has 415 red and blue balls. If there are 87 blue balls, what is the probability of randomly selecting a red ball to the nearest percent?

F 79%   H 21%
G 50%   J 20.9%

43. You are looking at new cars. The car lot has 20 cars on it. You want to test drive 5 of the cars. How many different possible combinations are there?

A 1   C 120
B 5   D 15,504

44. Find $_5P_4$.

F 120   H 5
G 24   J 1

Name _____ Date _____ Class _____

CHAPTER 9 Cumulative Test
Form C

1. A contestant on a game show will win the grand prize if he or she selects the correct door. The contestant has 7 possible doors to choose from. If the contestant randomly guesses, what is the probability that the contestant will select the correct door?
   A 0.14
   B 0.17
   C 0.83
   D 0.92

2. Simplify -5^4.
   F $-\frac{4}{5}$
   G 5
   H -625
   J $\frac{1}{625}$

3. Express $\frac{33}{4}$ as a percent.
   A $8\frac{1}{4}\%$
   B $8\frac{3}{4}\%$
   C $33\frac{1}{4}\%$
   D 825%

4. A lot in the country has 450 feet of frontage. How many yards is this?
   F 150 yd
   G 50 yd
   H 37.5 yd
   J 20 yd

5. Simplify $(-3)^{-3}$.
   A $-\frac{1}{27}$
   B -27
   C $\frac{1}{27}$
   D 27

6. Find the perimeter of the figure.

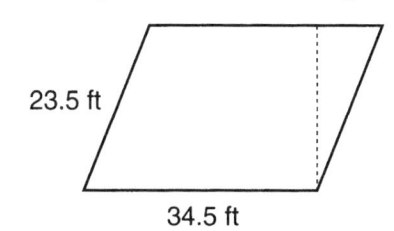

   23.5 ft
   34.5 ft

   F 810.8 ft
   G 116 ft
   H 58 ft
   J 57 ft

7. During the past month, $708.34, $430.11, $150.72, $230.75, and $360.00 were spent on repairs. What is the median amount spent on repairs?
   A $150.72
   B $360.00
   C $390.38
   D $375.98

8. Find the area of the figure.

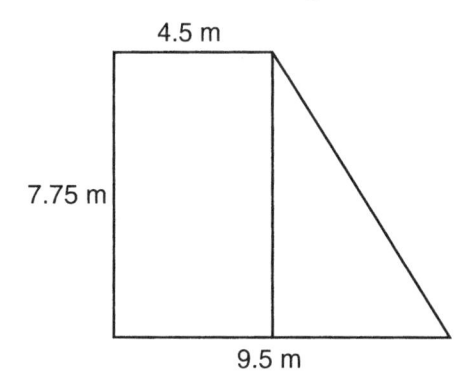

   4.5 m
   7.75 m
   9.5 m

   F 34.875 m²
   G 73.6 m²
   H 54.25 m²
   J 36.8125 m²

9. Find $\sqrt{111}$ to the nearest hundredth.
   A 10.52
   B 10.53
   C 10.54
   D 10.55

10. Solve the equation $3.5z = 31.5 - z$.
    F $z = \frac{1}{7}$
    G $z = 7$
    H $z = 24.5$
    J $z = 27$

11. A shipping carton measures 12 in. high by 6 in. wide by 4.5 in. deep. What is the volume of the carton if the width is tripled?
    A 8748 in³
    B 972 in³
    C 324 in³
    D 108 in³

Cumulative Test
Chapter 9 Form C, continued

12. Simplify $-5(-a - 7) + 10a$.
- F $5a - 35$
- G $-5a + 35$
- H $15a - 35$
- J $15a + 35$

13. Evaluate $5ab^3$ for $a = 2$ and $b = -3$.
- A -270
- B 270
- C 30
- D -30

14. What are the coordinates of point C?

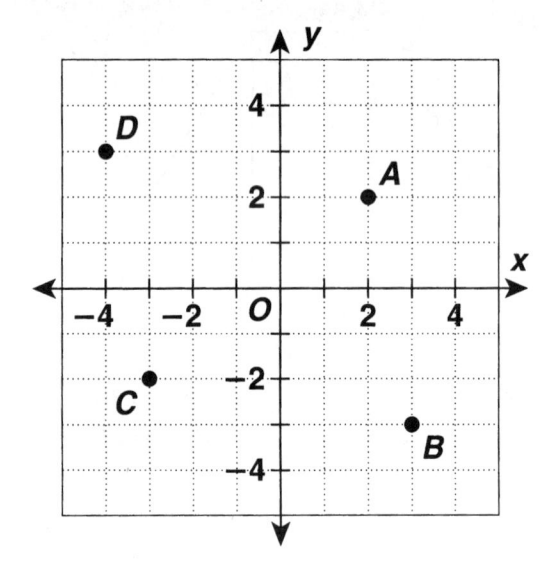

- F $(-2, -3)$
- G $(-3, -2)$
- H $(4, 3)$
- J $(-4, 3)$

15. Josh can walk one mile in $\frac{1}{3}$ of an hour. If he kept up this rate, how many hours would he take to walk 7 miles?
- A $7\frac{1}{3}$ hr
- B 4 hr
- C 3 hr
- D $2\frac{1}{3}$ hr

16. A small stepping stone has a diameter of 1.5 ft. What is the distance around the stepping stone? Use 3.14 for π.
- F 4.71 ft
- G 6 ft
- H 7.17 ft
- J 9.42 ft

17. Solve the proportion $\frac{x}{9} = \frac{4.1}{36.9}$.
- A $x = \frac{1}{4}$
- B $x = 1$
- C $x = 4.1$
- D $x = 9$

18. Your bowling scores for the last six games are:
119, 104, 113, 119, 119, and 129.
What is the mode?
- F 117
- G 165
- H 188
- J 119

19. Simplify using positive exponents:
$\left(\frac{5}{6}\right)\left(\frac{5}{6}\right)\left(\frac{5}{6}\right)\left(\frac{1}{5}\right)\left(\frac{1}{5}\right)\left(\frac{1}{5}\right)$.
- A $\left(\frac{5}{6}\right)^3\left(\frac{1}{5}\right)^3$
- B $\left(\frac{1}{5}\right)\left(\frac{5}{6}\right)^6$
- C $6\frac{5}{6}$
- D $6\frac{5}{6}$

20. A farmer currently farms 2000 acres of land. If he increases his acreage to 2700, what is the percent of increase in acreage?
- F 25.9%
- G 35%
- H 125.9%
- J 135%

21. Which statement says that 6 less than three times a number equals twice the number?
- A $6 - 3n = 2n$
- B $3n - 6 = 2n$
- C $6 - 3n = n$
- D $3(n - 6) = 2n$

22. Find all square roots of the number 576.
- F $24, -24$
- G 28.8
- H $18, 32$
- J 288

Name _____ Date _____ Class _____

CHAPTER 9 Cumulative Test
Form C, continued

23. Express 3.16×10^7 in standard notation.
    A 31,600
    B 3,160,000
    C 31,600,000
    D 316,000,000

24. Solve $55c + 30 = -135$.
    F $c = -3$
    G $c = 3$
    H $c = -50$
    J $c = 110$

25. Simplify $9 - (-12) \div (-6) + 9$.
    A -2
    B 12.5
    C -12
    D 16

26. Find the surface area of a sphere that has a diameter of 1 m. Use 3.14 for π.
    F 1.05 m^2
    G 3.14 m^2
    H 9.86 m^2
    J 12.56 m^2

27. Find the measure of an angle whose supplement is 3 more than 5 times the measure of its complement.
    A 18°
    B 33.8°
    C 67.5°
    D 68.25°

28. Susan is making trail mix. She purchased $3\frac{1}{5}$ lb of almonds, $5\frac{1}{2}$ lb of peanuts, and $1\frac{1}{6}$ lb of cashews. How many pounds of nuts did she purchase?
    F $9\frac{1}{10}$ lb
    G $9\frac{1}{2}$ lb
    H $9\frac{5}{6}$ lb
    J $9\frac{13}{15}$ lb

29. Which inequality is shown?

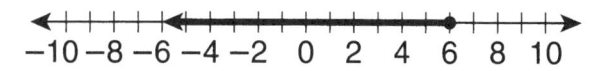

    (number line from -10 to 10, shaded for $x \leq 6$)

    A $4x - 6 \leq 12$
    B $-2x + 5 \geq -7$
    C $2x \leq 12 - x$
    D $-2x < 12$

30. How many sides does a trapezoid have?
    F 4
    G 6
    H 8
    J 10

31. What are two angles that both measure 30° called?
    A complementary angles
    B supplementary angles
    C congruent angles
    D right angles

32. A hot air balloon started at 800 feet, rose 50 feet, and then dropped 250 feet. What is the percent change from its original height?
    F $33\frac{1}{3}\%$
    G -25%
    H $33\frac{1}{4}\%$
    J 20%

33. Identify the figure.
    A acute triangle
    B obtuse triangle
    C isosceles triangle
    D equilateral triangle

34. Ms. Whipple is looking at a camera for her husband. When she looked at the camera originally, it had a price of $189.00. Two weeks later it was marked down to $126.99. What is the percent of markdown to the nearest tenth?
    F 32.8%
    G 48.8%
    H 62.01%
    J 67.2%

Cumulative Test
Form C, continued

35. Rich and Christine sent out 180 wedding invitations. 25 people responded a week later. What fraction and percent, to the nearest tenth of a percent, of responses had they received?

A $\frac{5}{12}$, 41.6% C $\frac{1}{9}$, 7.2%

B $\frac{15}{38}$, 39.4% D $\frac{5}{36}$, 13.9%

36. Find the measure of ∠CBF.

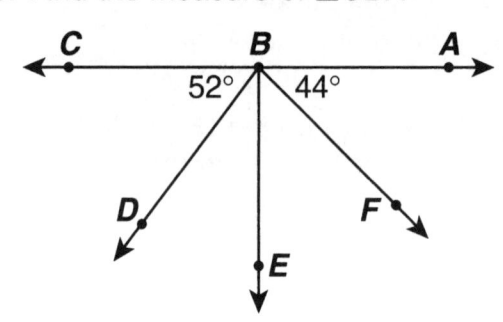

F 44°
G 108°
H 136°
J not enough information

37. Find the volume of the figure to the nearest tenth. Use 3.14 for π.

A 594.0 m³
B 1865.2 m³
C 2564.6 m³
D 7693.8 m³

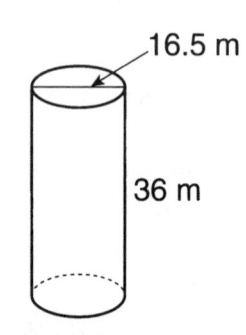

38. 120 people were asked about their favorite color. Blue was the favorite of 58 people. Identify the sample.

F 58 preferring blue
G 120 asked their favorite color
H all people at the school
J cannot be determined

39. If $P(A) = \frac{4}{5}$, what are the odds in favor of A not happening?

A 4:1 C 1:4
B 4:5 D 5:4

40. Sandra bought $150\frac{3}{4}$ feet of ribbon. What is $150\frac{3}{4}$ as a decimal?

F 15.75 H 150.3
G 1.575 J 150.75

41. Simplify $\frac{9!}{5!(9-5)!}$.

A 1 C 3024
B 126 D 15,120

42. By mistake someone has mixed up some nuts, bolts, and screws. You know there are a total of 400 in the bin. If there are 170 bolts and 200 screws, what is the probability of randomly selecting a nut?

F 0.075 H 7.5
G 0.75 J 75

43. You are re-hanging some posters in your room. You have 15 different posters, but you only want to hang 4. How many different possible combinations are there?

A 273 C 32,760
B 1365 D 5.45 × 10¹⁰

44. Find $_7P_4$.

F 6 H 210
G 35 J 840

Name _____ Date _____ Class _____

CHAPTER 10 Cumulative Test
Form A

1. Solve the system:
   $4x - 3y = -8$
   $x + 3y = 13$
   A (1, 4) **B** (−1, 4)

2. A sample space consists of 18 separate events that are equally likely. What is the probability of each?
   A 0 **C** 1
   B $\frac{1}{18}$ **D** 18

3. Select the graph of the solution set for $0.2x > 0.2$.

   A
   <-+-+-+-+-+-+-+-+-+-○->
   −10 −8 −6 −4 −2 0 2 4 6 8 10

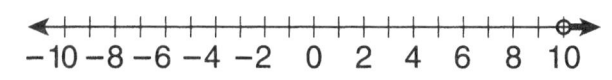

   B
   <-+-+-+-+-○←←←+-+-+-+->
   −10 −8 −6 −4 −2 0 2 4 6 8 10

   C
   <-+-+-+-+-+-○-+-+-+-+->
   −10 −8 −6 −4 −2 0 2 4 6 8 10

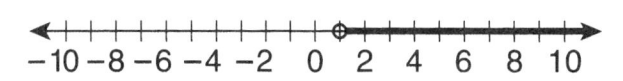

   D
   <-+-+-+-+-+-+-●-+-+-+->
   −10 −8 −6 −4 −2 0 2 4 6 8 10

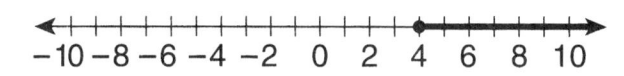

4. 2^3 means:
   A 2 • 2 • 2 **B** 2 • 3

5. Express 0.15 as a percent.
   A 150% **C** 0.15%
   B 15% **D** 0.0015%

6. If your window measures 60 inches, what is its length in feet?
   A 5 ft **B** 6.6 ft

7. A file contains 3300 bytes of data. What is this number in scientific notation?
   A 3.3×10^4 **C** 3.3×10^3
   B 3.3×10^2 **D** 33×10^{-3}

8. Find the missing length in the right triangle.

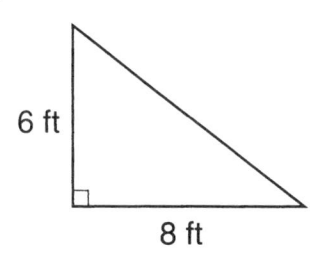

   A 10 ft **B** 14 ft

9. Simplify $-4 - (-14)$.
   A 10 **C** 18
   B −10 **D** −18

10. A scale model of a statue is 14 inches high by 3 inches wide. If the height of the actual statue is 14 feet, how wide is it?
    A 3 in. **C** 3 ft
    B 42 in. **D** 14 ft

11. A particular size battery is available in packages of 3 for $1.25 or packages of 4 for $1.75. Which is the better buy, and what is its price per battery?
    A 3 for $1.25; $0.42
    B 3 for $1.25; $0.44
    C 4 for $1.75; $0.42
    D 4 for $1.75; $0.44

Name _____ Date _____ Class _____

CHAPTER 10 Cumulative Test
Form A, continued

12. Which angles in the figure are complementary?

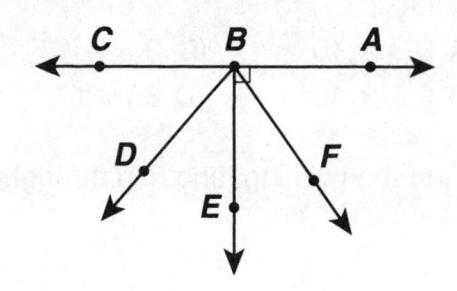

A $\angle ABF$ and $\angle FBE$
B $\angle ABE$ and $\angle EBC$

13. Solve $-10 + h = 10$.
A $h = 0$ C $h = 1$
B $h = 20$ D $h = 10$

14. The perimeter of a square flower bed is 8 feet. What is the length of one side?
A 2 ft   B 32 ft

15. What is "2.4 times a number" written as an algebraic expression?
A $2.4x$ B $2.4 + x$

16. Solve $\frac{3}{4}x = -\frac{3}{8}$.
A $x = -\frac{9}{32}$ B $x = -\frac{1}{2}$

17. Find the surface area of the figure.

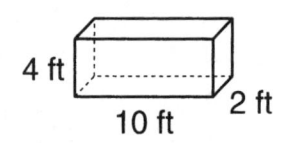

4 ft, 10 ft, 2 ft

A 16 ft^2 C 80 ft^2
B 68 ft^2 D 136 ft^2

18. Louis surveyed his friends to see how long, in minutes, it takes them to walk to school. What is the mean length of time of the data 7, 5, 11, 4, 8, and 13?
A 7.5 min   C 8 min
B 9.6 min   D 9 min

19. Find the square root and round to the nearest thousandth, if necessary.
$\sqrt{13}$
A 3.605   B 3.606

20. Find the value of -3^2.
A 9   B -9

21. You are making a quilt out of rectangles. One piece has dimensions of $1\frac{1}{3}$ in. by $2\frac{1}{6}$ in. What is the perimeter of this rectangle?
A $3\frac{2}{9}$ in.   C $4\frac{1}{3}$ in.
B $3\frac{1}{2}$ in.   D 7 in.

22. Choose the box-and-whisker plot that matches the data: 3, 5, 5, 5, 6, 9, 7, 7, 9, 8.

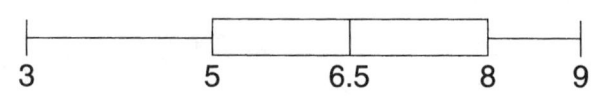

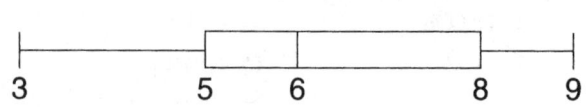

A (3, 5, 6.5, 8, 9)

B (3, 5, 6, 8, 9)

23. Solve $\frac{2}{5} + 2a = -\frac{2}{5}$.
A $a = -\frac{4}{5}$ B $a = -\frac{2}{5}$

Name _____ Date _____ Class _____

CHAPTER 10 Cumulative Test
Form A, continued

24. What is the surface area of the figure? Use 3.14 for π.
(2 m, 6 m)
A 18.8 m²
B 25.12 m²
C 37.7 m²
D 43.96 m²

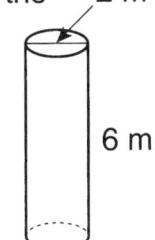

25. Find the measure of ∠ x.
A 115°
B 65°
C 90°
D 180°

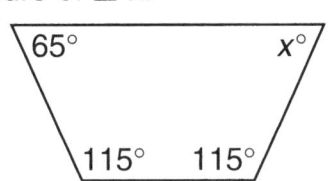

26. Identify the angle that is congruent to ∠2.

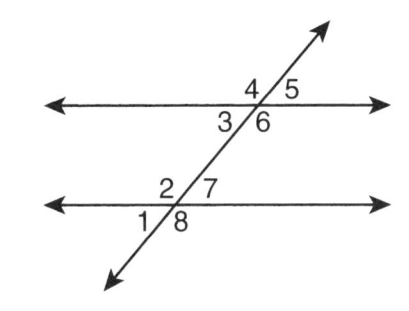

A ∠1   C ∠5
B ∠7   D ∠8

27. Evaluate the expression 2mn when m = −3 and n = 1.
A −7   C 7
B −6   D 6

28. Which name does not apply to the figure?
A square
B quadrilateral
C polygon
D parallelogram

29. Solve the inequality −y + 10 < y − 10.
A y ≤ 0   C y < 1
B y > 0   D y > 10

30. If $P(A) = \frac{1}{5}$, what are the odds against A happening?
A 1:6   C 4:1
B 1:5   D 5:1

31. Bagel Emporium conducted a survey and found that five out of every ten customers prefer plain bagels. What is this statistic as a percent?
A 50%   C 5%
B 10%   D 25%

32. On a map, $\frac{1}{4}$ in. equals 10 miles. How many miles does a map distance of $2\frac{1}{4}$ in. represent?
A $2\frac{1}{4}$ mi   C 20 mi
B $18\frac{1}{4}$ mi   D 90 mi

33. A mixture of nuts and candy contains 2 pounds of peanuts for every pound of candy. In 30 pounds of mix, how many pounds of candy are there?
A 10 lb   C 20 lb
B 15 lb   D 30 lb

34. Simplify the expression using positive exponents: $\frac{h^4}{h^3}$.
A $\frac{1}{h}$ C $\frac{1}{2h}$
B 4h   D h

Cumulative Test
Chapter 10 Form A, continued

35. Find the value of 3^{-2}.
   A $\frac{1}{9}$
   B -9
   C $-\frac{1}{9}$
   D 9

36. Combine like terms: $a + 2a - 5a$.
   A $-3a$
   B $-3a^3$
   C $-2a$
   D $-2a^3$

37. Sean wants to put a braid trim around a box that measures 14 inches by 12 inches. If the braid costs $0.10 per linear inch, how much will the decoration cost?
   A $2.60
   B $5.20

38. Find the area of the figure.
   A 7.5 in^2
   B 15 in^2

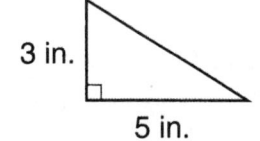

   3 in.
   5 in.

39. A camera that costs $200 wholesale is sold in a store for $233. What is the percent increase in price?
   A 33%
   B 16.5%
   C 14.2%
   D 66%

40. Find the total interest on a simple interest loan if the amount borrowed is $1000 at 5% for 3 years. Assume one payment is made at the end of the 3 years.
   A $1150.00
   B $150.00

Name _____ Date _____ Class _____

Chapter 10 Cumulative Test
Form B

1. Solve the system:
   $2x + 5y = 0$
   $x = -3y + 1$
   - **A** (−5, 2)
   - **B** (0, 0)
   - **C** (2, 5)
   - **D** (−2, 0)

2. A sample space consists of 84 separate events that are equally likely. What is the probability of each?
   - **F** 0
   - **G** $\frac{1}{84}$
   - **H** 1
   - **J** 84

3. Select the graph of the solution set for $0.6x + 2 > 8$.
   - **A** (number line, open circle at 10, shaded left)
   - **B** (number line, open circle at 4, shaded right)
   - **C** (number line, open circle at 2, shaded right)
   - **D** (number line, shaded between around 4 and 8)

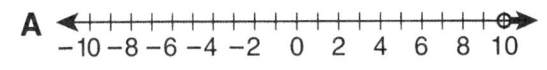

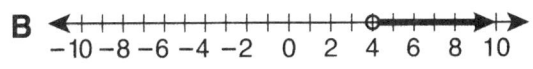

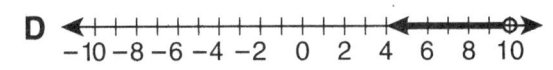

4. 16^4 means:
   - **F** 16 ÷ 4
   - **G** 16 • 4
   - **H** 16 • 16 • 16 • 16
   - **J** 16 • 16 • 16 • 16 • 16

5. Express 7.1 as a percent.
   - **A** 710%
   - **B** 71%
   - **C** 0.71%
   - **D** 0.0071%

6. Dennis measured a board to be 102 inches. How many feet long is the board?
   - **F** 34 ft
   - **G** 10.2 ft
   - **H** 8.5 ft
   - **J** 51 ft

7. A company produces 443,000 small appliances per year. What is this number in scientific notation?
   - **A** 4.43×10^6
   - **B** 44.3×10^4
   - **C** 4.43×10^5
   - **D** 4.43×10^{-5}

8. Find the missing length in the right triangle to the nearest tenth.
   - **F** 9.5 m
   - **G** 14.9 m
   - **H** 110.5 m
   - **J** 221 m

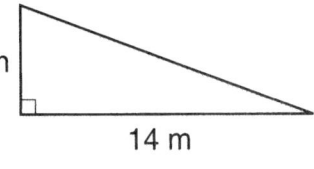

   (right triangle with legs 5 m and 14 m)

9. Simplify $-3 + (-15) - (-7)$.
   - **A** 5
   - **B** −5
   - **C** 11
   - **D** −11

10. A photograph that is 4 inches high by 6 inches wide is to be scaled to 12 inches wide. To keep the picture in proportion, how high will it need to be?
    - **F** 18 in.
    - **G** 5.3 in.
    - **H** 6 in.
    - **J** 8 in.

11. Foodmart has your favorite cereal on sale. A 28-ounce box sells for $1.40. Shop-n-Save has a 32-ounce box for $1.75. Which would be the better buy and price?
    - **A** the 28-oz box at $0.050 per oz
    - **B** the 32-oz box at $0.050 per oz
    - **C** the 28-oz box at $0.055 per oz
    - **D** the 32-oz box at $0.055 per oz

Copyright © by Holt, Rinehart and Winston.
All rights reserved.

Holt Pre-Algebra

Cumulative Test
Chapter 10 Form B, continued

12. Which angles in the figure are supplementary?

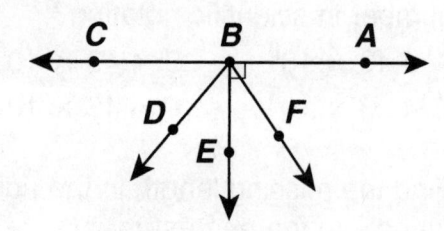

F $\angle ABE$ and $\angle CBE$
G $\angle ABE$ and $\angle FBE$
H $\angle ABF$ and $\angle EBF$
J $\angle ABF$ and $\angle CBD$

13. Solve $-14 + 3h = 13$.

A $h = 27$ C $h = -\frac{1}{3}$
B $h = 9$ D $h = 3$

14. The perimeter of the square addition to Mrs. Weagley's house is 100 feet. What is the length of one side?

F 15 ft   H 25 ft
G 20 ft   J 400 ft

15. What is "0.8 less than a number" written as an algebraic expression?

A $0.8 - x$ C $x - 0.8$
B $0.8 + x$ D $-0.8 - x$

16. Solve $\frac{16}{45}x = -\frac{4}{9}$.

F $x = \frac{5}{16}$ H $x = -\frac{20}{9}$
G $x = -\frac{5}{4}$ J $x = -\frac{16}{5}$

17. Find the surface area of the figure.

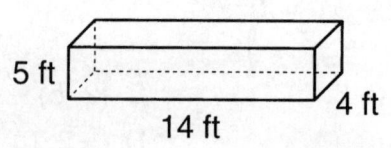

5 ft, 14 ft, 4 ft

A 280 ft^2 C 292 ft^2
B 118 ft^2 D 380 ft^2

18. What is the mean amount spent on taxi fares for a month in which you spent $25, $45, $10, $45, and $30?

F $30   H $31
G $35   J $45

19. Find the square root and round to the nearest thousandth, if necessary.
$\sqrt{34}$

A 5.828   C 5.836
B 5.831   D 34.000

20. Find the value of -5^3.

F -125 H 625
G 125   J $-15,625$

21. Find the perimeter of a rectangle measuring $2\frac{2}{3}$ in. by $2\frac{1}{5}$ in.

A $4\frac{3}{8}$ in.   C $5\frac{13}{15}$ in.
B $4\frac{13}{15}$ in.   D $9\frac{11}{15}$ in.

Name _____ Date _____ Class _____

CHAPTER 10 Cumulative Test
Form B, continued

22. Choose the box-and-whisker plot that matches the data: 43, 51, 52, 52, 69, 69, 71, 87, 65, 83.

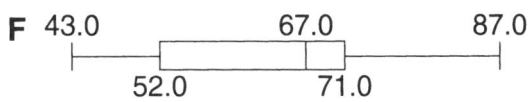

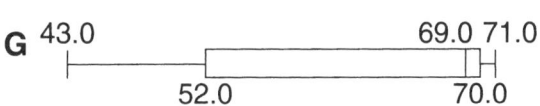

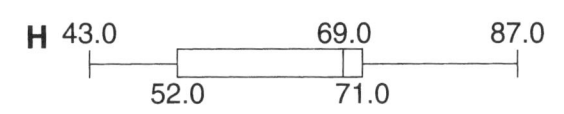

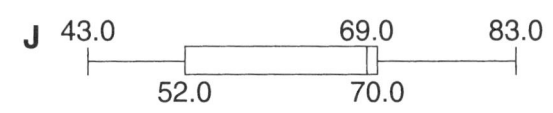

F 43.0 ... 52.0 ... 67.0 ... 71.0 ... 87.0

G 43.0 ... 52.0 ... 69.0 71.0 ... 70.0

H 43.0 ... 52.0 ... 69.0 ... 71.0 ... 87.0

J 43.0 ... 52.0 ... 69.0 ... 70.0 ... 83.0

23. Solve $\frac{2}{5} + \frac{3a}{5} = \frac{5}{4} - \frac{2a}{5}$.

A $a = \frac{33}{4}$ C $a = \frac{33}{20}$

B $a = \frac{17}{4}$ D $a = \frac{17}{20}$

24. What is the surface area of the figure? Use 3.14 for π.   7 ft   10 ft

F 384.65 ft²
G 219.8 ft²
H 296.73 ft²
J 747.32 ft²

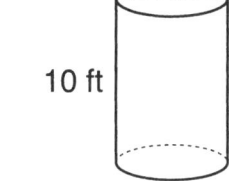

25. Find the measure of $\angle x$.

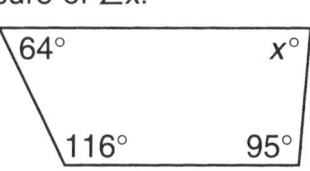

A 116°
B 113°
C 90°
D 85°

(64°, x°, 116°, 95°)

26. Identify one angle that is congruent to $\angle 8$.

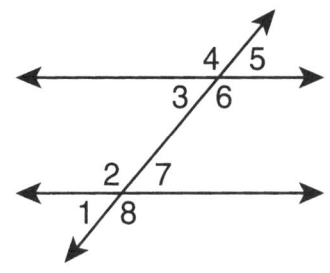

F $\angle 1$
G $\angle 7$
H $\angle 5$
J $\angle 6$

27. Evaluate the expression $5ab$ when $a = -5$ and $b = 3$.

A 3   C 75
B −60   D −75

28. Which name does not apply to the figure?

F trapezoid
G square
H rectangle
J parallelogram

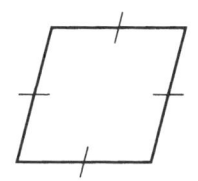

29. Solve the inequality $-10y + 15 < -15y + 20$.

A $y \leq 1$ C $y < 1$
B $y > 1$ D $y \geq 1$

30. If $P(A) = \frac{1}{6}$, what are the odds against A happening?

F 1:6   H 5:1
G 1:5   J 6:1

31. In a recent survey it was found that three out of every five pre-school children prefer singing over doing arts and crafts. Express this statistic as a percent.

A 60%   C 5%
B 30%   D 3%

259   Holt Pre-Algebra

Cumulative Test
Form B, continued

32. The scale on a blueprint drawing is such that $\frac{1}{4}$ in. equals 1 ft. If on the blueprint a living room measures $3\frac{3}{4}$ in. wide, then what is the actual width of the room?

F $3\frac{3}{4}$ ft H $12\frac{3}{4}$ ft
G 4 ft J 15 ft

33. A paint mixture contains 9 gallons of base for every gallon of color. In 230 gallons of paint, how many gallons of color are there?

A 23 gal C 115 gal
B 76 gal D 207 gal

34. Simplify the expression using positive exponents: $\frac{t^4}{t^8}$.

F $\frac{t}{t^2}$ H $\frac{1}{2t}$
G $\frac{1}{t^4}$ J t^4

35. Find the value of 9^{-2}.

A $\frac{1}{81}$ C $-\frac{1}{81}$
B -81 D $\frac{1}{18}$

36. Combine like terms: $4a^2 + 2a^2 - 9a^2$.

F $-3a^2$ H $-72a^2$
G $-3a^6$ J $-72a^6$

37. What will it cost to buy ceiling molding to go around a rectangular room that measures 11 feet by 10 feet, if the molding costs $3.31 per linear foot?

A $66.20 C $72.82
B $69.51 D $139.02

38. Find the area of the figure.

F 83 ft²
G 135 ft²
H 67.5 ft²
J 270 ft²

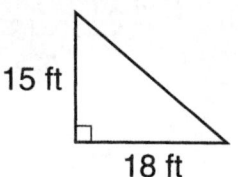

39. Last year Marta earned $356 per week. This year her salary is $371 per week. What is the percent of increase to the nearest tenth of a percent?

A 4.0% C 95.8%
B 4.2% D 96.0%

40. Find the total amount due on a simple interest loan if the principal is $800 with a rate of 8% for 5 years. Assume one payment is made at the end of the 5 years.

F $320.00 H $1120.00
G $864.00 J $2080.00

Name _____ Date _____ Class _____

CHAPTER 10 Cumulative Test
Form C

1. Solve the system:
   $x + 4y = 37$
   $6x + 3y = 75$
   - **A** (8, 8)
   - **B** (−7, 8)
   - **C** (9, 7)
   - **D** (−9, 8)

2. A sample space consists of 124 separate events that are equally likely. What is the probability of each?
   - **F** 0
   - **G** $\frac{1}{124}$
   - **H** 1
   - **J** 124

3. Select the graph of the solution set for $0.5x - 8 > -4 - 3.5x$.
   - **A** number line with closed dot at 6, shaded right
   - **B** number line with closed dot at −2, shaded left
   - **C** number line with open dot at 3, shaded right
   - **D** number line with open dot at −1, shaded left

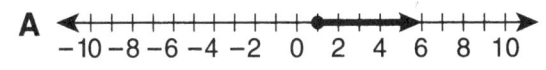

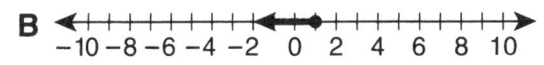

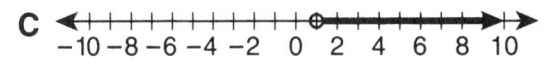

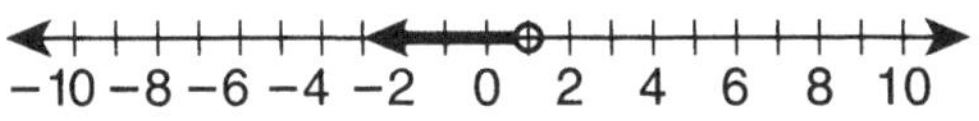

4. $\left(\dfrac{2}{3}\right)^4$ means:
   - **F** $\dfrac{2}{3} \div \dfrac{2}{3}$
   - **G** $\dfrac{2}{3} \cdot \dfrac{2}{3}$
   - **H** $\dfrac{2}{3} \cdot \dfrac{2}{3} \cdot \dfrac{2}{3} \cdot \dfrac{2}{3}$
   - **J** $\dfrac{2}{3} \cdot \dfrac{2}{3} \cdot \dfrac{2}{3} \cdot \dfrac{2}{3} \cdot \dfrac{2}{3}$

5. Express 0.00047 as a percent.
   - **A** 4.7%
   - **B** 0.047%
   - **C** 0.47%
   - **D** 0.0047%

6. You measured the height of a barn to be 234 inches. How many yards tall is the barn?
   - **F** 78 yd
   - **G** 23.4 yd
   - **H** 19.5 yd
   - **J** 6.5 yd

7. A science experiment produced 10,230,000 cells. What is this number in scientific notation?
   - **A** 1.023×10^7
   - **B** 1.023×10^6
   - **C** 1.23×10^7
   - **D** 1.23×10^6

8. Find the missing length in the right triangle to the nearest tenth. 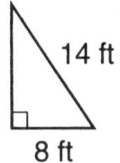
   - **F** 6.0 ft
   - **G** 11.5 ft
   - **H** 16.1 ft
   - **J** 132.0 ft

   (right triangle with hypotenuse 14 ft and leg 8 ft)

9. Simplify $-5(-15) - (-5)$.
   - **A** −15
   - **B** −35
   - **C** 70
   - **D** 80

10. A sketch of a decorative wall panel is 8.5 inches wide by 11 inches tall. If the finished panel is to be scaled to 5 feet wide, how high, to the nearest tenth of a foot, will it need to be to keep the same proportions?
    - **F** 5.5 ft
    - **G** 6.5 ft
    - **H** 42.5 ft
    - **J** 18.7 ft

Name _____ Date _____ Class _____

Cumulative Test
Chapter 10 Form C, continued

11. Walgreen's has bottled water advertised a $0.69 for a 20-ounce bottle and $1.20 for a 32-ounce bottle. Which would be the better buy and price?
  A the 20-oz bottle at $0.035 per oz
  B the 32-oz bottle at $0.035 per oz
  C the 20-oz bottle at $0.038 per oz
  D the 32-oz bottle at $0.038 per oz

12. Which angles in the figure are NOT supplementary?

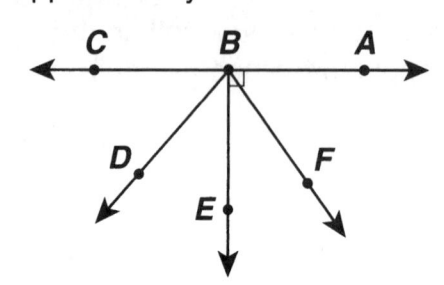

  F $\angle CBE$ and $\angle ABE$
  G $\angle CBF$ and $\angle ABF$
  H $\angle CBD$ and $\angle ABD$
  J $\angle CBD$ and $\angle EBD$

13. Solve $-10 + 3h = 10 - h$.
  A $h = 0$ C $h = -10$
  B $h = 5$ D $h = 10$

14. The area of a square outside patio is 49 square feet. What is the perimeter of the patio?
  F 7 ft   H 28 ft
  G 14 ft   J 49 ft

15. What is "0.8 less than the product of a number and 5" written as an algebraic expression?
  A $0.8 - 5x$ C $5x - 0.8$
  B $0.8 + x + 5$ D $0.8x - 5$

16. Solve $-\frac{27}{48}x = -\frac{3}{5}$.
  F $\frac{15}{16}$ H $-\frac{15}{16}$
  G $\frac{16}{15}$ J $-\frac{16}{15}$

17. Find the surface area of the figure.

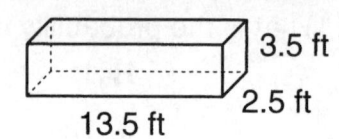

  3.5 ft
  2.5 ft
  13.5 ft

  A 19.5 ft^2 C 118.13 ft^2
  B 89.75 ft^2 D 179.5 ft^2

18. What is the mean cost of your lunch in a week in which you spend $7.50, $5.00, $5.61, $4.50, and $3.55?
  F $3.95   H $5.00
  G $5.23   J $7.50

19. Find the square root and round to the nearest hundredth, if necessary. $\sqrt{222}$
  A 14.89   C 14.91
  B 14.90   D 15.0

20. Find the value of -8^4.
  F -32 H 4096
  G 32   J -4096

21. What is the perimeter of a rectangle measuring $3\frac{1}{6}$ in. by $1\frac{3}{5}$ in.?
  A $1\frac{47}{48}$ in.   C $5\frac{1}{15}$ in.
  B $4\frac{4}{11}$ in.   D $9\frac{8}{15}$ in.

Name _____ Date _____ Class _____

CHAPTER 10 Cumulative Test
Form C, continued

22. Choose the box-and-whisker plot that matches the data: 53, 61, 62, 62, 79, 79, 81, 97, 75, 93.

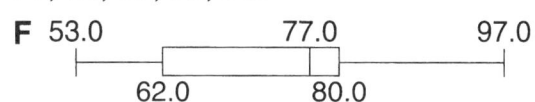

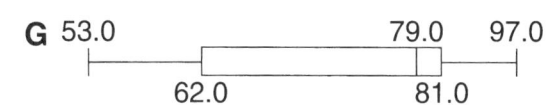

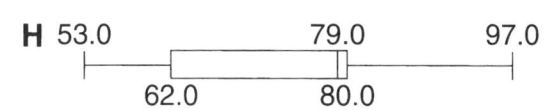

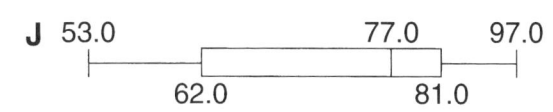

F 53.0 — 62.0 — 77.0 — 80.0 — 97.0

G 53.0 — 62.0 — 79.0 — 81.0 — 97.0

H 53.0 — 62.0 — 79.0 — 80.0 — 97.0

J 53.0 — 62.0 — 77.0 — 81.0 — 97.0

23. Solve $\frac{2}{9} + \frac{3}{5}b = \frac{5}{8} - \frac{2}{5}b + \frac{5}{8}$.

A $b = \frac{53}{36}$
B $b = -\frac{53}{36}$
C $b = \frac{37}{36}$
D $b = \frac{36}{72}$

24. What is the surface area of the figure? Use 3.14 for π and round to the nearest hundredth.

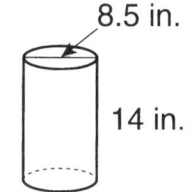

F 794.03 in^2
G 1201.05 in^2
H 487.09 in^2
J 747.32 in^2

(8.5 in., 14 in.)

25. Find the measure of $\angle x$.

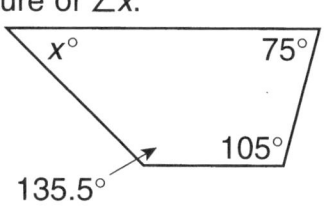

A 44.5°
B 45.5°
C 75°
D 15°

(x°, 75°, 105°, 135.5°)

26. Identify all angles congruent to $\angle 1$.

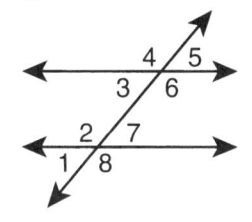

F $\angle 2, \angle 7, \angle 8$
H $\angle 3, \angle 5, \angle 7$
G $\angle 2, \angle 3, \angle 4$
J $\angle 5, \angle 6, \angle 7$

27. Evaluate the expression $6ab^2$ when $a = 4$ and $b = -3$.

A 216
B −216
C 72
D −72

28. Which name does not apply to the figure?

F prism
G polygon
H trapezoid
J quadrilateral

29. Solve the inequality $-9.5y + 14 < -6y - 7.3$. Round to the nearest tenth.

A $y \geq 6.1$
B $y \leq 6.1$
C $y > 6.1$
D $y < 6.1$

30. If $P(A) = \frac{1}{3}$, what are the odds against A happening?

F 4:3
G 3:2
H 2:3
J 2:1

31. A study showed that seven out of every eight people regularly brush their teeth before going to bed. Express this statistic as a percent.

A 62.5%
B 70%
C 87.5%
D 92%

263

Holt Pre-Algebra

Name _____ Date _____ Class _____

CHAPTER 10 **Cumulative Test**
Form C, continued

32. On a scale model of a garden $\frac{1}{4}$ inch equals 1 yard. If the model is $5\frac{3}{4}$ inch wide, then what is the actual width of the garden?

   F $1\frac{3}{16}$ yd H $5\frac{3}{4}$ yd
   G 6 yd J 23 yd

33. A punch recipe calls for 5 quarts of fruit juice for every quart of sparkling water. In 36 quarts of punch, how many quarts of fruit juice are there?

   A 6 qt C 36 qt
   B 30 qt D 41 qt

34. Simplify the expression using positive exponents: $\frac{t^4 u^5}{t^8 u}$.

   F $\frac{u}{t}$ H $\frac{u^5}{u^2}$
   G $\frac{u^4}{t^4}$ J ut

35. Find the value of $(-3)^{-3}$.

   A $-\frac{1}{27}$ C $\frac{1}{27}$
   B -27 D 27

36. Combine like terms: $3a^2 + 2b^2 - 3a^2$.

   F $-2ab$ H $7ab^2$
   G $2a^4$ J $2b^2$

37. A border for a flower bed costs $3.50 per foot for the first 10 feet and $2.75 per foot for every additional foot. How much will the border cost for a flower bed that measures 12 feet by 10 feet?

   A $156.00 C $121.00
   B $128.50 D $345.00

38. Find the area of the figure.

   F 4.75 ft²
   G 9.5 ft²
   H 9.75 ft²
   J 19.5 ft²

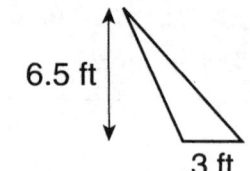

   6.5 ft
   3 ft

39. Dianne made 230 chocolate chip cookies and 125 butterscotch cookies last week. This week, she made 380 cinnamon raisin cookies. What is the percent increase in the number of cookies she baked to the nearest tenth of a percent?

   A 6.6% C 39.5%
   B 7.0% D 45.7%

40. Find the total amount due on a simple interest loan if the principal is $20,800 with a rate of 5.5% for 10 years. Assume one payment is made at the end of the 10 years.

   F $9360 H $32,240
   G $11,440 J $21,944

Name _____ Date _____ Class _____

Cumulative Test
Chapter 11 Form A

Select the best answer.

1. Which of the following equations is not linear?
   A $x + y^2 = 7$ B $5x = y + 2$

2. Solve the system of equations:
   $-3x + y = -3$
   $y = x - 3$.
   A $(0, -3)$ B $(-3, 0)$

3. Solve $4y + 1 = 8 + 2y$.
   A $y = -\frac{2}{7}$ B $y = \frac{7}{2}$

4. In Mrs. Trista's class there are 15 girls and 20 boys. Mrs. Trista randomly selects one student. What is the probability it will be a girl?
   A $\frac{3}{4}$ B $\frac{3}{7}$

5. Express 55% as a decimal.
   A 0.0055   C 0.55
   B 0.055   D 5.5

6. Which statement explains why the graph is misleading?

   100 books   200 books

   A The numbers are not realistic.
   B The area of the books distorts the comparison.

7. Solve for l: $P = 2w + 2l$.
   A $l = \frac{P - 2w}{2}$
   B $l = P - w$

8. Solve $\frac{1}{4}x = 1$.
   A $x = 4$
   B $x = 0.25$

9. Dilation of a figure changes _____.
   A size and shape
   B size only

10. $10 is 25% of what amount?
    A $2.50   C $40
    B $25   D $100

11. To determine an experimental probability, do _____.
    A multiple trials
    B a tree diagram

12. What scale factor relates a 5-inch scale model to a 30-foot statue?
    A 1: 72   C 6: 5
    B 1: 6   D 1 in. to 30 ft

13. Sue left a $2.00 tip for a meal that cost $15. To the nearest tenth, at what rate did she tip?
    A 22.5%   C 13.3%
    B 20%   D 6.6%

Name _____ Date _____ Class _____

CHAPTER 11 Cumulative Test
Form A, continued

14. Tessellations are patterns that _____.

 A are rotated
 B are shifted and rotated
 C cover a plane with no gaps
 D are changed in size and shape

15. Find the slope of the line that passes through (1, 2) and (2, 4).

 A 2
 B $\frac{1}{2}$

16. Calculate the means of the *x*- and *y*-coordinates for the data:

x	3	3	5	4	5
y	4	3	7	5	6

 A 4; 5
 B 4; 6

17. If a 3-inch cube is built from 1-inch cubes, what is the ratio of the corresponding volumes?

 A 3:1
 B 27:1

18. What is the slope-intercept form of the line passing through the points (−2, 5) and (6, 1)?

 A $y = -\frac{1}{2}x + 4$
 B $y = x - 2$

19. Express 220% as a decimal.

 A 0.22
 B 2.2
 C 22.0
 D 220.0

20. Choose the equation which means "*y* varies directly with *x*, and when *y* is 10, *x* is 5."

 A $10y = 5x$
 B $y = 5x$
 C $y = 2x$
 D $2 = x$

21. 50% of 201 is about what number?

 A 100
 B 50

22. The graph represents which inequality?

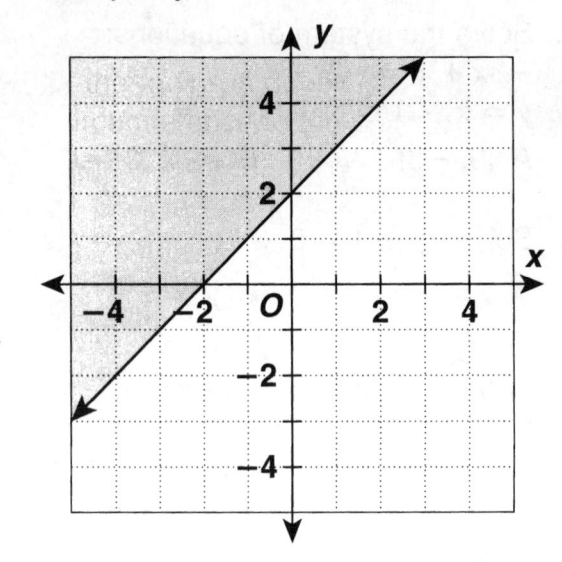

 A $y \leq -x + 3$
 B $2y < -2x - 5$
 C $y > -x + 2$
 D $y \geq x + 2$

23. Simplify $\frac{\sqrt{36}}{10}$.

 A $\frac{3}{5}$
 B $\frac{\sqrt{18}}{5}$

24. A child's large wooden cube measures 12 inches by 12 inches by 12 inches. Find the surface area of the cube.

 A 1728 in²
 B 864 in²

25. Find the unknown number in the proportion $\frac{6}{x} = \frac{24}{5}$.

 A $x = 1.25$
 B $x = 4$

26. In a three-dimensional figure, a(n) _____ is a flat surface.

 A vertex
 B edge
 C face
 D line

Name _____ Date _____ Class _____

CHAPTER 11 Cumulative Test
Form A, continued

27. Find the measure of the smaller angle.

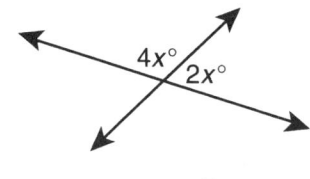

    A 60°   **B** 30°

28. What is the point-slope form of the line with slope 3 that passes through (2, −1)?

    A $y + 1 = 3(x - 2)$
    B $y - 1 = -3(x + 2)$
    C $y + 1 = 3(x + 2)$
    D $y - 1 = -3(x - 2)$

29. The two figures shown are in proportion. Find the value of x.

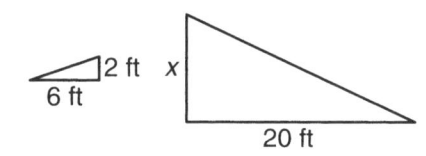

    A $6\frac{2}{3}$ ft   **B** 60 ft

30. The perimeter of a garden is to be at least 90 feet. If the width is 20 feet, then which statement correctly indicates the length?

    A $L \geq 70$ ft   **C** $L \geq 30$ ft
    B $L \geq 50$ ft   **D** $L \geq 25$ ft

31. Simplify $2 + 6 \cdot (-2)$.

    A −16   **B** −10

32. Express the ratio of 5 pieces of red candy to 20 total pieces of candy as a fraction in lowest terms.

    A $\frac{1 \text{ red}}{4 \text{ total}}$ **B** $\frac{3 \text{ red}}{4 \text{ total}}$

33. Find the perimeter of the parallelogram.

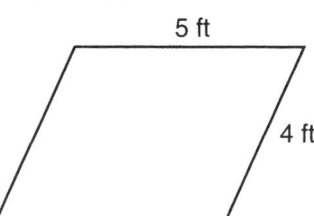

    A 40 ft
    B 20 ft
    C 18 ft
    D 9 ft

34. Find the average number of customer complaints over the past 8 days at Clara's Coffee Cart: 8, 2, 15, 15, 20, 15, 3, and 7.

    A 10.6   **B** 12.1

35. Find the surface area of the given figure. Use 3.14 for π.

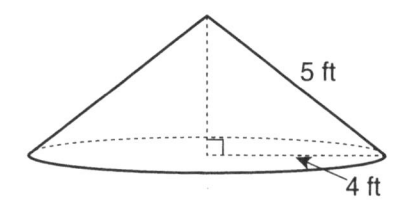

    A 251.2 ft²   **B** 113.04 ft²

36. Find the value of x in the parallelogram.

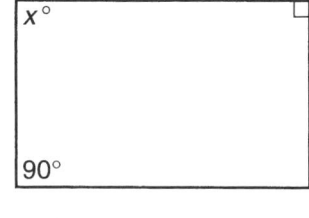

    A $x = 90$
    B $x = 180$
    C $x = 360$
    D not enough information

37. Express $(x)(x)(x)(x)(x)(x)(x)(x)$ using exponents.

    A x^8 **C** $8x$
    B $x + 8$ **D** 8^x

38. What is an equation for the line that is perpendicular to $x - y = 5$ and passes through (2, 2)?

    A $x + y = 4$ **B** $y - x = 4$

Cumulative Test
Chapter 11 Form A, continued

39. Find the measure of ∠CBD.

A 67°

B 23°

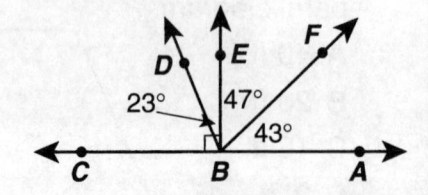

40. Rey flips a coin and spins a number spinner, then records the outcomes. These two events are _____.

A independent events

B dependent events

C random numbers

D permutations

41. Find the value of $_5P_4$.

A 120

B 1.25

C 5

D 1

42. Which of the graphs represents the inequality $x \geq -2$?

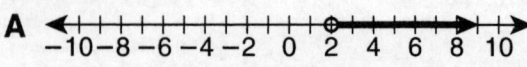

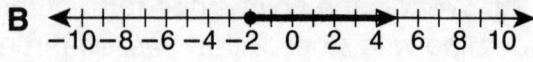

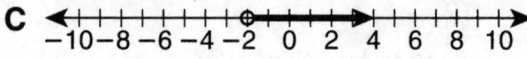

A, B, C, D

Name _____ Date _____ Class _____

Cumulative Test
Chapter 11 Form B

Select the best answer.

1. Which of the following equations is not linear?
   A $6x + 3y = 75$ C $y = 3$
   B $0.5x = y$ D $y = x^2$

2. Solve the system of equations:
   $x - 2y = 7$
   $-x + 3y = -14$.
   F $(-7, -7)$ H $(-7, -1)$
   G no solution   J $(7, 0)$

3. Solve $-3y + 9 = -7 + 9y$.
   A $y = -\frac{3}{4}$ C $y = \frac{3}{4}$
   B $y = \frac{4}{3}$ D $y = 3$

4. There are 87 green balls and 41 red balls in a bag. If one is randomly selected, what is the probability that it will be a green ball?
   F $\frac{87}{41}$ H $\frac{41}{128}$
   G $\frac{87}{128}$ J $\frac{1}{128}$

5. Express 3.2% as a decimal.
   A 0.0032   C 0.32
   B 0.032   D 32

6. Which statement explains why the graph is misleading?

   1980 = $1.00
   1990 = $0.63
   2000 = $0.48

   F The hands distract from the data.
   G The numbers are not realistic.
   H Nothing is wrong with the graph.
   J The area changes more than the amount.

7. Solve for h: $V = \frac{1}{3}Bh$.
   A $h = \frac{V}{3B}$ C $h = \frac{B}{3V}$
   B $h = \frac{3B}{V}$ D $h = \frac{3V}{B}$

8. Solve $-\frac{1}{8}x = 18$
   F $x = -144$ H $x = -2.25$
   G $x = 144$ J $x = 2.25$

9. Changing the size but not the shape of a figure is called _____.
   A dilation   C reflection
   B rotation   D translation

10. $75 is 20% of what amount?
    F $15   H $375
    G $90   J $1500

Cumulative Test
Chapter 11 Form B, continued

11. Repeating trials many times to try to determine the likelihood of an event is related to _____.
    A dependent events
    B theoretical probability
    C experimental probability
    D the Fundamental Counting Principle

12. What scale factor relates a 10-inch scale model to a 50-foot actual model?
    F 1:5
    G 1:60
    H 5:1
    J 1:6

13. The sales tax is $44.75 on a TV that sells for $895. What is the sales tax rate?
    A 40%
    B 20%
    C 5%
    D 4.5%

14. What is the name given to a repeating pattern of figures that completely covers a plane with no gaps or overlaps?
    F tessellation
    G transformation
    H correspondence
    J rotational symmetry

15. Find the slope of the line that passes through (3, −1) and (2, 3).
    A 2
    B −2
    C 4
    D −4

16. Calculate the means of the x- and y-coordinates for the data:

x	14	15	16	16	18
y	13	14	17	15	16

    F 15.8; 15
    G 5; 15
    H 16; 15
    J 10.8; 7.5

17. If a 4-inch cube is built from 1-inch cubes, what is the ratio of the corresponding volumes?
    A 4:1
    B 16:1
    C 24:1
    D 64:1

18. What is the slope-intercept form of the line passing through the points (−5, 0) and (2, −2)?
    F $y = \frac{5}{7}x - \frac{10}{7}$
    G $y = \frac{5}{7}x - \frac{10}{7}$
    H $y = \frac{2}{7}x - \frac{10}{7}$
    J $y = -\frac{2}{7}x - \frac{10}{7}$

19. Express 4% as a decimal.
    A 0.04
    B 0.40
    C 4.0
    D 40.0

20. Choose the equation which means "y varies directly with x, and y is 12 when x is 4."
    F $y = 3x$
    G $y = 12x$
    H $y = 12$
    J $12 = 3x$

21. 19 is about what percent of 82?
    A 25%
    B 35%
    C 40%
    D 50%

Cumulative Test
Form B, continued

22. The graph represents which inequality?

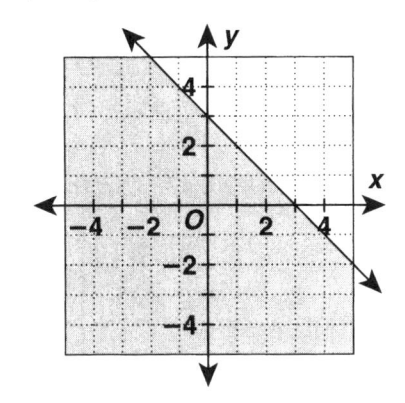

- F $y \leq -x + 3$
- G $y < -x + 3$
- H $y > -x + 3$
- J $y \geq -x + 3$

23. Simplify $\frac{\sqrt{125}}{20}$.

- A $\frac{\sqrt{25}}{5}$
- B $\frac{25}{4}$
- C $\frac{5}{2}$
- D $\frac{\sqrt{5}}{4}$

24. A briefcase measures 10 inches by 4.9 inches by 13.7 inches Find the surface area of the briefcase.

- F 671.3 in^2
- G 506.3 in^2
- H 253.1 in^2
- J 28.6 in^2

25. Find the unknown number in the proportion $\frac{5}{x} = \frac{20}{16}$.

- A $x = 40$
- B $x = 6.2$
- C $x = 4$
- D $x = 0.16$

26. In a three-dimensional figure, an edge is where two _____ meet.

- F vertices
- G curves
- H points
- J faces

27. Find the measure of the smaller angle.

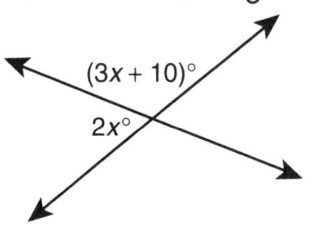

- A 112°
- B 90°
- C 68°
- D 34°

28. What is the point-slope form of the line with slope −2 that passes through (3, −1)?

- F $y + 1 = -2(x - 3)$
- G $y - 1 = -2(x - 3)$
- H $y + 1 = -2(x + 3)$
- J $y - 1 = -2(x + 3)$

29. The wheelchair ramp shown below is going to be rebuilt, with the proportions staying the same. How many feet will it rise if the new length is going to be 9 feet?

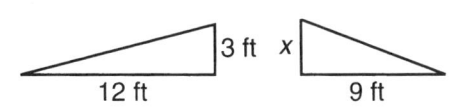

- A 36 ft
- B 4 ft
- C 3 ft
- D 2.25 ft

30. Andrea wants the area of her rectangular garden to be at least 36 square feet. If the garden is 4 feet wide, what must the length be?

- F $L \geq 9$ ft
- G $L > 9$ ft
- H $L \geq 14$ ft
- J $L \leq 14$ ft

31. Simplify $72 \div 8 + 8 \cdot (-4)$.

- A −23
- B −68
- C 41
- D −18

Name _____ Date _____ Class _____

CHAPTER 11 Cumulative Test
Form B, continued

32. Express the ratio of 11 cars to 55 people as a fraction in lowest terms.

F $\dfrac{1 \text{ car}}{5 \text{ people}}$ H $\dfrac{5 \text{ cars}}{1 \text{ person}}$

G $\dfrac{11 \text{ cars}}{55 \text{ people}}$ J $\dfrac{44 \text{ cars}}{55 \text{ people}}$

33. Find the perimeter of the parallelogram. 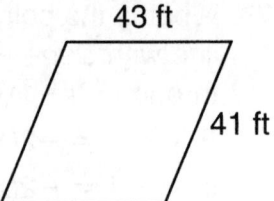
- A 168 ft
- B 127 ft
- C 125 ft
- D 84 ft

(43 ft, 41 ft)

34. Find the average number of phone calls made by a salesperson over the past 8 hours: 8, 21, 3, 19, 27, 41, 35, 32.
- F 24
- G 23.25
- H 24.5
- J 27

35. Find the surface area of the given figure. Use 3.14 for π. 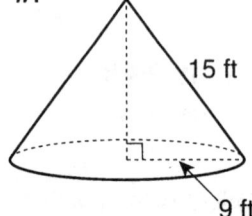
- A 3815.1 ft²
- B 1271.7 ft²
- C 678.24 ft²
- D 423.9 ft²

(15 ft, 9 ft)

36. Find the value of x in the parallelogram. 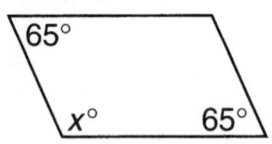
- F x = 65
- G x = 90
- H x = 115
- J x = 145

(65°, 65°, x°)

37. Express $(-6x)(-6x)(-6x)(-6x)$ using exponents.
- A $-6x^4$
- B $-24x$
- C $(-6x)^4$
- D $-(-6x)^4$

38. What is an equation for the line that is parallel to $3x - 2y = 6$ and passes through $(3, -1)$?
- F $2y - 3x = -11$
- G $2y + 3x = -7$
- H $-3x + 2y = 7$
- J $2x + 3y = 3$

39. Find the measure of $\angle DBF$. 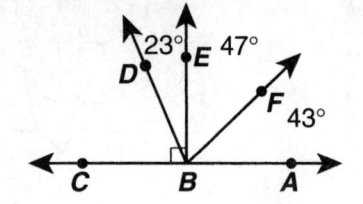
- A 203°
- B 70°
- C 43°
- D 47°

40. Claire draws a card from a deck, puts it back, and draws another card from the same deck. These two events are _____.
- F independent events
- G dependent events
- H random numbers
- J permutations

41. Find the value of $_8C_6$.
- A 20,160
- B 56
- C 28
- D 48

42. Which of the graphs represents the inequality $3x \geq 6$? 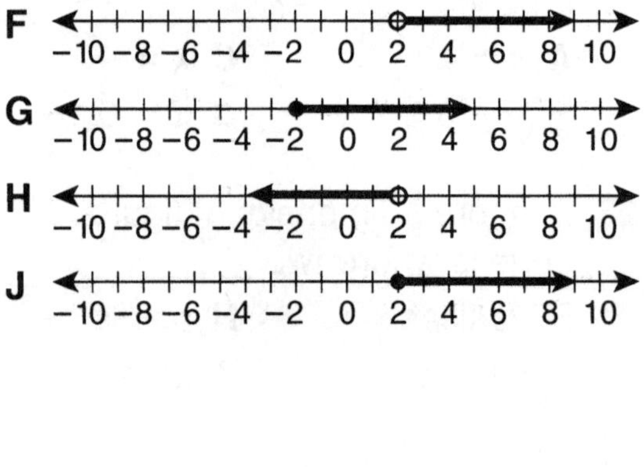
- F
- G
- H
- J

272   Holt Pre-Algebra

Name _____ Date _____ Class _____

CHAPTER 11 Cumulative Test
Form C

Select the best answer.

1. Which of the following equations is not linear?
   A $6x = 7$
   B $0.5x^2 = y$
   C $y = x + 4$
   D $y = x$

2. Solve the system of equations:
   $5x - 4y = 15$
   $-3x + 6y = -9$.
   F $(-1, 1)$
   G no solution
   H $(-1, 3)$
   J $(3, 0)$

3. Solve $-4y + 7 = -7 + 4y - 4$.
   A $y = -\frac{9}{4}$
   B $y = \frac{9}{4}$
   C $y = 2$
   D $y = \frac{4}{9}$

4. A regular deck of cards has four suits with 13 cards per suit. The deck is shuffled several times. If one card is randomly selected, what is the probability that it will be a heart or a four?
   F $\frac{4}{13}$
   G $\frac{17}{52}$
   H $\frac{1}{13}$
   J $\frac{1}{4}$

5. Express 0.82% as a decimal.
   A 0.0082
   B 0.082
   C 0.82
   D 8.2

6. Which statement explains why the graph is misleading?

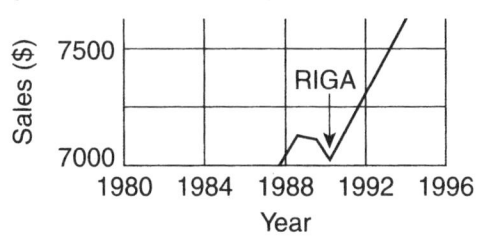

   F The scale is misleading.
   G It does not show this year's data.
   H The numbers are not realistic.
   J Nothing is wrong with the graph.

7. Solve for C: $F = \frac{9}{5}C + 32$.
   A $C = \frac{5}{9}(F + 32)$
   B $C = \frac{5}{9}(F - 32)$
   C $C = \frac{9}{5}F + 32$
   D $C = \frac{5}{9}F + 32$

8. Solve $-\frac{4}{3}x = 12$.
   F $x = -16$
   G $x = 16$
   H $x = -9$
   J $x = 9$

9. When a figure is dilated, it _____.
   A changes size but not shape
   B changes size and shape
   C is translated
   D is rotated

10. $3000 is 150% of what amount?
    F $1500
    G $2000
    H $3000
    J $4500

11. In experimental probability, the probability is determined by _____.
    A tree diagrams
    B repeated trials
    C mathematical calculations
    D Fundamental Counting Principle

Name _____ Date _____ Class _____

CHAPTER 11 Cumulative Test
Form C, continued

12. What scale factor relates a 12-inch scale model to a 72-foot actual model?
F 1:6
G 1:72
H 6:1
J $\frac{1}{12}$ in. to 1 ft

13. A softball bat, with tax, costs $37.17. If the sales tax is $2.17, what is the sales tax rate?
A 5.8%
B 7%
C 6.2%
D 4.5%

14. Which of the following figures will not form a tessellation?
F, G, H, J

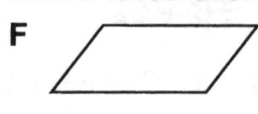

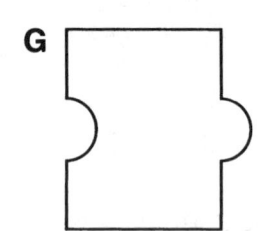

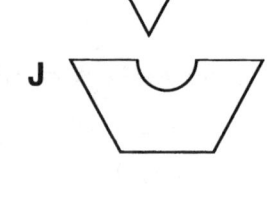

15. Find the slope of the line that passes through (−1, −3) and (−2, 2).
A −5
B $-\frac{1}{5}$
C −2
D $\frac{1}{3}$

16. Calculate the means of the x- and y-coordinates for the data:

x	114	115	116	116	118
y	113	114	117	115	116

F 115.8; 115
G 115; 115.8
H 116; 115
J 115.8; 117

17. If a 6-inch cube is built from 1-inch cubes, what is the ratio of the corresponding volumes?
A 6:1
B 18:1
C 36:1
D 216:1

18. What is the slope-intercept form of the line passing through the points (−2, 1) and (5, −2)?
F $y = -\frac{3}{7}x + 2$
G $y = \frac{3}{7}x - \frac{1}{7}$
H $y = -\frac{3}{7}x - \frac{1}{7}$
J $y = -\frac{3}{7}x + \frac{1}{7}$

19. Express 0.304% as a decimal.
A 0.304
B 0.00304
C 0.0034
D 0.0304

20. Choose the equation which means "y varies directly with x, and y is 6.5 when x is 2.5."
F $y = 4x$
G $y = 8x$
H $y = 6.5$
J $y = 2.6x$

21. 40% of 118.7 is about what number?
A 48
B 40
C 57
D 30

22. The graph represents which inequality? 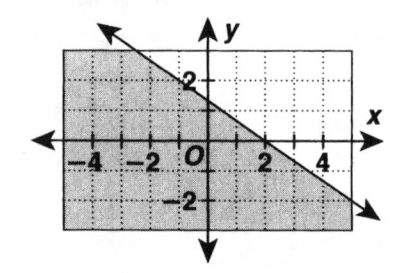
F $y \leq -3x + 2$
G $4y < -3x + 6$
H $4y \leq -3x + 6$
J $4y \geq -3x + 6$

Name _____ Date _____ Class _____

CHAPTER 11 Cumulative Test
Form C, continued

23. Simplify $\frac{\sqrt{49}}{14}$.

    A $\frac{\sqrt{7}}{2}$
    B $\frac{\sqrt{7}}{14}$
    C $\frac{7}{2}$
    D $\frac{1}{2}$

24. Serena is wrapping a gift for her brother's birthday. The package measures 20 inches × 14.5 inches × 23.7 inches. How much wrapping paper does she need to wrap the gift?

    F 6873.0 in²
    G 2215.3 in²
    H 1107.7 in²
    J 58.2 in²

25. Find the unknown number in the proportion $\frac{5}{x} = \frac{40}{15}$.

    A $x = 1.875$
    B $x = 3$
    C $x = 5$
    D $x = 8$

26. In a three-dimensional figure, a vertex is where _____ edges meet.

    F exactly two
    G exactly three
    G two or more
    J three or more

27. Find the measure of the angle.

    A 220°
    B 90°
    C 110°
    D 30°

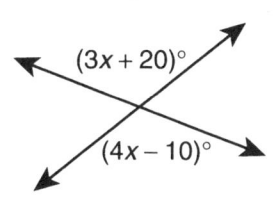

    $(3x + 20)°$
    $(4x - 10)°$

28. What is the point-slope form of the line with slope −3 that passes through (−4, −1)?

    F $y - 1 = -3(x - 4)$
    G $y + 1 = -3(x + 4)$
    H $y - 1 = -3(x + 4)$
    J $y - 1 = 3(x + 3)$

29. For the incline shown below, what measurement in the horizontal direction corresponds to 12 feet in the vertical direction?

    3 ft
    15 ft

    A 2.4 ft
    B 3 ft
    C 15 ft
    D 60 ft

30. The area of a triangle must be no greater than 75 cm². If the base is 20 cm, what must the height be?

    F $h \leq 3.25$ cm
    G $h \geq 3.25$ cm
    H $h \leq 7.5$ cm
    J $h < 7.5$ cm

31. Simplify $18 + 8 \div (-4) \cdot 7 - 3^2$.

    A 36.5
    B −19
    C −8.07
    D −5

32. Express the ratio of 24 balls for 80 cats as a fraction in lowest terms.

    F $\frac{3 \text{ balls}}{10 \text{ cats}}$
    G $\frac{24 \text{ balls}}{80 \text{ cats}}$
    H $\frac{10 \text{ balls}}{3 \text{ cats}}$
    J $\frac{80 \text{ balls}}{24 \text{ cats}}$

33. Find the perimeter of the figure created by the triangle and the parallelogram.

    30 ft
    25.5 ft →   25.5 ft
                25.5 ft

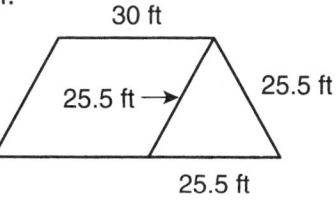

    A 765.0 ft
    B 249.8 ft
    C 111.0 ft
    D 136.5 ft

CHAPTER 11 Cumulative Test
Form C, continued

34. Over the past 9 days, 39, 41, 39, 39, 47, 61, 55, 46, and 52 people entered the hiking trail. Find the mean of the data set to the nearest hundredth.
  F 42.71
  G 46.56
  H 46.00
  J 22.00

35. Find the surface area of the given figure. Use 3.14 for π and round to the nearest tenth.
  A 351.7 ft²
  B 401.9 ft²
  C 452.2 ft²
  D 480.0 ft²

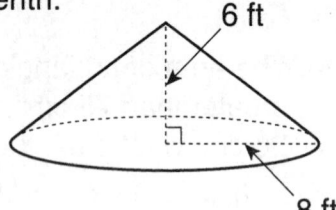

  6 ft
  8 ft

36. Find the value of x in the regular hexagon.
  F x = 120
  G x = 86.7
  H x = 60
  J x = 30

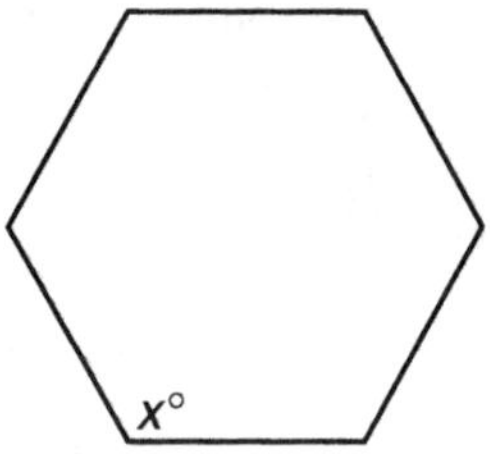

37. Simplify $(-ab)^3 (-ab)^2$.
  A $(-ab)^6$
  B $(-ab)^5$
  C $(ab)^5$
  D $(ab)^6$

38. What is an equation for the line that is perpendicular to $4x - 5y = 20$ and passes through $(-2, -3)$?
  F $4y - 5x = 7$
  G $4y - 5x = -2$
  H $5x + 4y = -22$
  J $-4x + 5y = 2$

39. Find the measure of $\angle CBF$.
  A 137°
  B 70°
  C 67°
  D 43°

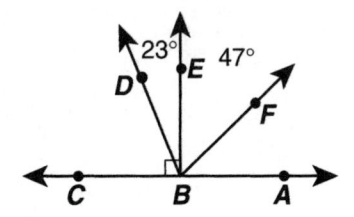

40. A bag contains 10 marbles, 3 of which are blue. Two random picks without replacement constitute _____.
  F independent events
  G dependent events
  H random numbers
  J permutations

41. Find the value of $_8P_6$.
  A 1.33
  B 48
  C 56
  D 20,160

42. Choose the graph which represents the inequality $2x + 4 < 3x + 2$.
  F 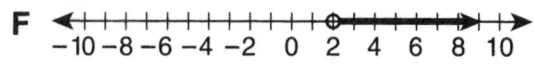
  G 
  H 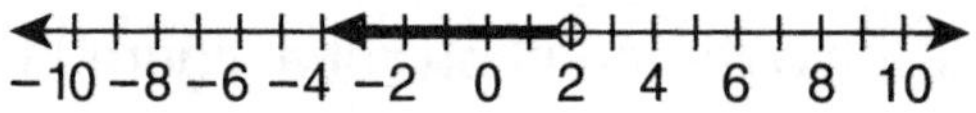
  J

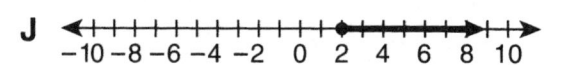

Name _____ Date _____ Class _____

Cumulative Test
Chapter 12 Form A

Select the best answer.

1. Each morning at Park Elementary, teachers take a count of how many students will be buying lunch. On Monday, the cook was told to prepare 340 lunches. If the 1st grade teacher counted 17 lunch buyers in her classroom, what percent of the lunches prepared were for first graders?
   A 5%　　　　B 20%

2. Identify the angle that is complementary to $\angle ABF$.

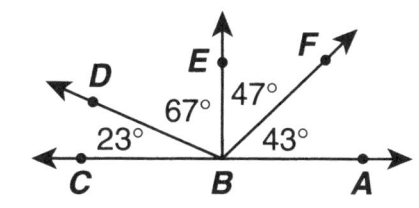

   A $\angle CBE$　　　C $\angle EBF$
   B $\angle CBD$　　　D $\angle FBC$

3. The diameter of a plate is 8.0 inches. What is the area of the flat upper surface? Use 3.14 for π.
   A 50.24 in^2　　　B 25.12 in^2

4. A scuba diver is swimming 3 meters below sea level then descends another 6 meters. What is the diver's new depth?
   A −3 m　　　　B −9 m

5. Find the slope of the line $2x - 3y = 16$.
   A $\frac{3}{2}$　　　　B $\frac{2}{3}$

6. Find the value of $_8C_3$.
   A 5　　　　C 120
   B 56　　　D 360

7. Solve the system of equations:
   $x + y = -3$
   $x - y = -5$.
   A (4, 2)　　　C (−4, 1)
   B (−5, 2)　　D no solution

8. Solve $6c - (5c - 1) = 2$.
   A $c = \frac{1}{11}$　　　B $c = 1$

9. Company A rents copy machines for $300 plus $0.05 per copy. Company B charges of $600 plus $0.01 per copy. For which number of copies is Company B's rate higher?
   A 7500 copies　　C 7000 copies
   B 8000 copies　　D 8500 copies

10. Solve the equation for s:
    $a + b = s + r$.
    A $s = a + b - r$　　C $s = \frac{a}{r} + b$
    B $s = r(a + b)$　　D $s = \frac{a+b}{r}$

11. Find the mean for the data set 49, 52, 52, 52, 74, 67, 55, 55.
    A 52　　　　B 57

12. Find $f(1)$ for $f(x) = 2(2x + 1)$.
    A 6　　　　B 5

Name _____ Date _____ Class _____

CHAPTER 12 Cumulative Test
Form A, continued

13. Express $3^4 3^4$ using positive exponents.
    A 9^{16} C 3^{16}
    B 9^8 D 3^8

14. Simplify $-4b - 4b$.
    A $-8b$ C $8b$
    B 0 D $8b^2$

15. Find the quotient $\frac{2}{3} \div \frac{1}{6}$.
    A $\frac{1}{9}$ B 4

16. Find the rule for the linear function.

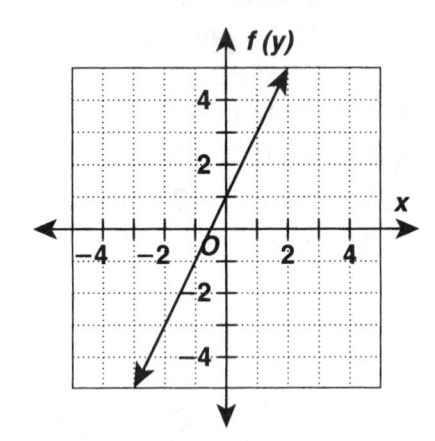

    A $f(x) = -(2x + 1)$
    B $f(x) = 2x + 1$

17. What is 10% of 500?
    A 500   C 5
    B 50   D 0.5

18. Convert 6 hours to seconds.
    A 360 s   C 10,800 s
    B 5400 s   D 21,600 s

19. Solve $-4(-2y) = -2$.
    A $y = -\frac{1}{4}$ C $y = \frac{1}{4}$
    B $y = -4$ D $y = 4$

20. The sum of 6 and four times a number is 38. What is the number?
    A 7   C 11
    B 8   D 16

21. Find the missing value in the proportion $\frac{1}{12} = \frac{x}{60}$.
    A $x = 60$ C $x = 5$
    B $x = 6$ D $x = 1$

22. Solve the inequality $5x < 45$.
    A $x < 9$ B $x > 9$

23. A particular substance has a half-life of 60 minutes. To the nearest thousandth, how much of a 200 mg sample is left after 10 hours?
    A 0.391 mg   B 0.195 mg

Cumulative Test
Form A, continued

24. Which angle is supplementary to ∠7?

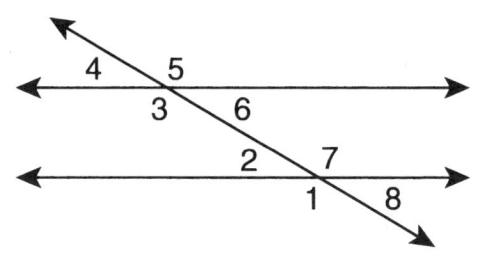

A ∠5 C ∠1
B ∠8 D ∠3

25. Find the perimeter of the figure.

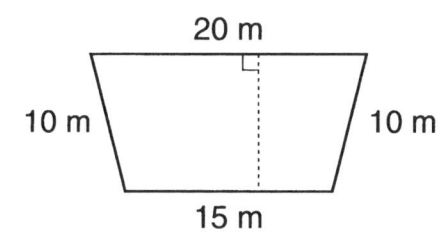

A 55 m B 175 m

26. Find the missing length in the right triangle to the nearest tenth.

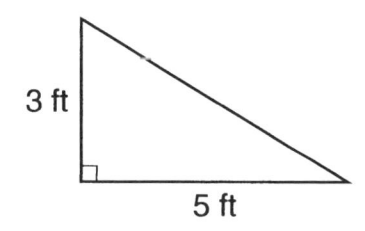

A 4.0 ft B 5.8 ft

27. If a costume maker takes 18 hours to sew 8 costumes, at what rate per hour are the costumes being made?

A $\frac{4}{9}$ costume/hour

B $\frac{4}{18}$ costume/hour

C $\frac{18}{8}$ costume/hour

D $\frac{9}{4}$ costume/hour

28. Find the y-intercept for $x + y = 5$.

A (2, 5) C (1, 0)
B (0, 3) D (0, 5)

29. Use the data in the table to find the inverse variation equation.

x	1	5	10	15
y	5	1	$\frac{1}{2}$	$\frac{1}{3}$

A $y = \frac{5}{x}$ B $y = \frac{1}{5x}$

30. Which ordered pair is a solution to the equation $y = -x + 1$?

A (−1, −3) C (−1, −2)
B (−1, 3) D (−1, 2)

31. What is the point-slope form equation for the line with a slope of 5 that passes through (2, 0)?

A $y = 5(x - 2)$ B $y = -5(x + 2)$

Name _____ Date _____ Class _____

CHAPTER 12 Cumulative Test
Form A, continued

32. How many sides does a regular hexagon have?
  A 4   C 6
  B 7   D 8

33. On a multiple-choice test, each question has 7 possible answers. What is the probability that a random guess on the first question will be correct?
  A 7   C $\frac{1}{7}$
  B 1   D 0

34. 60 is 80% of what number?
  A 750   C 48
  B 75   D 7.5

35. Simplify $\left(\frac{3}{7}\right)^2$.
  A $\frac{6}{14}$ B $\frac{9}{49}$

36. Solve $-\frac{1}{4}x = -23$.
  A $x = 27$ C $x = -88$
  B $x = -69$ D $x = 92$

37. Find the 5th term of the sequence defined by $a_n = 2(n + 1)$.
  A 12   B 10

38. Find the 8th term in the arithmetic sequence 3, 5, 7, 9, ...
  A 19   B 17

39. The price of a couch is reduced from $800 to $675. Find the rate of discount to the nearest tenth of a percent.
  A 15.6%   B 18.5%

40. A large carton in the shape of a cube measures 9 feet on a side. What is the volume of the carton?
  A 18 ft³   C 720 ft³
  B 81 ft³   D 729 ft³

41. Find the area of the triangle.

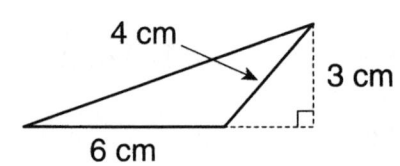

4 cm, 3 cm, 6 cm

  A 12 cm²   B 9 cm²

42. Express 51,000 in scientific notation.
  A 5.1×10^6 C 5.1×10^3
  B 5.1×10^5 D 5.1×10^4

Name _____ Date _____ Class _____

CHAPTER 12 Cumulative Test
Form B

Select the best answer.

1. On Saturday, the Henry County Fair had an attendance of 5625 people. The gates admitted 2645 adults, 2154 children, and the remainder senior citizens. About what percent were senior citizens?
   - **A** 7%
   - **B** 15%
   - **C** 24%
   - **D** 31%

2. Which angle is complementary to $\angle CBD$?

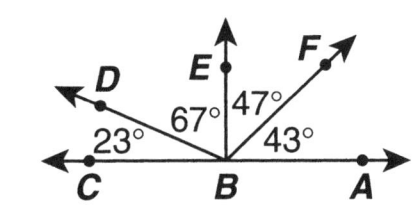

   - **F** $\angle ABF$
   - **G** $\angle FBA$
   - **H** $\angle ABE$
   - **J** $\angle DBE$

3. The diameter of a circle is 6 inches. What is the area? Use 3.14 for π.
   - **A** 3.5 in^2
   - **B** 18.84 in^2
   - **C** 28.26 in^2
   - **D** 28.3 in^2

4. An airplane flies at a cruising altitude of 2900 feet. It descends 1200 feet as it begins to reach its destination. As it approaches the airport, it descends an additional 1550 feet. What is the new altitude of the airplane?
   - **F** 1700 ft
   - **G** 1500 ft
   - **H** 300 ft
   - **J** 150 ft

5. Find the slope of the line $4x - 5y = 32$.
   - **A** $-\frac{5}{4}$
   - **B** $\frac{5}{4}$
   - **C** $-\frac{4}{5}$
   - **D** $\frac{4}{5}$

6. Find the value of $_8C_2$.
   - **F** 28
   - **G** 56
   - **H** 40,320
   - **J** (80,640

7. Solve the system of equations:
   $3x + 3y = 18$
   $2x - 2y = 4$
   - **A** $(-4, 2)$
   - **B** $(4, 2)$
   - **C** $(7, 2)$
   - **D** $(1, 11)$

8. Solve $7c + 6 = 2 + 3c$.
   - **F** $c = -\frac{2}{5}$
   - **G** $c = 1$
   - **H** $c = -1$
   - **J** $c = \frac{4}{5}$

9. A salesperson has two job offers. Company A is offering $200 weekly plus 10% commission on sales. Company B is offering $400 weekly plus 5% commission on sales. For what level of sales does Company A have the better offer?
   - **A** $2000
   - **B** $4000
   - **C** $4100
   - **D** $3900

Name _____ Date _____ Class _____

CHAPTER 12 **Cumulative Test**
Form B, continued

10. Solve the equation for h:
    $V = \frac{1}{3}\pi r^2 h$.

    F $h = \frac{V}{3}\pi r^2$ **H** $h = \frac{V}{\pi r}$

    G $h = \frac{3V}{\pi r^2}$ **J** $h = \frac{V}{\pi}r^2$

11. Find the median weight of a group of apples with individual weights of 4.7, 4.0, 6.2, 6.5, 6.1, 4.7, 4.0, 6.2, 6.3, 6.5, 4.7, 6.2, 6.5, and 6.0 ounces.

    A 4.7 oz **C** 5.6 oz
    B 6.5 oz **D** 6.15 oz

12. Find $f(3)$ for $f(x) = 2(x+1)^2$.

    F $2x^2 + 4x - 1$ **H** 22
    G 32 **J** 20

13. Express $(-4)^4(-4)^9$ using positive exponents.

    A $(16)^{36}$ **C** $(16)^{13}$
    B $(-4)^{13}$ **D** $(4)^{13}$

14. Simplify $-5n + 3n$.

    F $2n$ **H** $-2n$
    G $8n$ **J** $-8n$

15. Find the quotient $\frac{8}{15} \div \frac{4}{5}$.

    A $\frac{2}{3}$ **C** $\frac{2}{15}$
    B $\frac{32}{75}$ **D** $\frac{4}{15}$

16. Find the rule for the linear function.

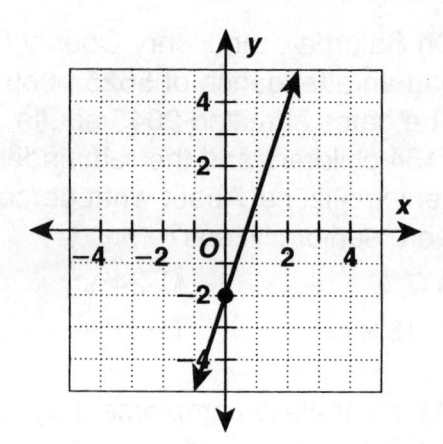

    F $f(x) = -3x - 2$ **H** $f(x) = 3x - 2$
    G $f(x) = 3x + 2$ **J** $f(x) = x - 2$

17. What is $6\frac{1}{5}\%$ of 80,000?

    A 50 **C** 4960
    B 496 **D** 49,600

18. Convert 4 weeks to minutes.

    F 80,640 min **H** 40,320 min
    G 57,600 min **J** 10,080 min

19. Solve $-12 = -3(-3y)$.

    A $y = \frac{4}{3}$ **C** $y = -\frac{4}{3}$
    B $y = 9$ **D** $y = -9$

20. The sum of twice a number and 28 equals 36 plus the number.

    F 21 **H** 8
    G 20 **J** -8

21. Find the missing value in the proportion $\frac{4}{8} = \frac{x}{24}$.

    A $x = 4$ **C** $x = 32$
    B $x = 12$ **D** $x = 256$

Holt Pre-Algebra

Name _____ Date _____ Class _____

CHAPTER 12 Cumulative Test
Form B, continued

22. Solve the inequality $-5x > 45$.
    F $x > 9$ H $x < 9$
    G $x < -5$ J $x < -9$

23. An investment of $2000 will double every 10 years. Which exponential function could be used to calculate the balance in x years?
    A $f(x) = 2000(2)^{x/10}$
    B $f(x) = 2000(2)^x$
    C $f(x) = 2000(x)^{10}$
    D $f(x) = x(10)^2$

24. Which angle is congruent to $\angle 4$?

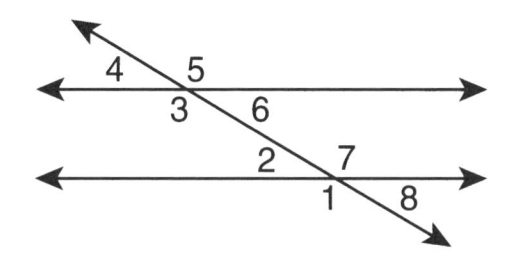

    F $\angle 6$ H $\angle 7$
    G $\angle 8$ J $\angle 2$

25. Find the perimeter of the figure.

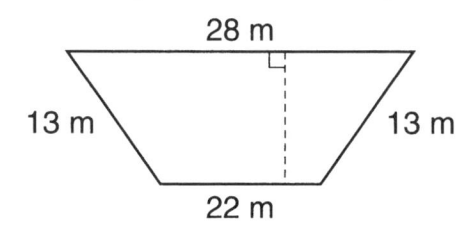

    28 m (top), 13 m (left), 13 m (right), 22 m (bottom)

    A 286 m   C 41 m
    B 76 m   D 28 m

26. Find the missing length in the right triangle. 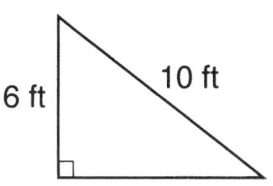
    F 7 ft
    G 8 ft
    H 9 ft
    J 10 ft

    (6 ft, 10 ft)

27. If you can type 1080 words in 40 minutes, then how many words per minute can you type?
    A 180 words/min   C 27 words/min
    B 31 words/min   D 9 words/min

28. Find the y-intercept for $3x + y = 0$.
    F (0, 0)   H (3, 0)
    G (−9, 0)   J (0, −9)

29. Use the data in the table to find the inverse variation equation.

x	15	30	45	60
y	10	5	$3\frac{1}{3}$	$2\frac{1}{2}$

    A $y = \dfrac{150}{x}$ C $y = \dfrac{x}{2}$
    B $y = 2x$ D $y = x + 15$

30. Which ordered pair is a solution to the equation $y = -x + 7$?
    F (4, 12)   H (4, 4)
    G (4, −3)   J (4, 3)

31. What is the point-slope form equation for the line with a slope of $-\frac{2}{3}$ passes through (0, −2)?

A $y - 2 = \frac{2}{3}x$ C $y - 3 = \frac{3}{2}x$

B $y + 2 = \frac{2}{3}x$ D $y + 2 = -\frac{2}{3}x$

32. How many sides does a regular octagon have?

F 4   H 6
G 7   J 8

33. A six-sided die is rolled. What is the probability of rolling a number less than 5?

A 4   C $\frac{2}{3}$
B $\frac{5}{6}$ D $\frac{1}{6}$

34. 560 is 140% of what number?

F 40   H 400
G 78.4   J 19,600

35. Simplify $\left(\frac{8}{3}\right)^2$.

A $21\frac{1}{3}$ C $\frac{8}{9}$
B $\frac{64}{9}$ D $\frac{3}{8}$

36. Solve $\frac{16}{45}x = -\frac{4}{9}$.

F $x = \frac{5}{16}$ H $x = -\frac{20}{9}$
G $x = -\frac{16}{5}$ J $x = -\frac{5}{4}$

37. Find the 5th term of the sequence defined by $a_n = \frac{n+1}{n+2}$.

A $\frac{1}{2}$ C $\frac{6}{7}$
B $\frac{5}{6}$ D $\frac{7}{8}$

38. Find the 13th term in the arithmetic sequence 2, 7, 12, 17, ...

F 62   H 57
G 60   J 55

39. A computer, priced at $1380, is marked down to $971.52. What is the rate of markdown to the nearest tenth of a percent?

A 22.7%   C 29.6%
B 23.4%   D 30.6%

40. Find the volume of a rectangular solid measuring 11 feet by 14 feet by 4 feet.

F 1331 ft³   H 154 ft³
G 616 ft³   J 29 ft³

41. Find the area of the triangle.

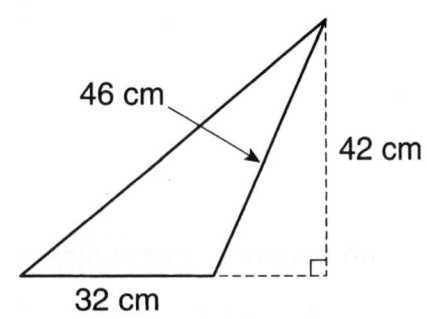

46 cm, 42 cm, 32 cm

A 1344 cm²   C 736 cm²
B 966 cm²   D 672 cm²

42. Express 680,000 in scientific notation.

F 6.8×10^5 H 6.8×10^{-5}
G 6.8×10^4 J 6.8×10^{-4}

Name _____ Date _____ Class _____

Cumulative Test
Chapter 12 Form C

Select the best answer.

1. Every morning Thrush's bakery makes 15 dozen cookies. The baker chooses to make peanut butter, chocolate chip, and oatmeal cookies. If she makes 42 peanut butter, 66 chocolate chip, and the rest oatmeal, what percentage of the cookies are oatmeal?
   - A 22%
   - B 40%
   - C 54%
   - D $63\frac{2}{3}$%

2. Identify the pair of angles that are supplementary.

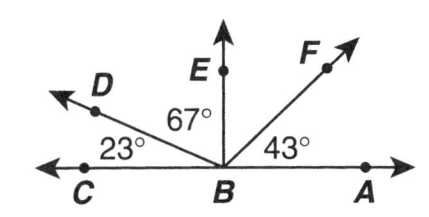

   (figure: rays from point B; ∠CBD = 23°, ∠DBE = 67°, ∠EBA = 43°; C and A are on a line, with D, E, F above)

   - F ∠CBD & ∠ABD
   - G ∠ABF & ∠CBD
   - H ∠DBE & ∠EBF
   - J ∠ABF & ∠FBE

3. The diameter of a discus is about 8.6 inches. To the nearest hundredth, what is the area of one flat surface of the discus? Use 3.14 for π.
   - A 27.00 in²
   - B 29.03 in²
   - C 58.06 in²
   - D 232.23 in²

4. On a cold winter night, the temperature at midnight is 6°F below zero. By five o'clock in the morning the temperature has dropped another four degrees. What is the temperature at this time?
   - F 10°F
   - G 2°F
   - H −10°F
   - J −12°F

5. Find the slope of the line $2x - 5y = -16$.
   - A $-\frac{5}{2}$
   - B $\frac{5}{2}$
   - C $-\frac{2}{5}$
   - D $\frac{2}{5}$

6. Find the value of $_{11}C_3$.
   - F 990
   - G 165
   - G 40,320
   - H 120,960

7. Solve the system of equations:
   $x + y = 14$
   $\frac{1}{3}x = \frac{1}{3}y + \frac{2}{3}$.
   - A (−8, 7)
   - B (7, 7)
   - C (8, 6)
   - D no solution

8. Solve $-6c + 9 = 3 + 4c + 3c$.
   - F $c = -\frac{2}{15}$
   - G $c = \frac{6}{13}$
   - H $c = -\frac{13}{6}$
   - J $c = \frac{13}{6}$

9. A car rental company has two rental rates. The first rate is $45 per day plus $0.15 per mile. The second rate is $90 per day plus $0.07 per mile. If you plan to rent for one day, what is the least number of miles you would need to drive to pay less by taking the second rate?
   - A 501 mi
   - B 282 mi
   - C 563 mi
   - D 205 mi

10. Solve the equation for B: $V = \frac{1}{3} Bh$.
    - F $B = \frac{V}{3h}$
    - G $B = \frac{3h}{V}$
    - H $B = \frac{3V}{h}$
    - J $B = \frac{h}{3V}$

Copyright © by Holt, Rinehart and Winston.
All rights reserved.

285

Holt Pre-Algebra

Cumulative Test
Chapter 12 Form C, continued

11. Find the mean speed of cars measured by radar to the nearest tenth. The individual speeds are 41.0, 43.5, 40.9, 44.3, 40.6, 43.4, 41.6, 40.7, 44.2, 41.0, 41.6, 44.3, 41.7, and 43.5 miles per hour.
- **A** 42.3 mph
- **B** 42.9 mph
- **C** 43.5 mph
- **D** 41.7 mph

12. Find $f(-1)$ for $f(x) = x(x^2 + 1)$.
- **F** 0
- **G** 2
- **H** −2
- **J** 3

13. Express $(-3)^7(-3)^8$ using positive exponents.
- **A** $(-3)^{15}$
- **B** $(3)^{15}$
- **C** $(9)^{56}$
- **D** $(9)^{15}$

14. Simplify $-14mn - (-7mn - 7mn)$.
- **F** $-7mn$
- **G** $-14mn$
- **H** $-28mn$
- **J** 0

15. Find the quotient $\frac{2}{7} \div \frac{1}{8}$.
- **A** $\frac{1}{28}$
- **B** $\frac{2}{7}$
- **C** $\frac{1}{5}$
- **D** $\frac{16}{7}$

16. Find the rule for the linear function.

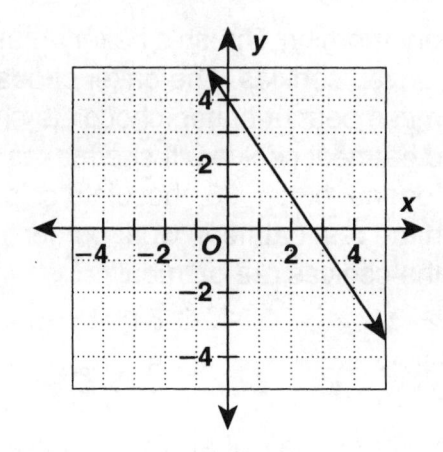

- **F** $f(x) = -2x + 4$
- **G** $f(x) = -\frac{3}{2}x + 4$
- **H** $f(x) = 3x + 4$
- **J** $(x) = \frac{3}{2}x + 4$

17. What is 190% of 3490?
- **A** 663
- **B** 6631
- **C** 66,310
- **D** 663,100

18. Convert 10,800 seconds to days.
- **F** 10 days
- **G** $\frac{1}{4}$ day
- **H** $\frac{1}{8}$ day
- **J** $\frac{1}{16}$ day

19. Solve $-18 = -2(-5y)$.
- **A** $y = -\frac{9}{5}$
- **B** $y = -10$
- **C** $y = \frac{9}{5}$
- **D** $y = 10$

20. A tree 7 feet tall grows at the rate of 2 feet per year. How many years will it take for it to be 15 feet tall?
- **F** 14 yr
- **G** 9 yr
- **H** 11 yr
- **J** 4 yr

Name _____ Date _____ Class _____

CHAPTER 12 Cumulative Test
Form C, continued

21. Find the missing value in the proportion $\frac{x}{72} = \frac{3}{12}$.

A $x = 3$
B $x = 9$
C $x = 18$
D $x = 216$

22. Solve the inequality $-10x < -100$.

F $x < 10$
G $x > 10$
H $x > -10$
J $-x > -10$

23. The half-life of caffeine in children is 3 hours. If a child consumes 40 mg of caffeine, how much is present after 4 hours? Round to the nearest tenth.

A 15.9 mg
B 15.0 mg
C 5.0 mg
D 2.5 mg

24. Which angles are congruent to ∠1?

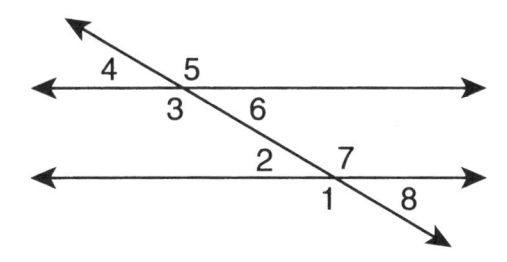

F ∠3, ∠5, ∠7
G ∠4, ∠5, ∠8
H ∠2, ∠3, ∠4
J ∠3, ∠6, ∠7

25. Find the perimeter of the trapezoid.

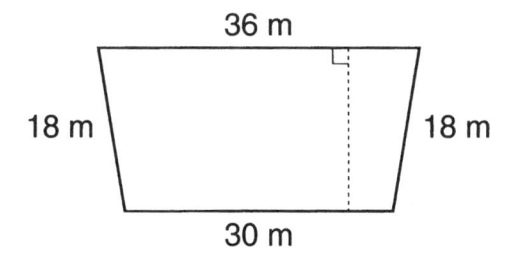

A 540 m
B 102 m
C 54 m
D 84 m

26. To the nearest tenth, find the missing length in the right triangle.

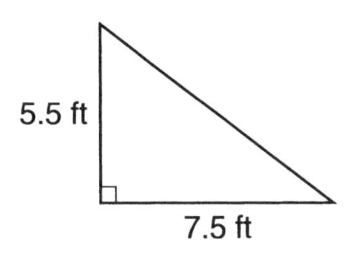

F 2.0 ft
G 4.0 ft
H 13 ft
J 9.3 ft

27. A machine can fill 4818 bags of candy in 0.6 hour. How many bags can be filled in an hour?

A 2891 bags
B 4810 bags
C 6883 bags
D 8030 bags

28. Find the y-intercept for $-2x + 3y = 6$.

F $(-2, 3)$
G $(-2, 0)$
H $(3, 0)$
J $(0, 2)$

29. Use the data in the table to find the inverse variation equation.

x	10	5	4	2
y	0.01	0.02	0.025	0.05

A $y = \frac{1}{10x}$
B $y = 10x$
C $y = \frac{1}{2x}$
D $y = 0.01x$

30. Which ordered pair is a solution to the equation $y = -7x - 33$?

F $(0, -6)$
G $(0, -33)$
H $(0, -40)$
J $(-6, 0)$

Cumulative Test
Chapter 12 Form C, continued

31. What is the point-slope form equation for the line with a slope of $-\frac{2}{3}$ that passes through (5, 2)?

A $y - 5 = -\frac{2}{3}(x - 2)$

B $y - 2 = -\frac{2}{3}(x - 5)$

C $y + 5 = \frac{2}{3}(x - 2)$

D $y + 2 = \frac{2}{3}(x - 5)$

32. How many sides does a regular heptagon have?

F 5   **H** 7
G 6   **J** 9

33. A bag contains four red marbles, three blue marbles, and five green marbles. What is the probability that a randomly selected marble is not blue?

A $\frac{1}{6}$ **C** $\frac{3}{4}$

B $\frac{1}{4}$ **D** $\frac{5}{6}$

34. $2\frac{1}{2}\%$ of what number is 68?

F 27,200   **H** 272
G 2720   **J** 1.7

35. Simplify $\left(\frac{5}{8}\right)^3$.

A $\frac{5}{512}$ **C** $1\frac{3}{5}$

B $\frac{125}{512}$ **D** $15\frac{5}{8}$

36. Solve $-\frac{16}{21}x = -\frac{12}{35}$.

F $x = \frac{9}{20}$ **H** $x = \frac{45}{16}$

G $x = \frac{16}{45}$ **J** $x = \frac{36}{5}$

37. Find the 5th term of the sequence defined by $a_n = 3\left(\frac{n}{n+1}\right)$.

A $\frac{12}{5}$ **C** $\frac{5}{6}$

B $\frac{5}{2}$ **D** $\frac{4}{5}$

38. Find the 25th term in the arithmetic sequence 1, 7, 13, 19, ...

F 151   **H** 145
G 150   **J** 144

39. A TV set costs $345 before tax. The sales tax is $17.25. What is the sales tax rate to the nearest tenth of a percent?

A 4.8%   **C** 5.0%
B 5.3%   **D** 0.5%

40. A rectangular carton measures 15 feet by 19 feet by 13 feet. What is the volume of the carton?

F 247 ft^3 **H** 3705 ft^3
G 285 ft^3 **J** 4275 ft^3

41. Find the area of the triangle.

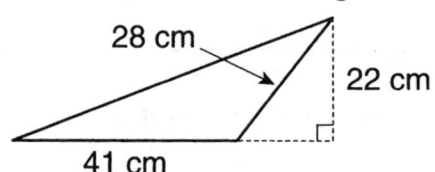

28 cm, 22 cm, 41 cm

A 1148 cm^2 **C** 451 cm^2
B 574 cm^2 **D** 902 cm^2

42. Express 29,000,000 in scientific notation.

F 2.9×10^6 **H** 2.9×10^7
G 2.9×10^8 **J** 2.9×10^9

Name _____ Date _____ Class _____

CHAPTER 13 Cumulative Test
Form A

Choose the best answer.

1. Evaluate the expression $5a - 3b$ for $a = 6$ and $b = 8$.
   A 18   **B** 6

2. Solve $6x = 84$.
   A $x = 14$ **B** $x = 78$

3. Simplify $(-2)^3 + (-3)^2$.
   A -17 **B** 1

4. What is $x^5 \cdot x^2 \cdot x$ written as one power?
   A x **C** x^7
   B x^2 **D** x^8

5. What is 3.62×10^{-2} in standard notation?
   F 0.00362   **H** 0.362
   G 0.0362   **J** 362

6. Simplify $\frac{18}{24}$.
   A $\frac{2}{3}$ **B** $\frac{3}{4}$

7. Add $\frac{5}{12} + \frac{1}{12}$. Give the answer in simplest form.
   A $\frac{1}{2}$ **B** $\frac{1}{4}$

8. Multiply $3.7(-2.4)$.
   A -8.88 **B** 1.3

9. Solve $h + 8.2 = 4.9$.
   A $h = 13.1$
   B $h = 3.7$
   C $h = -3.3$
   D $h = -4.7$

10. Evaluate $\sqrt{9} + 27$.
    A 6   **B** 30

11. Mr. Jetter's class made a stem-and-leaf plot showing the daily high temperatures in °F. How many stems does the stem-and-leaf plot have?

    Daily High Temperatures (°F)

Stem	Leaves
5	4 4 8 9 9 9 9
6	1 1 3 4 7 7 8
7	0 2 4 5 5 6 6 6

    Key: 6|3 means 63

    A 25 stems
    B 22 stems
    C 13 stems
    D 3 stems

12. Marisa's quiz scores were 7, 9, 10, 9, 8, and 10. Find the first quartile.
    A 8   **B** 9

13. What type of relationship is shown by the scatter plot?

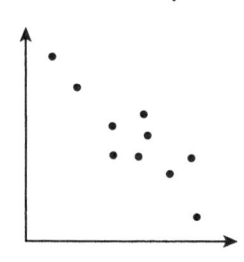

    A positive correlation
    B negative correlation

CHAPTER 13 Cumulative Test
Form A, continued

Use the figure for 14 and 15.

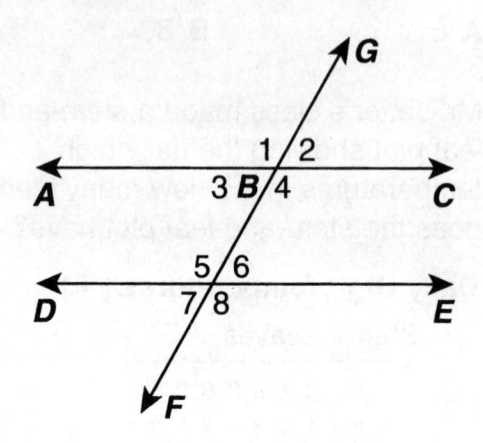

14. What type of angle is ∠ABC?
- A acute
- B obtuse
- C right
- D straight

15. Line AC is parallel to line DE. If m∠3 is 70°, what is m∠6?
- F 110°
- G 90°
- H 70°
- J 20°

16. Which quadrilateral could have its vertices at coordinates (4, 2), (7, 2), (7, 5), (4, 5)?
- A trapezoid
- B square

17. Which figure is symmetric?
- A 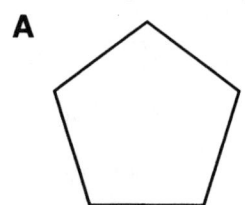
- B

18. What is the length of the hypotenuse of the triangle, to the nearest tenth?
- A 4.9 in. 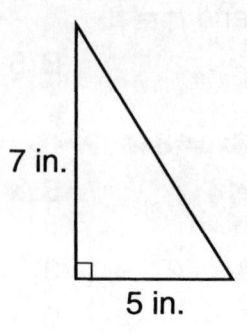
- B 8.6 in.

19. A round table has a diameter of 36 inches. What is the circumference of the top of the table? Use 3.14 for π.
- A 226.08 in.
- B 113.04 in.

20. A rectangular prism is 7 inches long, 5 inches wide, and 3 inches high. What is the volume of the prism?
- A 15 in^3
- B 105 in^3

21. Find the surface area of the pyramid. 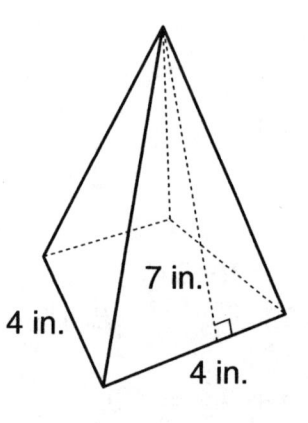
- A 72 in^2
- B 128 in^2

Cumulative Test — Chapter 13, Form A, continued

22. Which pair of ratios is proportional?

A $\frac{6}{10}$ and $\frac{10}{15}$

B $\frac{12}{18}$ and $\frac{8}{12}$

23. A car is traveling at the rate of 60 miles per hour. How far does the car travel in 1 minute?

A 3,600 mi   B 1 mi

24. On a scale drawing of a room, 1 inch represents 2 feet. In the scale drawing, the room is 7 inches long. How long is the actual room?

A 3.5 ft   C 14 ft
B 7 ft   D 84 ft

25. In a survey of 75 people, 60 people said they own a car. What percent of the people surveyed own a car?

F 8%   H 75%
G 60%   J 80%

26. Brandy and 2 friends went out to dinner. They left a 15% tip on a dinner check of $45. How much did they leave as a tip?

A $6.75   B $20.25

27. What is the probability of the spinner landing on a number less than 4?

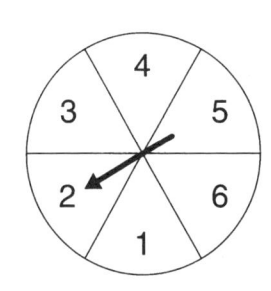

A $\frac{2}{3}$ B $\frac{1}{2}$

28. In how many ways can the letters of the word WORD be arranged?

A 24 ways   B 12 ways

29. A bag contains 5 green candles, 2 yellow candles, and 3 red candles. What are the odds of drawing a red candle from the bag?

A 3 to 10
B 3 to 7
C 10 to 3
D 7 to 3

30. Solve $6d - 24 = 3d - 3$.

A $d = 7$ B $d = -7$

31. Which ordered pair is a solution to this system of equations?

$$x + y = 14$$
$$x - y = 2$$

A (6, 8)   B (8, 6)

32. What is the slope of the line that passes through the points (−3, 5) and (1, 8)?

A $\frac{4}{3}$ B $\frac{3}{4}$

33. What is the y-intercept of the line shown by the equation $-3x + y = 6$?

A −3   B 6

CHAPTER 13 Cumulative Test
Form A, continued

34. Multiply $(y - 5)(y + 2)$.
  F $y^2 - 3y - 7$
  G $y^2 - 7y - 10$
  H $y^2 - 3y + 10$
  J $y^2 - 3y - 10$

35. Which inequality is shown on the graph?

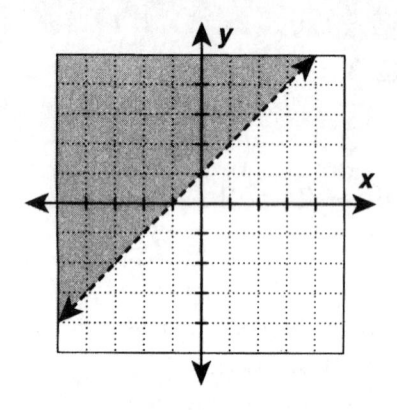

  A $y > x + 1$
  B $y < x + 1$

36. What are the next three terms in the sequence 1, 3, 6, 10, . . .?
  A 15, 21, 28
  B 11, 13, 16

37. What is the rule for the function table?

x	0	1	2	3
y	5	6	7	8

  A $y = 5x$
  B $y = x + 5$

38. Find $f(2)$ for $f(x) = x^2 - 3x + 7$.
  A $f(2) = 5$
  B $f(2) = -21$

Name _____ Date _____ Class _____

CHAPTER 13 Cumulative Test Form B

Choose the best answer.

1. Evaluate the expression $x(3 + y) - 4$ for $x = 7$ and $y = 1$.
   - A 70
   - B 24
   - C 18
   - D 0

2. Solve $\frac{a}{13} = 15$.
   - F $a = 205$
   - G $a = 195$
   - H $a = 185$
   - J $a = 28$

3. Simplify $(1 + 2 \cdot 4)^2$.
   - A 18
   - B 33
   - C 81
   - D 144

4. What is $7^0 \cdot 7^4 \cdot 7^2 \cdot 7^3$ written as one power?
   - F 7^0
   - G 7^4
   - H 7^9
   - J 7^{24}

5. What is 5.09×10^{-4} in standard notation?
   - A 50,900
   - B 5,090
   - C 0.00509
   - D 0.000509

6. Which decimal is equivalent to $\frac{9}{37}$?
   - F 0.937
   - G $0.\overline{243}$
   - H 0.243
   - J $0.\overline{234}$

7. Add $\frac{13}{15} + \frac{8}{15}$. Give the answer in simplest form.
   - A $\frac{7}{10}$
   - B $1\frac{2}{5}$
   - C $1\frac{2}{3}$
   - D $2\frac{1}{15}$

8. Evaluate $\left(2\frac{3}{4}\right)p$ for $p = \frac{2}{5}$.
   - F $1\frac{1}{10}$
   - G $2\frac{3}{10}$
   - H $3\frac{3}{20}$
   - J $6\frac{7}{8}$

9. Solve $2.3n = 0.92$.
   - A $n = 0.4$
   - B $n = 2.5$
   - C $n = 4$
   - D $n = 25$

10. Evaluate $\sqrt{81} - \sqrt{25}$.
    - F $\sqrt{56}$
    - G 16
    - H 14
    - J 4

11. Mr. Jetter's class made a stem-and-leaf plot showing the daily high temperatures in °F. How many leaves does the stem-and-leaf plot have?

    Daily High Temperatures (°F)

Stem	Leaves
5	4 4 8 9 9 9 9
6	1 1 3 4 7 7 8
7	0 2 4 5 5 6 6 6

    Key: 6|3 means 63

    - A 3 leaves
    - B 13 leaves
    - C 22 leaves
    - D 25 leaves

12. Armando's quiz scores were 7, 9, 10, 9, 8, 10, 9, and 8. Find the third quartile.
    - F 8
    - G 8.5
    - H 9
    - J 9.5

Cumulative Test
13 Form B, continued

13. What type of relationship is shown by the scatter plot?

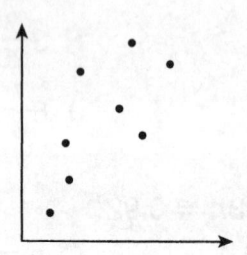

- **A** positive correlation
- **B** negative correlation
- **C** constant correlation
- **D** no correlation

Use the figure for 14 and 15.

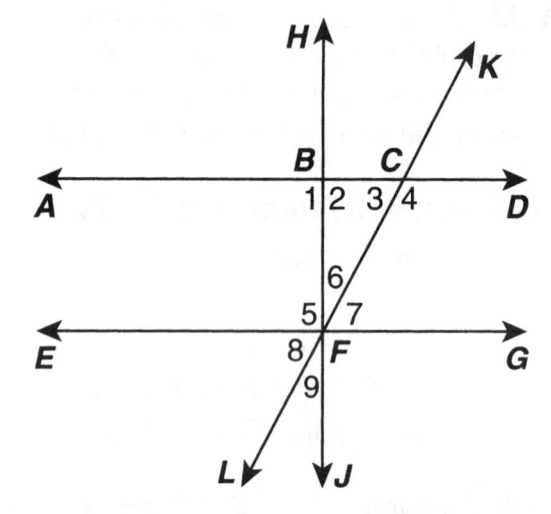

14. What type of angle is ∠GFC?
- **F** acute
- **G** obtuse
- **H** right
- **J** straight

15. Line AD is parallel to line EG. If m∠3 is 70°, what is m∠7?
- **A** 20°
- **B** 70°
- **C** 90°
- **D** 110°

16. Which quadrilateral could have its vertices at coordinates (3, 2), (7, 2), (1, 5), (5, 5)?
- **F** rectangle
- **G** square
- **H** parallelogram
- **J** trapezoid

17. Which of the following figures has line symmetry?

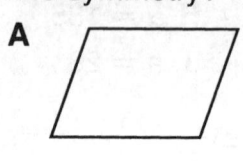

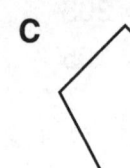

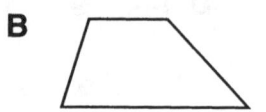

18. What is the length of the missing side of the triangle, to the nearest tenth?

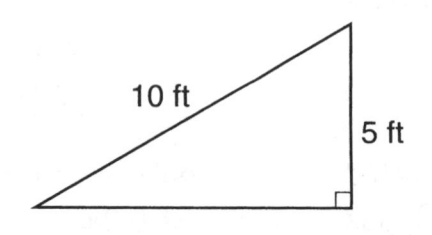

- **F** 5.0 ft
- **G** 8.6 ft
- **H** 8.7 ft
- **J** 11.2 ft

19. A round table has a diameter of 36 inches. What is the area of the top of the table? Use 3.14 for π.
- **A** 1,017.4 in^2
- **B** 508.7 in^2
- **C** 226.1 in^2
- **D** 113.0 in^2

Cumulative Test

CHAPTER 13 Form B, continued

20. A rectangular prism is 2 inches long, 4.5 inches wide, and 8 inches long. What is the volume of the prism?
- **F** 9 in³
- **G** 14.5 in³
- **H** 36 in³
- **J** 72 in³

21. Find the surface area of the pyramid.

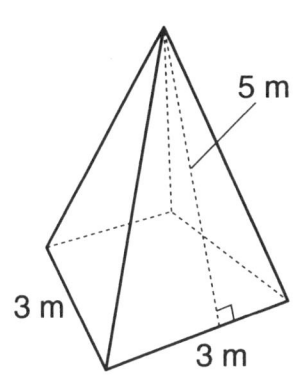

- **A** 15 m²
- **B** 31.5 m²
- **C** 39 m²
- **D** 69 m²

22. Which pair of ratios is proportional?
- **F** $\frac{9}{15}$ and $\frac{6}{9}$
- **G** $\frac{8}{12}$ and $\frac{6}{8}$
- **H** $\frac{15}{20}$ and $\frac{10}{16}$
- **J** $\frac{16}{24}$ and $\frac{10}{15}$

23. A painter can paint 1 foot of molding in 1 minute. How many inches of molding can he paint in one hour?
- **A** 720 in.
- **B** 144 in.
- **C** 60 in.
- **D** 12 in.

24. The distance between two cities on a map is 4 centimeters. The scale on the map says 1 cm = 45 km. What is the actual distance between the cities?
- **F** 11.25 km
- **G** 90 km
- **H** 162 km
- **J** 180 km

25. In a survey of 65 students, 80% said they make their bed every morning. How many of the students surveyed make their bed every morning?
- **A** 80 students
- **B** 65 students
- **C** 52 students
- **D** 15 students

26. Consuelo bought a sweater on sale at 15% off the regular price. If the regular price was $45.00, how much did she pay for the sweater?
- **F** $6.75
- **G** $38.25
- **H** $42.00
- **J** $51.75

27. What is the probability of the spinner landing on an odd number less than 6?

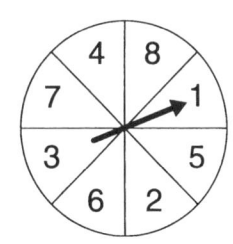

- **A** $\frac{5}{6}$
- **B** $\frac{5}{8}$
- **C** $\frac{1}{2}$
- **D** $\frac{3}{8}$

28. A pizza store offers a choice of 8 toppings. How many different pizzas are possible if you choose a pizza with 3 toppings?
- **F** 24
- **G** 48
- **H** 56
- **J** 336

29. A 1–6 number cube is rolled. What are the odds that the outcome is a multiple of 3?
- **A** 1 to 1
- **B** 1 to 2
- **C** 2 to 1
- **D** 1 to 3

Cumulative Test
Chapter 13 Form B, continued

30. Solve $7(b + 4) = 5b - 2$.
- F $b = 15$
- G $b = 13$
- H $b = -15$
- J no solution

31. Which ordered pair is a solution to this system of equations?
$$2x + y = 14$$
$$3x - 2y = 7$$
- A (4, 6)
- B (5, 8)
- C (5, 4)
- D (4, 5)

32. What is the slope of the line that passes through the points (−2, −5) and (1, 4)?
- F $\frac{1}{3}$
- G 3
- H $-\frac{1}{3}$
- J −3

33. What is the y-intercept of the line represented by $-3x + 2y = 6$?
- A 6
- B 3
- C 2
- D $\frac{3}{2}$

34. Multiply $(3y - 5)(y - 1)$.
- F $3y^2 - 8y + 5$
- G $3y^2 - 2y + 5$
- H $3y^2 - 8y - 6$
- J $3y^2 - 8y - 5$

35. Which inequality is shown on the graph?

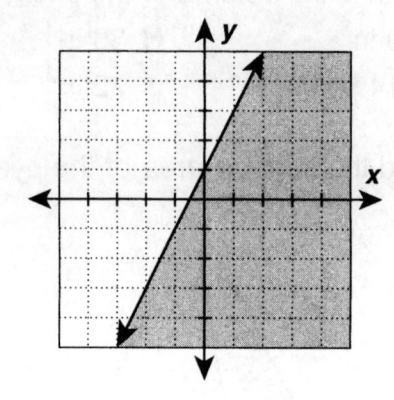

- A $y > 2x + 1$
- B $y < 2x + 1$
- C $y \geq 2x + 1$
- D $y \leq 2x + 1$

36. What are the next three terms in the sequence 5, 8, 12, 17, . . . ?
- F 23, 30, 38
- G 18, 19, 20
- H 23, 24, 25
- J 22, 27, 32

37. Which is the rule for the function table?

x	−1	0	1	2
y	4	7	10	13

- A $y = x + 5$
- B $y = 2x + 8$
- C $y = x + 11$
- D $y = 3x + 7$

38. Find $f(-2)$ for $f(x) = 3x^2 - x + 7$.
- F $f(-2) = 45$
- G $f(-2) = 21$
- H $f(-2) = 15$
- J $f(-2) = -3$

Name _____ Date _____ Class _____

CHAPTER 13 Cumulative Test
Form C

Choose the best answer.

1. Evaluate the expression $8ab - 3ac$ for $a = 2$, $b = 6$, and $c = 3$.
   - A 15
   - B 78
   - C 139
   - D 503

2. Solve $26n = 117$.
   - F $n = 4.05$
   - G $n = 4.13$
   - H $n = 4.5$
   - J $n = 91$

3. Simplify $(9 + 5)^3 - (9 - 5)^3$.
   - A 0
   - B 30
   - C 1,000
   - D 2,680

4. What is $\dfrac{(8^6 \cdot 8^4)}{8^2}$ written as one power?
   - F 8^8
   - G 8^5
   - H 8^2
   - J 8^1

5. What is 9.812×10^5 in standard notation?
   - A 0.00009812
   - B 9,812
   - C 98,120
   - D 981,200

6. Which number does **not** have the same value as the others?
   - F $\dfrac{75}{37}$
   - G $2.\overline{027}$
   - H $2\dfrac{1}{37}$
   - J $2.\overline{137}$

7. Evaluate $1 - y - \dfrac{17}{20}$ for $y = \dfrac{7}{20}$.
   - A $-\dfrac{1}{5}$
   - B $\dfrac{1}{10}$
   - C $\dfrac{1}{5}$
   - D $1\dfrac{1}{2}$

8. Multiply $\left(\dfrac{4}{5}\right)\left(1\dfrac{2}{3}\right)\left(\dfrac{15}{16}\right)$. Give the answer in simplest form.
   - F $\dfrac{1}{2}$
   - G $1\dfrac{1}{4}$
   - H $1\dfrac{1}{2}$
   - J $3\dfrac{1}{2}$

9. Solve $b - 0.39 = 1.61 - 2.04$.
   - A $b = 0.82$
   - B $b = 0.04$
   - C $b = -0.04$
   - D $b = -0.82$

10. Evaluate $\sqrt{121} - \sqrt{36 + 64}$.
    - F -3
    - G 1
    - H 3
    - J 13

11. Mrs. Wexler's class made a back-to-back stem-and-leaf plot showing the daily high and low temperatures in °F. How many leaves does the stem-and-leaf plot have?

Daily High and Low Temperatures (°F)

Leaves	Stem	Leaves
1 7 8 9	4	
0 0 1 2 4 4 6	5	4 4 8 9 9 9 9
9	6	1 1 3 4 7 7 8
0 1 3 5 5 5 8	7	0 2 4 5 5 6 6 6
1 2 3		

Key: 6|3 means 63

   - A 4 leaves
   - B 31 leaves
   - C 44 leaves
   - D 48 leaves

12. Hope's math test scores are 85, 95, 100, 90, 85, 100, 95, and 80. Find the third quartile.
    - F 85
    - G 95
    - H 97.5
    - J 100

Cumulative Test

Chapter 13 Form C, continued

13. What type of relationship is shown by the scatter plot?

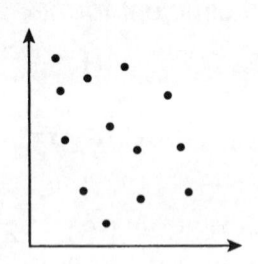

A positive correlation
B negative correlation
C constant correlation
D no correlation

Use the figure for 14 and 15.

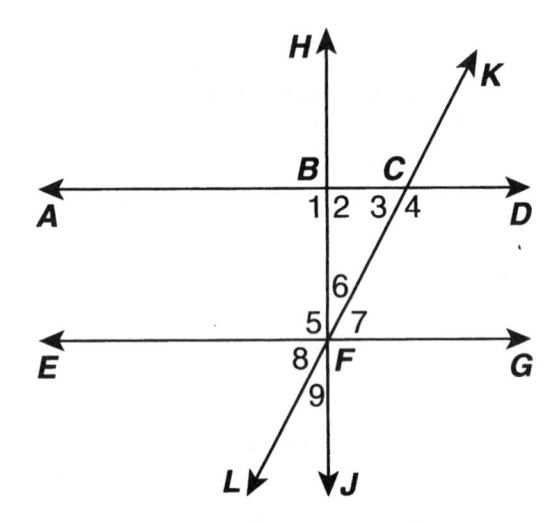

14. What type of angle is ∠KFE?
F obtuse
G straight
H right
J acute

15. Line AD is parallel to line EG. If m∠4 is 110°, what is m∠8?
A 20°
B 70°
C 90°
D 110°

16. Which quadrilateral could have its vertices at coordinates (4, 2), (9, 2), (2, 5), (9, 5)?
F rectangle
G square
H parallelogram
J trapezoid

17. Which of the following figures is **not** symmetric?

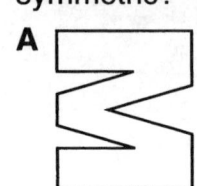

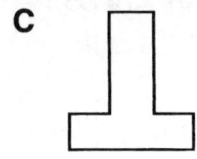

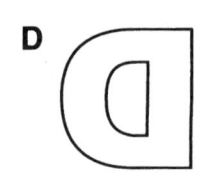

A M
B G
C T
D D

18. What is the length of the missing side of the triangle, to the nearest tenth?

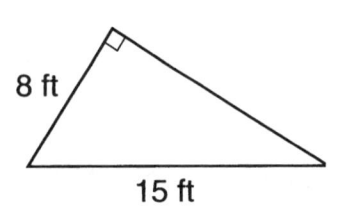

8 ft, 15 ft

F 7.0 ft
G 12.6 ft
H 12.7 ft
J 17.0 ft

19. A round table has a diameter of 3 feet. What is the area of the top of the table? Use 3.14 for π.
A 7.1 ft^2
B 9.4 ft^2
C 18.8 ft^2
D 28.3 ft^2

Holt Pre-Algebra

Name _____ Date _____ Class _____

CHAPTER 13 Cumulative Test
Test C, continued

20. A fish tank is 20.5 inches long, 12 inches wide, and 14 inches high. What is the volume of the fish tank?
    F 279 in³   H 1,066 in³
    G 287 in³   J 3,444 in³

21. Find the surface area of the cone to the nearest tenth. Use 3.14 for π.

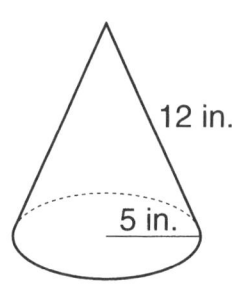

    12 in.
    5 in.

    A 188.4 in²   C 266.9 in²
    B 219.8 in²   D 314 in²

22. Which pair of ratios is **not** proportional?
    F $\frac{15}{24}$ and $\frac{21}{32}$ H $\frac{24}{40}$ and $\frac{9}{15}$
    G $\frac{30}{45}$ and $\frac{14}{21}$ J $\frac{16}{20}$ and $\frac{20}{25}$

23. A runner can jog 6 miles in 1 hour. How many feet does she jog in 1 minute?
    A 52.8 ft   C 5,280 ft
    B 528 ft   D 52,800 ft

24. The distance between two cities on a map is 1.25 inches. The scale on the map is 1 in. = 60 mi. What is the actual distance between the cities?
    F 48 mi   H 65 mi
    G 61.25 mi   J 75 mi

25. In a survey of 225 students, 175 of them said they drink a glass of milk every morning. To the nearest tenth of a percent, what percent of the students surveyed drink a glass of milk every morning?
    A 17.5%   C 77.7%
    B 22.5%   D 77.8%

26. The rate for defective light bulbs at a light bulb plant is 1.5%. If 3,000 light bulbs are produced per hour, how many defective light bulbs are produced in 24 hours?
    F 108 bulbs   H 1,080 bulbs
    G 450 bulbs   J 1,800 bulbs

27. What is the probability of the spinner landing on a prime number?

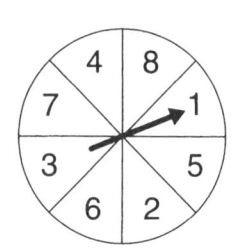

    A $\frac{1}{4}$ C $\frac{3}{8}$
    B $\frac{1}{2}$ D $\frac{5}{8}$

28. A pizza store offers a choice of 2 crusts and 8 toppings. How many different pizzas are possible if you choose a pizza with 3 toppings?
    F 16 pizzas   H 56 pizzas
    G 48 pizzas   J 112 pizzas

Cumulative Test

Chapter 13 Test C, continued

29. Two 1–6 number cubes are rolled. What are the odds that the sum of the 2 cubes is 7?
- **A** 1 to 5
- **B** 1 to 6
- **C** 5 to 1
- **D** 6 to 1

30. Solve $2.2p + 5.9 = 8.3 + 1.6p$.
- **F** $p = 4$
- **G** $p = 2$
- **H** $p = -4$
- **J** no solution

31. Which ordered pair is a solution to this system of equations?
$$y = 3x + 9$$
$$2x + 5y = 11$$
- **A** $(-2, 3)$
- **B** $(3, 0)$
- **C** $(3, 1)$
- **D** $(-2, 15)$

32. Which set of points represents a line that does **not** have a slope of -1?
- **F** $(3, -1), (-7, 9)$
- **G** $(-2, 2), (5, -5)$
- **H** $(-1, 0), (3, 4)$
- **J** $(-6, 3), (0, -3)$

33. What is the y-intercept of the line shown by the equation $4(x - 2y) = 3$?
- **A** $-\frac{1}{2}$
- **B** $-\frac{3}{8}$
- **C** $\frac{3}{8}$
- **D** $\frac{1}{2}$

34. Multiply $(3a - 5b)(3a + 5b)$.
- **F** $9a^2$
- **G** $9a^2 - 30ab + 25b^2$
- **H** $9a^2 - 25b^2$
- **J** $9a^2 + 25b^2$

35. Which inequality is shown on the graph?

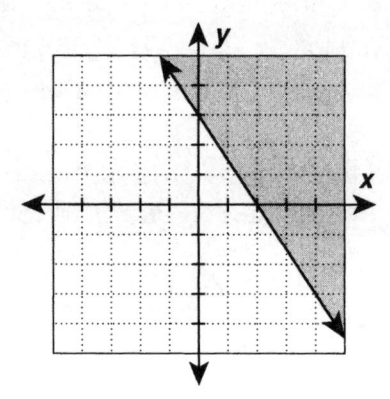

- **A** $y \leq -\left(\frac{3}{2}\right)x + 3$
- **B** $y \geq -\left(\frac{3}{2}\right)x + 3$
- **C** $y < -\left(\frac{3}{2}\right)x + 3$
- **D** $y > -\left(\frac{3}{2}\right)x + 3$

36. What are the next 3 terms in the sequence 13, 12, 10, 7, 3, . . .?
- **F** $-2, -8, -15$
- **G** $8, 14, 21$
- **H** $2, 1, 0$
- **J** $0, -4, -9$

37. What is the rule for the function table?

x	−5	−1	4	8
y	4	−4	−14	−22

- **A** $y = -3x + 2$
- **B** $y = -2x - 6$
- **C** $y = -5x + 1$
- **D** $y = -3x - 11$

38. Find $f(8)$ for $f(x) = \left(\frac{1}{2}\right)x^2 - \left(\frac{1}{4}\right)x + 8$.
- **F** $f(8) = 10$
- **G** $f(8) = 22$
- **H** $f(8) = 35$
- **J** $f(8) = 38$

Name _____ Date _____ Class _____

CHAPTER 14 Cumulative Test
Form A

Choose the best answer.

1. Which is the algebraic expression for the word phrase *6 more than twice a number* n?
   - **A** $6n + 2$
   - **B** $2n + 6$

2. Find the solution for $x - 7 = 24$.
   - **A** $x = 31$
   - **B** $x = 17$

3. Evaluate $9 - k$ for $k = -20$.
   - **A** -29
   - **B** -11
   - **C** 11
   - **D** 29

4. Solve $g + 18 \leq -45$.
   - **F** $g \leq -27$
   - **G** $g \leq 27$
   - **H** $g \leq -63$
   - **J** $g \leq 63$

5. Evaluate 2^4.
   - **A** 8
   - **B** 16

6. What is 0.45 expressed as a fraction in simplest form?
   - **A** $\frac{9}{20}$
   - **B** $\frac{9}{25}$

7. Add $4.32 + (-7.19)$.
   - **A** -3.13
   - **B** -2.87

8. Divide $\frac{9}{10} \div \frac{3}{5}$.
   - **A** $1\frac{1}{2}$
   - **B** $\frac{27}{50}$

9. Solve $d + \frac{1}{3} < \frac{5}{6}$.
   - **A** $d > \frac{1}{2}$
   - **B** $d > 1\frac{1}{6}$
   - **C** $d < \frac{1}{2}$
   - **D** $d < 1\frac{1}{6}$

10. What is $\sqrt{44}$ rounded to the nearest tenth?
    - **A** 6.6
    - **B** 6.7

For 11–12, use the stem-and-leaf plot, which shows the heights in inches of the boys on a middle school basketball team.

Basketball Team Heights

Stems	Leaves
6	4 5 7 8 9
7	0 1 1 3

11. What is the mode of the heights?
    - **A** 69
    - **B** 69.5
    - **C** 70
    - **D** 71

12. What is the first quartile?
    - **A** 66
    - **B** 67

301

Holt Pre-Algebra

Name _____ Date _____ Class _____

CHAPTER 14 Cumulative Test
Form A, continued

Use the figure for 13 and 14.

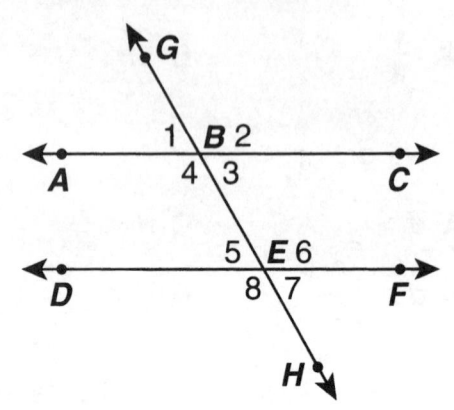

13. Which angle is obtuse?

    A ∠ABE   B ∠ABG

14. Line AC is parallel to line DF. If m∠5 is 50°, what is m∠3?

    A 130°   C 50°
    B 90°   D 40°

15. What is the sum of the measures of the angles of a pentagon?

    F 900°   H 360°
    G 540°   J 180°

16. Identify the type of translation.

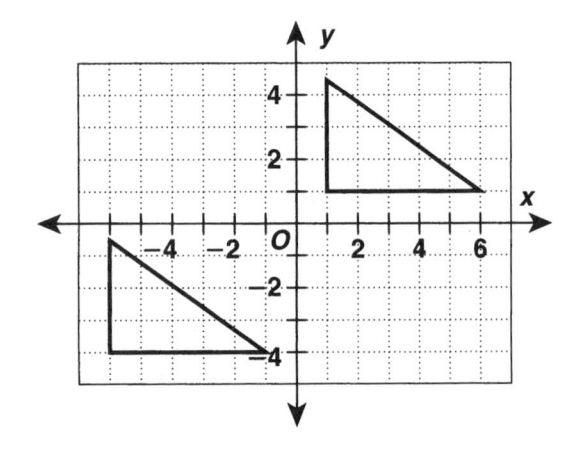

    A translation   C reflection
    B rotation   D tessellation

17. What is the area of a figure with vertices (3, 1), (3, 5), and (5, 5)?

    A 8 square units
    B 4 square units

18. A round table has a radius of 18 inches. What is the area of the top of the table to the nearest tenth? Use 3.14 for π.

    A 113.0 in^2 B 1,017.4 in^2

19. A rectangular pyramid has a base with an area of 36 square centimeters and a height of 7 centimeters. What is the volume of the pyramid?

    A 252 cm^3 B 84 cm^3

20. Which is the better buy?

    A a 28-oz bottle of ketchup for $1.29
    B a 40-oz bottle of ketchup for $1.89

21. Solve the proportion $\frac{5}{m} = \frac{15}{24}$.

    A $m = 3$ B $m = 8$

22. Which scale reduces the size of the actual object?

    A 1 ft to 6 in.   B 1 in. to 6 ft

23. Michael bought a CD on sale for $12. The sale price was 75% of the regular price. What is the regular price of the CD?

    A $9.00   C $12.75
    B $12.25   D $16.00

Copyright © by Holt, Rinehart and Winston.
All rights reserved.

302

Holt Pre-Algebra

Name _____ Date _____ Class _____

CHAPTER 14 Cumulative Test
Form A, continued

24. Jamie borrowed $3,000 from a bank for 2 years at a simple interest rate of 6% per year. How much interest will she pay on the loan?
    A $360 **B** $3,360

25. A baseball team won 14 of its last 21 games. What is the probability that the team will win its next game?
    A $\frac{1}{2}$ **B** $\frac{2}{3}$

26. How many combinations of 3 different flowers can you choose from 5 different flowers?
    A 10 different combinations
    B 15 different combinations

27. A bag contains 5 green candles, 2 yellow candles, and 3 red candles. What is the probability of picking a red candle and then a green candle from the bag without replacement?
    A $\frac{3}{20}$ **B** $\frac{1}{6}$

28. Which graph represents the solution to the inequality, $4x - 9 > 7$?
    A ← −4 −2 0 2 4 →
    B ← −4 −2 0 2 4 →

29. Solve $P = 2a + b$ for b.
    A $b = P - 2a$ **B** $b = 2a - P$

30. What is the equation of the line that passes through the points (3, 5) and (1, 1)?
    A $y = -2x + 11$ **B** $y = 2x - 1$

31. Which equation gives the direct variation if $y = 8$ when $x = 2$?
    A $y = 4x$ **B** $y = \left(\frac{1}{4}\right)x$

32. Which inequality is shown on the graph?

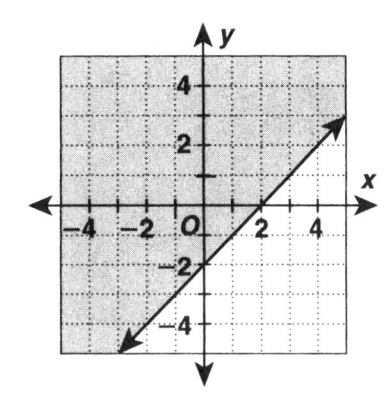

    A $y \geq x - 2$ **B** $y \leq x - 2$

33. What is the ninth term in the sequence, 3, 6, 9, …?
    A 27 **B** 30

34. What is the rule for the function table?

x	−1	0	1	2
y	3	5	7	9

    A $y = 2x - 1$ **B** $y = 2x + 5$

Cumulative Test
CHAPTER 14 Form A, continued

35. Find $f(3)$ for $f(x) = 4x^2 - 1$.
  A $f(3) = 35$ **B** $f(3) = 143$

36. What is the degree of the polynomial $x + 3x^2 - 9x$?
  A 2   **B** 3

37. Add $(7m + 3) + (2m^2 - 3m - 5)$.
  A $6m^2 + 2$ **C** $2m^2 + 4m - 2$
  B $2m^2 + 7m - 5$ **D** $4m^2$

38. Multiply $(x^4y^2)(2x^2y^3)$.
  F $2(xy)^{48}$ **H** $2x^8y^6$
  G $(2xy)^{48}$ **J** $2x^6y^5$

39. Which set is a finite set?
  A {whole numbers greater than 50}
  B {whole numbers less than 50}

40. Which statement is true about the conjunction P and Q?
  A P and Q must both be true if the conjunction is true.
  B P and Q must both be false if the conjunction is false.

41. What is the degree of vertex B?

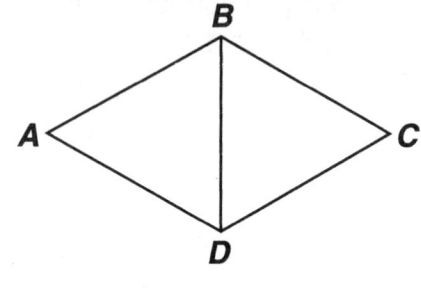

  A 2   **B** 3

Name _____ Date _____ Class _____

Cumulative Test
Chapter 14 — Form B

Choose the best answer.

1. Which is the algebraic expression for the word phrase *5 less than the product of 8 and* p?
   - **A** $5 - 8p$
   - **B** $8p - 5$
   - **C** $5p - 8$
   - **D** $p + 8 - 5$

2. Find the solution for $x - 26 = 78$.
   - **F** $x = 3$
   - **G** $x = 52$
   - **H** $x = 104$
   - **J** $x = 2,028$

3. Evaluate $w - (-17)$ for $w = 8$.
   - **A** -25
   - **B** -9
   - **C** 9
   - **D** 25

4. Solve $29 \geq b - 17$.
   - **F** $b \leq 46$
   - **G** $b \geq 46$
   - **H** $b \leq 12$
   - **J** $b \geq 12$

5. Evaluate $(-8)^2$.
   - **A** -64
   - **B** -16
   - **C** 16
   - **D** 64

6. What is 0.125 as a fraction in simplest form?
   - **F** $\frac{1}{125}$
   - **G** $\frac{1}{80}$
   - **H** $\frac{1}{8}$
   - **J** $1\frac{1}{4}$

7. Add $6.85 + 3.3$.
   - **A** 10.15
   - **B** 9.88
   - **C** 8.88
   - **D** 7.18

8. Divide $16 \div 2\frac{2}{3}$.
   - **F** 6
   - **G** 12
   - **H** $13\frac{1}{3}$
   - **J** $42\frac{2}{3}$

9. Solve $3n \leq -\frac{7}{9}$.
   - **A** $n \leq -3\frac{6}{7}$
   - **B** $n \leq -\frac{7}{27}$
   - **C** $n \geq -\frac{7}{27}$
   - **D** $n \geq -3\frac{6}{7}$

10. What is $\sqrt{176}$ rounded to the nearest tenth?
    - **F** 13
    - **G** 13.2
    - **H** 13.3
    - **J** 14

For 11–12, use the stem-and-leaf plot, which shows the points scored by a middle-school football team during the season.

Football Points Scored

Stems	Leaves
0	3 7 7 7
1	2 3 4 4 9
2	1 4

11. What is the mode of the scores?
    - **A** 7
    - **B** 11
    - **C** 12.8
    - **D** 13

12. What is the third quartile?
    - **F** 14
    - **G** 19
    - **H** 21
    - **J** 24

Name _____ Date _____ Class _____

CHAPTER 14 Cumulative Test
Form B, continued

Use the figure for 13 and 14.

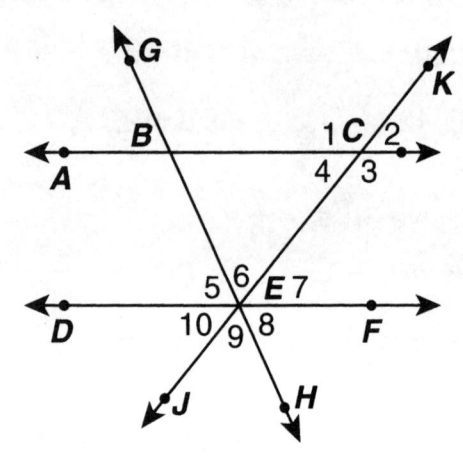

13. Which angle is supplementary to ∠ABE?
    A ∠GBC
    B ∠BCE
    C ∠EBC
    D ∠DEJ

14. Line AC is parallel to line DF. If m∠DEC is 150°, what is m∠1?
    F 210°
    G 180°
    H 150°
    J 30°

15. What is the sum of the measures of the angles of an octagon?
    A 180°
    B 1,080°
    C 1,440°
    D 1,800°

16. Identify the type of translation.

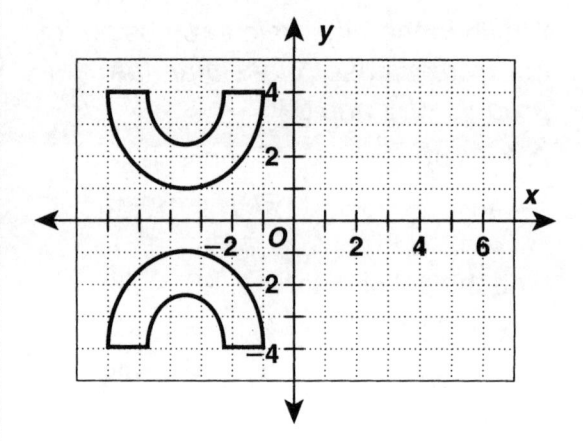

    F translation
    G dilation
    H reflection
    J tessellation

17. What is the area of a figure with vertices (1, 1), (8, 1), and (5, 5)?
    A 28 square units
    B 21 square units
    C 14 square units
    D 11 square units

18. A tree trunk has a radius of 11 inches. What is the circumference of the tree trunk to the nearest tenth? Use 3.14 for π.
    F 34.5 in.
    G 69.1 in.
    H 138.2 in.
    J 379.9 in.

19. The base of a cone has a radius of 6 centimeters. The cone is 7 centimeters tall. What is the volume of the cone to the nearest tenth? Use 3.14 for π.
    A 260 cm^3
    B 263.7 cm^3
    C 263.8 cm^3
    D 264.0 cm^3

Copyright © by Holt, Rinehart and Winston.
All rights reserved.

Holt Pre-Algebra

Name _____ Date _____ Class _____

Cumulative Test
CHAPTER 14 Form B, continued

20. Which is the best buy?
    F an 8-oz can of peaches for $0.59
    G a 12-oz can of peaches for $0.89
    H a 20-oz can of peaches for $1.49
    J a 32-oz can of peaches for $2.29

21. Solve the proportion $\frac{5}{8} = \frac{a}{20}$.
    A $a = 2.5$ C $a = 15$
    B $a = 12.5$ D $a = 17$

22. Which scale enlarges the size of the actual object?
    F 1 m to 100 cm   H 1 cm to 10 m
    G 10 mm to 1 cm   J 10 m to 1 cm

23. Alyssa is selling bags of peanuts to raise money for a fundraiser. So far, she has sold 36 bags of peanuts. This is 45% of her goal. What is Alyssa's goal?
    A 16 bags of peanuts
    B 52 bags of peanuts
    C 80 bags of peanuts
    D 81 bags of peanuts

24. Noah borrowed $7,500 for 2 years at a simple interest rate of 6.5% per year. How much interest will he pay on the loan?
    F $97.50   H $975.00
    G $487.50   J $8,475.00

25. Mario tossed a coin 40 times, and it landed heads up 24 times. What is the experimental probability that the coin will land heads up on the next toss?
    A $\frac{2}{3}$ C $\frac{1}{2}$
    B $\frac{3}{5}$ D $\frac{2}{5}$

26. How many different 3-topping pizzas are possible if there are 10 toppings from which to choose?
    F 30 pizzas   H 120 pizzas
    G 60 pizzas   J 720 pizzas

27. A bag contains 5 green marbles, 4 yellow marbles, and 3 red marbles. What is the probability of picking a red marble and then a yellow marble from the bag without replacement?
    A $\frac{1}{11}$ C $\frac{7}{12}$
    B $\frac{1}{12}$ D $\frac{20}{33}$

28. Which graph represents the solution for the inequality $3x + 9 < 6$?
    F ← ← ← ← ○ + + + + + → (at −1)
    G ← + + + ○ + + + + + → (at −1)
    H ← + + + + + + ● + + → (at 1)
    J ← + + + + ○ + + + + → (at −1)

29. Solve $y = 2x + 5$ for x.
    A $x = 5 - 2y$ C $x = \frac{(y-5)}{2}$
    B $x = 2y - 5$ D $x = 2(y - 5)$

30. What is the equation of the line that passes through the points $(-2, 5)$ and $(1, -1)$?
    F $y = 3x + 11$ H $y = -2x + 1$
    G $y = 2x + 9$ J $y = 3x - 4$

31. Which equation gives the direct variation if $y = 8$ when $x = -2$?
    A $y = -4x$ C $y = 4x$
    B $y = \left(\frac{1}{4}\right)x$ D $y = -\left(\frac{1}{4}\right)x$

Name _____ Date _____ Class _____

CHAPTER 14 Cumulative Test
Form B, continued

32. Which inequality is shown on the graph?

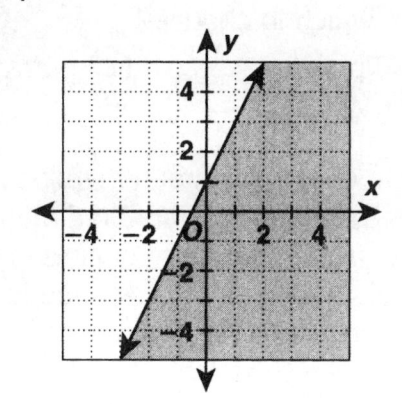

F $y > 2x + 1$ **H** $y \geq 2x + 1$
G $y < 2x + 1$ **J** $y \leq 2x + 1$

33. What is the tenth term in the sequence 5, 8, 11, 14, …?
A 29   **C** 35
B 32   **D** 38

34. What is the rule for the function table?

x	−1	0	1	2
y	7	5	3	1

F $y = -7x$ **H** $y = x + 5$
G $y = -2x + 5$ **J** $y = -3x + 7$

35. Find $f(-2)$ for $f(x) = 2x^2 + 4x + 1$.
A $f(-2) = -15$ **C** $f(-2) = 17$
B $f(-2) = 1$ **D** $f(-2) = 25$

36. What is the degree of the polynomial $2x + 3x^2 - 4x^5$?
F 2   **H** 4
G 3   **J** 5

37. Add $(3m^2 - 5m) + (3m - 5m^2)$.
A $-4m^2$ **C** $-2m^2 - 2m$
B 0   **D** $-4m$

38. Multiply $3z(2z^4 - 4z^3)$.
F $6z^5 - 12z^4$ **H** $6z^4 - 12z^3$
G $5z^5 - z^4$ **J** $5z^4 - z^3$

39. Which set is an infinite set?
A {integers less than 5}
B {positive integers less than 5}
C {integers between −5 and 5}
D {integers with absolute values less than 5}

40. Which statement is **not** true about the conjunction P and Q?
F Both P and Q must be true if the conjunction is true.
G Only P must be true if the conjunction is true.
H If both P and Q are false, the conjunction is false.
J The conjunction is false if P is false.

41. What is the degree of vertex A?

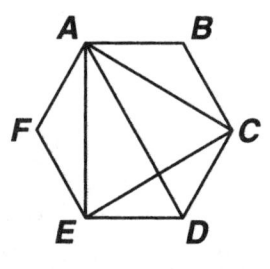

A 3   **C** 5
B 4   **D** 6

Name _____ Date _____ Class _____

CHAPTER 14 Cumulative Test
Form C

Choose the best answer.

1. Which is the algebraic expression for the word phrase *half the difference of* n *and 15*?
   A $\left(\frac{1}{2}\right)(n+15)$ C $\left(\frac{1}{2}\right)(15-n)$
   B $\left(\frac{1}{2}\right)(n-15)$ D $\left(\frac{1}{2}\right)n-15$

2. Find the solution for $3.9 + y = 11.7$.
   F $y = 3$ H $y = 8.2$
   G $y = 7.8$ J $y = 15.6$

3. Evaluate $t - (-17) - 9$ for $t = -24$.
   A -50 C -2
   B -16 D 2

4. Solve $3k + 11 > -55$.
   F $k < -15$ H $k > -22$
   G $k < 22$ J $k > 55$

5. Evaluate $-(-9)^3$.
   A 729 C -27
   B 27 D -729

6. What is -2.325 expressed as a fraction in simplest form?
   F $-2\frac{13}{400}$ H $-2\frac{3}{8}$
   G $-2\frac{3}{25}$ J $-2\frac{13}{40}$

7. Add $-8.95 + (-4.093) + 6.43$.
   A 1.573 C -6.613
   B 11.287 D -19.473

8. Divide $-12\frac{2}{3} \div 9\frac{1}{2}$.
   F $-\frac{3}{4}$ H $-1\frac{1}{3}$
   G $-1\frac{1}{6}$ J $-120\frac{1}{3}$

9. Solve $5n \le \frac{1}{4} - \frac{7}{8}$.
   A $n \le -3\frac{1}{8}$ C $n \le -\frac{1}{8}$
   B $n \le -\frac{9}{40}$ D $n \le 3\frac{1}{8}$

10. What is $\sqrt{276.8}$ rounded to the nearest tenth?
    F 16.0 H 16.7
    G 16.6 J 17.0

For 11 and 12, use the stem-and-leaf plot, which shows the daily high temperatures (in °F) for 10 days.

Daily High Temperatures (in °F)

Stems	Leaves
8	8 9
9	2 3 4 6 9
10	0 1 4

11. What is the mode of the temperatures?
    A 95 C 104
    B 95.6 D no mode

12. What is the third quartile?
    F 92 H 100
    G 95 J 101.5

Name _____ Date _____ Class _____

Cumulative Test
Chapter 14 Form C, continued

Use the figure for 13 and 14.

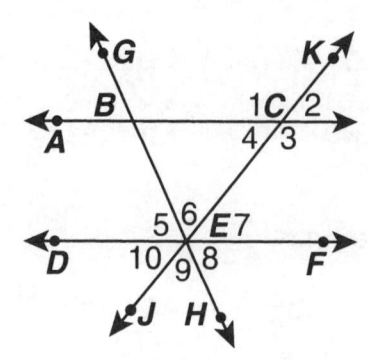

13. Line AC is parallel to line DF. Which angle is **not** supplementary to ∠DEB?
    A ∠GEF
    B ∠ABE
    C ∠DEH
    D ∠JEF

14. Line AC is parallel to line DF. If m∠CEF is 45°, what is m∠1?
    F 325°
    G 180°
    H 135°
    J 45°

15. What is the measure of each angle of a regular decagon?
    A 100°
    B 144°
    C 180°
    D 1,440°

16. Identify the type of transformation.

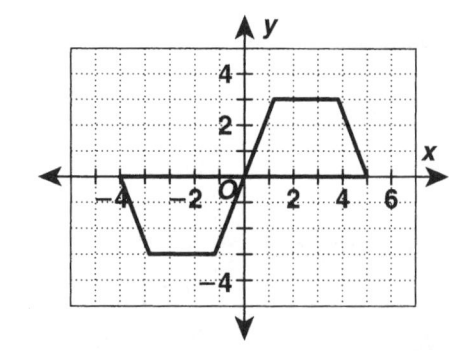

    F translation
    G 90° clockwise rotation
    H reflection
    J 180° rotation

17. What is the area of a figure with vertices (−5, −3), (2 −3), (−3, 1), and (0, 1)?
    A 10 square units
    B 14 square units
    C 20 square units
    D 40 square units

18. The wheel of a bicycle has a diameter of 24 inches. About how far does the bicycle go if the wheel revolves 20 times? Use 3.14 for π.
    F 126 ft
    G 252 ft
    H 480 ft
    J 1,507 ft

19. The base of a rectangular pyramid has a length of 6.3 meters and a width of 5.9 meters. The height of the pyramid is 7.2 meters. What is the volume of the pyramid to the nearest tenth?
    A 89.2 m^3
    B 90 m^3
    C 267.6 m^3
    D 270.0 m^3

20. Which is the best buy?
    F a 9.5-oz box of crackers for $1.59
    G a 14.5-oz box of crackers for $2.39
    H a 22.5-oz box of crackers for $3.79
    J a 33.5-oz box of crackers for $5.59

21. Solve the proportion $\frac{2.5}{1.8} = \frac{15}{m}$.
    A m = 10.5
    B m = 10.8
    C m = 27
    D m = 67.5

Cumulative Test

Chapter 14 Test C, continued

22. Which scale does **not** preserve the size of the actual object?
F 18 ft to 6 yd
G 6 yd to 2 ft
H 24 in. to 2 ft
J 72 in. to 2 yd

23. Maggie has saved $84 for a new racing bicycle. This is 37.5% of the price. How much does the bicycle cost?
A $31.50
B $121.50
C $224
D $252

24. Tomás took out a new car loan with an annual simple interest rate of 2.9% for 36 months. At the end of the loan he had paid $1,305 in interest. What was the principal of the loan?
F $13,050
G $13,695
H $15,000
J $18,305

25. Marcia tossed 2 number cubes 75 times and found that 30 times the sum was less than or equal to 7. What is the probability that the sum will be less than or equal to 7 on the next toss?
A $\frac{1}{3}$
B $\frac{2}{5}$
C $\frac{1}{2}$
D $\frac{3}{5}$

26. How many 5-player starting squads can be formed from a basketball team of 15 players?
F 75 squads
G 225 squads
H 2,250 squads
J 3,003 squads

27. A bag contains 5 green marbles, 3 yellow marbles, and 2 red marbles. What is the probability of picking three green marbles from the bag without replacement?
A $\frac{1}{8}$
B $\frac{1}{12}$
C $\frac{3}{50}$
D $\frac{3}{100}$

28. Which graph represents the solution of the inequality $3x + 9 \geq 1 - x$?

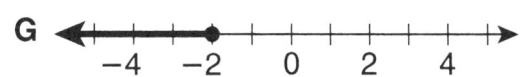

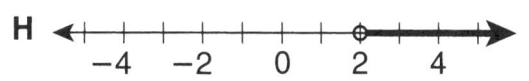

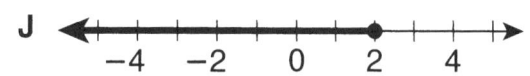

29. Solve $a^2 + b^2 = c^2$ for b.
A $b = \sqrt{c^2 - a^2}$
B $b = a^2 - c^2$
C $b = c^2 - a^2$
D $b = \sqrt{a^2 - c^2}$

30. What is the equation of the line that passes through the points $(3, -5)$ and $(-5, 11)$?
F $y = 3x + 11$
G $y = 2x + 9$
H $y = 3x - 4$
J $y = -2x + 1$

31. What equation gives the direct variation if $y = -2$ when $x = 8$?
A $y = -4x$
B $y = \left(\frac{1}{4}\right)x$
C $y = 4x$
D $y = -\left(\frac{1}{4}\right)x$

Cumulative Test

Chapter 14 Test C, continued

32. What inequality is shown on the graph?

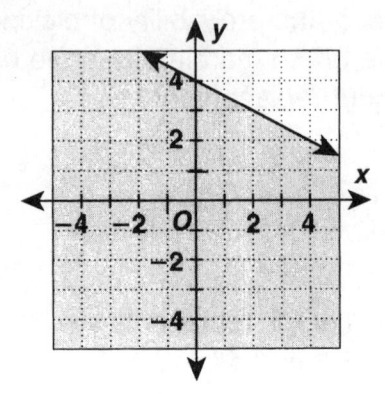

F $y \leq -\left(\frac{1}{2}\right)x + 4$ H $y \leq \left(\frac{1}{2}\right)x + 4$

G $y \leq \left(\frac{1}{2}\right)x - 4$ J $y \leq -\left(\frac{1}{2}\right)x + 4$

33. What is the eleventh term in the sequence 50, 46, 42, 38, …?
A 10
B 6
C 0
D −10

34. What is the rule for the function table?

x	−4	−1	2	5
y	−18	−9	0	9

F $y = x - 2$ H $y = 3(x - 2)$
G $y = 3(2 - x)$ J $y = 3x - 2$

35. Find $f(-5)$ for
$f(x) = \left(\frac{1}{2}\right)x^2 - \left(\frac{1}{2}\right)x + 5$.

A $f(-5) = 12$ C $f(-5) = 13\frac{3}{4}$
B $f(-5) = 12\frac{1}{2}$ D $f(-5) = 20$

36. What is the degree of the polynomial $\left(\frac{1}{2}\right)x^2 + 5x^3 - 4x$?

F $\frac{1}{2}$ H 2
G 1   J 3

37. Add $(3m^2 + mn) + (3mn - m^2)$.
A $4m^2 + 4mn$ C $2m^2 - 4mn$
B $2m^2 + 4mn$ D $m^2 + 4mn$

38. Multiply $(-3abc^2)(-2a^2b + 4ac^2)$.
F $6a^3b^2c^2 - a^2bc^4$
G $6a^3b^2c^2 - 12a^2bc^4$
H $6a^3b^2c^2 + 12a^2bc^4$
J $-6a^5b^3c^6$

39. Which set is an infinite set?
A {rational numbers between −10 and 10}
B {whole numbers between −10 and 10}
C {integers between −10 and 10}
D {natural numbers between −10 and 10}

40. What conditions are required to make the disjunction P or Q false?
F Both P and Q are true.
G P is true and Q is false.
H P is false and Q is true.
J Both P and Q are false.

41. What is the degree of vertex D?

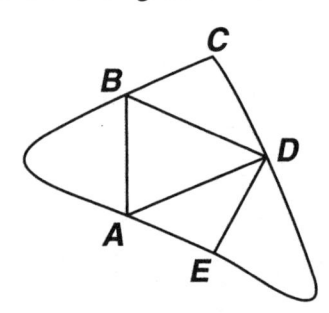

A 5   C 3
B 4   D 2

Name _____ Date _____ Class _____

End of Course Assessment

Select the best answer.

1. Which of the following is a correct statement?
   A $0.03 > 30\%$
   B $0.03 = \frac{3}{10}$
   C $0.30 < \frac{3}{10}$
   D $30\% = \frac{3}{10}$

2. Valleywood Golf Course offers classes Monday through Saturday. The Saturday classes last 75 minutes while the weekday classes last 45 minutes. If they offered 705 minutes of classes last week, what was the maximum number of weekday classes?
   F 9
   G 14
   H 12
   J 17

3. Find the area of a trapezoid with the dimensions $b_1 = 9$, $b_2 = 16$, $h = 13.4$.
   A 335 sq units
   B 223.4 sq units
   C 189.2 sq units
   D 167.5 sq units

4. A car salesman makes $445 per week plus commission. If during the week he sells a new van for $34,960 and earns a 2.5% commission on the sale, how much money did he earn for the week?
   F $429
   G $874
   H $885
   J $1319

5. What is the value of $f(-3)$ for the function $f(x) = 4.7x + 1.6$?
   A -18.3
   B -15.7
   C -12.5
   D 15.7

6. If $\angle A$ and $\angle B$ are supplementary, and $m\angle A = 57°$, what is $m\angle B$?
   F 33°
   G 123°
   H 213°
   J 303°

7. At a school festival, a colored chip is randomly drawn out of a bag, and replaced. The table below shows the results of 50 draws. Estimate the probability of choosing a purple chip.

Outcome	Blue	Red	Green	Purple	Gold
Draws	7	11	6	12	14

   A 12%
   B 18%
   C 24%
   D 88%

8. What are the coordinates of point K?
   F $(-2, 4)$
   G $(-2, 4)$
   H $(2, -4)$
   J $(2, -4)$

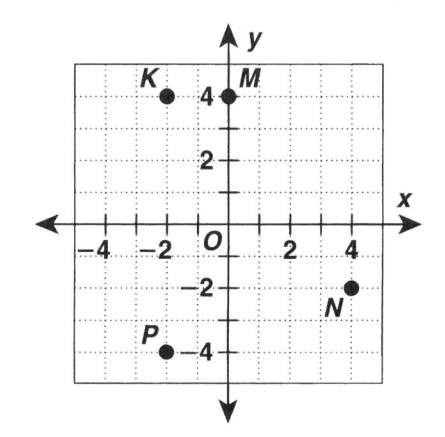

9. Solve for k, $\frac{4}{5}k + \frac{7}{10} = \frac{13}{15}k - \frac{3}{5}$
   A $k = \frac{1}{10}$
   B $k = 1\frac{1}{2}$
   C $k = 19\frac{1}{2}$
   D $k = 21\frac{2}{5}$

313

Holt Pre-Algebra

Name _____ Date _____ Class _____

End of Course Assessment

10. What is the slope of a line that is perpendicular to the line that passes through the points (4, 7) and (9, 3)?

F $-\dfrac{4}{5}$ H $\dfrac{4}{5}$

G $-\dfrac{5}{4}$ J $\dfrac{5}{4}$

11. Evaluate $\dfrac{3 + k + 9}{2}$ for $k = -16$.

A −2 C −4
B 2 D 14

12. In a school survey, three homerooms are chosen and ten students from each homeroom are randomly chosen to complete the survey. Identify the sampling method.

F Population H Systematic
G Random J Stratified

13. Give the 8th term in the sequence the numerator of each fraction is 1.

$6\dfrac{1}{6}, 6\dfrac{1}{3}, 6\dfrac{1}{2}, \ldots$

A $6\dfrac{2}{3}$ C 7

B $6\dfrac{5}{6}$ D $7\dfrac{1}{3}$

14. On a drawing with a scale of $\dfrac{1}{8}$ in.: 1 ft, a window is $\dfrac{3}{4}$ in. wide. What is the actual width of the window?

F 5.5 ft H 6 ft
G 5.75 ft J 6.25 ft

15. What is the equation of direct variation given that y is 16 when x is −2?

A $y = -8x$ C $y = 8x$
B $y = -\dfrac{1}{8}x$ D $x = -8y$

16. The graph represents which inequality?

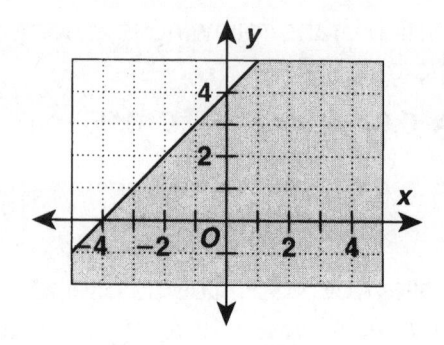

F $y \le x + 4$ H $y < x + 4$
G $y = 4x$ J $y > x + 4$

17. Combine like terms
$7a + 4b - 3a - 2b$.

A $4a + 2b$ C $4a - 2b$
B $11a - b$ D $7ab$

18. Find the mean, median, and mode of the data set. 5, 8, 6, 5, 8, 4, 8, 4

F 6, 5.5, 8 H 8, 5.5, 8
G 5.5, 6, 6 J 6, 5.5, 5

19. Simplify $19 + (3 \cdot 2^4)$.

A 38 C 912
B 67 D 1315

20. What is the square root of 429 to the nearest tenth?

F 8.6 H 214.5
G 20.7 J 184,041

21. When starting a family vacation, the Singler's odometer in their van read 15,674.7. At the end of the trip the odometer read 16,495.2, how far did the Singler's travel?

A 32,169.9 mi C 820.5 mi
B 1117.2 mi D −820.5 mi

Name _____ Date _____ Class _____

End of Course Assessment

22. Caleb and Drew are playing a game with a pair of dice. Caleb needs a sum of 5 or greater to win. What is his probability of winning on his next turn?

    F $\frac{5}{6}$ H $\frac{1}{6}$

    G $\frac{2}{5}$ J $\frac{2}{3}$

23. Wesley invested $6500 in a mutual fund at a yearly rate of $3\frac{3}{4}\%$. If he has earned $975 in interest, how long has the money been invested?

    A 2.5 years   C 4 years
    B 3 years   D 4.5 years

24. If a pool table measures 4 ft by 8 ft, what is the length from the back edge of the top left pocket to the bottom right pocket to the nearest tenth?

    F 80 ft   H 6.9 ft
    G 24.3 ft   J 8.9 ft

25. What is the sum of the interior angles of the polygon?

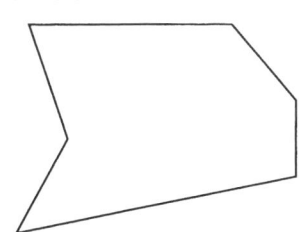

    A 180°   C 540°
    B 360°   D 720°

26. Evaluate $2x + 5y$ for $x = 12$ and $y = 6$.

    F 72   H 54
    G 25   J 42

27. Solve $\frac{3}{18} = \frac{a}{30}$.

    A 6   C 15
    B 5   D 90

28. Your favorite brand of cereal comes in four different sized boxes. Which size is the best deal?

Toasted Almond Crunch	
15 oz	$2.39
18 oz	$2.87
24 oz	$3.79
32 oz	$5.10

    F 15 oz   H 24 oz
    G 18 oz   J 32 oz

29. The line has what type of slope?

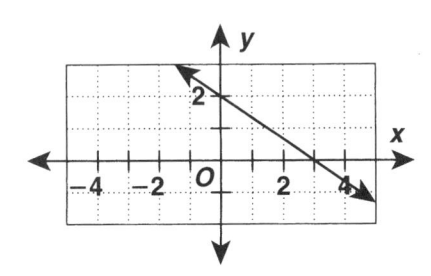

    A positive   C zero
    B negative   D undefined

30. Which is the number 0.0000042 in scientific notation?

    F 0.42×10^{-7} H 4.2×10^{-6}
    G 4.2×10^{6} J 4.2×10^{-5}

31. Which ordered pair is a solution of the system of equations?
    $y = 3x + 1$ $y = 5x - 3$

    A (2, 3)   C (1, 2)
    B (0, 1)   D (2, 7)

Name _____ Date _____ Class _____

End of Course Assessment

32. For a graduation party Mrs. Kennedy prepares a meat tray. If she buys 3 pounds each of ham and turkey, 2 pounds of roast beef, and 1 pound of salami, how much will the meat cost?

Today's Specials	
Roast Beef	$6.29 lb
Ham	$3.29 lb
Salami	$3.89 lb
Turkey	$4.79 lb

F $18.26 H $40.71
G $31.13 J $47.26

33. If Hannah purchased a 5-day parking pass for $36.25, how much did she pay per day?
A $6.75   C $7.00
B $6.85   D $7.25

34. A school festival has sold 760 tickets. If this is 95% of their goal, how many more tickets do they need to sell to reach 100% of their goal?
F 22   H 58
G 40   J 800

35. Find the perimeter of the figure.

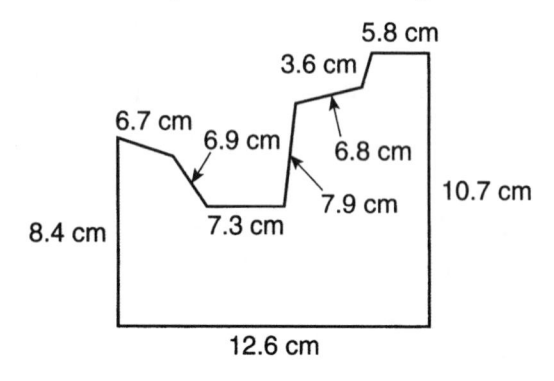

A 69.8 cm   C 73.1 cm
B 70 cm   D 76.7 cm

36. Jasmin is walking a group of 6 dogs. In how many different orders can the dogs enter her house in single file to get a drink of water?
F 21   H 480
G 40   J 720

37. How many times did the temperature rise one degree per hour?

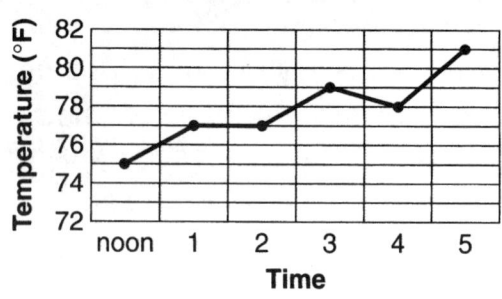

A none   C 2
B 1   D 3

38. Give the 9th term in the sequence 2, −6, 18, −54 . . .
F −39,366   H 1459
G −4374   J 13,122

39. What is the range and the first and third quartiles of the data set?
14, 16, 12, 15, 22, 18, 16, 10, 12
A 22, 12, 17   C 12, 12, 17
B 22, 15, 17   D 12, 12, 15

40. Find the volume. Use 3.14 for π. 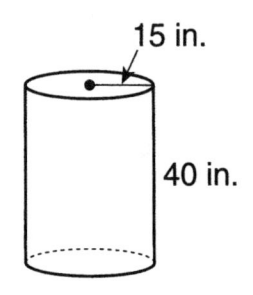
F 1200 in^3
G 1884 in^3
H 3768 in^3
J 28,260 in^3

316   Holt Pre-Algebra

Inventory Test

Select the best answer for questions 1–40.

1. Twice a number, increased by 28 equals 48 plus the number. What is the number?
   A 21
   B) 20
   C 18
   D −18

2. An airplane flies at a cruising altitude of 2900 ft. It descends 1200 ft as it begins to reach its destination. As it approaches the airport, it descends an additional 1400 ft. What is the new altitude of the airplane?
   F 1700 ft
   G 1500 ft
   H) 300 ft
   J 150 ft

3. If you can type 1,240 words in 40 minutes, how many words per minute can you type?
   A 180 words/min
   B) 31 words/min
   C 27 words/min
   D 9 words/min

4. Solve the inequality $-5x < -45$.
   F) $x > 9$
   G $x < -5$
   H $x < 9$
   J $x < -9$

5. Evaluate the expression $-|7 - y|$ for $y = -6$.
   A −1
   B) −13
   C 1
   D 13

6. Which is equivalent to $(-4)^4(-4)^9$?
   F $(16)^{36}$
   G) $(-4)^{13}$
   H $(16)^{13}$
   J $(-4)^{36}$

7. Express 680,000 using scientific notation.
   A) 6.8×10^5
   B 6.8×10^4
   C 6.8×10^{-5}
   D 6.8×10^{-4}

8. Simplify $6 \times 1\frac{9}{10}$.
   F $6\frac{9}{10}$
   G $\frac{25}{11}$
   H) $11\frac{2}{5}$
   J $\frac{79}{10}$

9. Divide $\frac{6}{7} \div \frac{4}{5}$.
   A $\frac{3}{28}$
   B $\frac{24}{35}$
   C $\frac{14}{17}$
   D) $\frac{15}{14}$

10. Which is equivalent to $9\sqrt{36 + 64}$?
    F) 90
    G 126
    H 30
    J 23

11. A potential candidate for mayor wants to know whether she is positioned to launch a successful campaign. 25 residents chosen at random from each of 10 randomly identified city neighborhoods are asked about their opinions on 10 key issues. Identify the sampling method used.
    A biased
    B random
    C systematic
    D stratified

12. To the nearest tenth, find the mean weight of a group of apples with individual weights of 4.7, 4.0, 6.2, 6.5, 6.1, 4.7, 4.0, 6.2, 6.3, 6.5, 4.7, 6.2, 6.5, and 6.0 ounces.
    F 4.7 oz
    G 6.5 oz
    H) 5.6 oz
    J 6.2 oz

13. Six employees of a small company were asked the distance they traveled to work, to the nearest tenth of a mile. Find the range of the distances: 2.7, 5.9, 1.7, 4.7, 6.9, 3.5.
    A 0.8 mi
    B 1.7 mi
    C) 5.2 mi
    D 5.9 mi

14. Identify the pair of angles that are complementary.

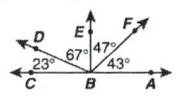

    F $\angle CBE$ & $\angle CBD$
    G $\angle ABF$ & $\angle CBD$
    H $\angle DBE$ & $\angle EBF$
    J) $\angle ABF$ & $\angle FBE$

15. Which angle is congruent to $\angle 5$?

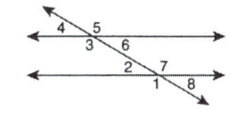

    A $\angle 6$
    B $\angle 8$
    C) $\angle 7$
    D $\angle 2$

16. A decagon has ten sides. What is the sum of the interior angles in a decagon?
    F) 1,440°
    G 1,800°
    H 2,160°
    J Unknown. The interior angle formula only applies to regular polygons.

17. The diameter of an ice-hockey puck is 3.0 inches. To the nearest tenth, what is the area of the flat upper surface? Use 3.14 for π.
    A 3.5 in²
    B 9.4 in²
    C) 7.1 in²
    D 28.3 in²

18. Find the missing length in the right triangle.

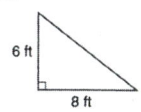

    F 7 ft
    G 8 ft
    H 9 ft
    J) 10 ft

19. Find the volume of the pyramid.

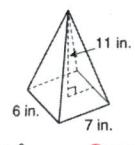

    A 462 in³
    B 242 in³
    C) 154 in³
    D 14 in³

20. To estimate the number of fish in a lake, biologists catch, mark, and release 50 fish. One week later, they catch 33 fish and find that 15 of them are marked. What is their best estimate for the total number of fish in the lake?
    F) 110
    G 73
    H 83
    J 60

21. Convert 8 weeks to minutes.
    A) 80,640 min
    B 57,600 min
    C 40,320 min
    D 10,080 min

22. Find the missing value in the proportion $\frac{4}{8} = \frac{x}{64}$.
    F $x = 4$
    G $x = 12$
    H) $x = 32$
    J $x = 256$

23. An interior decorator is making a scale drawing of a room. If the scale is 2 in. : 3 ft, how wide is the drawing of 7.5-ft bay window?
    A) 5 in.
    B 11.25 in.
    C 15 in.
    D 2.5 in.

24. What is $6\frac{1}{5}$% of 8000?
    F 500
    G) 496
    H 4,960
    J 49,600

25. 56 is 140% of what number?
    A) 40
    B 78.4
    C 400
    D 19,600

26. A computer, priced at $1,380, is marked down to $1,057. What is the rate of markdown?
    F 22.7%
    G) 23.4%
    H 29.6%
    J 30.6%

27. A six-sided die is rolled twice. What is the probability of rolling the same number twice in a row?
    F $\frac{1}{6}$
    G) $\frac{1}{36}$
    H $\frac{2}{3}$
    J $\frac{1}{3}$

28. Find the value of $_{10}C_2$.
    A 90
    B) 45
    C 40,320
    D 80,640

29. In a certain town, 4% of the population commutes to work by bicycle. If a person is randomly selected from the town, what are the odds against selecting someone who commutes by bicycle?
    A 24:25
    B) 24:1
    C 1:25
    D 1:24

30. Solve $-7c + 6 = 2 + 8c$.
    F $c = \frac{15}{4}$
    G $c = -\frac{15}{4}$
    H) $c = \frac{4}{15}$
    J $c = 8$

31. Find the inequality that corresponds to the given number line.

    A) $2x + 3 < 11$
    B $2x + 3 > 11$
    C $2x + 3 \geq 11$
    D $2x + 3 \leq 11$

32. Solve the formula $A = \frac{bh}{2}$ for b.
    F $b = \frac{A}{2h}$
    G $b = \frac{Ah}{2}$
    H) $b = \frac{2A}{h}$
    J $b = \frac{h}{2A}$

33. Solve the system of equations:
    $2x + 2y = 18$
    $6x + y = 39$
    A $(-6, 2)$
    B) $(6, 3)$
    C $(7, 2)$
    D $(-2, 11)$

34. Find the rule for the linear function.

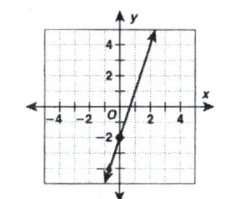

    F) $f(x) = 3x - 2$
    G $f(x) = 3x + 2$
    H $f(x) = -3x - 2$
    J $f(x) = x - 2$

35. What is the point-slope form of the equation for the line with a slope of $\frac{2}{3}$ that passes through (0, 2).
    A) $y - 2 = \frac{2}{3}x$
    B $y + 2 = \frac{2}{3}x$
    C $y - 3 = \frac{3}{2}x$
    D $y + 2 = -\frac{2}{3}x$

36. Given that y varies directly with x, find the equation of direct variation if y is 10 when x is 5.
    F $y = x + 5$
    G) $y = 2x$
    H $y = \frac{50}{x}$
    J $y = 0.5x$

37. Find the 12th term in the arithmetic sequence 2, 7, 12, 17, ...
    A 62
    B 60
    C) 57
    D 55

38. Complete the ordered pairs in the table for the equation $y = -x + 8$.

x	3	8	0.5
y	??	??	??

    F 15, 0, 8.5
    G 5, −2, 2.5
    H 3, 8, 0.5
    J) 5, 0, 7.5

39. A colony of bacteria triples in size every 4 days. If there were originally 27 cells, how many cells are there after 4 weeks?
    A 729
    B 2187
    C 19,683
    D) 59,049

40. Find $f(4)$ for the quadratic function $f(x) = 3x^2 - x - 5$.
    F 3
    G 15
    H) 39
    J 135

317   **Holt Pre-Algebra**

Chapter 1 Quiz — Section A

Choose the best answer.

1. Evaluate $2x + 8$ for $x = 5$.
   A 10   C 28
   B 18   D 32

2. Evaluate $2.5r + 12$ for $r = 6$.
   A 24   **C 27**
   B 162   D 44

3. Evaluate $4a + 7c$ for $a = 5$ and $c = 3$.
   A 118   C 30
   B 41   D 19

4. Which algebraic expression represents "4 times the sum of 12 and b"?
   A $4 + 12 + b$ **C $4(12 + b)$**
   B $4 + (12 + b)$ D $4(12 - b)$

5. Which algebraic expression represents "3 less than the sum of 5 and r"?
   A $(5 \cdot r) - 3$ C $3 - (5 + r)$
   B $(5 + r) - 3$ D $(5 - r) - 3$

6. Which value of z is the solution for the equation $43 - z = 18$?
   A $z = 25$ C $z = 13$
   B $z = 15$ D $z = 61$

7. Which value of t is the solution for the equation $9.45 = t + 3.7$?
   A $t = 13.15$ **C $t = 5.75$**
   B $t = 6.2$ D $t = 6.12$

8. What is the value of k for this equation: $\frac{k}{8} = 12$?
   A $k = 96$ C $k = 66$
   B $k = 20$ D $k = \frac{3}{2}$

9. What is the value of m for this equation: $4m - 15 = 33$?
   A $m = 48$ C $m = 60$
   B $m = 52$ **D $m = 12$**

10. Which inequality is represented by this graph?

    A $x + 3 > 10$ **C $y + 5 > 15$**
    B $t - 3 \le 22$ D $5 > \frac{w}{2}$

11. Simplify $3(2x - 5) + 2x$.
    A $6x - 15 + 2x$ C $7x - 5$
    B $8x - 15$ D $6x - 15$

12. Solve $6d + 4 + 5d - 2d = 58$.
    A $d = 13d$ **C $d = 6$**
    B $d = 10$ D $d = 13d + 4$

Chapter 1 Quiz — Section B

Choose the best answer.

1. Which ordered pair is a solution for $y = 2x + 8$?
   A (1, 20)   C (3, 15)
   B (4, 12)   **D (4, 16)**

2. Which ordered pair is a solution for $y = 5x - 3$?
   A (17, 4)   **C (4, 17)**
   B (5, 28)   D (3, 18)

3. The price of a 5-line newspaper ad is $8, plus $1.50 for each additional line. The equation for price p for buying an ad is $p = 8 + 1.50l$. What is the cost of an ad with 4 extra lines?
   A $14   C $11.50
   B $13   D $9.50

4. What ordered pair is missing from the table below for $y = 2x - 5$?

x	$2x$	y	(x, y)
8	2(8)	11	(8, 11)
9	2(9)	13	(9, 13)
10	2(10)	15	(,)
11	2(11)	17	(11, 17)

   A (10, 15)   C (10, 14)
   B (10, 12)   D (15, 11)

Use this coordinate plane for questions 5–7.

5. Identify the coordinates for point B.
   A $(-2, -3)$ **C $(-2, 3)$**
   B $(2, 3)$ D $(2, 3)$

6. Identify the coordinates for point A.
   A $(3, 3)$ C $(2, 2)$
   B $(2, 3)$ D $(-2, -3)$

The table shows how many seconds it takes for a garage door to open.

Time (seconds)	Door Opening (feet)
0	0
2	1
4	2
6	3
8	4
10	5
12	6
14	7
16	8

7. How many seconds does it take for the garage door to be half-way up?
   A 4 seconds   C 6 seconds
   B 8 seconds   D 16 seconds

Chapter 2 Quiz — Section A

Add.

1. $6 + (-9)$
   A -15 C 3
   B -3   D 15

Evaluate the expression.

2. $11 + d + (-4)$ for $d = -6$
   A -9 **C 1**
   B -1 D 9

Subtract.

3. $-12 - (-4)$
   A -16 C 8
   B -8   D 16

Multiply or divide.

4. $-7(-2)$
   A -14 C 5
   B -5 **D 14**

5. $\frac{-9(11)}{-3}$
   A -33 **C 33**
   B 27   D 99

Solve.

6. $-17 + v = 3$
   A $v = -14$ C $v = 14$
   B $v = 3$ **D $v = 20$**

7. $8r = -48$
   A $r = -8$ C $r = 6$
   B $r = -6$ D $r = 384$

8. $\frac{h}{6} = -7$
   A $h = -42$ C $h = 13$
   B $h = -13$ D $h = 42$

9. $4 - p = 12$
   A $p = -16$ C $p = 8$
   B $p = -8$ D $p = 24$

10. $\frac{k}{5} < -2$
    A $k < -10$ C $k < 7$
    B $k < -7$ D $k < 10$

Chapter 2 Quiz — Section B

Write using exponents.

1. $n \cdot n \cdot n \cdot n \cdot n$
   A n^3 **C n^5**
   B n^4 D n^6

Evaluate.

2. $(-3)^4$
   A -81 C 12
   B -12 **D 81**

Simplify.

3. $14 + (-2 - (-2)^3)$
   A -4 **C 20**
   B 4   D 24

4. $(6 \cdot 2)^2 + 4$
   A 28   C 164
   B 148   D 576

Multiply or divide. Write the product as one power.

5. $4^3 \cdot 4^0 \cdot 4^2 \cdot 4^5$
   A 4^9 C 4^{11}
   B 4^{10}   D 256^{10}

6. $\frac{f^7}{g^3}$
   A fg^4 C $\frac{f^4}{g}$
   B fg^{10} **D Cannot simplify**

Simplify.

7. 10^{-4}
   A 0.00001   C 1000
   B 0.0001   D 10,000

8. $\frac{6^5}{6^3}$
   A 1   **C 36**
   B 12   D 216

Write the number in standard notation.

9. 1.42×10^{-2}
   A 0.00142   C 1.42
   B 0.0142   D 142

Write the number in scientific notation.

10. 0.000721
    A 0.721×10^{-5} **C 7.21×10^{-4}**
    B 7.21×10^{-5} D 721×10^{-4}

318   Holt Pre-Algebra

Chapter 3 Quiz — Section A

Choose the best answer.

1. Simplify the fraction $\frac{42}{90}$.
   A $\frac{22}{45}$ C $\frac{9}{15}$
   (B) $\frac{7}{15}$ D $\frac{16}{30}$

Add or subtract. Express in simplest form.

2. $\frac{7}{10} + \frac{9}{10}$
   F $1\frac{1}{5}$ H $1\frac{1}{3}$
   G $1\frac{1}{5}$ (J) $1\frac{3}{5}$

3. $\frac{8}{9} - \frac{2}{9}$
   A $\frac{6}{9}$ C $\frac{10}{9}$
   (B) $\frac{2}{3}$ D $1\frac{1}{9}$

4. Multiply $-0.02(32.7)$.
   (F) -0.654 H 0.654
   G -6.54 J 6.54

5. Evaluate $3\frac{1}{5}n$ for $n = \frac{3}{4}$.
   A $1\frac{5}{9}$ C $3\frac{1}{9}$
   (B) $2\frac{2}{5}$ D $3\frac{3}{20}$

6. Divide $3\frac{3}{5} \div 6$. Express in simplest form.
   (F) $\frac{3}{5}$ H $\frac{108}{5}$
   G $18\frac{3}{5}$ J $6\frac{11}{5}$

Add or subtract.

7. $\frac{6}{7} - \frac{4}{5}$
   A $1\frac{23}{25}$ C 1
   B $\frac{1}{12}$ (D) $\frac{2}{35}$

8. $\frac{3}{8} + \frac{1}{7}$
   (F) $\frac{29}{56}$ H $\frac{18}{56}$
   G $-\frac{4}{15}$ J $-\frac{3}{56}$

9. Solve $x + \frac{4}{15} = \frac{2}{3}$.
   A $x = \frac{14}{15}$ (C) $x = \frac{2}{5}$
   B $x = -\frac{6}{15}$ D $x = -\frac{14}{15}$

10. Solve $3.7 + y > 4.9$.
    F $y < -1.2$ H $y < 1.2$
    G $y > -1.2$ (J) $y > 1.2$

Chapter 3 Quiz — Section B

Choose the best answer.

Simplify each expression.

1. $\sqrt{9 + 16}$
   A 4 C 7
   (B) 5 D 25

2. $\sqrt{64} - \sqrt{36}$
   F 10 (H) 2
   G 0 J 5.29

3. $11 + \sqrt{25}$
   A 36 C 12
   B 6 (D) 16

Use a calculator to find each square root. Round to the nearest tenth.

4. $\sqrt{132}$
   F 11.2 H 11.8
   (G) 11.5 J 14.9

5. $-\sqrt{47}$
   (A) -6.9 C 6.4
   B -6.2 D 6.9

6. $\sqrt{56.4}$
   A 5.7 (C) 7.5
   B 6.2 D 8.4

7. What kind of number is 0?
   (F) rational
   G irrational
   H not a real number
   J negative number

8. What kind of number is $-\sqrt{2}$?
   A rational
   (B) irrational
   C not a real number
   D natural number

Chapter 4 Quiz — Section A

Choose the best answer.

1. Identify the population and sample in the following example. A convenience store owner wants to know which soda-pop is the most popular in his store. He surveys 30 random customers and notes which soda they buy.
   A All people; customers who buy soda
   B Customers who buy soda; people who shop at the store
   (C) People who shop at the store; customers who are surveyed
   D People who shop at the store; customers who buy soda

2. Identify the sampling method used in the following example. A door-to-door poll-taker stops at every third house.
   F Random
   (G) Systematic
   H Stratified
   J Not a sample method

3. Which set of numbers represent the data values in the stem-and-leaf plot?

Stem	Leaves
5	1 4
6	0 5 8
7	
8	7 7 9

   A 51, 54, 60, 64, 68, 87, 87, 89
   B 51, 54, 60, 65, 68, 68, 87, 87
   C 50, 54, 60, 65, 68, 87, 87, 89
   (D) 51, 54, 60, 65, 68, 87, 87, 89

4. Find the mean, median, and mode of the data set: 15, 7, 9, 12, 21, 11, 13, 12, 18
   F mean: 12, median: 12, mode: 13.1
   (G) mean: 13.1, median: 12, mode: 12
   H mean: 12, median: 13.1, mode: 12
   J mean: 13.5, median: 12, mode: 12

5. Find the range and the first and third quartiles of the data set: 42, 33, 47, 50, 37, 51, 49, 35, 52, 48, 53
   A range: 18, 1st: 42, 3rd: 50
   B range: 20, 1st: 40, 3rd: 49
   (C) range: 20, 1st: 37, 3rd: 51
   D range: 18, 1st: 38, 3rd: 51

6. Use the box-and-whiskers graph to find the third quartile.

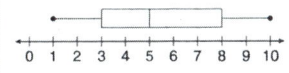

   F 5
   (G) 8
   H 10
   J 9

Chapter 4 Quiz — Section B

Choose the best answer.

1. What is the frequency of the data value 8 in the data set below?
   9, 7, 8, 6, 7, 6, 5, 9, 8, 6, 9, 8, 7
   A 0 C 1
   B 2 (D) 3

2. Use the following line graph to estimate the number of units sold for the month of August. 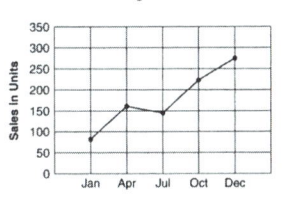
   F 150 H 140
   (G) 175 J 200

3. What is the interval used in the following histogram? 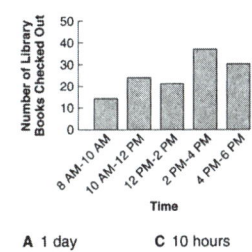 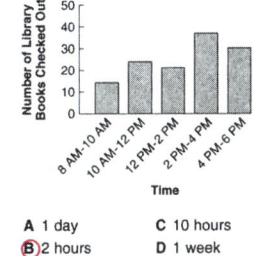
   A 1 day C 10 hours
   (B) 2 hours D 1 week

4. Which of the following statements is not misleading?
   F A little league baseball player hits the ball 55% percent of the time, while a professional player hits the ball 33% of the time. Little-leaguers are better hitters than professionals.
   G A basketball player makes 75% of her shots during a single game. She is obviously a great shooter.
   (H) A hockey team scores no points during one game. The best the team can do is tie the game.
   J The kicker for the school football team scored more points than any other player on the team for the season. He is the best player on the team.

5. Which of the following data sets has a negative correlation?
   (A) The total distance driven and the amount of tread on the tires.
   B The number of miles driven and the amount of gas used.
   C The number of passengers in a car and the number of driver licenses.
   D The speed of the car and the rotation rate of its tires.

Holt Pre-Algebra

CHAPTER 5 Quiz — Section A

Choose the best answer.

1. Which of the following is not true for the figure below?

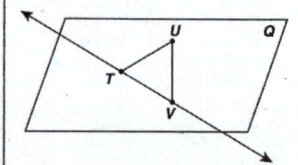

   A plane *TUV*, point *T*, point *U*, point *V*
   B plane Q, point *T*, point *U*, point *V*
   C $\overline{TQ}$, $\overline{TU}$, $\overline{UV}$ ✓
   D $\overline{TV}$, $\overline{TU}$, $\overline{UV}$

2. If ∠F and ∠G are vertical angles and m∠F = 63°, find m∠G.
   F 180°   H 117°
   G 63° ✓   J 100°

3. Which of the following is true for the figure below where line *e* ∥ line *f*? 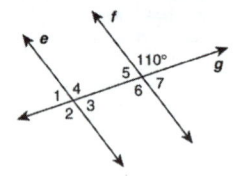
   A ∠2 ≅ ∠7
   B m∠3 = 70° ✓
   C line *f* ⊥ line *g*
   D m∠6 = 70°

4. If the congruent angles of an isosceles triangle each measure 41°, what is the measure of the third angle?
   F 41°   H 59°
   G 98° ✓ J 118°

5. A regular polygon that has six sides is a hexagon. What is the angle measure of a hexagon?
   A 120° ✓   C 720°
   B 60°   D 90°

6. Which set of terms applies to the figure below? 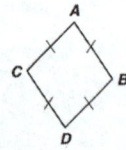
   F quadrilateral, parallelogram, square
   G quadrilateral, parallelogram, rhombus ✓
   H parallelogram, trapezoid, rectangle
   J quadrilateral, trapezoid, rhombus

7. What is the slope of the line that passes through the points (−3, 7) and (5, −1)?
   A −1 ✓   C 1
   B 3   D −3

CHAPTER 5 Quiz — Section B

Choose the best answer.

1. Identify the correct congruence statement for the pair of polygons shown below. 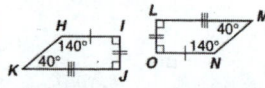
   A trapezoid HIJK ≅ trapezoid LMNO
   B trapezoid KJIH ≅ trapezoid MLON ✓
   C trapezoid IJKH ≅ trapezoid ONML
   D trapezoid JIHK ≅ trapezoid ONML

2. triangle ABC ≅ triangle EDF. Find *f*. 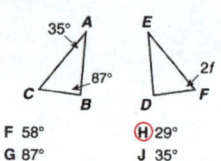
   F 58°   H 29° ✓
   G 87°   J 35°

3. Identify which transformation is represented in the following figure. 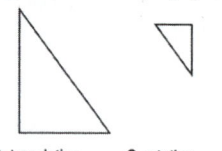
   A translation   C rotation
   B reflection   D none of these ✓

4. A square has vertices T(1, 2), U(4, 2), V(4, 5), W(1, 5). If the square is rotated clockwise 90 degrees around (0, 0), what is the coordinate of W?
   F (5, −1) ✓   H (2, −4)
   G (5, −4)   J (2, −1)

5. Which of the following is not true about tessellations?
   A The angles at the vertices add to 360°.
   B Tessellations cover an entire plane.
   C A regular tessellation uses two or more regular polygons. ✓
   D There are no gaps or overlaps in a tessellation.

CHAPTER 6 Quiz — Section A

Choose the best answer.

1. Find the perimeter of the figure below. 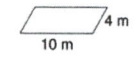
   A 40 m   C 28 m ✓
   B 10 m   D 4 m

2. Find the area of the figure with vertices (−1, −1), (3, −1), (3, 4), and (−1, 4).
   F 18 units²   H 10 units²
   G 20 units² ✓ J 9 units²

3. Find the perimeter of the figure below. 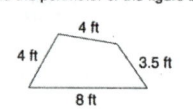
   A 19.5 ft ✓   C 24 ft
   B 21 ft   D 18 ft

4. Find the area of the figure with vertices (0, 3), (3, −1), and (−2, −1).
   F 20 units²   H 5 units²
   G 10 units² ✓ J 4 units²

5. Find the length of the hypotenuse of a right triangle that has legs of lengths 12 in. and 9 in.
   A 17 in.   C 21 in.
   B 13 in.   D 15 in. ✓

6. Find the circumference of a circle that has a radius of 7 cm. Use 3.14 for π.
   F 153.86 cm   H 43.96 cm ✓
   G 114.74 cm   J 21.98 cm

7. Find the area of a circle with center (2, 3) that passes through (2, 7). Use 3.14 for π.
   A 50.24 units² ✓   C 25.12 units²
   B 12.56 units²   D 41.36 units²

8. A merry-go-round that has a radius of 20 ft has a duration per ride of 15 revolutions. How far does a person travel who rides on the merry-go-round? Use 3.14 for π.
   F 600 ft   H 942 ft
   G 18,840 ft   J 1884 ft ✓

CHAPTER 6 Quiz — Section B

Choose the best answer.

1. A flat surface of a three-dimensional figure is a(n) _____.
   A edge   C point
   B vertex D face ✓

2. Find the volume of a rectangular prism with base 2 units by 3 units and height 5 units.
   F 15 units³   H 62 units³
   G 30 units³ ✓ J 20 units³

3. Find the volume of the figure below to the nearest tenth of a unit using 3.14 for π.
   A 2373.8 cm³ ✓   C 791.3 cm³
   B 1017.4 cm³   D 9495.4 cm³

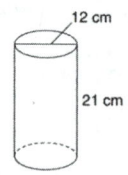

4. Find the volume of the figure below. 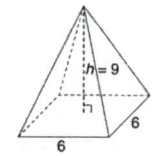
   F 324 units³   H 108 units³ ✓
   G 144 units³   J 236 units³

5. Find the surface area of the figure below. Use 3.14 for π.
   A 1055.04 in²   C 326.56 in²
   B 427.04 in² ✓   D 653.12 in²

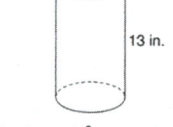

6. A regular square pyramid has a base with sides that measure 4 ft and a slant height of 7 ft. Find the surface area of the pyramid.
   F 37.3 ft²   H 128 ft²
   G 56 ft²   J 72 ft² ✓

7. Find the surface area of a cone with diameter 6 cm and slant height 12 cm. Use 3.14 for π.
   A 339.12 cm²   C 113.04 cm²
   B 141.30 cm² ✓ D 452.16 cm²

8. Find the volume of a sphere with radius 11 m to the nearest tenth of a unit using 3.14 for π.
   F 1519.8 m³   H 3271.1 m³
   G 2696.5 m³   J 5572.5 m³ ✓

9. Find the surface area of a sphere with diameter 34 in. to the nearest tenth of a unit using 3.14 for π.
   A 20,569.1 in²   C 14,519.4 in²
   B 3629.8 in² ✓   D 6112.96 in²

Chapter 7 Quiz — Section A

Choose the best answer.

1. Find two ratios that are equivalent to the ratio $\frac{15}{6}$.
   A $\frac{30}{10}, \frac{10}{3}$
   C $\frac{30}{18}, \frac{5}{2}$
   B $\frac{45}{18}, \frac{5}{2}$ ✓
   D $\frac{45}{10}, \frac{10}{3}$

2. Which conversion is correct?
   F 3 yd = 36 in.
   G 1 day = 1440 sec
   H 2 mi = 10,560 ft ✓
   J 200 km = 2,000,000 cm

3. A thunder storm produced 114 lightning strikes in $1\frac{1}{2}$ hours. What was the unit rate of lightning strikes?
   A 76 strikes per hour ✓
   B 228 strikes per hour
   C 114 strikes per hour
   D 57 strikes per hour

4. The typical speed of sound is 760 miles per hour. If a plane is traveling exactly at the speed of sound, it is said to be at Mach 1. What is the Mach of a plane, to the nearest tenth, traveling 960 miles per hour?
   F 1.5
   G 1.7
   H 1.3 ✓
   J 0.8

5. If an object is traveling 30 feet per second, what is its speed in miles per hour to the nearest whole number? There are 5280 feet in a mile.
   A 20 mph ✓
   B 55 mph
   C 25 mph
   D 32 mph

6. What is the value of a in the proportion $\frac{a}{24} = \frac{6}{16}$?
   F $a = 64$
   G $a = 3$
   H $a = 8$
   J $a = 9$ ✓

Chapter 7 Quiz — Section B

1. A rectangle with dimensions 7 × 13 is enlarged by 15%. What are the new dimensions?
   A 10.5 × 19.5
   B 1.05 × 1.95
   C 8.05 × 14.95 ✓
   D 5.95 × 11.05

2. A figure has vertices (−13, 13), (26, 52), (39, 39). What would be the new coordinates of the vertices if the image were reduced by a scale factor of 1.3 with the origin as the center of dilation?
   F (−16.9, 16.9), (33.8, 67.6), (50.7, 50.7)
   G (−10, 10), (20, 40), (30, 30) ✓
   H (10, 10), (−20, 40), (−30, 30)
   J (16.9, 16.9), (33.8, 67.6), (50.7, 50.7)

3. A picture that is originally 820 pixels long and 410 pixels tall is to be scaled to 600 pixels long. To the nearest pixel, how tall is the new picture?
   A 560
   B 1200
   C 190
   D 300 ✓

4. An advertisement on a billboard measures 22 ft long and 8 ft high. If the ad is transferred to the side of a bus and is 30 in. long, how tall is the new ad, to the nearest inch?
   F 11 in. ✓
   G 9 in.
   H 10 in.
   J 12 in.

5. If two towns are 3 inches apart on a map and in real dimensions they are 15 miles apart, what is the scale of the map?
   A 1 in.:15 mi
   B 1 in.:5 mi ✓
   C 3 in.:1 mi
   D 1.5 in.:3 mi

6. What scale factor would reduce an object by 33%?
   F 24 in.:1 yd ✓
   G 1 ft:66 ft
   H 10 mm:1 cm
   J 2 m:5 m

7. A common scale for do-it-yourself airplane models is 1:48. The F-117A Stealth Fighter is 63 feet, 9 inches long. To the nearest inch, how long would a model of this plane be?
   A 9 in.
   B 12 in.
   C 16 in. ✓
   D 13 in.

Chapter 8 Quiz — Section A

Choose the best answer.

1. What is the percent of the unknown value p represented on the number line below?

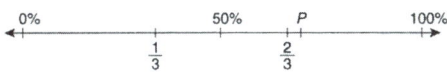

   0% — 50% — P — 100%
   $\frac{1}{3}$ — $\frac{2}{3}$

   A 66%
   B 25%
   C 70% ✓
   D 60%

2. What is 40% written as a fraction?
   F $\frac{1}{25}$
   G $\frac{2}{5}$ ✓
   H $\frac{3}{8}$
   J $\frac{3}{5}$

3. What is $\frac{5}{8}$ written as a decimal?
   A 0.625 ✓
   B 0.375
   C 0.5
   D 0.58

4. 54 is what percent of 150?
   F 50%
   G 54%
   H 42%
   J 36% ✓

5. Find 5% of 356.
   A 71.2
   B 17.8 ✓
   C 35.6
   D 20

6. 33 is 220% of what number?
   F 15 ✓
   G 72.6
   H 22
   J 100

7. The length of the Nile river is 6693 km. The Amazon river is 96% as long as the Nile. To the nearest kilometer, what is the length of the Amazon?
   A 6693 km
   B 6256 km
   C 6425 km ✓
   D 6021 km

Chapter 8 Quiz — Section B

Choose the best answer.

1. What is the percent increase from 19 to 27, to the nearest percent?
   A 30%
   B 42% ✓
   C 22%
   D 46%

2. If a stock started the week at $23.04 per share, and at the end of the week it had lost 8.5% of its value, to the nearest penny what was the final price of the stock?
   F $1.96
   G $20.96
   H $21.08 ✓
   J $27.56

3. Estimate 19% of 25.
   A about 5 ✓
   B about 4
   C about 6
   D about 3

4. 80 is about what percent of 228?
   F 66%
   G 33% ✓
   H 80%
   J 50%

5. A car salesman was able to sell a car for $12,500, earning a commission of 5%. How much was his commission?
   A $716
   B $423
   C $500
   D $625 ✓

6. The sales tax in Alex's city is 7.33%. He bought a video game system for $299, and 2 games at $49 apiece. What was his total bill, to the nearest dollar?
   F $397
   G $335
   H $426 ✓
   J $468

7. A computer program used to require 450 clock cycles to process a certain input file. After the programmer optimized the program, it only took 60 clock cycles to process the same file. To the nearest percent, what was the percent decrease?
   A 87% ✓
   B 72%
   C 60%
   D 89%

CHAPTER 9 Quiz — Section A

Choose the best answer.

1. A recent insurance industry survey discovered that 25% of teenage drivers will have an accident within 6 months of receiving their license. What is the probability that a teenage driver will not get into an accident within 6 months of receiving his or her license?
   A 60%
   B 25%
   C 75%
   D 33%

2. Eight teams meet annually for a basketball tournament. The table below shows each teams' probability of winning. Team 6 is from River City North; team 7 is from River City South; and team 8 is from River City East. What is the probability that a team not from River City will win the tournament?

Team	Probability
1	0.07
2	0.13
3	0.05
4	0.14
5	0.16
6	0.18
7	0.21
8	0.06

   F 37.5%
   G 55%
   H 45%
   J 62.5%

3. There are five finalists at the Edgewater Kennel Show. The German Shepherd has a $\frac{1}{6}$ chance of winning the show. The Boxer is $\frac{1}{2}$ as likely to win as the German Shepherd, and the Black Lab, Poodle, and Border Collie have an equal chance of winning. What is the probability of the Border Collie's winning the show?
   A $\frac{1}{3}$
   B $\frac{1}{4}$
   C $\frac{1}{5}$
   D $\frac{1}{6}$

4. Kitty watches the birds fly by the windows. She notices that out of 52 birds, 17 are finches. Estimate to the nearest percent the probability that the next bird flying by will be a finch.
   F 25%
   G 33%
   H 50%
   J 75%

5. A baseball player's chance of getting a hit is 0.29. Using the following table, estimate the chances the baseball player has of getting a hit at least 2 out of 5 times.
   86 58 52 79 19 86 31 87 72 91
   47 94 18 39 47 32 66 67 89 93
   26 44 61 27 34 22 43 54 56 12
   74 83 21 96 11 25 16 52 23 30
   52 73 74 02 98 58 28 19 32 68
   A 40%
   B 25%
   C 33%
   D 50%

CHAPTER 9 Quiz — Section B

Choose the best answer.

1. What is the probability when rolling two dice that the total shown on the dice is 2 or 3?
   A $\frac{1}{6}$
   B $\frac{1}{4}$
   C $\frac{1}{36}$
   D $\frac{1}{12}$

2. In a weekend bingo game Agnes has a 5% chance of winning the pot. Her friend Dorothy bought more cards than her and has an chance 8% of winning. What is the probability that Agnes or Dorothy will win the pot?
   F 13%
   G 5%
   H 8%
   J 3%

3. In computer languages, an identifier is a word the programmer uses to name a variable, function, or label. Some languages allow all letters, the underscore character '_', and the digits 0-9 to be used in identifiers, but the first character cannot be a digit. If the language is not case sensitive, i.e., upper and lower case letters are the same, how many possible 4-character identifiers are there?
   A 1,874,161
   B 1,213,056
   C 1,367,631
   D 531,441

4. Simplify 7!
   F 5040
   G 720
   H 40,320
   J 120

5. In a yacht race with 12 boats, how many ways out of the total number of boats can participants finish 1st, 2nd, and 3rd?
   A 220
   B 479,001,600
   C 1320
   D 12

6. Three cards are chosen at random from a standard deck of 52 cards. If each card is replaced before the next card is drawn, are the events dependent or independent? And, what is the probability of drawing a five, then a red card, and then a face card?
   F dependent; $\frac{3}{338}$
   G dependent; $\frac{42}{2197}$
   H independent; $\frac{3}{338}$
   J independent; $\frac{42}{2197}$

7. The probability of a student passing a certain multiple-choice test if he or she guesses at all of the answers is $\frac{3}{20}$. What are the odds that the student will pass?
   A 3:20
   B 3:17
   C 17:3
   D 3:23

CHAPTER 10 Quiz — Section A

Choose the best answer.

1. Chris has a job selling newspapers every morning. He receives a base pay of $10 per day and also gets 15 cents for every paper he sells. One morning he made $16.45. How many papers did he sell?
   A 52
   B 43
   C 64
   D 39

2. Solve 18s + 22 = −14.
   F s = −2
   G s = −3
   H s = 3
   J s = 2

3. Solve $\frac{n}{2} + 7 = 22$.
   A n = 58
   B n = 30
   C n = 26
   D n = 15

4. Solve 9y − 6 + y + 8 = 42.
   F y = 10
   G y = 5
   H y = 4
   J y = 8

5. Solve $\frac{h}{2} + \frac{h}{5} = 14$.
   A h = 16
   B h = 38
   C h = 24
   D h = 20

6. Solve 12x + 15 = 24 − 6x.
   F $x = 6\frac{1}{2}$
   G $x = \frac{1}{2}$
   H $x = 2\frac{1}{6}$
   J $x = 1\frac{1}{2}$

7. Solve $\frac{3w}{2} + \frac{1}{2} = w + 4$.
   A w = 7
   B w = 4
   C w = 8
   D w = 3

CHAPTER 10 Quiz — Section B

Choose the best answer.

1. A school booster club is selling T-shirts. The silk-screening company is selling shirts to the boosters for $3 each, plus a one-time fee of $150 to cover supply costs. If the boosters want to make at least $300 in profits and they sell a shirt for $12, how many shirts do they need to sell?
   A 35 or more
   B 60 or more
   C 25 or more
   D 50 or more

2. Solve 5x − 2 < 13.
   F x < 2
   G x < 7
   H x < 3
   J x < 15

3. Solve −14 > 2x + 4.
   A x < −5
   B x < −9
   C x > 6
   D x > 8

4. Solve 8x − 5 − 4x ≤ 3.
   F x ≤ 2
   G x ≥ 2
   H x ≤ −2
   J $x \geq -\frac{1}{2}$

5. Solve $\frac{3x}{4} - \frac{1}{2} \geq \frac{5}{8}$.
   A $x \geq \frac{5}{3}$
   B $x \geq \frac{3}{2}$
   C x ≥ 2
   D x ≥ 3

6. Solve t − 3u + 5 = v for t.
   F t = v − 3u + 5
   G t = v + 3u + 5
   H t = v −3u − 5
   J t = v + 3u − 5

7. Solve A = πr² for r.
   A $r = \frac{\sqrt{A}}{\pi}$
   B $r = \frac{A}{\pi}$
   C $r = \sqrt{\frac{A}{\pi}}$
   D $r = \frac{A}{\sqrt{\pi}}$

8. Identify the solution for the system
   x + 2y = 11
   3x + y = 8.
   F (1, 2)
   G (1, 5)
   H (3, 7)
   J (2, 2)

CHAPTER 11 Quiz — Section A

Choose the best answer.

1. Which of the following equations is not linear?
   A $y = x + 6$
   (C) $y = 2x^2$
   B $y = -8$
   D $y = 9x$

2. The formula for converting temperature from Celsius to Fahrenheit is $f = \frac{9}{5}c + 32$ where c is the temperature in Celsius. If the temperature is 21° Celsius, what is the temperature in degrees Fahrenheit?
   F 45.5°F
   H 21.0°F
   G 72.3°F
   (J) 69.8°F

3. What is the slope of the line that passes through the two points (1, 7) and (4, 6)?
   (A) $-\frac{1}{3}$
   C $\frac{3}{2}$
   B $-\frac{2}{3}$
   D $\frac{1}{2}$

4. What is the slope of the line in the figure below?

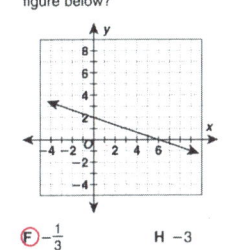

   (F) $-\frac{1}{3}$
   H -3
   G $\frac{1}{2}$
   J -2

5. Identify the line that is perpendicular to the line passing through the points (−5, 0) and (9, −7).
   A $5x - y = 7$
   (C) $-2x + y = -6$
   B $x + 6y = 4$
   D $-3x + 4y = 8$

6. Identify the x- and y-intercepts for the line $8x + 4y = 16$.
   F $x = 1, y = 5$
   H $x = 8, y = 4$
   (G) $x = 2, y = 4$
   J $x = 0, y = 0$

7. Express the equation $2x - 6y = 8$ in slope-intercept form.
   A $y = 2x - 6$
   C $y = \frac{1}{2}x - \frac{3}{2}$
   (B) $y = \frac{1}{3}x - \frac{4}{3}$
   D $y = 3x - 4$

8. Find the equation that has a slope of $-\frac{1}{8}$ and passes through the point (0, −4).
   F $y = -\frac{1}{4}x - 8$
   H $y = \frac{1}{4}x - 4$
   G $y = \frac{1}{8}x$
   (J) $y = -\frac{1}{8}x - 4$

CHAPTER 11 Quiz — Section B

Choose the best answer.

1. The data set below represents a direct variation. Identify the constant of proportionality.

x	1	2	3	4	5	6	7
y	$\frac{9}{2}$	9	$\frac{27}{2}$	18	$\frac{45}{2}$	27	$\frac{63}{2}$

   A $\frac{2}{9}$
   (C) $\frac{9}{2}$
   B 9
   D 2

2. Find the equation of direct variation, given that y varies directly with x, and x is 36 when y is 99.
   F $y = 99x$
   H $y = 36x$
   (G) $y = \frac{11}{4}x$
   J $y = \frac{4}{11}x$

3. Which inequality is represented in the graph below?

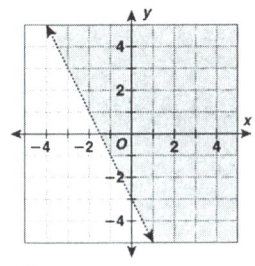

   (A) $y > -2x - 3$
   C $y < -2x - 3$
   B $y \le -2x - 3$
   D $y \ge -2x - 3$

4. Which inequality is represented in the graph below?

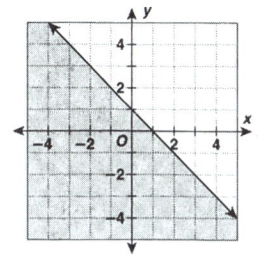

   F $y \ge -x + 1$
   H $y < 2x + 1$
   G $y < -x - 1$
   (J) $y \le -x + 1$

5. Which inequality represents the following scenario? A newspaper stand in a city sells paper A for 35 cents and paper B for 50 cents. To be profitable, the stand needs to sell at least $50 worth of newspapers per day.
   A $35A + 50B > 50$
   (B) $7A + 10B \ge 1000$
   C $35A + 50B \ge 500$
   D $7A + 10B > 50$

6. Find a line of best fit for the data in the table below.

x	−5	−3	0	1	3	5	9
y	1	3	2	3	5	6	6

   F $y = -\frac{1}{2}x + 3$
   H $y = 2x + 3$
   (G) $y = \frac{1}{2}x + 3$
   J $y = 2x - 3$

CHAPTER 12 Quiz — Section A

Choose the best answer.

1. Identify the common difference for the following arithmetic sequence: 52, 49, 46, 43, ...
   A 3
   (C) −3
   B −5
   D 7

2. Find the 18th term for the arithmetic sequence where $a_1 = 6$ and $d = 3$.
   F 60
   H 54
   (G) 57
   J 51

3. Find the next three terms for the sequence 7, 20, 33, 46, 59, ...
   A 75, 91, 107
   C 64, 69, 74
   B 66, 73, 80
   (D) 72, 85, 98

4. Identify the common ratio for the following geometric sequence: $\frac{1}{3}, \frac{2}{9}, \frac{4}{27}, \frac{8}{81}, \ldots$
   (F) $\frac{2}{3}$
   H $\frac{1}{3}$
   G $\frac{1}{6}$
   J $\frac{2}{9}$

5. Find the 6th term for the following geometric sequence: 1, 3, 9, 27, ...
   A 81
   C 30
   (B) 243
   D 729

6. Find the next three terms for the sequence −1, 2, −4, 8, −16, ...
   F 32, 64, 128
   H −32, 64, −128
   G 18, 20, 22
   (J) 32, −64, 128

7. Find the next three terms for the sequence 0, 2, 6, 12, 20, ...
   A 25, 30, 40
   (C) 30, 42, 56
   B 22, 26, 32
   D 28, 30, 32

8. Find the first three terms of the sequence for which the first term is 2 and each successive term is 1 more than twice the previous term.
   F 2, 4, 8
   H 2, 4, 6
   (G) 2, 5, 11
   J 2, 3, 5

9. Find the first three terms of the sequence defined by $a_n = \frac{n-1}{n}$.
   A $\frac{1}{2}, \frac{2}{3}, \frac{3}{4}$
   C $\frac{1}{2}, \frac{3}{4}, \frac{5}{6}$
   B $0, 2, \frac{3}{2}$
   (D) $0, \frac{1}{2}, \frac{2}{3}$

CHAPTER 12 Quiz — Section B

Choose the best answer.

1. Which relationship below is not a function?
   A $y = 5x$
   (C) $x = 3$
   B $y = x^2$
   D $y = 6$

2. Evaluate the function $f(x) = -x + 4$ for $x = -6$.
   F −6
   (H) 10
   G −2
   J 4

3. Identify the equation for the function represented in the graph below.

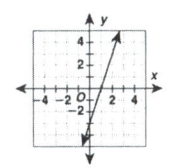

   A $y = x + 4$
   C $y = x^2 - 5$
   B $y = \frac{1}{3}x - 3$
   (D) $y = 3x - 3$

4. Identify the equation for the function represented in the graph below.

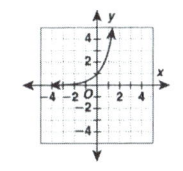

   (F) $y = 3^x$
   H $y = 2^x$
   G $y = x^3$
   J $y = 3x$

5. The radioactive isotope carbon-14 is the element used in the common technique of carbon dating to determine the age of an organic material. The half-life of carbon-14 is about 5500 years. If a sample contains 30 mg of carbon-14, how much would remain after 16,500 years?
   A 3 mg
   (C) 3.75 mg
   B 2.75 mg
   D 2 mg

6. Identify the equation for the function represented in the graph below.

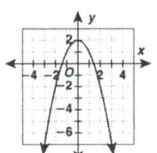

   F $y = -x + 2$
   H $y = x^2 + 2$
   G $y = x + 2$
   (J) $y = -x^2 + 2$

7. Find the constant of proportionality for the inverse variation represented in the table below.

x	−9	−5	−1	1	3	5	9
y	$-\frac{1}{3}$	$-\frac{3}{5}$	−3	3	1	$\frac{3}{5}$	$\frac{1}{3}$

   (A) 3
   C −3
   B −2
   D 2

323 Holt Pre-Algebra

Chapter 13 Quiz Form A

Choose the best answer.

1. Classify $6y^5 + 5x$.
   A monomial
   B binomial
   C trinomial
   D none of the above

2. Classify $7mn^{0.6}$.
   A monomial
   B binomial
   C trinomial
   D none of the above

3. Find the degree of $5y^4 + y^2 + 7$.
   A 4
   B 5
   C 7
   D 12

4. Which polynomial has a degree of 7?
   A $2x^5 + 4x^2$
   B $4f^6 + 3f^7 + f$
   C $2x + 5x^4$
   D $4c + 20c^{0.7}$

5. Identify like terms in the polynomial $6x + 5y^3 - 6 + 2y^3$.
   A 6, 6x
   B y^3
   C $5y^3; 2y^3$
   D no like terms

6. Simplify $7n^2 + 7p^2 - 4n^2 + p^2$.
   A $11n^2 - 8p^2$
   B $3n^2 + 8p^2$
   C $3n^2 + 7n^2$
   D $11n^2 - 7p^2$

7. Simplify $9(5x^4 + 7x)$.
   A $45x^4 + 7x$
   B $45x^{36} + 63$
   C $45x^4 + 63x$
   D $108x^5$

8. Simplify $3(3x^2 + 5y) - 7x^2$.
   A $2x^2 + 5y$
   B $16x^2 + 15y$
   C $2x^2 + 15y$
   D $30x^2 + 15y$

Chapter 13 Quiz Form B

Choose the best answer.

1. Add $(7a^2 - 5b) + (-4a^2 + 2b + 8)$.
   A $11a^2 + 7b + 8$
   B $3a^2 - 3b + 8$
   C $3a^2 + 3b^2 + 8$
   D $3a^2 + 7b + 8$

2. Add $(5xy^2 + 3x - 2y) + (3xy^2 + y - 3)$.
   A $8xy^2 + 3x + y - 3$
   B $8xy^2 + 3x - y - 3$
   C $8xy^4 + 3x - 2y - 3$
   D $15xy^2 + 3x - 2y - 3$

3. Find the opposite of $-3hj^3 + 5hj - 4$.
   A $-3hj^3 - 5hj + 4$
   B $3hj^3 + 5hj + 4$
   C $3hj^3 - 5hj + 4$
   D $4 - 5hj + -3hj^3$

4. Subtract $(x^4 - 4x^2) - (4x - x^2 + 7)$.
   A $x^4 + 4x^2 - 4x - x^2 - 7$
   B $x^4 + 5x^4 - 4x - 7$
   C $x^4 + 4x^2 + 4x + 7$
   D $x^4 - 3x^2 - 4x - 7$

5. Multiply $(4m^4n^4)(3m^3n)$.
   A $7m^{12}n^4$
   B $12m^{12}n^4$
   C $12m^7n^4$
   D $12m^7n^5$

6. Multiply $3y^3z(6y^4 - 5z^2)$.
   A $9y^7z - 2y^3z^3$
   B $18y^7z - 15y^3z^3$
   C $18y^7z^5 - 15z^2$
   D $18y^{12} - 15y^3z^2$

7. Multiply $(a + 5)(a + 7)$.
   A $a^2 + 7a + 5a + 12$
   B $a^2 + 35a + 35$
   C $a^2 + 12a^2 + 35$
   D $a^2 + 12a + 35$

Chapter 14 Quiz Form A

Choose the correct answer.

1. cats ☐ {pets}
   A $\notin$
   B $\in$
   C $\subset$
   D $\varnothing$

2. $3x + \left(\frac{12}{x}\right) + y^3$ ☐ {polynomials}
   A $\notin$
   B $\in$
   C $\subset$
   D $\cap$

3. Describe the set: $H = \{odd\ integers\}$.
   A infinite
   B finite
   C $\notin$ {all numbers}
   D $\not\subset$ {all numbers}

4. Which of the following is true about sets M and N? $M = \{0, 4, 8, 9\}$ $N = \{-4, 4, 7\}$
   A $M \cap N = \{\}$
   B $M \cup N = \{4\}$
   C $M \cap N = \{4\}$
   D $M \cap N = \{-4\}$

5. What is the union of sets T and R? $T = \{3, 7, 11\}$ $R = \{prime\ numbers\}$
   A $T \cup R = T$
   B $T \cup R = R$
   C $T \cup R = \{\}$
   D $T \cap R = R$

6. What is the intersection of sets E and F? $E = \{2, 8, 104\}$ $F = \{even\ numbers\}$
   A $E \cup F = F$
   B $E \cap F = F$
   C $E \cup F = \{\}$
   D $E \cap F = E$

7. Identify $X \cap Y$ in the Venn diagram.

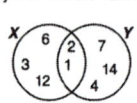

   A $\{3, 6, 12\}$
   B $\{4, 7, 14\}$
   C $\{1, 2\}$
   D $\{\}$

Chapter 14 Quiz Form B

Choose the best answer.

For 1 and 2, use the following disjunction:
P: The person must be 8 years old.
Q: The person must be at least 44 inches tall.

1. James is 7 years old and 45 inches tall.
   A P: False; Q: True; P or Q: False
   B P: True; Q: False; P or Q: True
   C P: False; Q: True; P or Q: True
   D P: False; Q: False; P or Q: False

2. Star is 9 years old and 48 inches tall.
   A P: True; Q: False; P or Q: False
   B P: True; Q: True; P or Q: True
   C P: False; Q: True; P or Q: True
   D P: False; Q: False; P or Q: True

3. In the statement, "You will be healthier if you exercise every day," what is the term that describes, "exercise every day?"
   A hypothesis
   B conclusion
   C conjunction
   D deduction

Use the figure for 4 and 5.

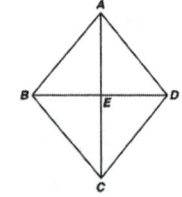

4. What is the degree of point E?
   A 1 degree
   B 2 degrees
   C 3 degrees
   D 4 degrees

5. Name the figure.
   A Euler circuit
   B Königsberg Bridge
   C Hamiltonian circuit
   D Euler path

Holt Pre-Algebra

Chapter Test
Form A

Evaluate each expression for the given values of the variables.

1. $3x + 2y$ for $x = 8$ and $y = 6$
   36

2. $13m - 2n$ for $m = 3$ and $n = 4$
   31

3. $5(k + 8) - 2m$ for $k = 6$ and $m = 3$
   64

Write an algebraic expression for each word phrase.

4. 5 more than twice a number p
   $2p + 5$

5. 7 times the sum of h and 12
   $7(h + 12)$

6. 4 less than the sum of g and 15
   $(g + 15) - 4$

Solve.

7. $z + 18 = 54$
   $z = 36$

8. $m - 4.5 = 12$
   $m = 16.5$

Write an equation, then solve.

9. The depth of Lake Superior is 1330 feet. This is 407 feet deeper than Lake Michigan. How deep is Lake Michigan?
   $d + 407 = 1330$; $d = 923$ ft

10. The length of Lake Ontario is 193 miles, which is 48 miles less than the length of Lake Huron. How many miles long is Lake Huron?
    $\ell - 48 = 193$; $\ell = 241$ mi

Solve and check.

11. $\frac{n}{8} = 12$
    $n = 96$

12. $7k = 91$
    $k = 13$

Write an equation, then solve.

13. The Smith family spends an average of $450 monthly for groceries. Groceries account for $\frac{1}{8}$ of their monthly costs. How much are their monthly costs?
    $\frac{1}{8}m = \$450$; $m = \$3600$

14. Janice spent $325 on clothes. This was 5 times the amount spent for her 5-year-old brother. How much money was spent on clothes for her brother?
    $5c = \$325$; $c = \$65$

Solve and graph.

15. $x + 8 > 12$ ____ **$x > 4$**

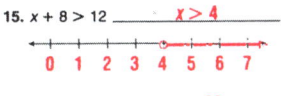

16. $z - 5 \leq 15$ ____ **$z \leq 20$**

Form A, continued

Solve and graph.

17. $4x \geq 52$ ____ **$x \geq 13$**

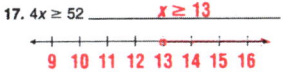

18. $\frac{k}{5} < 12$ ____ **$k < 60$**

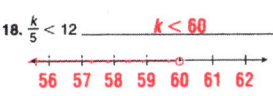

Simplify.

19. $4(z + 8) - z$ **$3z + 32$**

20. $2(4y + 6) - 3y$ **$5y + 12$**

21. $3(x - 5) + 8x$ **$11x - 15$**

Solve.

22. $4y + 3y = 63$
    $y = 9$

23. $10g - 4g = 54$
    $g = 9$

24. $12t - 4t = 96$
    $t = 12$

Determine whether the ordered pair is a solution of the given equation.

25. $y = x + 9$; (7, 16)
    (7, 16) is a solution

26. $y = 2x + 8$; (5, 28)
    (5, 28) is not a solution

Give the coordinates of each point identified on the coordinate plane.

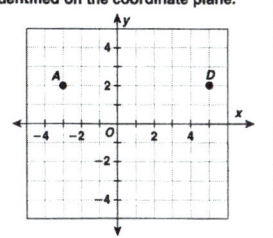

27. A **$(-3, 2)$**

28. D **$(5, 2)$**

Study the table and use it to answer the following questions.

Rocket Experiment with Water Pressure

Water (ounces)	Height of Rocket (feet)
4	22
8	50
10	60
16	91

29. How much water was used to launch the rocket 50 feet?
    8 ounces

30. If twice as much water is used, will the rocket soar twice as far? Explain.
    Not always; possible answer: with 8 oz the rocket went 50 ft, but with 16 oz it did not go 100 ft.

Chapter Test
Form B

Evaluate each expression for the given values of the variables.

1. $6x + 3y$ for $x = 7$ and $y = 8$
   66

2. $1.8s - 5p$ for $s = 9$ and $p = 2$
   6.2

3. $j(5 + t) - 8$ for $j = 8$ and $t = 6$
   80

Write an algebraic expression for each word phrase.

4. 8 more than the product of z and 8
   $8z + 8$

5. 4 less than the quotient of x and 8
   $\frac{x}{8} - 4$

6. 7 less than the sum of m and 12
   $(m + 12) - 7$

Solve and check.

7. $n + 36 = 154$
   $n = 118$

8. $4.9 = m - 2.6$
   $m = 7.5$

Write an equation, then solve.

9. The average weight of an elephant is 5450 kg. This is 4270 kg more than the average weight of giraffe. What is the average weight of a giraffe?
   $w + 4270 = 5450$; $w = 1180$ kg

10. The giraffe sleeps 180 minutes per day, which is 60 minutes less than an elephant sleeps each day. How long does the elephant sleep each day?
    $s - 60 = 180$; $s = 240$ min

Solve and check.

11. $\frac{m}{15} = 18$
    $m = 270$

12. $23y = 92$
    $y = 4$

Write an equation, then solve.

13. A koala bear eats about 2.5 pounds of eucalyptus leaves each day. This is about $\frac{1}{10}$ of his total body weight. What does a koala bear weigh?
    $\frac{1}{10}w = 2.5$; $w = 25$ lb

14. The hummingbird beats its wings 5400 beats per minute. This is 30 times faster than a stork's wing beats per minute. How many wing beats per minute does a stork make?
    $30b = 5400$; $b = 180$

Solve and graph.

15. $12 + x \geq 20$ ____ **$x \geq 8$**

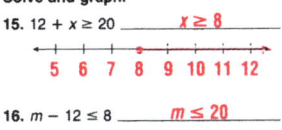

16. $m - 12 \leq 8$ ____ **$m \leq 20$**

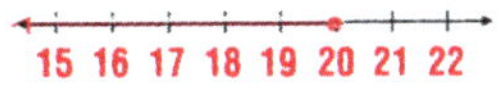

Form B, continued

Solve and graph.

17. $13y \leq 104$ ____ **$y \leq 8$**

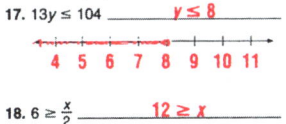

18. $6 \geq \frac{x}{2}$ ____ **$12 \geq x$**

Simplify.

19. $7(2b - 8)$
    $14b - 56$

20. $4(4x - 4) + 2x$
    $18x - 16$

21. $3(4b + 3) - 5$
    $12b + 4$

Solve.

22. $24k - 7k = 51$ **$k = 3$**

23. $45 = 3y + 6y$ **$y = 5$**

24. $\frac{z}{12} = 9$ **$z = 108$**

Determine whether the ordered pair is a solution of the given equation.

25. $y = 7x + 7$; (4, 30)
    (4, 30) is not a solution

26. $y = 4x + 12$; (3, 24)
    (3, 24) is a solution

Give the coordinates of each point identified on the coordinate plane.

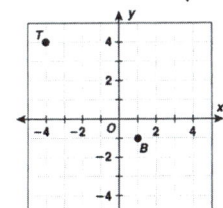

27. T **$(-4, 4)$**

28. B **$(1, -1)$**

Study the table and use it to answer the following questions.

Filling the Bath Tub

Time (minutes/seconds)	Height of Bath Water
1:00	4 inches
2:00	7.5 inches
3:00	10.2 inches
4:00	11.6 inches
4:23	12 inches

29. During which minute does the most water enter the bathtub as it is being filled?
    First minute

30. After 2 minutes of filling the tub, how many more inches of water must be added for the tub to be filled?
    4.5 inches

Chapter 1 Chapter Test — Form C

Evaluate each expression for $x = 2.5$, $y = 12$, and $z = 4$.

1. $2y - 2x$ — **19**
2. $z(4 + x) + 8$ — **34**
3. $5xyz$ — **600**

Write an algebraic expression for each word phrase.

4. 5 less than the product of 8 and k — **$8k - 5$**
5. twice the sum of k and 8 — **$2(k + 8)$**
6. half the sum of 12 and t — **$\frac{12 + t}{2}$ or $\frac{1}{2}(12 + t)$**

Solve and check.

7. $2.8 + t = 9.4$ — **$t = 6.6$**
8. $18 = d - 5$ — **$d = 23$**

Write an equation, then solve.

9. The world's largest cherry pie weighs 37,740 pounds. This is 7625 pounds heavier than the largest apple pie. How heavy is the largest apple pie?
   $c + 7625 = 37{,}740$; $c = 30{,}115$ lb

10. The world's largest lollipop weighs 2220 pounds, which is 10,126 pounds lighter than the largest popsicle. What is the weight of the largest popsicle?
    $p - 10{,}126 = 2{,}220$; $p = 12{,}346$ lb

Solve and check.

11. $\frac{n}{6} - 4 = 8$ — **$n = 72$**
12. $13x + 14 = 40$ — **$x = 2$**

Write an equation, then solve.

13. The population of Nevada is close to 1.6 million, which is about $\frac{1}{5}$ of Michigan's population. What is the estimated population of Michigan?
    $\frac{1}{5}m = 1.6$; $m = 8$ million people

14. There are about 1.2 million people in the U.S. who speak Chinese. This is about 3 times the number of people who speak Greek. How many people speak Greek?
    $3c = 1.2$; $c = 0.4$ million or 400,000 people

Solve and graph.

15. $3.2 + x \geq 9$ — **$x \geq 5.8$**
16. $8 < x - 3$ — **$x > 11$**

Chapter 1 Chapter Test — Form C, continued

Solve and graph.

17. $3x + 8 \geq 41$ — **$x \geq 11$**
18. $6 < \frac{a}{4}$ — **$a > 24$**

Simplify.

19. $6(3h + 9) - 4h$ — **$14h + 54$**
20. $6(5y - 7) + 8y$ — **$38y - 42$**
21. $4(2y - 3) + 5y$ — **$13y - 12$**

Solve.

22. $12g + 13g = 450$ — **$g = 18$**
23. $45 = 3y + 6y$ — **$y = 5$**
24. $42 = 9k - 2k$ — **$k = 6$**

Determine whether the ordered pair is a solution of the given equation.

25. $y = 17x - 3$; (2, 31) — **(2, 31) is a solution**
26. $y = 4x - 12$; (8, 4) — **(8, 4) is not a solution**

27. Give the coordinates point C, point T and point A, on the coordinate plane.
    $C(-6, 3)$; $T(-2, 1)$; $A(-6, -2)$

28. Connect all 3 points to form a geometric figure. Identify the figure formed.
    triangle

Office Deliveries 5-Story Building

Floor #	# of Deliveries	Time (minutes)
1	2	11
2	6	20
3	3	13
4	1	4
5	3	15

29. On which floors were 3 deliveries made? — **3rd floor and 5th floor**
30. Do 6 deliveries take twice as long as 3 deliveries? Why or why not?
    No; possible answer: fewer packages to deliver.

Chapter 2 Chapter Test — Test A

Add.

1. $6 + (-1)$ — **5**
2. $-9 + (-5)$ — **-14**
3. $-6 + 6$ — **0**

Evaluate each expression for the given value of the variable.

4. $10 + j$ for $j = -3$ — **7**
5. $n + (-2)$ for $n = 5$ — **3**
6. $s + 6$ for $s = -6$ — **0**

Subtract.

7. $-9 - 2$ — **-11**
8. $2 - (-7)$ — **9**
9. $-12 - (-5)$ — **-7**

Evaluate each expression for the given value of the variable.

10. $4 - g$ for $g = -2$ — **6**
11. $-16 - t$ for $t = 4$ — **-20**
12. $n - (-6)$ for $n = 5$ — **11**

Multiply or divide.

13. $5(-7)$ — **-35**
14. $-2(-12)$ — **24**
15. $\frac{2^5}{2^2}$ — **2^3 or 8**

Simplify.

16. $8(-10 + 7)$ — **-24**
17. $9(-1 + 4)$ — **27**
18. $\frac{-4(5)}{-4}$ — **5**

Solve.

19. $-9 + d = 23$ — **$d = 32$**
20. $4h = -28$ — **$h = -7$**
21. $\frac{k}{9} = -3$ — **$k = -27$**

Solve and graph. (2-5)

22. $t - 1 \leq 4$ — **$t \leq 5$**

Chapter 2 Assessment — Test A, continued

Solve and graph.

23. $\frac{y}{4} < -2$ — **$y < -8$**
24. $4c \geq 12$ — **$c \geq 3$**

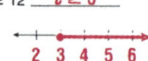

Write using exponents.

25. $3 \cdot 3 \cdot 3$ — **3^3**
26. $(-n) \cdot (-n)$ — **$(-n)^2$**
27. $k \cdot k \cdot k \cdot k$ — **k^5**

Simplify.

28. 2^3 — **8**
29. $(-4)^3$ — **-64**
30. 5^4 — **625**
31. $(8 - 7)^2$ — **1**
32. $9 + (-5)^2$ — **34**
33. $(1 \cdot 6)^2$ — **36**

Multiply or divide. Write the product or quotient as one power.

34. $n^1 \cdot n^4 \cdot n^7$ — **n^{12}**
35. $4^3 \cdot 4^2 \cdot 4^0 \cdot 4^1$ — **4^6**
36. $\frac{u^7}{u^2}$ — **u^5**

Simplify the powers of 10.

37. 10^{-3} — **0.001**
38. 10^{-2} — **0.01**
39. 10^{-6} — **0.000001**

Simplify.

40. $10^{-3} \cdot 10^5$ — **100**
41. 1.6×10^{-2} — **0.016**
42. $\frac{2^3}{2^5}$ — **$\frac{1}{4}$ or 0.25**

Write each number in standard notation.

43. 3.2×10^3 — **3200**
44. 1.75×10^{-3} — **0.00175**
45. 9.46×10^0 — **9.46**

Write each answer in scientific notation.

46. In one year, Americans made 445 billion phone calls. Write this number in scientific notation.
    4.45×10^{11}

47. A bookstore orders a shipment of books. The books weigh 3.2 lb each. How much will the shipment of 100 books weigh?
    3.2×10^2 lb

Chapter 2 Assessment — Test B

Add.
1. 7 + (−8) **−1**
2. −18 + (−5) **−23**
3. −6 + 15 **9**

Evaluate each expression for the given value of the variable.
4. 14 + j for j = −7 **7**
5. n + (−9) for n = 15 **6**
6. s + 7 − (−3) for s = −6 **4**

Subtract.
7. −9 − 22 **−31**
8. 10 − (−7) **17**
9. −18 − (−6) **−12**

Evaluate each expression for the given value of the variable.
10. 9 − g for g = −15 **24**
11. −26 − t for t = 9 **−35**
12. n − (−14) for n = 1 **15**

Multiply or divide.
13. 25(−7) **−175**
14. −28(−12) **336**
15. $\frac{8^9}{8^3}$ **8^6 or 262,144**

Simplify.
16. 8(−13 − 7) **−160**
17. −9(−16 + 4) **108**
18. $\frac{-3(14)}{-6}$ **7**

Solve.
19. 19 + d = 53 **d = 34**
20. 14h = −98 **h = −7**
21. $\frac{k}{81} = -3$ **k = −243**

Solve and graph.
22. t − 15 ≤ −9 **t ≤ 6**

Chapter 2 Assessment — Test B, continued

Solve and graph.
23. $\frac{y}{8} < -2$ **y < −16**
24. 7c ≥ 49 **c ≥ 7**

Write using exponents.
25. 7 · 7 · 7 · 7 **7^4**
26. (−k) · (−k) · (−k) **$(-k)^3$**
27. t · t · t · t · t **t^5**

Simplify.
28. 7^4 **2401**
29. $(-9)^3$ **−729**
30. 12^2 **144**
31. $(8 + (-7))^8$ **1**
32. 19 + (−5) + $(-3)^3$ **−13**
33. $(5 + 7 \cdot 4)^2 + 7$ **1096**

Multiply or divide. Write the product or quotient as one power.
34. $n^{11} \cdot n^{11} \cdot n^{11}$ **n^{33}**
35. $4^6 \cdot 4^2 \cdot 4^0 \cdot 4^1$ **4^9**
36. $\frac{u^{16}}{u^{14}}$ **u^2**

Simplify the powers of 10.
37. 10^{-6} **0.000001**
38. 10^{-4} **0.0001**
39. 10^{-10} **0.0000000001**

Simplify.
40. $10^{-5} \cdot 10^{11}$ **1,000,000**
41. 2.49×10^{-2} **0.0249**
42. $\frac{6^3}{6^5}$ **$\frac{1}{36}$ or $0.27\overline{7}$**

Write each number in standard notation.
43. 1.42×10^6 **1,420,000**
44. 3.56×10^{-4} **0.000356**
45. 5.12×10^0 **5.12**

Write each answer in scientific notation.
46. The thinnest commercial glass is 0.000984 in. thick. The glass on an aquarium is 1000 times as thick. How thick is the glass?
9.84×10^{-1} in.
47. A pet store buys a truckload of dog food. The bags weigh 50 lb each. How much will all 1000 bags ordered weigh?
5×10^4 lb

Chapter 2 Assessment — Test C

Add.
1. 27 + (−38) **−11**
2. −48 + (−25) **−73**
3. −46 + 25 **−21**

Evaluate each expression for the given value of the variable.
4. 84 + j for j = −28 **56**
5. n + (−69) for n = 75 **6**
6. s + 27 − (−93) for s = −16 **104**

Subtract.
7. −89 − 32 **−121**
8. 46 − (−15) **61**
9. −58 − (−66) **8**

Evaluate each expression for the given value of the variable.
10. 35 − g for g = −26 **61**
11. −86 − t for t = 45 **−131**
12. n − (−64) for n = 29 **93**

Multiply or divide.
13. 525(−4) **−2100**
14. −108(−14) **1512**
15. $\frac{17^8}{17^3}$ **17^5 or 1,419,857**

Simplify.
16. 26(−23 − 19) **−1092**
17. −14(−36 + 14) **308**
18. $\frac{-13(20)}{4}$ **−65**

Solve.
19. −126 = 39 − d **d = 165**
20. 74h = −518 **h = −7**
21. $\frac{4h}{-7} = -64$ **h = 112**

Solve and graph.
22. t − 25 ≤ −19 **t ≤ 6**

Chapter 2 Assessment — Test C, continued

Solve and graph.
23. $\frac{-y + 12}{4} > 8$ **y < −20**
24. 16c ≥ 112 **c ≥ 7**

Write using exponents.
25. 17 · 17 · 17 · 17 **17^4**
26. (−t) · (−t) · (−t) **$(-t)^3$**
27. −5 · 5 · 5 · s · s · b · b · b · b **$-5^3 s^2 b^4$**

Simplify.
28. 34^3 **39,304**
29. $(-12)^3$ **−1728**
30. 18^4 **104,976**
31. $(9 + (-6))^8$ **6561**
32. 19 + $(-5)^3$ **−106**
33. $(13 \cdot 2)^2 + 56$ **732**

Multiply or divide. Write the product or quotient as one power.
34. $(n^2)^2 \cdot n^{15} \cdot n^{11} \cdot n^{19}$ **n^{49}**
35. $7 \cdot 7^2 \cdot 7 \cdot 6^0$ **7^4**
36. $\frac{u^{45}}{u^{14}}$ **u^{31}**

Simplify the powers of 10.
37. 10^{-11} **0.00000000001**
38. 10^{-7} **0.0000001**
39. 10^{-4} **0.0001**

Simplify.
40. $(13 - 3)^{15} (6 + 4)^{-9}$ **1,000,000**
41. 3.567×10^{-3} **0.003567**
42. $\frac{12^3}{12^5}$ **$\frac{1}{144}$ or 0.00694**

Write each number in standard notation.
43. 5.234×10^6 **5,234,000**
44. 1.9×10^{-5} **0.000019**
45. 6.48×10^0 **6.48**

Write each answer in scientific notation.
46. The mass of a small insect is 0.0000569 g. How much would 100 of the insects weigh?
5.69×10^{-3} g
47. A pet store buys a truckload of dog food. The bags weigh 25 lb each. How much will all 10,000 bags weigh?
2.5×10^5 lb

Chapter 3 Chapter Test — Form A

Simplify.

1. $\frac{6}{9}$ — $\frac{2}{3}$
2. $-\frac{25}{75}$ — $-\frac{1}{3}$
3. $\frac{12}{28}$ — $\frac{3}{7}$

Write each decimal as a fraction in simplest form.

4. 0.39 — $\frac{39}{100}$
5. 0.24 — $\frac{6}{25}$

Write each fraction as a decimal.

6. $\frac{7}{10}$ — 0.7
7. $\frac{1}{5}$ — 0.2

Add or subtract.

8. $\frac{3}{5} - \frac{1}{5}$ — $\frac{2}{5}$
9. $-1.4 + 1.9$ — 0.5

Evaluate each expression for the given value of the variable.

10. $2.9 + j$ for $j = 1.3$ — 4.2
11. $4 + n$ for $n = -\frac{2}{3}$ — $3\frac{1}{3}$

Multiply. Write each answer in simplest form.

12. $\frac{3}{5}\left(\frac{1}{2}\right)$ — $\frac{3}{10}$
13. $5.3(4.1)$ — 21.73
14. $-0.2(2.7)$ — -0.54

Evaluate $\frac{1}{6}y$ for each value of y.

15. $y = 4$ — $\frac{2}{3}$
16. $y = \frac{1}{9}$ — $\frac{1}{54}$
17. $y = \frac{2}{13}$ — $\frac{1}{39}$

Divide. Write each answer in simplest form.

18. $\frac{9}{16} \div \frac{1}{2}$ — $1\frac{1}{8}$
19. $9 \div \frac{3}{4}$ — 12

Divide.

20. $0.36 \div 0.24$ — 1.5
21. $9.12 \div 0.5$ — 18.24

Evaluate $\frac{2}{x}$ for each value of x.

22. $x = 0.4$ — 5
23. $x = 0.2$ — 10

Add or subtract.

24. $\frac{1}{2} + \frac{1}{6}$ — $\frac{2}{3}$
25. $3\frac{1}{6} + 1\frac{7}{12}$ — $4\frac{3}{4}$

Evaluate each expression for the given value of the variable.

26. $\frac{1}{8} + x$ for $x = \frac{1}{4}$ — $\frac{3}{8}$
27. $x - \frac{4}{5}$ for $x = \frac{1}{5}$ — $-\frac{3}{5}$

Solve.

28. $m - 1.6 = 5.2$ — $m = 6.8$
29. $\frac{2}{9}m = \frac{2}{3}$ — $m = 3$
30. $3.0h = 12.0$ — $h = 4$

Solve.

31. $y - \frac{1}{8} \geq \frac{5}{8}$ — $y \geq \frac{3}{4}$
32. $7 + y > 8.2$ — $y > 1.2$
33. $3a < \frac{1}{6}$ — $a < \frac{1}{18}$

Simplify each expression.

34. $\sqrt{2 + 2}$ — 2
35. $\sqrt{36} - \sqrt{4}$ — 4
36. $\sqrt{16} + 2$ — 6

Use a calculator to find each value. Round to the nearest tenth.

37. $\sqrt{196}$ — 14.0
38. $\sqrt{12}$ — 3.5
39. $-\sqrt{29}$ — -5.4

State if the number is rational, irrational, or not a real number.

40. 0.91 — rational
41. -5 — rational
42. $0.\overline{3}$ — rational

43. Jessie bought 6 carnations that were $0.39 each. How much did she spend? — $2.34

44. There is a fourth of a pie left. Your mom says you can have half of it. How much of the pie is that? — $\frac{1}{8}$

Chapter 3 Chapter Test — Form B

Simplify.

1. $\frac{21}{30}$ — $\frac{7}{10}$
2. $-\frac{45}{120}$ — $-\frac{3}{8}$
3. $\frac{15}{17}$ — already simplified

Write each decimal as a fraction in simplest form.

4. -3.46 — $-3\frac{23}{50}$
5. -0.07 — $-\frac{7}{100}$

Write each fraction as a decimal.

6. $\frac{7}{25}$ — 0.28
7. $-\frac{25}{5}$ — -5

Add or subtract.

8. $\frac{7}{23} - \frac{9}{23}$ — $-\frac{2}{23}$
9. $-1.4 + 0.72$ — -0.68

Evaluate each expression for the given value of the variable.

10. $8 + n$ for $n = -\frac{1}{4}$ — $7\frac{3}{4}$
11. $\frac{4}{12} + f$ for $f = -\frac{9}{12}$ — $-\frac{5}{12}$

Multiply. Write each answer in simplest form.

12. $\frac{3}{4}\left(\frac{1}{8}\right)$ — $\frac{3}{32}$
13. $-5\left(1\frac{5}{6}\right)$ — $-9\frac{1}{6}$
14. $4.73(3.1)$ — 14.663

Evaluate $5\frac{2}{3}y$ for each value of y.

15. $y = 4$ — $22\frac{2}{3}$
16. $y = -\frac{1}{9}$ — $-\frac{17}{27}$
17. $y = \frac{7}{17}$ — $2\frac{1}{3}$

Divide. Write each answer in simplest form.

18. $\frac{5}{6} \div \frac{9}{16}$ — $1\frac{13}{27}$
19. $6\frac{1}{3} \div 3\frac{1}{2}$ — $1\frac{17}{21}$

Divide.

20. $1.9 \div 0.05$ — 38
21. $7.15 \div 1.3$ — 5.5

Evaluate $\frac{8}{x}$ for each value of x.

22. $x = 2.5$ — 3.2
23. $x = 0.04$ — 200

Add or subtract.

24. $\frac{2}{3} + \frac{4}{5}$ — $1\frac{7}{15}$
25. $2\frac{5}{6} - 1\frac{3}{10}$ — $1\frac{8}{15}$

Evaluate each expression for the given value of the variable.

26. $\frac{1}{8} + x$ for $x = -4\frac{1}{8}$ — -4
27. $n - 3\frac{5}{6}$ for $n = -\frac{1}{15}$ — $-3\frac{9}{10}$

Solve.

28. $m - 5.6 = -0.9$ — $m = 4.7$
29. $\frac{3}{7}m = -\frac{6}{7}$ — $m = -2$
30. $3.7h = 0.74$ — $h = 0.2$

Solve.

31. $y - \frac{1}{8} \geq \frac{3}{4}$ — $y \geq \frac{7}{8}$
32. $5.5 + y > 2.3$ — $y > -3.2$
33. $6a < -\frac{5}{8}$ — $a < -\frac{5}{48}$

Simplify each expression.

34. $\sqrt{7 + 9}$ — 4
35. $\sqrt{36} - \sqrt{81}$ — -3
36. $\sqrt{25} + 24$ — 29

Use a calculator to find each value. Round to the nearest tenth.

37. $\sqrt{136}$ — 11.7
38. $-\sqrt{34}$ — -5.8
39. $\sqrt{82.9}$ — 9.1

State if the number is rational, irrational, or not a real number.

40. $\sqrt{27}$ — irrational
41. $-\sqrt{\frac{81}{16}}$ — rational
42. $\sqrt{\frac{1}{9}}$ — rational

43. If you buy 2.5 pounds of hamburger and it costs $2.10 per pound, how much does the package cost? — $5.25

44. A book of stamps contains 20 stamps. If you used one fourth of them, how many did you use? — 5

Chapter Test
3 Form C

Simplify.

1. $\frac{75}{204}$ — $\frac{25}{68}$
2. $-\frac{108}{320}$ — $-\frac{27}{80}$
3. $\frac{14}{53}$ — already simplified

Write each decimal as a fraction in simplest form.

4. -4.3125 — $-4\frac{5}{16}$
5. $0.0\overline{3}$ — $\frac{1}{30}$

Write each fraction as a decimal.

6. $\frac{9}{16}$ — 0.5625
7. $-\frac{19}{40}$ — -0.475

Add or subtract.

8. $\frac{5}{136} - \frac{9}{136}$ — $-\frac{1}{34}$
9. $-17.67 + 25.13$ — 7.46

Evaluate each expression for the given value of the variable.

10. $54 + n$ for $n = -\frac{7}{16}$ — $53\frac{9}{16}$
11. $\frac{2}{9} + f$ for $f = -\frac{8}{9}$ — $-\frac{2}{3}$

Multiply. Write each answer in simplest form.

12. $13\frac{4}{7}\left(\frac{3}{5}\right)$ — $8\frac{1}{7}$
13. $11\left(4\frac{5}{12}\right)$ — $48\frac{7}{12}$
14. $-0.47(92.8)$ — -43.616

Evaluate $16\frac{2}{5}y$ for each value of y.

15. $y = 4$ — $65\frac{3}{5}$
16. $y = \frac{1}{16}$ — $1\frac{1}{40}$
17. $y = \frac{5}{29}$ — $2\frac{24}{29}$

Divide. Write each answer in simplest form.

18. $13\frac{3}{4} \div 1\frac{2}{3}$ — $8\frac{1}{4}$
19. $6\frac{2}{3} \div 7\frac{1}{2}$ — $\frac{8}{9}$

Divide.

20. $0.512 \div 0.08$ — 6.4
21. $2002 \div 5.2$ — 385

Chapter Test
3 Form C, continued

Evaluate $\frac{24}{x}$ for each value of x.

22. $x = 0.9$ — $26.\overline{6}$
23. $x = 0.012$ — 2000

Add or subtract. Write each answer in simplest form.

24. $\frac{4}{5} + \frac{3}{8}$ — $1\frac{7}{40}$
25. $6\frac{5}{8} - 1\frac{9}{10}$ — $4\frac{29}{40}$

Evaluate each expression for the given value of the variable.

26. $\frac{2}{3} + x$ for $x = \frac{7}{12}$ — $1\frac{1}{4}$
27. $n - 5\frac{5}{6}$ for $n = -4\frac{7}{8}$ — $-10\frac{17}{24}$

Solve.

28. $m + 25.6 = -0.19$ — $m = -25.79$
29. $\frac{4}{17}m = -\frac{8}{17}$ — $m = -2$
30. $23.7h = 16.59$ — $h = 0.7$

Solve.

31. $y - \frac{1}{2} \geq \frac{7}{8}$ — $y \geq 1\frac{3}{8}$
32. $43.6 + y > 32.3$ — $y > -11.3$
33. $-3a < -13\frac{4}{5}$ — $a > 4\frac{3}{5}$

Simplify each expression.

34. $\sqrt{460 + 24}$ — 22
35. $\sqrt{121} - \sqrt{196}$ — -3
36. $\sqrt{324} + 324$ — 342

Use a calculator to find each value. Round to the nearest tenth.

37. $\sqrt{1436}$ — 37.9
38. $-\sqrt{284}$ — -16.9
39. $\sqrt{982.9}$ — 31.4

State if the number is rational, irrational, or not a real number.

40. $\sqrt{11}$ — irrational
41. $\sqrt{576}$ — rational
42. $\sqrt{5\frac{1}{16}}$ — rational

43. The area of a rectangle is its length times its width. Find the area if the length is 2.3 feet and the width is 0.7 foot.

1.61 feet^2

44. Find the area of a rectangle if the length is $2\frac{2}{3}$ inches and the width is $\frac{4}{9}$ inches.

$1\frac{5}{27}$ in^2

Chapter Test
4 Form A

Identify the population, sample, and sampling method.

1. Every 15th student eating in the school cafeteria was asked his or her favorite dessert.

All students eating in the school cafeteria; every 15th student; systematic.

2. 300 households in Sylvania were selected by random digit dialing to respond to a survey.

All households in Sylvania with telephones; 300 households called; random.

3. Ten video stores in a large city were randomly selected and 50 customers were randomly selected from each store and asked what was their favorite video.

People going to the video stores in the city; those asked the questions; stratified.

Organize the data to make a stem-and-leaf plot.

4. The following are test scores for a math class.

Test Scores			
98	85	70	93
85	74	98	81
89	78	100	85

Stem	Leaves
7	0 4 8
8	1 5 5 5 9
9	3 8 8
10	0

Use the data. 10, 25, 18, 15, 21, 20, 12 Round to the nearest tenth.

5. Find the mean. — 17.3
6. Find the median. — 18
7. Find the mode. — none
8. Find the range. — 15
9. Find the third quartile. — 21
10. Find the first quartile. — 12
11. Make a box-and-whisker plot.

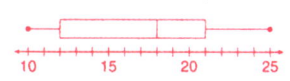

Use the data. Round to the nearest tenth. 8, 5, 5, 8, 6, 5, 3, 4

12. Find the mean. — 5.5
13. Find the median. — 5
14. Find the mode. — 5
15. Find the range. — 5
16. Find the third quartile. — 7
17. Find the first quartile. — 4.5
18. Make a box-and-whisker plot.

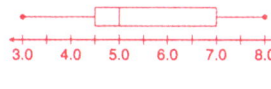

Chapter Test
4 Form A, continued

Use the data.
21, 21, 19, 22, 24, 19, 21, 24, 21, 24, 24, 22, 19, 22, 21, 19, 24, 21

19. Make a frequency table.

Scores	Frequency
19	4
21	6
22	3
24	5

20. Make a bar graph.

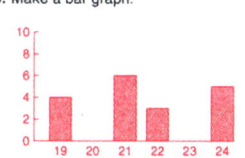

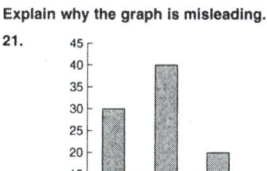

21. Explain why the graph is misleading.

Since the vertical axis does not start at 0, it looks as though 3 times as many students score between 70–79 than score between 90–99.

Use the data to work with scatter plots.

Test Score	70	85	60	95	70	80	75	85	90
Hours Studied	5	6	3	7	4	5	5	5	7

22. Use the data to construct a scatter plot.

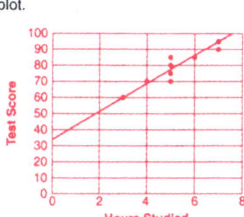

23. Draw a line of best fit on the scatter plot you drew for problem 22.

The line is shown on the above scatter plot.

24. Does the data set have a positive, negative, or no correlation? Explain.

Positive correlation. As hours of study increase, so do test scores.

Chapter Test
Form B

Identify the population, sample, and sampling method.

1. 250 students in an elementary school were randomly selected and asked what was their favorite color.

   All students in the school; 250 students; random.

2. Every twentieth visitor to the local zoo was asked a series of questions about the animal exhibits.

   All visitors to the zoo; those asked the questions; systematic.

3. Five grocery stores in a city were randomly selected and 100 customers were randomly surveyed from each store as to their favorite flavor of ice cream.

   Grocery store customers; those surveyed; stratified.

4. Make a back-to-back stem-and-leaf plot.

   Test Scores for 2 Math Classes

Class #1		Class #2		
98	85	70	93	88 94 82 100
85	74	98	81	88 100 84 100
89	78	100	85	88 94 94 71
89	93	85	81	94 82 82 94

Leaves	Stem	Leaves
0 4 8	7	1
1 1 5 5 5 5 9 9	8	2 2 2 4 8 8 8
3 3 8 8	9	4 4 4 4 4
0	10	0 0 0

Use the data. 98, 95, 89, 85, 89, 90

5. Find the mean. **91**
6. Find the median. **89.5**
7. Find the mode. **89**
8. Find the range. **13**
9. Find the third quartile. **95**
10. Find the first quartile. **89**
11. Make a box-and-whisker plot.

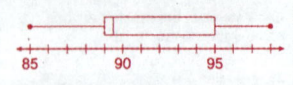

Use the data. Round to the nearest tenth. 28, 25, 28, 22, 28, 29, 23, 24

12. Find the mean. **25.9**
13. Find the median. **26.5**
14. Find the mode. **28**
15. Find the range. **7.0**
16. Find the third quartile. **28.0**
17. Find the first quartile. **23.5**
18. Make a box-and-whisker plot.

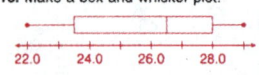

Chapter Test
Form B, continued

Use the test score data to answer the questions.
38,15,18,17,28,29,22,24,34,20,35,31,25,33,14

19. Make a frequency table with an interval of 10.

Scores	Frequency
10–19	4
20–29	6
30–39	5

20. Make a histogram.

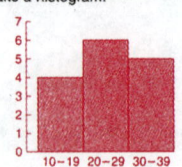

21. Explain why the graph is misleading.

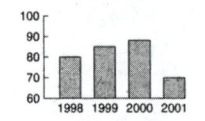

    Since the vertical axis does not start at 0, the bookings for 2000 appear to be almost 3 times as much as for 2001.

Use the data to work with scatter plots.

Test Score	55	75	50	80	95	90	85	80	85	90
Hours Watching TV per Week	22	15	25	10	7	9	10	6	15	7

22. Use the data to construct a scatter plot. Draw a line of best fit.

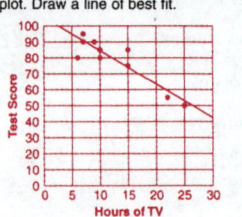

23. Draw a line of best fit on the scatter plot you drew for problem 22.

    The line is shown on the above scatter plot.

24. Does the data set have a positive, negative, or no correlation? Explain.

    Negative correlation. As hours of TV watching increase, test scores decrease.

Chapter Test
Form C

Identify the population, sample, and sampling method.

1. Seven elementary schools in a large city were randomly selected and then 100 students were randomly selected from each school. Each student was asked how much time each week he spends on homework.

   All students in the city's elementary schools; 100 students from each school for 700 students selected; stratified.

2. A manufacturer of automobile tires wants to check his best brand of tires for wear. He randomly selects 10 tires and puts them on the testing machine.

   All of the best brand of tires; 10 tires tested; random.

3. A local grocery store wants to know from how far away people come to shop at the store. Every 5th customer is asked how far he or she drove to get to the store.

   All people shopping at the grocery store; those asked the questions; systematic.

4. Make a back-to-back stem-and-leaf plot.

   Test Scores for 2 Math Classes

Class #1				Class #2			
78	75	76	97	85	92	87	95
81	84	78	83	84	94	87	100
82	98	95	85	87	82	94	94
89	96	75	83	74	100	82	77

Leaves	Stem	Leaves
8 8 6 5 5	7	4 7
9 5 4 3 3 2 1	8	2 2 4 5 7 7 7
8 7 6 5	9	2 4 4 4 5
	10	0 0

Use the data. 198, 195, 189, 185, 189, 190

5. Find the mean. **191**
6. Find the median. **189.5**
7. Find the mode. **189**
8. Find the range. **13**
9. Find the third quartile. **195**
10. Find the first quartile. **189**
11. Make a box-and-whisker plot.

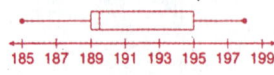

Chapter Test
Form C, continued

Use the data. 1128, 507, 1634, 989, 1350, 1275, 1647, 1301, 1035

12. Find the mean. **1207.3**
13. Find the median. **1275**
14. Find the mode. **none**
15. Find the range. **1140**
16. Find the third quartile. **1492**
17. Find the first quartile. **1012**
18. Make a box-and-whisker plot.

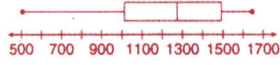

Use the test score data.
138, 115, 128, 117, 128, 129, 122, 124, 134, 120, 135, 131, 125, 133, 114

19. Make a frequency table with an interval of 10.

Scores	Frequency
110–119	3
120–129	7
130–139	5

20. Make a histogram

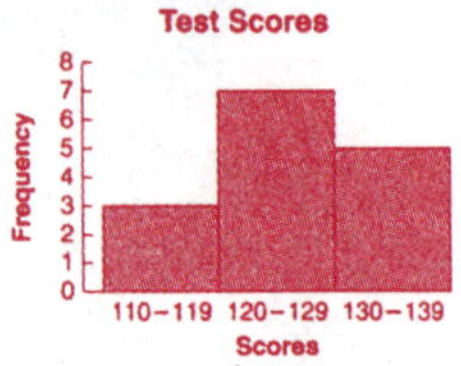

21. The graph shows the number of cars and trucks on a highway. Explain why the graph is misleading.

    The size of the cars versus the size of the trucks distorts the comparison.

Test Scores	85	80	67	89	74	91	81	79	96	79
Hours of Exercise	3.0	8.0	3.0	8.5	1.0	8.0	5.5	9.5	2.0	2.5

22. Use the data above to construct a scatter plot. Draw a line of best fit.

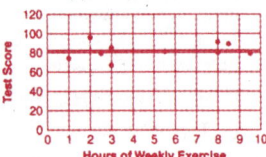

23. Draw a line of best fit on the scatter plot you drew for problem 22.

    The line is shown on the above scatter plot.

24. Does the data set have a positive, negative, or no correlation? Explain.

    No correlation. Hours of exercise does not affect test scores.

Chapter 5 Chapter Test — Form A

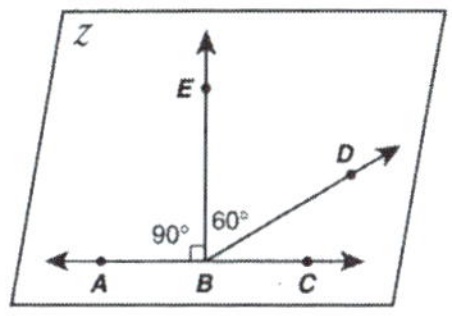

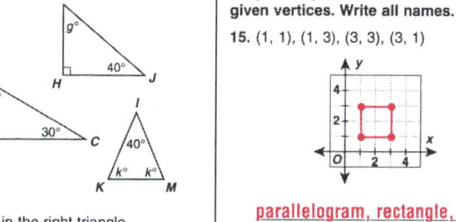

1. Name one point in the figure. **any of point A, point B, point C, point D, or point E**
2. Name a line in the figure. **$\overleftrightarrow{AC}$**
3. Name a plane in the figure. **Z, or any 3 noncollinear points can name the plane**
4. Name one line segment in the figure. **any of $\overline{AB}, \overline{AC}, \overline{BC}, \overline{BE}, \overline{BD}$**
5. Name one ray in the figure. **any of $\overrightarrow{BA}, \overrightarrow{BC}, \overrightarrow{BE}, \overrightarrow{BD}, \overrightarrow{AB}, \overrightarrow{AC}, \overrightarrow{CA}, \overrightarrow{CB}$**
6. Name one angle congruent to ∠7. **any of ∠2, ∠3, ∠6** 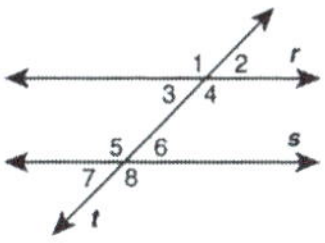
7. Which line is the transversal? **t**
8. If m∠6 is 40°, what is m∠5? **140°**
9. Find g in the right triangle. **50°** 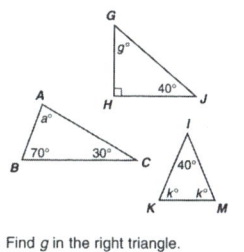
10. Find a in the acute triangle. **90°**
11. Which triangle is an isosceles triangle? **△KIM**

Find the sum of the angle measures.

12. pentagon **540°**
13. triangle **180°**
14. rectangle **360°**

Graph the quadrilaterals with the given vertices. Write all names.

15. (1, 1), (1, 3), (3, 3), (3, 1) — **parallelogram, rectangle, rhombus, square**
16. (2, 3), (6, 3), (6, 0), (2, 0) — **parallelogram, rectangle**

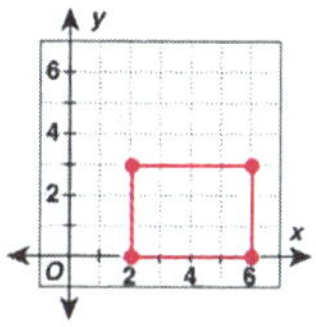

Write a congruence statement.

17. **△ABC ≅ △DEF**
18. **rectangle ABCD ≅ EFGH**

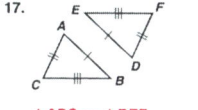

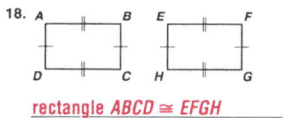

Identify each as a translation, rotation, reflection, or none of these.

19. **translation**
20. **reflection**

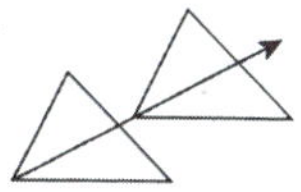

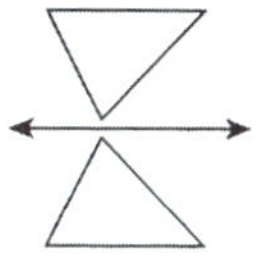

21. Complete the figure. The dashed line is the line of symmetry.
22. (figure)
23. Create a tessellation with the given figure.

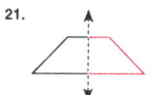

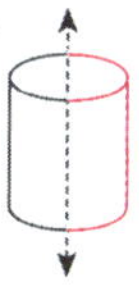

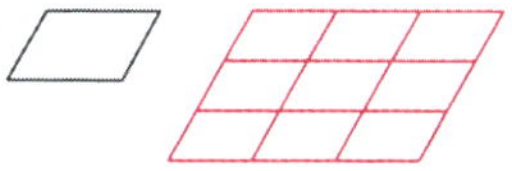

Chapter 5 Chapter Test — Form B

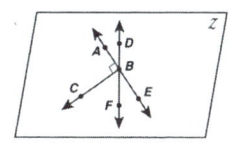

1. Name three points in the figure. **any three of point A, point B, point C, point D, point E, or point F**
2. Name a line in the figure. **$\overleftrightarrow{AE}$ or $\overleftrightarrow{DF}$**
3. Name a plane in the figure. **Z or any 3 non-collinear points can name the plane.**
4. Name two line segments in the figure. **any two of $\overline{AB}, \overline{AE}, \overline{BE}, \overline{BD}, \overline{DF}, \overline{BF},$ or $\overline{BC}$**
5. Name two rays in the figure. **any two of $\overrightarrow{AB}, \overrightarrow{BA}, \overrightarrow{BE}, \overrightarrow{DB}, \overrightarrow{BD}, \overrightarrow{FD}, \overrightarrow{BC},$ or $\overrightarrow{BF}$**
6. Name two angles congruent to ∠3. **any two of ∠2, ∠7, ∠6** 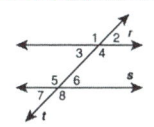
7. Which line is the transversal? **t**
8. If m∠6 is 40°, what is m∠7? **40°**
9. Find e in the obtuse triangle. **120°** 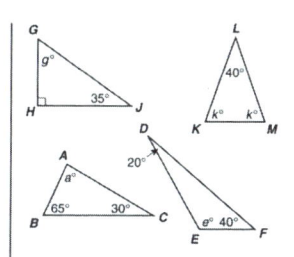
10. Find the unknown angle measures in the isosceles triangle. **70°**
11. Name one acute triangle. **△ABC or △KLM**

Find the sum of the angle measures.

12. hexagon **720°**
13. heptagon **900°**
14. octagon **600°**

Graph the quadrilaterals with the given vertices. Write all names.

15. (2, 2), (6, 2), (6, −1), (2, −1) — **parallelogram, rectangle**
16. (−2, 4), (−3, 1), (1, 1), (2, 4) — **parallelogram**

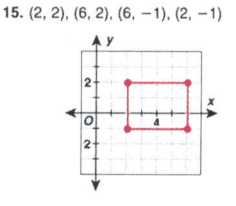

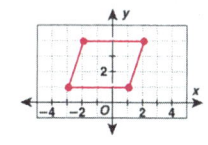

Write a congruence statement.

17. **pentagon ABCDE ≅ FGHJK**
18. **△ABC ≅ △DEF**

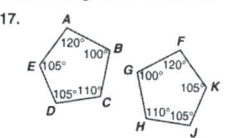

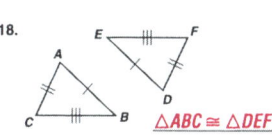

Identify each as a translation, rotation, reflection, or none of these.

19. **rotation**
20. **translation**

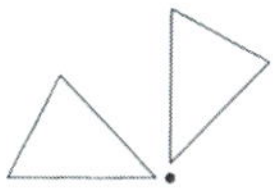

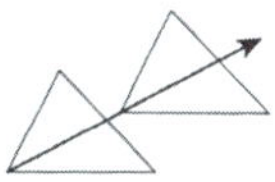

21. Complete the figure. The dashed line is the line of symmetry.
22. (figure)
23. Create a tessellation with the given figure.

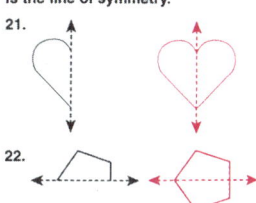

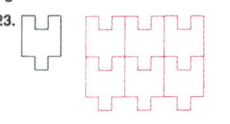

Chapter 5 — Chapter Test, Form C

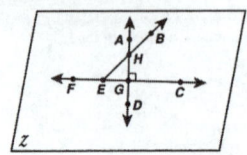

1. Name all points in the figure. **point A, point B, point C, point D, point E, point F, point G, and point H**
2. Name all lines in the figure. $\overline{AH}$, $\overline{AG}$, $\overline{AD}$, $\overline{DA}$, $\overline{GA}$, $\overline{HA}$, $\overline{FE}$, $\overline{FG}$, $\overline{FC}$, $\overline{CF}$, $\overline{GF}$, $\overline{EF}$
3. Name a plane in the figure. **Z, or any 3 noncollinear points can name the plane.**
4. Name eight line segments in the figure. **any eight of $\overline{AH}$, $\overline{AG}$, $\overline{AD}$, $\overline{BH}$, $\overline{BE}$, $\overline{CG}$, $\overline{CE}$, $\overline{CF}$, $\overline{DG}$, $\overline{DH}$, $\overline{EG}$, $\overline{EF}$, $\overline{EH}$, $\overline{FG}$, or $\overline{GH}$**
5. Name all acute angles in the figure. **∠AHB, ∠EHG, and ∠HEG** 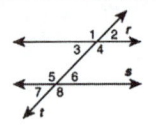
6. Name all angles congruent to ∠8. **∠5, ∠1, ∠4**
7. Which line is the transversal? **t**
8. If m∠7 is 35°, what is m∠4? **145°**

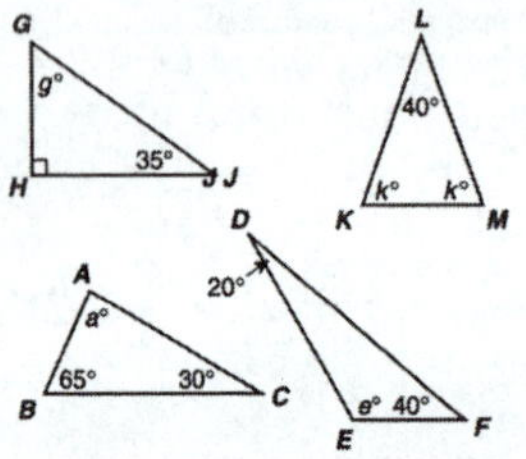

9. Find a in the acute triangle. **85°**
10. Find e in the obtuse triangle. **120°**
11. Which triangles are scalene triangles? **△DEF, △ABC, and △GHJ**

Find the sum of the angle measures.
12. hexagon **720°**
13. heptagon **900°**
14. octagon **1080°**

Graph the quadrilaterals with the given vertices. Write all names.
15. (−3, 0), (−2, −2), (−3, −4), (−4, −2) — **parallelogram, rhombus**
16. (−4, 3), (0, 3), (1, 1), (−3, 1) — **parallelogram**

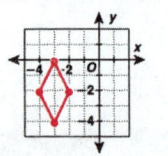

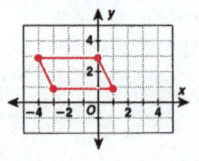

Write a congruence statement.
17. 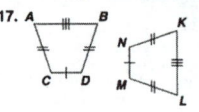 **quadrilateral ABDC ≅ KLMN**
18. 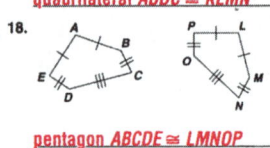 **pentagon ABCDE ≅ LMNOP**

Identify each as a translation, rotation, reflection, or none of these.
19. 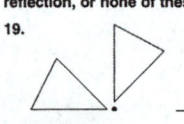 **rotation**
20. **reflection**

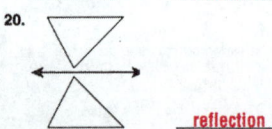

21. Complete the figure. The dashed line is the line of symmetry.
22. 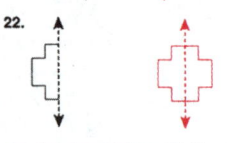
23. Create a tessellation with the given figure.

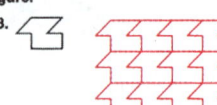

Chapter 6 — Chapter Test, Form A

Find the perimeter of each figure.
1. 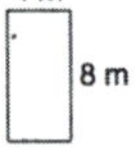 **24 m**
2. 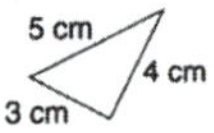 **12 cm**

Graph and find the area of each figure with the given vertices.
3. (0, 0), (0, 4), (3, 4), (3, 0) — **12 units²**
4. (1, 0), (3, 4), (5, 0) — **8 units²**

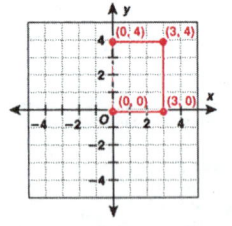

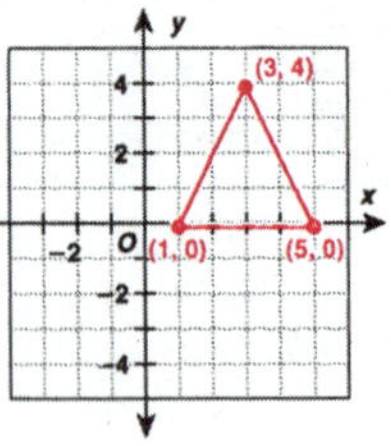

5. Find the length of the hypotenuse. 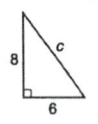 **c = 10**
6. Find the unknown side. **b = 12**

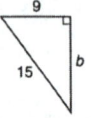

Find the circumference and area of each circle, both in terms of π and to the nearest tenth of a unit using 3.14 for π.
7. circle with radius 7 in. **14π in., 44.0 in.; 49π in², 153.9 in²**
8. circle with diameter 16 cm **16π cm, 50.2 cm; 64π cm², 201.0 cm²**

9. Use isometric dot paper to sketch a rectangular box that is 4 units long, 2 units wide, and 3 units tall. **Possible answer:**

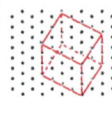

10. Sketch a one-point perspective drawing of a cube. **Possible answer:**

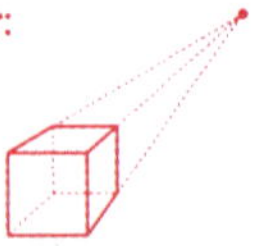

Find the volume of each figure to the nearest tenth. Use 3.14 for π.
11. 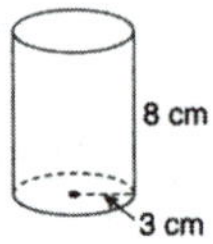 **226.1 cm³**
12. 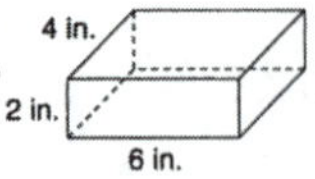 **48.0 in³**
13. 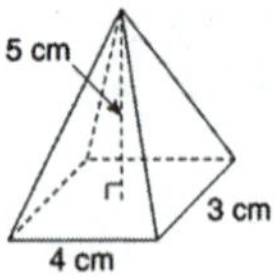 **20.0 cm³**
14. 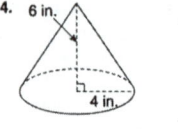 **100.5 in³**

Find the surface area of each figure to the nearest tenth. Use 3.14 for π.
15. the figure in Exercise 11 **207.2 cm²**
16. the figure in Exercise 12 **88.0 in²**
17. 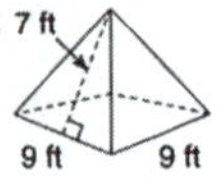 **207.0 ft²**
18. 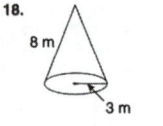 **103.6 m²**

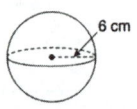

19. Find the surface area of the sphere, both in terms of π and to the nearest tenth of a unit using 3.14 for π. **144π cm², 452.2 cm²**
20. Find the volume of the sphere, both in terms of π and to the nearest tenth of a unit using 3.14 for π. **288π cm³; 904.3 cm³**

Chapter 6 Chapter Test — Form B

Find the perimeter of each figure.

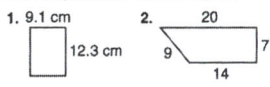

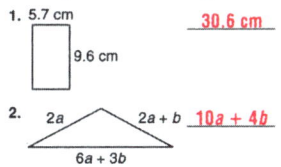

1. 9.1 cm, 12.3 cm — **42.8 cm**

2. 20, 9, 7, 14 — **50 units**

Graph and find the area of each figure with the given vertices.

3. (−2, 1), (0, 4), (5, 4), (3, 1) — **15 units²**

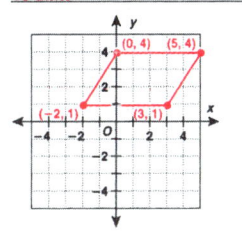

4. (−3, 1), (−1, 5), (2, 1) — **10 units²**

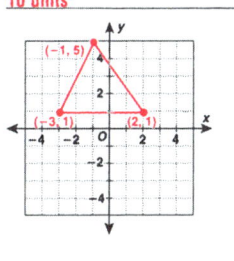

5. Find the length of the hypotenuse to the nearest tenth. (legs 6 and 3) — **6.7**

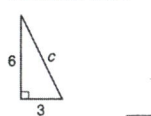

6. Find the unknown side to the nearest tenth. (hypotenuse 8, leg 5) — **6.2**

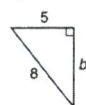

Find the circumference and area of each circle, both in terms of π and to the nearest tenth of a unit using 3.14 for π.

7. circle with radius 15.2 in. — **30.4π in., 95.5 in.; 231.0π in², 725.5 in²**

8. circle with diameter 28.6 cm — **28.6π cm, 89.8 cm; 204.5π cm², 642.1 cm²**

9. Use isometric dot paper to sketch a cube 5 units on each side. Possible answer: [cube sketch]

10. Sketch a one-point perspective drawing of a triangular box. Possible answer: [sketch]

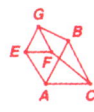

Find the volume of each figure to the nearest tenth. Use 3.14 for π.

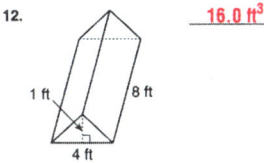

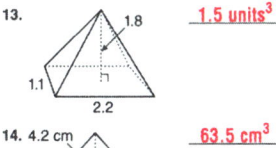

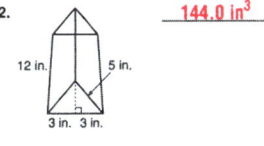

11. 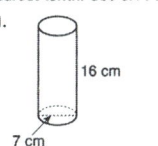 cylinder, r = 7 cm, h = 16 cm — **615.4 cm³**

12. triangular prism, 1 ft, 4 ft, 8 ft — **16.0 ft³**

13. pyramid, 1.1, 1.8, 2.2 — **1.5 units³**

14. 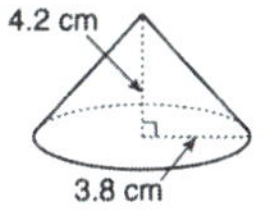 cone, 4.2 cm, 3.8 cm — **63.5 cm³**

Find the surface area of each figure to the nearest tenth. Use 3.14 for π.

15. the figure in Exercise 11 — **428.6 cm²**

16. a rectangular prism 7 in. by 4 in. by 5 in. — **166.0 in²**

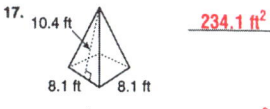

17. pyramid, 10.4 ft, 8.1 ft, 8.1 ft — **234.1 ft²**

18. 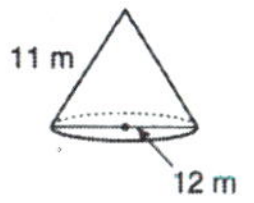 cone, 11 m, 12 m — **320.3 m²**

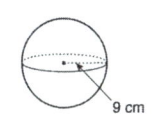

19. Find the surface area of the sphere, both in terms of π and to the nearest tenth of a unit using 3.14 for π. (r = 9 cm) — **324π cm², 1017.4 cm²**

20. Find the volume of the sphere, both in terms of π and to the nearest tenth of a unit using 3.14 for π. — **972π cm³; 3052.1 cm³**

Chapter 6 Chapter Test — Form C

Find the perimeter of each figure.

1. 5.7 cm, 9.6 cm — **30.6 cm**

2. 2a, 2a + b, 6a + 3b — **10a + 4b**

Graph and find the area of each figure with the given vertices.

3. (−3, 3), (−3, 0), (−2, 0), (−2, −2), (1, −2), (1, 0), (3, 0), (3, 2), (0, 2), (0, 3) — **21 units²**

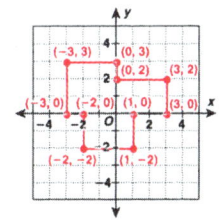

4. (−3, −2), (−2, 2), (3, 2), (4, −2) — **24 units²**

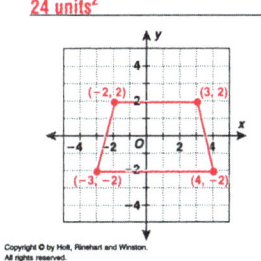

5. Find the length of the hypotenuse to the nearest hundredth. (legs 4 and 5) — **c ≈ 6.40**

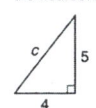

6. Find the unknown side to the nearest hundredth. (hypotenuse 12, leg 7) — **a ≈ 9.75**

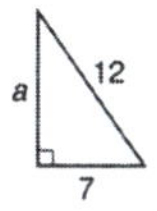

Find the circumference and area of each circle, both in terms of π and to the nearest tenth using 3.14 for π.

7. circle whose circumference is one-sixth the circumference of a circle with radius 18 in. — **6π in., 18.8 in.; 9π in², 28.3 in²**

8. circle whose area is four times the area of a circle with diameter 10 cm — **20π cm, 62.8 cm; 100π cm², 314.0 cm²**

9. Use isometric dot paper to sketch a rectangular box with a base 5 units long by 3 units wide and a height of 2 units. [sketch]

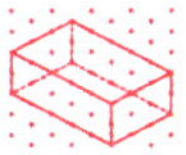

10. Sketch a two-point perspective drawing of a square box. Possible answer: [sketch]

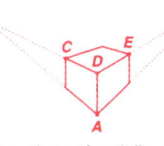

Find the volume of each figure to the nearest tenth. Use 3.14 for π.

11. cylinder, 18.4 cm, 12.2 cm — **2149.8 cm³**

12. prism, 12 in., 5 in., 3 in., 3 in. — **144.0 in³**

13. 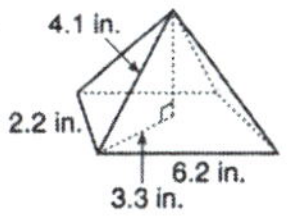 pyramid, 4.1 in., 2.2 in., 6.2 in., 3.3 in. — **11.1 in³**

14. 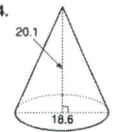 cone, 20.1, 18.6 — **1819.6 units³**

Find the surface area of each figure to the nearest tenth. Use 3.14 for π.

15. the figure in Exercise 11 — **938.5 cm²**

16. the figure in Exercise 12 — **216.0 in²**

17. 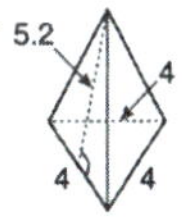 5.2, 4, 4, 4 — **38.1 units²**

18. the figure in Exercise 14 — **918.3 units²**

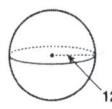

19. Find the surface area of the sphere, both in terms of π and to the nearest tenth of a unit using 3.14 for π. (r = 12) — **576.0π units², 1808.6 units²**

20. Find the volume of the sphere, both in terms of π and to the nearest tenth of a unit using 3.14 for π. — **2304π units³; 7234.6 units³**

Chapter 7 Chapter Test — Form A

Find two ratios that are equivalent to each given ratio.

1. $\frac{5}{10}$ Possible answer: $\frac{1}{2}, \frac{10}{20}$
2. $\frac{4}{6}$ Possible answer: $\frac{2}{3}, \frac{8}{12}$
3. $\frac{16}{4}$ Possible answer: $\frac{4}{1}, \frac{8}{2}$
4. $\frac{21}{27}$ Possible answer: $\frac{7}{9}, \frac{14}{18}$

Simplify to tell whether the ratios form a proportion.

5. $\frac{4}{12}$ and $\frac{2}{8}$ — **no**
6. $\frac{1}{2}$ and $\frac{4}{8}$ — **yes**
7. $\frac{1}{3}$ and $\frac{2}{6}$ — **yes**

Find the unit price for each offer and tell which is the better buy.

8. 20-oz box of cereal for $3.80; 15-oz box of cereal for $3.15
   The 20-oz box costs $0.19/oz and the 15-oz box costs $0.21/oz. The 20-oz box is a better buy.

9. 10 blank CDs for $2.50; 15 blank CDs for $3.00
   10 CDs cost $0.25 per CD and 15 CDs cost $0.20 per CD. 15 CDs is a better buy.

Find the appropriate factor for each conversion.

10. kilometers to meters
    $\frac{1000 \text{ m}}{1 \text{ kilometer}}$

11. gallons to quarts
    $\frac{4 \text{ quarts}}{1 \text{ gallon}}$

12. A car travels 5 miles in 6 minutes. What is its speed in miles per hour?
    50 mi/h

13. A home improvement store sells 2355 feet of wire per day. How many yards of wire does the store sell per day?
    785 yards

Solve each proportion.

14. $\frac{3}{9} = \frac{x}{3}$ **$x = 1$**
15. $\frac{32}{t} = \frac{4}{1}$ **$t = 8$**
16. $\frac{12}{22} = \frac{18}{p}$ **$p = 33$**

Dilate each figure by the given scale factor with the origin as the center of dilation.

17. Triangle with vertices $A(1, 2)$, $B(4, 1)$, $C(4, 4)$, scale factor = 2

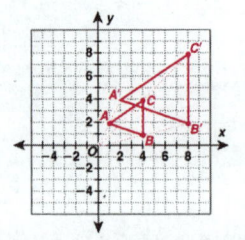

18. Triangle with vertices $A(4, 2)$, $B(8, 2)$, $C(6, 6)$, scale factor $\frac{1}{2}$

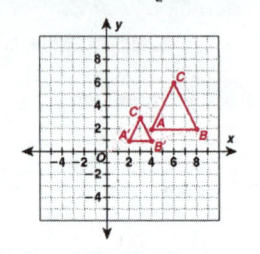

Use the properties of similar figures to answer each question.

19. Rectangle A has length 11 m and width 7 m. Rectangle B has length 33 m and width 21 m. Are rectangles A and B similar?
    yes

20. A soccer field for 13-year olds measures 50 yards wide by 100 yards. 8-year olds play on a similar soccer field that is 20 yards wide. How long is the field for 8-year olds?
    40 yards

21. Julia's room is 4 in. long on a scale drawing. If her room is actually 16 ft long, what is the scale?
    1 in.:4 ft

22. A drawing of a 78-foot long building was built using a scale of 1 in.:8 ft. What is the length of the drawing?
    $9\frac{3}{4}$ in.

Tell whether each scale reduces, enlarges, or preserves the size of the actual object.

23. 1 cm:12 m **reduces**
24. 6 ft:10 in. **enlarges**
25. 1 km: 1000 m **preserves**

A 3-in. cube is built from small cubes, each 1 in. on a side. Compare the following values.

26. the side lengths
    Larger cube is 3 times as long as smaller cube.

27. the volumes
    The volume of the larger cube is 27 times that of the smaller cube.

Chapter 7 Chapter Test — Form B

Find two ratios that are equivalent to each given ratio.

1. $\frac{2}{9}$ Possible answer: $\frac{4}{18}, \frac{6}{27}$
2. $\frac{50}{15}$ Possible answer: $\frac{10}{3}, \frac{100}{30}$
3. $\frac{11}{17}$ Possible answer: $\frac{22}{34}, \frac{33}{51}$
4. $\frac{18}{16}$ Possible answer: $\frac{9}{8}, \frac{36}{32}$

Simplify to tell whether the ratios form a proportion.

5. $\frac{4}{26}$ and $\frac{2}{13}$ — **yes**
6. $\frac{18}{60}$ and $\frac{3}{10}$ — **yes**
7. $\frac{5}{25}$ and $\frac{15}{50}$ — **no**

Find the unit price for each offer and tell which is the better buy.

8. 20 blank CDs for $2.79; 12 blank CDs for $1.20
   20 CDs cost $0.14 per CD and 12 CDs cost $0.10 per CD. 12 CDs is the better buy.

9. 6 paperback books for $19.00; 8 paperback books for $26.00
   6 books cost $3.17 per book and 8 books cost $3.25 per book. 6 books is the better buy.

Find the appropriate factor for each conversion.

10. months to years
    $\frac{1 \text{ yr}}{12 \text{ months}}$

11. pounds to ounces
    $\frac{16 \text{ oz}}{1 \text{ lb}}$

12. A woodworker can put together 2 wood toy trains per day. How many trains could the woodworker make in 8 weeks?
    112

13. A store sells 8-ounce packages of mushrooms for $1.29. What is the cost of 3 pounds of mushrooms?
    $7.74

Solve each proportion.

14. $\frac{4}{10} = \frac{y}{20}$ **$y = 8$**
15. $\frac{r}{0.32} = \frac{3}{2}$ **$r = 0.48$**
16. $\frac{11}{q} = \frac{5}{2}$ **$q = 4.4$**

Dilate each figure by the given scale factor with the origin as the center of dilation.

17. Triangle with vertices $A(2, 2)$, $B(8, 4)$, $C(4, 8)$, scale factor = $\frac{1}{4}$

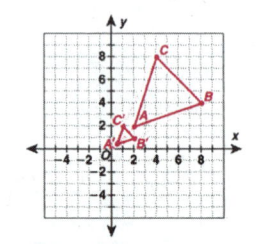

18. Quadrilateral with vertices $A(2, 2)$, $B(8, 2)$, $C(8, 4)$, $D(2, 6)$, scale factor 1.5

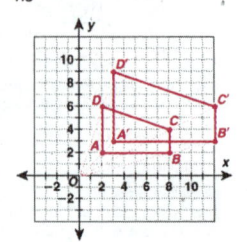

19. Jess made two picture frames. One frame is 8 inches by 9 inches. The other frame is 12 inches by 22 inches. Are the frames similar?
    no

20. Lisa is having a 5 in. by 7 in. photo made into a similar poster. If the poster is 2 ft wide, how long will it be?
    33.6 in. or 2.8 ft

21. If the scale is 1 cm:8 m, how tall is a drawing of a 654-m skyscraper?
    81.75 cm

22. A drawing of an airplane hangar was made using a scale of 1 in.:20 ft. If the hangar is actually 250 feet wide, how wide is the drawing?
    $12\frac{1}{2}$ in.

Tell whether each scale reduces, enlarges, or preserves the size of the actual object.

23. 15 ft:1 in. **enlarges**
24. 10 mm:1 cm **preserves**
25. 2 m:10 km   **reduces**

An 8-cm cube is built from small cubes, each 1 cm on a side. Compare the following values.

26. the side lengths
    Larger cube is 8 times as long as smaller cube

27. the volumes
    The volume of the larger cube is 512 times that of the smaller cube.

Chapter 7 Chapter Test — Form C

Find two ratios that are equivalent to each given ratio.

1. $\frac{5}{9}$ Possible answer: $\frac{10}{18}, \frac{15}{27}$
2. $\frac{2}{12}$ Possible answer: $\frac{1}{6}, \frac{3}{18}$
3. $\frac{15}{6}$ Possible answer: $\frac{5}{2}, \frac{10}{4}$
4. $\frac{14}{8}$ Possible answer: $\frac{7}{4}, \frac{21}{12}$

Simplify to tell whether the ratios form a proportion.

5. $\frac{6}{16}$ and $\frac{9}{24}$ **yes**
6. $\frac{36}{28}$ and $\frac{10}{7}$ **no**
7. $\frac{21}{27}$ and $\frac{7}{8}$ **no**

Find the unit price for each offer and tell which is the better buy.

8. $7.98 for a 3-pound ham; $10.84 for a 5-pound ham

The 3-lb ham costs $2.66 per lb and the 5-lb ham costs $2.17 per pound. The 5-pound ham is the better buy.

9. A 12-ounce drink for $1.40; a 20-ounce drink for $2.70.

The 12-oz drink costs $0.12 per ounce and the 20-oz drink costs $0.14 per ounce. The 12-ounce drink is the better buy.

Find the appropriate factor for each conversion.

10. weeks to hours
$\frac{168 \text{ hours}}{1 \text{ week}}$

11. miles to yards
$\frac{1760 \text{ yd}}{1 \text{ mi}}$

12. A car wash cleans automobiles at a rate of 35 per hour. How many cars do they clean in an 8-hour day?
280 automobiles

13. Zara biked for 2 hours at an average rate of 15 meters per second. How many kilometers did she bike?
108 km

Solve each proportion.

14. $\frac{6}{4} = \frac{x}{5}$ $x = 7.5$
15. $\frac{33}{t} = \frac{4}{1}$ $t = 8.25$
16. $\frac{12}{1.5} = \frac{40}{p}$ $p = 5$

Chapter 7 Chapter Test — Form C, continued

Identify the scale factor used in each dilation.

17. **2**

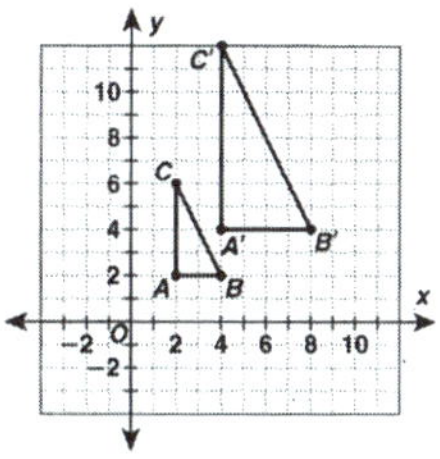

18. **0.5**

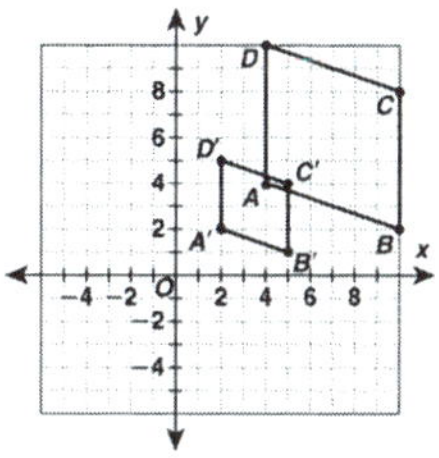

Use the properties of similar figures to answer each question.

19. Bart is using two different size triangles to make a tile design. One has sides of 4 in., 6 in., and 7 in. The other has sides of 12 in., 18 in., and 20 in. Are the two triangles similar? Explain.
No; $\frac{6}{18} \neq \frac{7}{20}$

20. The two triangles are similar. Use the scale factor to solve for x. 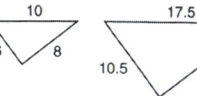
$x = 14$

21. A scale drawing of a rectangular swimming pool is 6 in. by 10.5 in. If the scale is 0.25 in.:1 ft, what is the perimeter of the actual pool?
132 ft

22. If the scale of a drawing is 2 in.:35 ft, how long would a 49-foot fence be in the drawing?
2.8 in.

Find the scale factor and tell whether it reduces, enlarges, or preserves the size of the actual object.

23. 1 cm:12 m **1:1200, reduces**
24. 2 ft:10 in. **12:5, enlarges**
25. 8 in.:32 ft **1:48, reduces**

For each cube, a reduced scale model is built using a scale factor of 0.25. Find the length of the model and the number of 1-cm cubes used to build it.

26. a 16-cm cube
4 cm, 64 1-cm cubes

27. a 48-cm cube
12 cm, 1728 1-cm cubes

Chapter 8 Chapter Test — Form A

Find the missing ratio or percent equivalent for each letter on the number line.

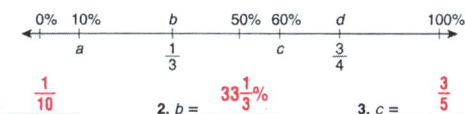

1. $a = \frac{1}{10}$
2. $b = 33\frac{1}{3}\%$
3. $c = \frac{3}{5}$

4. Write $\frac{7}{10}$ as a percent.
70%

5. Write 40% as a fraction.
$\frac{2}{5}$

6. What percent of 175 is 28?
16%

7. What percent of 305 is 122?
40%

8. 9.6 is 15% of what number?
64

9. 40% of what number is 114?
285

10. Estimate 25% of 203.
about 50

11. Estimate 12% of 80.
about 8

12. About 35 acres of a 125-acre farm are planted with corn. What percent of the farm's fields are planted with corn?
28%

13. Jake runs the 100-meter dash in track. Erika runs a race 400% the distance of Jake's race. How long is Erika's race?
400 m

14. Derek can carry 65% of his weight in his backpack while camping. If his backpack weighs 88.4 pounds, how much does Derek weigh?
136 pounds

15. Kersten has 12 postcards from New York City. This is 30% of her total postcard collection. How many postcards does she have in her collection?
40 postcards

16. Find the percent increase or decrease from 25 to 20.
20% decrease

Chapter 8 Chapter Test — Form A, continued

Give answers to the nearest percent.

17. A toy store sold a toy train set for $49.95 last year. This year, the same train set costs $52.50. What was the percent increase of the cost of the train set?
5%

18. The president of a small company made $72,000 last year. She cut her salary this year to $60,000 because the company was not doing as well. What was the percent decrease in her salary to the nearest percent?
17%

19. A rain jacket costs $52. It is on sale for 20% off. Estimate the discount on the jacket.
about $10

20. In a state with a sales tax rate of 6%, Alex bought paper for his printer for $17.99. How much was his sales tax to the nearest penny?
$1.08

21. Theo earned $3045 over the summer as a lifeguard. Of this, $669.90 was withheld for taxes. What percent of his income was withheld?
22%

22. Clarence earns a 10% commission on sales plus a $200 weekly salary. In one week, his sales totaled $2100. What was his total pay that week?
$410

23. Shannon invested $5000 in a bond. Her total simple interest on her investment after 3 years was $1200. What was the yearly interest rate on her investment?
8%

24. Tess borrowed $10,500 from the bank to buy a car. The length of her loan is 4 years with a simple interest rate of 7.5%. How much will she pay in interest if she pays off the loan at the end of the 4 years?
$3150

25. Stephen deposited $2500 in a savings account that earned an annual simple interest rate of 4%. When he closed his account, he had $3000. How long did he have his account?
5 years

Chapter 8 Test — Form B

Find the missing ratio or percent equivalent for each letter on the number line.

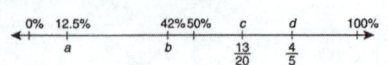

Number line: 0%, 12.5%, 42%, 50%, c, d, 100% with a, b, 13/20, 4/5

1. $a = \dfrac{1}{8}$
2. $b = \dfrac{21}{50}$
3. $c = 65\%$

4. Write $\dfrac{17}{25}$ as a percent.
   68%

5. Write 29% as a decimal.
   0.29

6. What percent of 67 is 134?
   200%

7. 48 is what percent of 192?
   25%

8. 12.5 is 20% of what number?
   62.5

9. 120% of what number is 98.4?
   82

10. Estimate 20% of 198.
    about 40

11. Estimate 3015 out of 8999 as a percent.
    about $33\dfrac{1}{3}\%$

12. Mille Lacs county covers 574 square miles of Minnesota. If Minnesota is 79,610 square miles, what percent of Minnesota is Mille Lacs county to the nearest tenth of a percent?
    0.7%

13. Elsa is 56 inches tall. Her brother's height is 65% of Elsa's height. How tall is her brother?
    36.4 in.

14. A store sold 156 winter coats during a sale. If this represented 60% of their total inventory, how many coats did they have before the sale?
    260

15. Monica has 64 stamps commemorating Olympic games in her collection. This is 16% of her total collection. How many stamps does she have in her collection?
    400

16. On sale, a sweater was reduced from $40 to $32. Find the percent of decrease.
    20%

Form B, continued

Give answers to the nearest percent.

17. A toy store sold a toy train set for $49.79 last year. This year, the same train set costs $51.29. What was the percent increase of the cost of the train set?
    3%

18. A small company had profits of $550,000 last year. This year, their profits were only $484,000. What was the percent decrease in their profits?
    12%

19. A pair of shoes are on sale for 25% off. They normally cost $42.95. Estimate the discount on the shoes.
    about $10

20. In a state with a sales tax rate of 6.5%, Thomas bought a new DVD player for $256.99 and a DVD movie for $24.99. How much is the sales tax on his purchase to the nearest penny?
    $18.33

21. Kirk earned $4147 over the summer working as a waiter. $829.40 was taken out for taxes. What percent of his income was withheld for taxes?
    20%

22. Justin works as a car salesman where he earns 8% commission on his sales and no weekly salary. What will his weekly sales have to be to earn $3280 for the week?
    $41,000

23. Karen deposited $8500 in a college savings account for her grandson that earns an annual simple interest rate of 6.5%. What will be the total amount in the account in 10 years?
    $14,025

24. Sadie borrowed $3500 from a bank at an annual simple interest rate for a home remodeling project. After 4 years, she repaid the bank $4200. What was the interest rate of the loan?
    5%

25. If Aisha deposits $2250 in a savings account that earns 3.5% annual simple interest, how long must she keep the money in the account for its total to reach $2722.50?
    6 yr

Chapter 8 Test — Form C

Gia's Budget (pie chart): 1/4 Savings, 30% Rent, 1/10 Spending Money, 20% Utilities, 3/20 Food

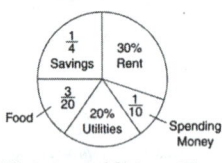

1. What percent of Gia's monthly income goes into savings?
   25%

2. What fraction of her income is used to pay for utilities?
   $\dfrac{1}{5}$

3. What percent of her income is used to buy food?
   15%

4. Write $\dfrac{8}{25}$ as a percent.
   32%

5. Write 6% as a fraction.
   $\dfrac{3}{50}$

6. What percent of 2950 is 531?
   18%

7. What percent of 460 is 621?
   135%

8. 51.45 is 35% of what number?
   147

9. 23.5% of what number is 42.3?
   180

10. Estimate 25% of 58.5.
    about 15

11. Estimate 110% of 89.75.
    about 99

12. Enrique has read 67.5% of a 354-page book. To the nearest page, how many pages has he read?
    239

13. Sophia needs to earn at least 25% of her college tuition before her parents will help with the rest. Tuition for next year is $3410. How much does Sophia need to earn?
    $852.50

14. An artist is designing a statue of a man standing next to a horse. The man will be 8.5 feet tall and the horse will be 130% the height of the man. How tall will the horse be?
    11.05 ft

15. If 114 out of 355 students in the eighth grade play instruments, what percent of eighth grade students do not play instruments? Round your answer to the nearest percent.
    68%

Form C, continued

Give answers to the nearest hundredth of a percent.

16. The number of children taking swimming lessons at the community pool last summer was 274. The number registered for lessons this summer is 292. What is the percent increase?
    6.57%

17. A pair of Kelly's new pants shrunk in the wash. The inseam was 37.25 inches before they were washed and 36.375 inches after they were washed. What is the percent decrease in the length of the pants?
    2.35%

18. During a sale, the price of a DVD was decreased by 50%. By what percent must the sale price be increased to restore the original price?
    200%

19. The total area of the state of Florida is 58,560 square miles. The total land area of Florida is 54,252 square miles. Estimate what percent of Florida's total area is water.
    about 8%

20. Janine is paid $200 a week plus 12.5% commission on sales. What were her weekly sales if she earned $950?
    $6000

21. Shaneece made $42,500 last year. Isak made $35,450 last year. If they are both taxed at 27.5% of the amount over $27,050, how much more tax did Shaneece pay than Isak?
    $1938.75

22. 19.5% of Alisha's paycheck is withheld each week for taxes. If she earns $568 per week, how much money is her check written for?
    $457.24

23. A credit union advertises that if you put $3000 in a certificate of deposit, you can earn $1800 in ten years. What yearly simple interest rate does the certificate of deposit offer?
    6%

24. Salim borrowed $12,375 for 4 years at an annual simple interest rate of 7.5%. If he makes one payment at the end of the loan, how much interest will he have to repay?
    $3712.50

25. Todd had to pay $1147.50 of interest on a loan with an annual simple interest rate of 8.5% that he had for 5 years. Assuming he makes one payment at the end of the loan, what was the principal of the loan?
    $2700

Chapter 9 Chapter Test Form A

An experiment consists of drawing 4 balls from a bag and counting the number of red balls. The table gives the probability of each outcome.

Number of Red Balls	0	1	2	3	4
Probability	0.025	0.36	0.27	0.282	0.063

1. What is the probability of drawing fewer than 2 red balls?

   0.385

2. What is the probability of drawing more than 1 red ball?

   0.615

3. Erin's soccer team has won 17 of their 20 games this season. What is the probability that they will win their next game?

   0.85 or 85%

4. A researcher conducted a survey of 324 high school students and found that 54 of them were enrolled in advanced chemistry. What is the probability that a randomly selected student is enrolled in advanced chemistry?

   $\frac{1}{6}$

Use the table of random numbers to simulate each situation. Use at least 10 trials for each simulation.

33	35	71	65	22	33	04	35	56	99
63	41	51	27	76	48	30	84	63	20
57	62	81	29	54	61	35	22	35	44
62	61	22	24	35	12	73	42	64	46
33	20	52	77	88	15	73	82	19	97
31	96	04	29	74	36	44	42	38	26
53	14	76	41	98	83	53	64	15	91
24	83	42	19	61	12	52	62	28	32

5. Katy hits the ball 58% of the time she bats. Estimate the probability that she will hit the ball at least 3 times in her next 4 at bats.

   possible answer: 40%

6. Sandra hits a golf ball over 120 yards on her first drive 67% of the time. Estimate the probability that she will hit the ball over 120 yards at least 4 times in the next 7 times she drives.

   possible answer: 90%

An experiment consists of spinning a spinner with equal chances of landing on one of four colors – red, blue, green, or yellow.

7. What is the probability of landing on blue or yellow?

   $\frac{1}{2}$

Chapter 9 Chapter Test Form A, continued

8. The PIN numbers for a cash card at a bank contain four digits 1–9. All codes are equally likely. Find the number of possible PIN numbers.

   6561

9. The flavors at an ice cream shop are chocolate, vanilla, mint, and strawberry. The cone choices are waffle or sugar. Describe all of the different ice cream cone options available.

   SC, SV, SM, SS, WC, WV, WM, WS

10. 6 swimmers are competing in the 100-yard butterfly. In how many different orders can all of the swimmers finish the race?

    720

11. Find the number of different 4-person teams that can be made from 14 people.

    1001

12. There are 3 apples, 5 oranges, and 2 tangerines in a bowl of fruit. Two pieces of fruit are chosen at random and not replaced. What is the probability of choosing an apple and then an orange?

    $\frac{1}{6}$

13. 5 red dice and 5 blue dice are put into a bag. What is the probability that when two dice are taken out, one is blue and one is red?

    $\frac{5}{9}$

14. There is a jar with 10 nickels and 5 dimes. If two coins are chosen at random, what is the probability of choosing first a nickel and then a dime?

    $\frac{5}{21}$

15. The odds of winning a door prize at a birthday party are 1:12. What is the probability of winning a door prize?

    $\frac{1}{13}$

16. Six people called a radio talk show: two lawyers, two doctors, a veterinarian, and an accountant. What is the probability that a randomly selected caller is not a doctor?

    $\frac{2}{3}$

17. Jared has collected 30 contest game pieces. Of those, 4 were winning pieces. What are the odds against winning a prize?

    13 to 2

Chapter 9 Chapter Test Form B

Trisha made an educated guess on four of the multiple choice questions on her drivers license examination. The table gives the probability of each possible result.

Number of Correct Answers	Probability
0	0.025
1	0.271
2	0.359
3	0.282
4	0.063

1. What is the probability of getting 3 or more correct?

   0.345

2. What is Trisha's probability of failing this part of the exam (getting fewer than 2 correct)?

   0.296

3. A ball was randomly drawn from a bag and then replaced. In 300 experiments, a green ball was chosen 58 times, a red ball was chosen 118 times, a yellow ball was chosen 99 times, and a blue ball was chosen 25 times. What is the probability of choosing a yellow ball?

   0.33 or 33%

4. A researcher conducted a survey of 425 high school students and found that 356 of them planned to attend college. Estimate to the nearest percent the probability that a randomly selected student plans to attend college.

   0.84 or 84%

Use the table of random numbers to simulate each situation. Use at least 10 trials for each simulation.

33	35	71	65	22	33	04	35	56	99
65	41	51	27	76	48	30	84	63	20
57	62	81	29	74	61	35	22	35	44
42	61	22	24	35	12	73	42	69	46
33	20	52	57	88	15	73	82	19	97
31	96	04	29	74	36	48	42	38	26
53	14	76	41	18	43	53	68	15	91
24	83	42	19	61	12	52	62	28	32

5. In a city in Alaska, snow falls on 64% of winter days. Estimate the probability that it will snow at least 6 out of 7 days during a week in January.

   possible answer: 20%

6. At a pizza place, about 45% of the customers order pepperoni on their pizza. Estimate the probability that at least 5 out of the next 6 customers will order pepperoni.

   possible answer: 10%

An experiment consists of rolling a fair eight-sided die.

7. What is the probability of rolling a 3, 4, or 5?

   $\frac{3}{8}$

Chapter 9 Chapter Test Form B, continued

8. Student ID codes at a university contain two letters followed by two digits 0–9 and then another two letters. All codes are equally likely. Find the number of possible student ID codes.

   45,697,600

9. A store sells three different styles of fleece jackets in 8 different colors. Each style can also be purchased with or without a hood. How many different versions of fleece jackets does the store sell?

   48

10. Maureen has 7 different plants that she wants to plant in a row in her flower bed. How many different ways can she arrange the plants?

    5040

11. If Maureen decides she only has room for 3 of the 7 plants in her flower bed, how many different selections of 3 plants does she have to choose from?

    35

12. Karleen made a snack mix by mixing 30 raisins, 45 peanuts, and 15 marshmallows. If she randomly selects two pieces from her snack mix, what is the probability they will both be peanuts?

    $\frac{22}{89}$

13. A fair die is rolled twice. What is the probability of getting 4 the first time but not the second?

    $\frac{5}{36}$

14. There is a jar with 10 nickels, 5 dimes, and 6 quarters. If three coins are chosen at random, what is the probability of choosing all quarters?

    $\frac{2}{133}$

15. If the odds of winning a free meal at a restaurant are 1:356, what is the probability of winning?

    $\frac{1}{357}$

16. A class ordered 7 pizzas. 3 are pepperoni, 2 are mushroom, and 2 are cheese. The boxes are not labeled. What is the probability that the first box opened will not contain a pepperoni pizza?

    $\frac{4}{7}$

17. If 5 of the first 175 customers that arrive at a boat sale will win free lifejackets, what are the odds against winning?

    34 to 1

Chapter 9 Chapter Test — Form C

Outcome	Probability
1	0.071
2	0.271
3	0.314
4	0.282
5	0.062

1. What is the probability of outcome 1 or 2 occurring?

   0.342

2. What is the probability of neither outcome 1 nor outcome 4?

   0.647

3. Among 692 patients who were tested for a certain disease, 295 tested negative, 104 tested positive, and 293 had inconclusive results. Estimate to the nearest thousandth the probability of a patient's testing negative.

   0.426 or 42.6%

Sport	Number of Students
Swimming	28
Basketball	31
Football	45
Tennis	24

4. 165 students were polled about the sports they participate in. Estimate to the nearest thousandth the probability that a randomly polled student does not participate in any sports.

   0.224 or 22.4%

Use the table of random numbers to simulate each situation. Use at least 10 trials for each simulation.

12	77	64	83	08	46	32	03	19	60
42	97	43	64	44	23	82	66	34	15
16	52	23	30	43	55	69	93	32	05
25	78	81	57	38	21	99	73	33	57
22	43	51	27	34	26	44	62	48	34
21	96	11	25	26	22	83	24	26	72
12	95	46	13	52	23	54	45	01	47
64	11	84	35	36	41	59	15	76	28

5. Mr. Guanara gives a pop quiz about 25% of the days his class meets. What is the probability that there will be a pop quiz at least 3 out of the next 6 times his class meets.

   possible answer: 20%

6. An inspector finds that about 3% of computer motherboards assembled are defective. What is the probability that 1 of the next 8 motherboards will be defective?

   possible answer: 20%

Three fair coins with sides marked A and B are tossed.

7. Find each probability. $P(ABA)$

   $\dfrac{1}{8}$

Chapter 9 Chapter Test — Form C, continued

8. Raffle tickets at a school have 6 letters followed by 2 digits 0–9. All possible raffle tickets are equally likely. Find the number of possible raffle tickets.

   30,891,577,600

9. At a sandwich shop, you can choose from 5 different kinds of bread, 7 different meats, 10 different cheeses, and 6 different sandwich spreads. How many different kinds of sandwiches are offered?

   2100

10. 14 people are lined up to purchase movie tickets. How many different ways can they be lined up?

    87,178,291,200

11. How many ways can a teacher choose the first, second, third, and fourth students to give a presentation from a class of 21?

    143,640

12. A group of 11 professionals includes 3 lawyers, 1 accountant, 2 doctors, 4 teachers, and 1 salesperson. What is the probability to the nearest thousandth that a committee of 2 randomly chosen people will consist of 2 teachers?

    0.109

13. Grace and her brothers are drawing straws to see who has to vacuum the living room. There are 8 long straws and 2 short straws. They will draw until someone gets a short straw. What is the probability that 3 long straws and then a short straw will be drawn?

    $\dfrac{2}{15}$

14. There is a jar with 9 pennies, 10 nickels, 5 dimes, and 6 quarters. If two coins are chosen at random, what is the probability of choosing first a nickel and then a dime?

    $\dfrac{5}{87}$

15. The probability of winning 3rd prize in a contest is $\dfrac{1}{500}$. What are the odds of winning 3rd prize?

    1:499

16. A batch of 16 vials of medicine includes 5 vials of painkiller, 2 vials of arthritis medication, 3 vials of medication that inhibits blood clotting, and 6 vials of a medication for high blood pressure. What is the likelihood that a randomly selected vial contains a medication related to blood or blood circulation?

    $\dfrac{9}{16}$

17. 8 students each have an equal chance of being selected for student council president. What are the odds against being selected?

    7:1

Chapter 10 Chapter Test — Form A

Solve.

1. $4t + 5 = 13$ — $t = 2$
2. $3h - 2 = 1$ — $h = 1$
3. $-2.1a + 1.3 = 5.5$ — $a = -2$
4. $\dfrac{x}{2} + 1 = 4$ — $x = 6$
5. $5p + 2p - 3 = 11$ — $p = 2$
6. $-3b - 6 + 7b = 6$ — $b = 3$
7. $\dfrac{s}{3} - \dfrac{2}{3} = \dfrac{1}{3}$ — $s = 3$
8. $\dfrac{x}{2} + \dfrac{x}{6} = 2$ — $x = 3$
9. $x - 1 = x + 7$ — **no solution**
10. $5t + 6 = 2t - 3$ — $t = -3$
11. $4(g + 1) = 2g - 6$ — $g = -5$
12. $\dfrac{a}{3} + \dfrac{a}{3} - 1 = a$ — $a = -3$

Solve and graph.

13. $3c - 2 \geq 4$ — $c \geq 2$
14. $-1 < 5z + 4$ — $-1 < z$
15. $t - 3t - 1 > 7$ — $t < -4$

Solve each equation for the indicated variable.

16. Solve $a + b + 3 = 5$ for b.

    $b = 2 - a$

17. Solve $P = 2(l + w)$ for l.

    $l = \dfrac{P - 2w}{2}$

18. Solve $d = r \cdot t$ for t.

    $t = \dfrac{d}{r}$

19. Solve $y = x^2$ for x.

    $x = \sqrt{y}$

Solve each equation for y and graph.

20. $y + 2x = 4$ — $y = -2x + 4$

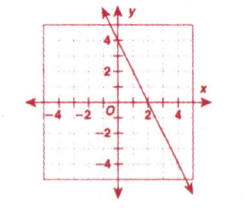

Chapter 10 Chapter Test — Form A, continued

21. $y - 3x = -2$ — $y = 3x - 2$

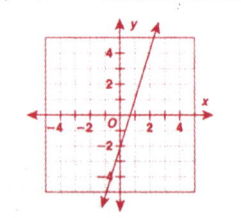

Determine if the ordered pair is a solution of the system of equations.

22. $(1, 3)$; $y = 2x - 3$, $y = x + 2$ — **no**
23. $(3, 6)$; $y = x + 3$, $y = 3x - 3$ — **yes**

Solve each system of equations.

24. $y = x - 5$, $y = 2x + 1$ — $(-6, -11)$
25. $y = x + 4$, $y = -x + 2$ — $(-1, 3)$
26. $y = 3x + 4$, $y = 2x + 5$ — $(1, 7)$
27. $x + y = 5$, $x - y = 7$ — $(6, -1)$

Write and solve an equation, inequality, or system of equations to answer the question.

28. Jeff bought 5 loaves of bread that were each the same price. He used coupons worth $2.60. Angie bought 3 loaves of bread, without using coupons. They paid the same total amount. What was the price of each loaf?

    $5x - 2.60 = 3x$; $x = \$1.30$

29. Maria has $50 to spend and would like to buy some music CDs. She also has a $10 gift certificate. What is the greatest amount that each CD can cost if Maria wants to buy 4 CDs? Assume all the CDs are the same price.

    $4x - 10 \leq 50$; $x \leq \$15$

30. Two numbers have a sum of 10, and a difference of 4. Find the two numbers.

    $x + y = 10$, $x - y = 4$; $x = 7$ and $y = 3$

Chapter Test — Form B

Solve.

1. $8t + 5 = 37$ — $t = 4$
2. $\frac{-b}{2} - 10 = 5$ — $b = -30$
3. $-5.9a - 5.5 = 12.2$ — $a = -3$
4. $\frac{g+3}{5} = 9$ — $g = 42$
5. $-5p - 3p + 4 = 36$ — $p = -4$
6. $\frac{4c}{5} - \frac{3c}{5} + \frac{1}{5} = \frac{-2}{5}$ — $c = -3$
7. $\frac{x}{2} + \frac{5x}{6} = 4$ — $x = 3$
8. $\frac{2y}{7} + \frac{4y}{7} - \frac{3}{14} = \frac{1}{14}$ — $y = \frac{1}{3}$
9. $-5x - 9 = -2x + 6$ — $x = -5$
10. $3t + 6 = 3t - 14$ — no solution
11. $5(g - 3) = 2g + 3$ — $g = 6$
12. $\frac{8a}{3} + \frac{4a}{3} - 1 = a + 5$ — $a = 2$

Solve and graph.

13. $4d - 5 > 11$ — $d > 4$ 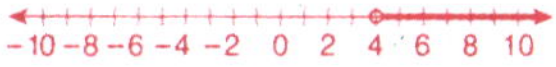
14. $-1 < \frac{k}{4} + \frac{1}{2} + \frac{k}{2}$ — $-2 < k$
15. $-5s - 9 - 3s \geq 15$ — $s \leq -3$

Solve each equation for the indicated variable.

16. Solve $P = a + b + c$ for b. — $b = P - a - c$
17. Solve $A = \frac{1}{2}rp$ for p. — $p = \frac{2A}{r}$
18. Solve $y = 2x + 3$ for x. — $x = \frac{y-3}{2}$
19. Solve $A = \pi r^2$ for r. — $r = \sqrt{\frac{A}{\pi}}$

Solve each equation for y and graph.

20. $2y - 3x = 4$ — $y = \frac{3x}{2} + 2$
21. $2y - 5x = 8$ — $y = \frac{5x}{2} + 4$

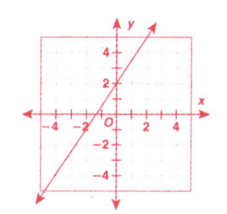

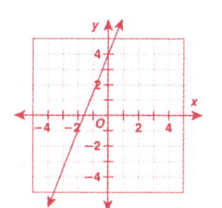

Determine if the ordered pair is a solution of the system of equations.

22. $(-1, 4); y = -5x - 1, y = 2x + 6$ — yes
23. $(2, 7); y = 3x + 1, y = -x - 8$ — no

Solve each system of equations.

24. $y = x + 2, y = 3x + 6$ — $(-2, 0)$
25. $y = -3x + 7, y = -2x - 5$ — $(12, -29)$
26. $x + y = 8, x - y = 12$ — $(10, -2)$
27. $2x - 3y = 5, 4x + y = 3$ — $(1, -1)$

Write and solve an equation, inequality, or system of equations to answer the question.

28. Olivia bought 5 greeting cards that were each the same price. She used coupons worth $2.50. Alexandra bought 4 cards and used a $1.00 coupon. Each girl paid the same amount. What was the price of each card?

    $5x - 2.5 = 4x - 1; x = \$1.50$

29. Brandon is planning a surprise party for Ashley. He has $25 to spend for drinks and a cake. He knows that the cake will cost $18 and drinks will cost $0.50 per person. What is the greatest number of people he can invite to the party?

    $0.50x + 18 \leq 25; x \leq 14$ people

30. Two numbers have a sum of 39. The larger number is 9 less than twice the smaller number. Find the two numbers.

    $x + y = 39, x = 2y - 9; x = 23$ and $y = 16$

Chapter Test — Form C

Solve.

1. $\frac{3}{4} - \frac{5t}{6} = \frac{5}{12}$ — $t = \frac{2}{5}$
2. $-6.8h + 15.3 = -39.1$ — $h = 8$
3. $\frac{8-a}{12} = -6$ — $a = -64$
4. $\frac{12+x}{5} + \frac{4}{5} = -20$ — $x = -116$
5. $12p + 8 - 7p - 3 = -20$ — $p = -5$
6. $\frac{3b}{4} - \frac{2}{3} + \frac{5b}{12} = -3$ — $b = -2$
7. $\frac{2y}{27} - \frac{2}{9} + \frac{y}{3} = -\frac{5}{17}$ — $y = -\frac{3}{17}$
8. $\frac{x+4}{5} - \frac{1}{3} + \frac{4x}{5} = \frac{1}{2}$ — $x = \frac{1}{30}$
9. $2(x - 5) + \frac{3x}{4} = 4x$ — $x = -8$
10. $-(t + 5) - 4 = -2(t - 6)$ — $t = 21$
11. $3(4g + 1) + 5 = 12g - 7$ — no solution
12. $-4\left(2a - \frac{1}{2}\right) + 5 = 2\left(4a - \frac{1}{2}\right)$ — $a = \frac{1}{2}$

Solve and graph.

13. $\frac{3n}{11} - 2 > 4$ — $n > 22$ 
14. $-\frac{x}{2} < \frac{x}{4} + \frac{3}{8}$ — $x > -\frac{1}{2}$ 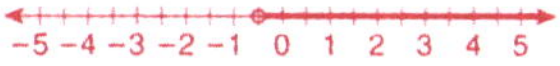
15. $2\left(\frac{v}{8} + \frac{1}{4}\right) \leq \frac{v+1}{6}$ — $v \leq -4$

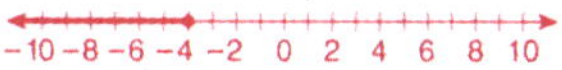

Solve each equation for the indicated variable.

16. Solve $C = \frac{5}{9}(F - 32)$ for F. — $F = \frac{9}{5}C + 32$
17. Solve $A = \frac{1}{2}h(b_1 + b_2)$ for b_1. — $b_1 = \frac{2A}{h} - b_2$
18. Solve $S = 6rs + 6sh$ for r. — $r = \frac{S - 6sh}{6s}$
19. Solve $V = \frac{1}{3}\pi r^2 h$ for r. — $r = \sqrt{\frac{3V}{\pi h}}$

Solve each equation for y and graph.

20. $5y + 2x = 15$ — $y = -\frac{2x}{5} + 3$ 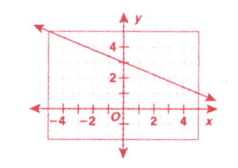
21. $y - 4 = -2(x + 3)$ — $y = -2x - 2$

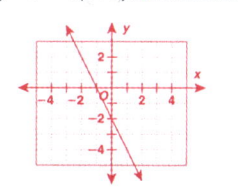

Determine if the ordered pair is a solution of the system of equations.

22. $(-1, 1); 2y - 3x = 5, x + 5y = 4$ — yes
23. $(3, 5); 8x - 4y = 4, 7y - 3x = -26$ — no

Solve each system of equations.

24. $y = 2x + 5, x = -2y - 5$ — $(-3, -1)$
25. $2y = -10x + 40, -6x + 4y = -24$ — $(4, 0)$
26. $2x - 5y = -19, -3x + 4y = 18$ — $(-2, 3)$
27. $8x + 9y = -18, 7x - y = 2$ — $(0, -2)$

Write and solve an equation, inequality, or system of equations to answer the question.

28. Two sisters left at the same time to travel home. Susanna drove at 65 miles per hour and started 9 miles farther from home than Rachel. Rachel drove at 55 miles per hour. They reached home at the same time. How far was each girl from home?

    $\frac{r}{55} = \frac{r+9}{65}$; $r = 49.5$ miles, $s = 58.5$ miles

29. Gerri received grades of 83, 94, 76, and 89 on four tests. What is the lowest average score she can afford to get on her next two tests to end up with an overall average of 90?

    $\frac{83 + 94 + 76 + 89 + 2x}{6} \geq 90$; $x = 99$

30. The difference of two numbers is 18. Their sum is 13 more than 3 times the smaller number. Find the two numbers.

    $x - y = 18, x + y = 3y + 13$; $x = 23$ and $y = 5$

339 Holt Pre-Algebra

Chapter 11 Chapter Test — Form A

Graph each equation and tell whether it is linear.

1. $y = x^2$ **Not linear**

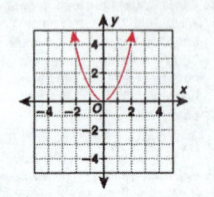

2. $y = x - 5$ **Linear**

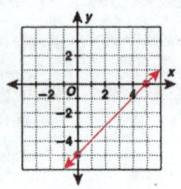

3. Find the slope of the line that passes through the points (5, 4) and (3, 1).
   $\dfrac{3}{2}$

Tell whether the lines passing through the given points are parallel or perpendicular.

4. line 1: (3, 0) and (2, 1); line 2: (4, 5) and (3, 6)
   parallel

Find the x-intercept and y-intercept of each line. Use the intercepts to graph the equation.

5. $y = -x + 5$ **5; 5**

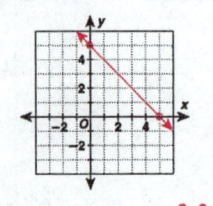

6. $x - y = -3$ **−3; 3**

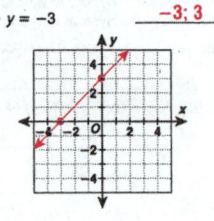

7. Write $x + y = 9$ in slope-intercept form and then find the slope and y-intercept.
   $y = -x + 9;\ -1;\ 9$

Chapter 11 Chapter Test — Form A, continued

8. Write the equation of the line that passes through the points (1, 9) and (2, 6) in slope-intercept form.
   $y = -3x + 12$

9. Identify a point the line for $y - 4 = 5(x - 3)$ passes through and identify the slope of the line.
   Possible answer: (3, 4); 5

10. Write the point-slope form of the equation for a line with a slope −2 that passes through the point (−1, 1).
    $y - 1 = -2(x + 1)$

Find each equation of direct variation, given that y varies directly with x.

11. y is 10 when x is 2
    $y = 5x$

12. y is −3 when x is $\tfrac{1}{2}$
    $y = -6x$

Graph each inequality.

13. $y > x - 1$

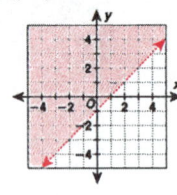

14. $y \leq x + 2$

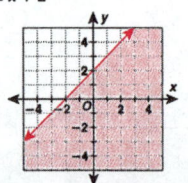

Use a line of best fit to answer the question.

15. Find a line of best fit for the data. Use the equation of the line to predict the y-value for x = 8.

x	1	3	4	5	7
y	5	8	11	14	17

Possible answer: $y = 2x + 3$; y = 19

16. A mechanic charges $20 plus $40 for each hour he works on a car. The equation $y = 40x + 20$ represents the total amount he charges for working x hours. If he works 5 hours on a car, how much should he charge?
    $220

Chapter 11 Chapter Test — Form B

Graph each equation and tell whether it is linear.

1. $y = 2x + 1$ **Linear**

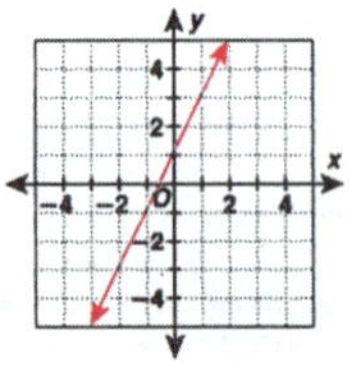

2. $y = x^2 - 5$ **Not linear**

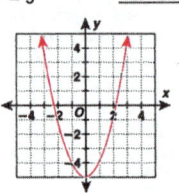

3. Find the slope of the line that passes through the points (−3, 4) and (5, −1).
   $-\dfrac{5}{8}$

Tell whether the lines passing through the given points are parallel or perpendicular.

4. line 1: (7, −2) and (3, 1); line 2: (−8, 2) and (−5, 6)
   perpendicular

Find the x-intercept and y-intercept of each line. Use the intercepts to graph the equation.

5. $3y = 2x - 3$ $\dfrac{3}{2};\ -1$

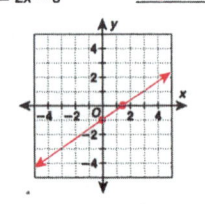

6. $x - 2y = -4$ **−4; 2**

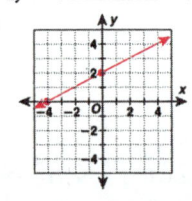

7. Write $5x - y = 8$ in slope-intercept form and then find the slope and y-intercept.
   $y = 5x - 8;\ 5;\ -8$

Chapter 11 Chapter Test — Form B, continued

8. Write the equation of the line that passes through the points (−2, 4) and (−4, 10) in slope-intercept form.
   $y = -3x - 2$

9. Identify a point the line for $y + 5 = -\tfrac{2}{3}(x - 3)$ passes through and identify the slope of the line.
   Possible answer: (3, −5); $-\tfrac{2}{3}$

10. Write the point-slope form of the equation for a line with a slope of −5 that passes through the point (−4, 3).
    $y - 3 = -5(x + 4)$

Find each equation of direct variation, given that y varies directly with x.

11. y is −3 when x is 12
    $y = -\tfrac{1}{4}x$

12. y is 6 when x is 9
    $y = \tfrac{2}{3}x$

Graph each inequality.

13. $y > 2x - 1$

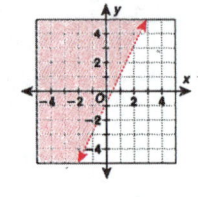

14. $y \leq \tfrac{1}{2}x + 2$

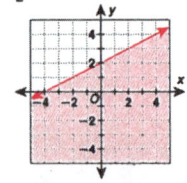

Use a line of best fit to answer the question.

15. The temperature one day in January dropped during a cold front. The data are shown below. Find a line of best fit for the data. Use the equation of the line to predict the temperature after 6 hours.

Hours	0	1	2	3	4
Temperature	40	35	32	29	27

Possible answer: $y = -3.2x + 39$; about 20 degrees

16. A hiker walking down a mountain decreases his altitude by 20 feet each minute he hikes. The equation $y = -20x + 8000$ represents his altitude after hiking x minutes. If he hikes 30 minutes, what is his altitude?
    7400 ft

Chapter 11 Chapter Test Form C

Graph each equation and tell whether it is linear.

1. $y = -\frac{1}{3}x - 2$ **Linear**

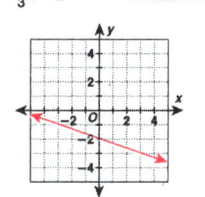

2. $y = 2x^2 - 3$ **Not linear**

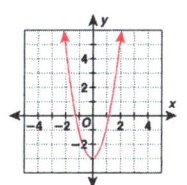

3. Find the slope of the line that passes through the points $(-7, -5)$ and $(4, -3)$.
   $\frac{2}{11}$

Tell whether the lines passing through the given points are parallel or perpendicular.

4. line 1: $(8, -4)$ and $(-2, 3.5)$;
   line 2: $(-4, -3)$ and $(6, -10.5)$
   parallel

Find the x-intercept and y-intercept of each line. Use the intercepts to graph the equation.

5. $-\frac{1}{3}y = -\frac{1}{2}x + 1$ **2; −3**

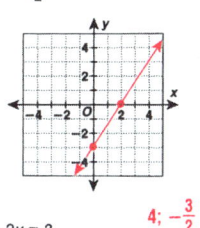

6. $\frac{3}{4}x - 2y = 3$ **4; $-\frac{3}{2}$**

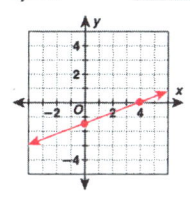

7. Write $5(4x - 2y) = 15$ in slope-intercept form and then find the slope and y-intercept.
   $y = 2x - \frac{3}{2}$; 2; $-\frac{3}{2}$

Chapter 11 Chapter Test Form C, continued

8. Write the equation of the line that passes through the points $(-10, 6)$ and $(15, 1)$ in slope-intercept form.
   $y = -\frac{1}{5}x + 4$

9. Identify a point the line for $y + 7 = -\frac{3}{5}(x - 4)$ passes through and identify the slope of the line.
   Possible answer: $(4, -7)$; $-\frac{3}{5}$

10. Write the point-slope form of the equation for a line with a slope of $-\frac{1}{7}$ that passes through the point $\left(-21, \frac{1}{2}\right)$.
    $y - \frac{1}{2} = -\frac{1}{7}(x + 21)$

Find each equation of direct variation, given that y varies directly with x.

11. y is -63 when x is 81
    $y = -\frac{7}{9}x$

12. y is $\frac{3}{5}$ when x is $\frac{1}{2}$
    $y = \frac{6}{5}x$

Graph each inequality.

13. $\frac{1}{4}y > -\frac{1}{6}x + \frac{1}{2}$

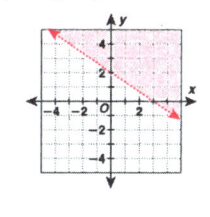

14. $-3x + 4y \leq 12$

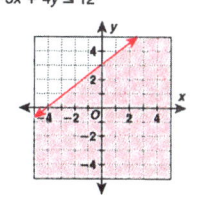

Use a line of best fit to answer the question.

15. The value of a car decreases as it is driven more miles. The data are shown below. Find a line of best fit for the data. Use the equation to predict the value of the car at 24,000 miles.

Miles	Value of Car
10,000	$14,500
12,000	$14,150
14,000	$13,815
16,000	$13,650
18,000	$13,500

Possible answer: $y = -0.125x + 15,673$; about $12,673

16. In order to rent an apartment, Vik makes an initial payment of $1535, and each month he must pay $750. The equation $y = 750x + 1535$ represents the amount of money Vik has paid after x months. After 2 years, how much rent money has Vik paid?
    $19,535

Chapter 12 Chapter Test Form A

Determine if each sequence could be arithmetic. If so, give the common difference and find the specified term.

1. $1, 3, 5, 7, 9, \ldots$; 8th term.
   yes; 2; 15

2. $1, 3, 9, 27, 81, \ldots$; 7th term.
   no

Determine if each sequence could be geometric. If so, give the common ratio and find the specified term.

3. $64, 32, 16, 8, \ldots$; 6th term.
   yes; $\frac{1}{2}$; 2

4. $1, -1, 1, -1, \ldots$; 7th term.
   yes; -1; 1

Use the first and second differences to find the next three terms in the sequence.

5. $1, 2, 4, 7, 11, \ldots$
   16, 22, 29

Find the first five terms of the sequence specified by the rule.

6. $a_n = \frac{2}{n}$
   $2, 1, \frac{2}{3}, \frac{1}{2}, \frac{2}{5}$

Determine if each relationship represents a function.

7.
x	1	2	3	4
y	6	−2	6	8

yes

8.

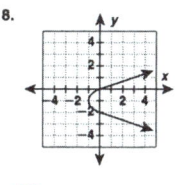

no

Write the rule for each linear function.

9.
x	−1	0	1	2
y	8	10	12	14

$y = 2x + 10$

10.

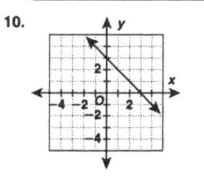

$y = -x + 3$

Chapter 12 Chapter Test Form A, continued

Complete the table for each exponential function and use it to graph the function.

11. $f(x) = 3\left(\frac{1}{2}\right)^x$

x	−1	0	1	2
y	6	3	$\frac{3}{2}$	$\frac{3}{4}$

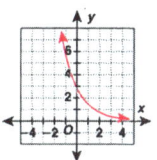

Complete the table for each quadratic function and use it to make a graph.

12. $f(x) = -x^2 + 2x$

x	−2	−1	0	1	2	3
y	−8	−3	0	1	0	−3

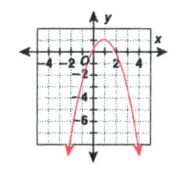

13. $f(x) = (x - 2)(x + 2)$

x	−2	−1	0	1	2
y	0	−3	−4	−3	0

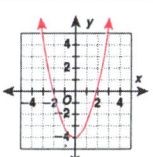

Tell whether the relationship is an inverse variation.

14. The table below shows the distance driven in a given time.

Time (hours)	2	4	5	7
Miles Driven	120	240	300	420

no

Graph the inverse variation.

15. $f(x) = \frac{9}{x}$

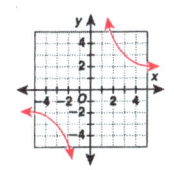

16. The height of a ball thrown horizontally from the top of a 75-meter tower is given by the function $f(t) = -5t^2 + 75$. What is the height after 2 seconds?
    55 m

Chapter 12 Chapter Test Form B

Determine if each sequence could be arithmetic. If so, give the common difference and find the specified term.

1. $2, \frac{7}{3}, \frac{8}{3}, 3, \frac{10}{3}, \ldots$; 11th term.

   yes; $\frac{1}{3}$; $\frac{16}{3}$

2. $-5, 6, 17, 28, 39, \ldots$; 20th term.

   yes; 11; 204

Determine if each sequence could be geometric. If so, give the common ratio and find the specified term.

3. $625, 125, 25, 10, \ldots$; 9th term.

   no

4. $-\frac{1}{2}, 2, -8, 32, \ldots$; 7th term.

   yes; -4; -2048

Use the first and second differences to find the next three terms in the sequence.

5. $6, 11, 15, 18, 20, \ldots$

   21, 21, 20

Find the first five terms of the sequence specified by the rule.

6. $a_n = \frac{n^2 + 1}{n}$

   $2, \frac{5}{2}, \frac{10}{3}, \frac{17}{4}, \frac{26}{5}$

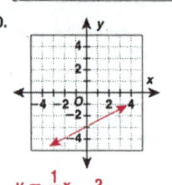

Determine if each relationship represents a function.

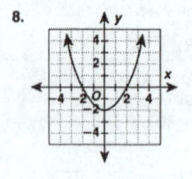

7.
x	0	4	0	9
y	6	9	5	-1

no

8.

yes

Write the rule for each linear function.

9.
x	-2	1	4	6
y	15	3	-9	-17

$y = -4x + 7$

10.

$y = \frac{1}{2}x - 3$

Chapter 12 Chapter Test Form B, continued

Create a table for each function and use it to graph the function.
Possible answer:

11. $f(x) = -2\left(\frac{1}{3}\right)^x$

x	-1	0	1	2
y	-6	-2	$-\frac{2}{3}$	$-\frac{2}{9}$

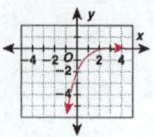

12. $f(x) = x^2 - 2x + 3$

x	-1	0	1	2	3
y	6	3	2	3	6

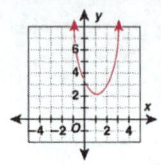

13. $f(x) = \frac{1}{2}(x - 2)(x + 1)$

x	-3	-2	-1	0	1	2	3
y	5	2	0	-1	-1	0	2

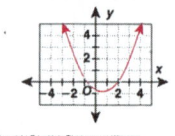

Tell whether the relationship is an inverse variation.

14. The table below shows the number of items purchased as a function of the cost per item.

Cost per Item ($)	7	10	12	15	20
Items Purchased	60	42	35	28	21

yes

Graph the inverse variation.

15. $f(x) = -\frac{4}{x}$.

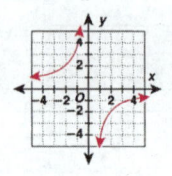

16. Ms. Suarez wants to earn $150 in interest over a 3-year period from a savings account. The principal she must deposit varies inversely with the interest rate of the account. If the interest rate is 0.08, she must deposit $625. If the interest rate is 0.064, how much must she deposit?

$781.25

Chapter 12 Chapter Test Form C

Determine if each sequence could be arithmetic. If so, give the common difference and find the specified term.

1. $-\frac{2}{7}, \frac{6}{7}, 2, \frac{20}{7}, \frac{27}{7}, \ldots$; 21th term.

   no

2. $-\frac{11}{3}, -\frac{2}{3}, \frac{7}{3}, \frac{16}{3}, \frac{25}{3}, \ldots$; 33rd term.

   yes; 3; $\frac{277}{3}$

Determine if each sequence could be geometric. If so, give the common ratio and find the specified term.

3. $1, \frac{1}{2}, \frac{1}{3}, \frac{1}{4}, \ldots$; 11th term.

   no

4. $2, 16, 128, 512, \ldots$; 9th term.

   no

Use the first and second differences to find the next three terms in the sequence.

5. $-\frac{18}{5}, -\frac{11}{5}, -\frac{3}{5}, \frac{6}{5}, \frac{16}{5}, \ldots$

   $\frac{27}{5}, \frac{39}{5}, \frac{52}{5}$

Find the first five terms of the sequence specified by the rule.

6. $a_n = \frac{n(n+1)(n+2)}{6}$

   1, 4, 10, 20, 35

Determine if each relationship represents a function.

7.
x	-2	6	11	17
y	11	17	-2	6

yes

8.

yes

Write the rule for each linear function.

9.
x	6	8	2	-4
y	-19	-29	1	31

$y = -5x + 11$

10.

$y = -2x - 3$

Chapter 12 Chapter Test Form C, continued

Create a table for each function and use it to graph the function.
Possible answer:

11. $f(x) = -\frac{1}{2}\left(\frac{1}{3}\right)^x$

x	-2	-1	0	1	2
y	$-\frac{9}{2}$	$-\frac{3}{2}$	$-\frac{1}{2}$	$-\frac{1}{6}$	$-\frac{1}{18}$

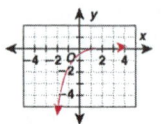

12. $f(x) = \frac{1}{2}x^2 + x - 2$

x	-3	-2	-1	0	1	2
y	$-\frac{1}{2}$	-2	$-\frac{5}{2}$	-2	$-\frac{1}{2}$	2

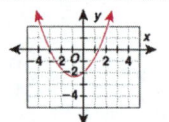

13. $f(x) = \frac{1}{2}(x - 3)(x + 2)$

x	-3	-2	-1	0	1	2	3
y	3	0	-2	-3	-3	-2	0

Tell whether the relationship is an inverse variation.

14. The table below shows the speed of a car in relation to time needed to cover a given distance.

Time (min)	10	15	24	30	50
Speed (kph)	60	40	25	20	12

yes

Graph the inverse variation.

15. $f(x) = \frac{-8}{(3x)}$.

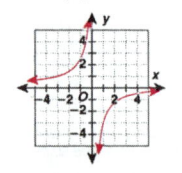

16. The resistance of a 30-m piece of wire varies inversely with the square of the diameter. If the diameter of the wire is 2 mm, it has a resistance of 0.2 ohms. What is the resistance of a wire with a diameter of 0.4 mm?

5 ohms

Chapter 13 Chapter Test Form A

Classify each expression as monomial, binomial, trinomial, or not a polynomial.

1. $5x^2$
   monomial
2. $9y^2$
   monomial
3. $3a^2 + b^2$
   binomial
4. $7x^2 + x^3 - 8x$
   trinomial

Find the degree of each polynomial.

5. $3x^2 + 4x$
   2
6. $4n^3 + 2n^8$
   8
7. $9a^5 - 3a^3 - 5a^2$
   5
8. $k^3 + k^4 + k^7 + k$
   7

Identify the like terms in each polynomial.

9. $4n^2 + 3 + 5n^2$
   $4n^2$ and $5n^2$
10. $6x^3 - 5x^2 + 5x^3$
    $6x^3$ and $5x^3$

Simplify.

11. $7y^2 + 3y - 4y^2$
    $3y^2 + 3y$
12. $4n^2 + 8 + 8n^2$
    $12n^2 + 8$
13. $5t^2 + t^2 - 3s^2$
    $6t^2 - 3s^2$

Add.

14. $(6x + 2) + (5x + 2)$
    $11x + 4$
15. $(8n + 8) + (6n - 5)$
    $14n + 3$
16. $(4x + 2x^2) + (4x^2 + 2x)$
    $6x + 6x^2$
17. $(5n^2 + 3n) + (2n^2 - 2n)$
    $7n^2 + n$
18. $(8a^3 - 6b) + (9b - 4a^3)$
    $4a^3 + 3b$
19. $(2x - 2x^3) + (8x^3 + 4x)$
    $6x + 6x^3$
20. $(7xy + 4x) + (3xy - 2x)$
    $10xy + 2x$

Find the opposite of each polynomial.

23. $5n^3$
    $-5n^3$
24. $6x^4y^4$
    $-6x^4y^4$

Subtract.

27. $5x - (2x - 4x^2)$
    $3x + 4x^2$
28. $(6n + 4n^2) - (2n^2 + 4n)$
    $2n + 2n^2$
29. $(4x^2 + 2y^2) - (3x^2 - 9y^2)$
    $x^2 + 11y^2$
30. $(8y^2 - 4) - (-5y^2 - 8)$
    $13y^2 + 4$
31. $(7n^2 - 5n) - (-7n + 5n^2)$
    $2n^2 + 2n$
32. $(9x + 7 + 10x^4) - (2x + 3 - 5x^4)$
    $7x + 4 + 15x^4$
33. $(4n - 2n^2 + 6a^2) - (-8n^2 - 6a^2)$
    $4n + 6n^2 + 12a^2$

Multiply.

34. $(3x^2)(5x^2)$
    $15x^4$
35. $(2n^2m^3)(3n^3m^2)$
    $6n^5m^5$
36. $5x(4x^2 + 8)$
    $20x^3 + 40x$
37. $9x^2(2x - 6x^3)$
    $18x^3 - 54x^5$
38. $3ab^2(2a^3b^3 + 2ab + 5b^2)$
    $6a^4b^5 + 6a^2b^3 + 15ab^4$
39. $(n + 3)(n + 4)$
    $n^2 + 7n + 12$
40. $(x + 2)(y + 7)$
    $xy + 7x + 2y + 14$
41. $(n + 6)(n - 6)$
    $n^2 - 36$
42. $(y + 5)(y - 4)$
    $y^2 + y - 20$
43. $(2a + b)(a - 4b)$
    $2a^2 - 7ab - 4b^2$
44. A rectangle has a length of $2w + 3$ and a width of $w - 5$. Write and simplify an expression for the area of the rectangle.
    $2w^2 - 7w - 15$

Chapter 13 Chapter Test Form B

Classify each expression as monomial, binomial, trinomial, or not a polynomial.

1. $6a^2 - 4a$
   binomial
2. $\left(\frac{1}{7}\right)x^5y^3$
   monomial
3. $8p^2 + p^{2.2}$
   not a polynomial
4. $4m^2n^2 + 5n^3 - 4n$
   trinomial

Find the degree of each polynomial.

5. $8x^5 + 6x^4$
   5
6. $-6a^3 + 8a - 2a^5$
   5
7. $5 + 3m^4 - 2m^7 + m^8$
   8
8. $b^3 - b^9 - 3 - 3b^2$
   9

Identify the like terms in each polynomial.

9. $t^2 + 9s^3 - 5t^4 + 8 + 4s^3$
   $9s^3$ and $4s^3$
10. $4a^3b^2 - 5a^2b^2 + a^3b^2$
    $4a^3b^2$ and a^3b^2

Simplify.

11. $v^2 + 6v - 8v^2 + 4v^2$
    $-3v^2 + 6v$
12. $-4x^2 + 9 + 5x^2$
    $x^2 + 9$
13. $5m^2n^2 + n^3 - 3m^2n^2 + 7n^3$
    $2m^2n^2 + 8n^3$

Add.

14. $(13x - 4) + (5x + 5)$
    $18x + 1$
15. $(6a + 9) + (6a^2 - 5a - 3)$
    $a + 6a^2 + 6$
16. $(3x + 2x^3 - 7) + (4x^3 - 2x + 7)$
    $x + 6x^3$
17. $(6z^4 + 7y^3 + 4) + (2y^3 - z^4 - 3)$
    $5z^4 + 9y^3 + 1$
18. $(7r^3 + 7s + 7r^2) + (4s - 4r^3)$
    $3r^3 + 11s + 7r^2$
19. $(5x - 2x^3 - 9) + (8x^3 - 3x - 5)$
    $2x + 6x^3 - 14$
20. $(7ab + 6b^2 - 5a) + (ab - 3a)$
    $8ab + 6b^2 - 8a$

Find the opposite of each polynomial.

23. $9a^5b^5c^3$
    $-9a^5b^5c^3$
24. $9y^3 - 6x$
    $-9y^3 + 6x$

Subtract.

27. $8b - (8a^2 + 7b - 5)$
    $-8a^2 + b + 5$
28. $(4x + 5x^2) - (3x^2 + x)$
    $3x + 2x^2$
29. $(x^3 - 5x + 2x^2) - (3x - 3x^3 + 9)$
    $4x^3 - 8x + 2x^2 - 9$
30. $(6y^2z^6 + 3yz - 4z^2) - (-5yz - 9)$
    $6y^2z^6 + 8yz - 4z^2 + 9$
31. $(n^4 + 9n^3m^7 - 5n) - (-6n - n^3m^7)$
    $n^4 + 10n^3m^7 + n$
32. $(6x - 6 + x^3) - (4 - 2x - 5x^3)$
    $8x - 10 + 6x^3$
33. $(-5s + 7s^2 + 4r^2 - 4) - (-3s^2 - 6s)$
    $s + 10s^2 + 4r^2 - 4$

Multiply.

34. $(5x^2)(2y^7)(4x^3)(2y^3)$
    $80x^5y^{10}$
35. $(9a^2b^5)(6ab^4)$
    $54a^3b^9$
36. $6n(3n^2 - 7)$
    $18n^3 - 42n$
37. $-4ab(5a^2 - 7b^3)$
    $-20a^3b + 28ab^4$
38. $6st^2(7s^3t^3 + st - 5t^2)$
    $42s^4t^5 + 6s^2t^3 - 30st^4$
39. $(a - 3)(y + 5)$
    $ay + 5a - 3y - 15$
40. $(x + 4)(x + 5)$
    $x^2 + 9x + 20$
41. $(b + 6)^2$
    $b^2 + 12b + 36$
42. $(y + 5)(y - 4)$
    $y^2 + y - 20$
43. $(5y - 7)^2$
    $25y^2 - 70y + 49$
44. A square has a side length of $3e + 4$. Write and simplify an expression for the area of the square.
    $9e^2 + 24e + 16$

Chapter 13 Chapter Test — Form C

Classify each expression as monomial, binomial, trinomial, or not a polynomial.

1. $9x^2 - \dfrac{7}{x^2}$
   not a polynomial

2. 88
   monomial

3. $5n^{11} + m^2 - nm$
   trinomial

4. $6a^2b^2 + \left(\dfrac{1}{2}b\right)n^{3.1} + 8n$
   not a polynomial

Find the degree of each polynomial.

5. $-2x + 8x$
   1

6. $-10n^8 + 10n^9 + n^8$
   8

7. $4 + 5y^3 - 8y^2 - 4y^9$
   9

8. $35j^4 + 28j^{12} - 15j^7$
   12

Identify the like terms in each polynomial.

9. $11x^2 + 11y^2 - t^4 + 9y^2 + 4x^{22}$
   $11y^2$ and $9y^2$

10. $g^2h^2 - 8gh^2 + g^2h + 8g^2h^2$
    g^2h^2 and $8g^2h^2$

Simplify.

11. $3(5x^2 + 3y) - 4x^2 + 4y^2$
    $11x^2 + 9y + 4y^2$

12. $5n^2 - 10m^3 - 4n^2 + m^3 + n^2$
    $2n^2 - 9m^3$

13. $3(3a^2b^2 + b^3) - 3a^2b^2 + 2(7a^3 + 4b^3)$
    $6a^2b^2 + 11b^3 + 14a^3$

Add.

14. $(3x^3 - 4x) + (-2x + 5 - x^3)$
    $2x^3 - 6x + 5$

15. $(9a + 5b^2) + (3a^2 - 2a) + (6b^2 - 5a^2)$
    $7a + 11b^2 - 2a^2$

16. $(4x + x^3) + (4x^3 - x + 7) + (3x + 9x^3)$
    $6x + 14x^3 + 7$

17. $(2a^3 + 6b^3 + 7) + (5b^3 - a^3b^4 - 3a^3)$
    $-a^3 + 11b^3 - a^3b^4 + 7$

18. $(6m^3n^3 + 2m + 12m^3n^3) + (-8m - n^3)$
    $18m^3n^3 - 6m - n^3$

19. $(xy^3 - 2x^3 - 9xy^3) + (8xy^3 - 2x^3 - 2)$
    $-4x^3 - 2$

20. $(7cd + 6d^2) + (cd - 3c) + (-8d - d^2)$
    $8cd + 5d^2 - 3c - 8d$

Chapter 13 Chapter Test — Form C, continued

Find the opposite of each polynomial.

23. $12x^8yz^3 - 7xy + 23$
    $-12x^8yz^3 + 7xy - 23$

24. $-11a^3 - 6ab^3 + 12a^4b^3$
    $11a^3 + 6ab^3 - 12a^4b^3$

Subtract.

27. $(5b + 2a^2) - (9a^2 + 12b - 8)$
    $-7b - 7a^2 + 8$

28. $(4n^2 + 10n^5m^5 - 14n) - (3n - 8n^5m^5)$
    $4n^2 + 18n^5m^5 - 17n$

29. $(6r^4s^4 + 11r^2 + 9r^2s^2) - (14r^2s^2 - 5r^2)$
    $6r^4s^4 + 16r^2 - 5r^2s^2$

30. $(12y^2z - 9yz - 4z^5) - (-5yz - 9y^2z)$
    $21y^2z - 4yz - 4z^5$

31. $(9s - 13s^2 - r^2 + 4) - (5r^2 + s - 18s^2)$
    $8s + 5s^2 - 6r^2 + 4$

32. $(13rs^8 + 9s^4 - 3r^6s^2) - (-4r^6s^2 + 9s^4)$
    $13rs^8 + r^6s^2$

33. $(6yz^5 + 16yz - 4z^2) - (5yz^5 - 9z^2 + z)$
    $yz^5 + 16yz + 5z^2 - z$

Multiply.

34. $8x^2(5x^2y^2 + 9x^3y^3)$
    $40x^4y^2 + 72x^5y^3$

35. $(-rs^5)(7r^5s)$
    $-7r^6s^6$

36. $6x^5y(6x^2y^3 - 7xy^4 + 12x^3)$
    $36x^8y^4 - 42x^7y^5 + 72x^9y$

37. $13fg(-9g^2 - 5f^3 + f^5g^2)$
    $-117fg^3 - 65f^4g + 13f^6g^3$

38. $10x^2y^2(8xy - 5xy^3 + 4x^3y)$
    $80x^3y^3 - 50x^3y^5 + 40x^5y^3$

39. $(9x + 2)(x + 5)$
    $9x^2 + 47x + 10$

40. $(2x + 4)(4x - 6)$
    $8x^2 + 4x - 24$

41. $(5a - 4)^2$
    $25a^2 - 40a + 16$

42. $(6y + x)(y - 8x)$
    $6y^2 - 47xy - 8x^2$

43. $(8r - 6s)(4r - 10s)$
    $32r^2 - 104rs + 60s^2$

44. A parallelogram has a base length of $5p - 1$ and a height of $2p + 3$. Write and simplify an expression for the area of the parallelogram.
    $10p^2 + 13p - 3$

Chapter 14 Chapter Test — Form A

Use the correct symbol to make each statement true. Choose $\in$ or $\notin$.

1. 13 ☐ {prime numbers}
   $\in$

2. pentagon ☐ {parallelograms}
   $\notin$

Tell whether the first set is a subset of the second set. Use the correct symbol.

3. N = {positive integers}; R = {whole numbers}
   $N \subset R$

4. E = {circles}; G = {plane figures}
   $E \subset G$

Find the intersection of the sets.

5. A = {0, 1, 2, 3}; B = {2, 3, 4, 5, 6}
   {2, 3}

6. R = {−2, −1, 0}; T = {0, 1, 2, 3, 4}
   {0}

For 16–20, find the union of the sets.

7. M = {0, 1, 2, 3}; N = {4, 5, 6}
   {0, 1, 2, 3, 4, 5, 6}

8. G = {multiples of 4}; H = {factors of 8}
   {multiples of 4, 1, 2}

For 9–11, use the Venn diagram.

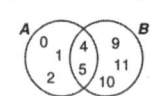

Set A: {0, 2, 4, 5} Set B: {4, 5, 9, 10, 11}

9. Identify the intersection of sets A and B.
   $A \cap B = \{4, 5\}$

10. Identify the union of sets A and B.
    $A \cup B = \{0, 1, 2, 4, 5, 9, 10, 11\}$

11. True or False: $A \subset B$.
    False

12. Make a truth table for the conjunction P and Q.
    P: A number is a multiple of 5.
    Q: A number is a multiple of 2.

P	Q	P and Q
T	T	T
T	F	F
F	T	F
F	F	F

13. Complete the truth table.
    P: Today is Monday.
    Q: There is school today.

P	Q	P and Q	P or Q
T	T	T	T
T	F	**F**	T
F	T	**F**	**T**
F	F	F	**F**

Chapter 14 Chapter Test — Form A, continued

For 14–15, name the hypothesis and the conclusion in each statement.

14. If you sleep, you will feel rested.
    H: You sleep.
    C: You feel rested.

15. In order to feel well, eat a good breakfast.
    H: You eat a good breakfast.
    C: You feel well.

16. Make a conclusion, if possible, from the deductive argument. If a rectangle has 4 equal sides, it is a square. Rectangle $ABCD$ has 4 equal sides.
    Rectangle $ABCD$ is a square.

For 17–19, use the graph.

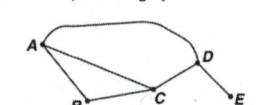

17. Find the degree of each vertex.
    A: 3, B: 2, C: 3,
    D: 3, E: 1

18. Is the graph connected? How do you know?
    The graph is connected because there is a path from each vertex to another vertex.

19. Determine whether the graph can be traversed through an Euler circuit. If your answer is yes, describe an Euler circuit in the graph.
    No; Not all vertices have even degrees.

For 20–21, use the graph.

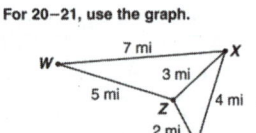

(W to X: 7 mi; W to Z: 5 mi; X to Z: 3 mi; X to Y: 4 mi; Z to Y: 2 mi)

20. A Hamiltonian circuit must pass through every vertex. How is it different from an Euler circuit?
    It is not necessary to traverse every edge.

21. What is the length of the shortest Hamiltonian circuit that begins and ends at W?
    18 mi

Chapter 14 — Chapter Test, Form B

Insert the correct symbol to make each statement true. Choose $\in$ or $\notin$.

1. $x^2 - 2x = 0$ ☐ {quadratic equations}
   $\in$

2. Delaware ☐ {United Nations}
   $\notin$

Determine whether the first set is a subset of the second set. Use the correct symbol.

3. N = {natural numbers}; R = {real numbers}
   $N \subset R$

4. P = {polygons}; T = {triangles}
   $P \not\subset T$

Find the intersection of the sets.

5. $A = \{2, 4, 6, 8\}$; $B = \{6, 8, 10, 12\}$
   $\{6, 8\}$

6. J = {negative integers}; $K = \{-4, -3, -2, -1, 0, 1\}$
   $\{-4, -3, -2, -1\}$

For 16–20, find the union of the sets.

7. G = {multiples of 2}; H = {multiples of 4}
   {multiples of 2}

8. X = {quadrilaterals}; Y = {polygons}
   {polygons}

For 9–11, use the Venn diagram.

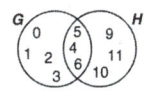

9. Identify the intersection of sets G and H.
   $G \cap H = \{4, 5, 6\}$

10. Identify the union of sets G and H.
    $G \cup H = \{0, 1, 2, 3, 4, 5, 6, 9, 10, 11\}$

11. Identify a subset in the diagram.
    no subsets

12. Make a truth table for the conjunction P and Q.
    P: A number is an even number.
    Q: A number is a factor of 90.

P	Q	P and Q
T	T	T
T	F	F
F	T	F
F	F	F

13. Complete the truth table.
    P: Today is Saturday.
    Q: There is practice today.

P	Q	P and Q	P or Q
T	T	T	T
T	F	F	T
F	T	F	T
F	F	F	F

For 14–15, identify the hypothesis and the conclusion in each conditional.

14. If you study, the test will be easy.
    H: You study.
    C: The test is easy.

15. To be healthy, eat well.
    H: You eat well.
    C: You will be healthy.

16. Make a conclusion, if possible, from the deductive argument.
    If a quadrilateral has 4 right angles, it is a rectangle. Quadrilateral $ABCD$ has angle measures of 90°, 90°, 120°, and 60°.
    No conclusion possible.

For 17–19, use the graph.

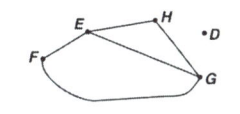

17. Find the degree of each vertex.
    D: 0, E: 3, F: 2,
    G: 3, H: 2

18. Is the graph connected? Explain.
    No; There is no path to vertex D.

19. Determine whether the graph can be traversed through an Euler circuit. If your answer is yes, describe an Euler circuit in the graph.
    No; Not all vertices have even degrees.

For 20–21, use the graph.

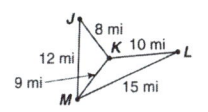

20. Describe how a Hamiltonian circuit differs from an Euler circuit.
    Both pass through every vertex, but an Euler circuit also must pass over every edge.

21. Determine the length of the shortest Hamiltonian circuit beginning at L.
    45 mi

Chapter 14 — Chapter Test, Form C

Insert the correct symbol to make each statement true. Choose $\in$ or $\notin$.

1. $y = x^3 - 3x$ ☐ {quadratic functions}
   $\notin$

2. Ohio River ☐ {rivers in the United States}
   $\in$

Determine whether the first set is a subset of the second set. Use the correct symbol.

3. P = {perimeter}; A = {area}
   $P \not\subset A$

4. $V = \{x^2, 3x^3, 2x\}$; M = {monomials}
   $V \subset M$

Find the intersection of the sets.

5. $A = \{x \mid x < 9\}$; $B = \{x \mid x > 12\}$
   { } (empty set)

6. C = {first 5 prime numbers}; D = {first 10 positive integers}
   $\{2, 3, 5, 7\}$

Find the union of the sets.

7. G = {multiples of 2}; H = {factors of 12}
   {1, 3, multiples of 2}

8. W = {square roots of 16, 36, 64}; Z = {squares of 2, 4, 6, 8}
   $\{4, 6, 8, 16, 36, 64\}$

For 9–11, use the Venn diagram.

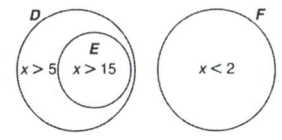

9. Identify the union of D and F.
   $D \cup F = \{x \mid x > 5$ or $x < 2\}$

10. Identify the intersections.
    $D \cap E = \{x \mid x > 15\}$
    $D \cap F = \emptyset$
    $E \cap F = \emptyset$

11. Identify any subsets.
    $E \subset D$

12. Make a truth table for the disjunction P or Q.
    P: A figure is a quadrilateral.
    Q: A figure is a regular polygon.

P	Q	P or Q
T	T	T
T	F	T
F	T	T
F	F	F

13. Complete the truth table.
    P: Ann is on the basketball team.
    Q: Ann plays the violin.

P	Q	P and Q	P or Q
T	T	T	T
T	F	F	T
F	T	F	T
F	F	F	F

For 14–15, identify the hypothesis and the conclusion in each conditional.

14. Work hard, and you will be rewarded.
    H: You work hard.
    C: You will be rewarded.

15. Plants need water to grow.
    H: Don't water your plants.
    C: Your plants won't grow.

16. Make a conclusion, if possible, from the deductive argument.
    If a rectangle has sides of 8 ft and 12 ft, then its area is 96 ft^2. Rectangle $WXYZ$ has an area of 96 ft^2.
    No conclusion; The rectangle could be 4 ft by 24 ft.

For 17–19, use the graph.

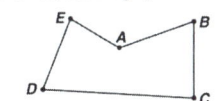

17. Find the degree of each vertex.
    A: 2, B: 2, C: 2, D: 2, E: 2

18. Is the graph connected? Explain.
    Yes; There is a path from each vertex to another vertex.

19. Determine whether the graph can be traversed through an Euler circuit. If your answer is yes, describe an Euler circuit in the graph.
    The graph represents an Euler circuit; D, E, A, B, C, D

For 20–21, use the graph.

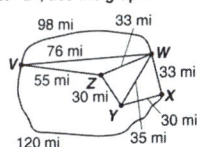

20. Determine the length of the shortest Hamiltonian circuit beginning at V.
    224 mi

21. Determine the length of the longest Hamiltonian circuit beginning at X.
    311 mi

Chapter 1 Performance Assessment
Algebra Toolbox

Demonstrate your knowledge by giving a clear, concise solution to the problems presented. Be sure to include all important information.

1. Draw a graph that is representative of the following statements.

   - A bus stops to pick up students for school.

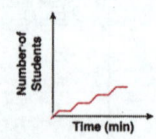

   - On a very hot day in July you enter your hot house and turn on the air conditioner.

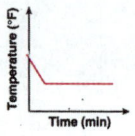

   Think of a situation that could be modeled by the graph.

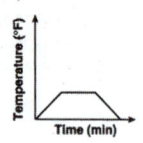

2. Label the axes of the graph and then write several sentences describing the graph.

   Possible answer: Turning on an oven, baking a cake and then turning the oven off to cool down.

Chapter 2 Performance Assessment Teacher Support
Chapter 2

Ten Famous Mathematicians

Euclid (325–265 B.C.) Best known for his book *The Elements*	Al-Khwarizmi (A.D. 780–850) Considered one of the first users of Algebra.
Omar Khayyam (A.D. 1048–1131) The first mathematician to deal with every type of cubic equation that yields a positive square root	Archimedes (287–212 B.C.) Best known for his practical applications to physics topics such as the lever and water displacement.
Plato (429–348 B.C.) Known to the world as a Greek philosopher	Mandelbrot (A.D. 1924–) The father of fractal geometry
Pythagoras (570–500 B.C.) Established a school in Croton, in Southern Italy and one of the most interesting figures in the history of mathematics.	Thales (624–547 B.C.) He is considered the "Father of Geometry" for his early discoveries and proofs of geometric relationships.
Hypatia (A.D. 370–415) Known as the first woman mathematician in history	Descartes (A.D. 1596–1650) Inventor of analytic geometry

1. Create a timeline showing the birthdates of the mathematicians.

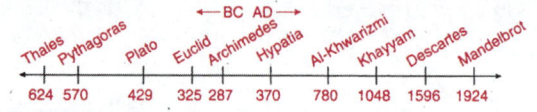

   ← BC AD →
   Thales Pythagoras Plato Euclid Archimedes Hypatia Al-Khwarizmi Khayyam Descartes Mandelbrot
   624 570 429 325 287 370 780 1048 1596 1924

2. Explain how a timeline is different than a number line. What do A.D. and B.C. represent?

   Possible answer: The timeline does not have a zero point. A.D. represents positive numbers and B.C. represents negative numbers.

3. List the mathematicians in order from youngest to oldest by age at death.

   Mandelbrot is still alive. Hypatia (45); Descartes (54); Euclid (60); Pythagoras and Khwarizmi (70); Archimedes (75); Thales (77); Plato (81); Khayyam (83)

Chapter 3 Performance Assessment
Rational and Real Numbers

The large outer square represents one whole unit. It has been divided into smaller pieces labeled A to K.

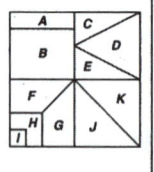

1. Decide what each fraction of the whole unit each piece represents.

   A $\frac{1}{16}$ E $\frac{1}{16}$ I $\frac{1}{64}$

   B $\frac{3}{16}$ F $\frac{3}{32}$ J $\frac{1}{8}$

   C $\frac{1}{16}$ G $\frac{3}{32}$ K $\frac{1}{8}$

   D $\frac{1}{8}$ H $\frac{3}{64}$

2. Write a brief explanation of how you determined the fraction for each of the following pieces: **Possible answers:**

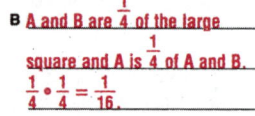

   B **A and B are $\frac{1}{4}$ of the large square and A is $\frac{1}{4}$ of A and B.**
   $\frac{1}{4} \cdot \frac{1}{4} = \frac{1}{16}$

   I **F, G, H, and I is $\frac{1}{4}$ of the large square and H and I is $\frac{1}{4}$ of that. I is $\frac{1}{4}$ of H and I.**
   $\frac{1}{4} \cdot \frac{1}{4} \cdot \frac{1}{4} = \frac{1}{64}$

   F **E, G, H, and I are $\frac{1}{4}$ of the large square and F and G is $\frac{3}{4}$ of that. F is $\frac{1}{2}$ of F and G.**
   $\frac{1}{2} \cdot \frac{3}{4} \cdot \frac{1}{4} = \frac{3}{32}$

   J **J and K are $\frac{1}{4}$ of the large square and J is $\frac{1}{2}$ of that.**
   $\frac{1}{2} \cdot \frac{1}{4} = \frac{1}{8}$

3. Design your own fraction square and give the fractions for the pieces.

   Answers will vary.

Chapter 4 Performance Assessment
Collecting, Displaying, and Analyzing Data

Park	One-Day Ticket Price	Number of Roller Coasters
Thrill Rides	$42.00	15
Flying Coasters Park	$44.99	15
Family Fun	$37.50	11
Gooseberry Farm	$48.00	10
Ride Monster	$39.99	9
Discovery Mountain	$34.95	8
Fun Town	$25.00	5

1. Make two bar graphs to represent the data. On one graph show the one-day ticket prices at each park. On the other, show the number of roller coasters.

2. Find the mean, median, and mode of the ticket price data.

   Mean **$38.92**   Median **$39.99**   Mode **none**

3. Find the mean, median, and mode of the number of roller coaster data.

   Mean **10.43**   Median **10**   Mode **15**

4. Make a scatter plot of ticket price vs the number of roller coasters and draw a trend line.

   Scatter plot should show a positive trend.

Chapter 5 Performance Assessment
Plane Geometry

Complete the tasks given below.

1. Graph two parallel lines and two perpendicular lines. What are the rules for determining whether two lines are parallel or perpendicular?

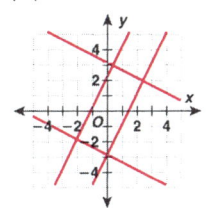

   Two parallel lines have the same slope. The product of the slopes of two perpendicular lines is −1.

2. Plot the points (0, 0), (4, 0), (3, 3) and (1, 3) on the coordinate grid. Connect the points in the order given and connect the last point to the first point.

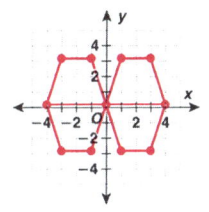

3. Rotate the original figure 180 degrees. Show the result of the transformation of the coordinate grid.

   (0, 0), (−4, 0), (−3, −3), (−1, −3)

4. Reflect the original figure across the x-axis. Show the result of the transformation of the coordinate grid.

   (0, 0), (4, 0), (3, −3), (1, −3)

5. Translate the original figure 4 units to the left. Show the result of the transformation of the coordinate grid.

   (−4, 0), (0, 0), (−1, 3), (−3, 3)

Chapter 6 Performance Assessment
Perimeter, Area, and Volume

1. Your soccer club wants to sell frozen yogurt to raise money for new uniforms. The club has the choice of the two different size containers shown below. Each container costs the club the same amount. The club plans to charge customers $2.50. Which container should the club buy? Explain.

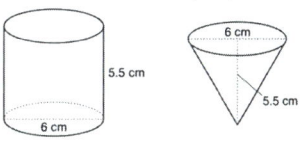

   Possible answer: Volumes: cylinder is 155.43 cm^3, cone is 51.81 cm^3. The soccer club would make more money using the cone-shaped container because it's volume is smaller.

2. A new movie theater is going to sell popcorn. The manager has the choice of the three different size containers shown below. The manager plans to charge $4.75 for a container of popcorn.

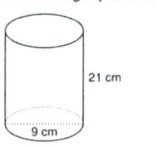

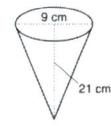

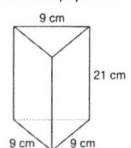

   a. Which container would you choose as the manager of the movie theater? Explain.

   The volumes: cylinder is 1335.29 cm^3, the cone is 445.10 cm^3, the prism is 736.55 cm^3. The cone has the smallest volume, so it would use less popcorn.

   b. If all of the containers are made using the same material, which container would cost the least to make?

   Possible answer: Surface areas: cylinder is 657.05 cm^2, cone is 303.47 cm^2, prism is 602.07 cm^2. The cone would cost the least.

Chapter 7 Performance Assessment
Ratios and Similarity

Complete the tasks given below.

The diagram shown below is an architect's plan for a bedroom area. Using a ruler, make the necessary measurements to answer the questions that follow. Give the actual measurements.

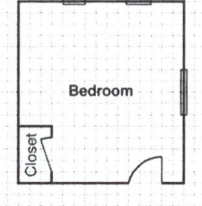

1 unit = 1 ft

1. How many feet wide is the door leading into the bedroom?

   3 ft

2. How many feet wide is the door leading into the closet?

   3 ft

3. How wide is the widest part of the bedroom?

   16 ft

4. What are the dimensions of the bedroom?

   15 ft by 16 ft

5. How many feet of baseboard will be needed to go around the bedroom, including the inside of the closet? (Hint: baseboard does not go in the doorways.)

   77 ft

6. If baseboard costs $2.35 per linear foot, what is the total cost for baseboard?

   $180.95

7. How many square yards of carpeting will be needed for the bedroom and closet? (round to the nearest yard)

   27 yd^2

8. If carpet costs $27.99 per square yard, what is the total cost for carpeting?

   $755.73

9. What are the widths of the windows?

   2 ft, 2.25 ft, and 4 ft

10. What are the dimensions of the closet?

    3 ft by 5 ft

11. If the walls are 8 feet high, what is the total area of the walls and ceiling, including the inside and outside of the closet? Subtract 95 square feet for windows and doors.

    769 ft^2

12. If one gallon of paint covers 300 square feet, how many gallons of paint will be needed to paint the walls and ceiling excluding windows and doors? Round up to the nearest gallon.

    3 gal

Chapter 8 Performance Assessment
Percents

You have been hired by Pizza Palace to do a new marketing campaign. The campaign should show that Pizza Palace has the best food.

The table shows the results of a survey to find out which restaurant people liked best.

Restaurant	Response
Taco Town	128
Pizza Palace	248
Mr. Cluck's Chicken	28
Burger Barn	12
Salad Haven	184

1. What percent of the people surveyed chose each of the five restaurants? Round to the nearest hundredth of a percent.

Restaurant	Percent
Taco Town	21.33%
Pizza Palace	41.33%
Mr. Cluck's Chicken	4.67%
Burger Barn	2%
Salad Haven	30.67%

2. In your presentation to Pizza Palace you want to give them a colorful representation of the results of the survey. You have chosen to make a circle graph. Calculate the size of each sector. Round to the nearest whole degree.

Restaurant	Size of Sector
Taco Town	77°
Pizza Palace	149°
Mr. Cluck's Chicken	17°
Burger Barn	7°
Salad Haven	110°

3. Construct the circle graph.

   Favorite Eating Places

4. What is your recommendation to Pizza Palace?

   Possible answer: Since Pizza Palace has the largest market share, customers must think it has the best food.

---

347   Holt Pre-Algebra

CHAPTER 9 Performance Assessment
Probability

Complete the tasks given below.

1. Which spinner was most likely the spinner that produced the following data set? Explain your answer.
   Data Set: yellow, green, green, yellow, yellow, yellow, green, yellow, yellow, yellow

   Spinner A   Spinner B   Spinner C

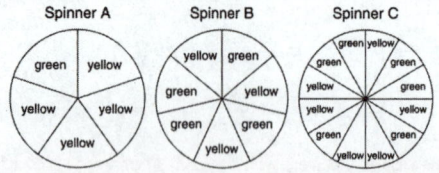

   Yellow occurred $\frac{7}{10}$ or 70% of the time. Spinner A has yellow 80%; spinner B has yellow 43%; spinner C has yellow 50%. Spinner A is most likely the spinner.

2. Show several ways to find the number of 3-letter combinations in the word EQUAL. Write your answers in an organized list.
   Possible Answers: $\frac{5!}{3!2!} = \frac{5 \times 4 \times 3 \times 2 \times 1}{3 \times 2 \times 1 \times 2 \times 1} = 10$

   EQU, EQA, EQL, EUA, EUL, EAL, QUA, QUL, QAL, and UAL

3. Use a random number table and write the first fifty numbers as single digits.
   Example: 43215 would be 4, 3, 2, 1, 5.
   A. How many 5's did you get? **Possible answer: 5**
   B. How many numbers larger than 6 did you get? **Possible answer: 15**
   C. How many odd numbers did you get? **Possible answer: 25**
   D. If you selected the next fifty random numbers, how many 5's would you expect to get? **Possible answer: 5**
   E. How many odd numbers would you expect to get? **Possible answer: 25**

CHAPTER 10 Performance Assessment
More Equations and Inequalities

Complete the tasks given below.

1. Solve each of the equations shown below. Describe the steps needed.
   A. $3x - 4 = 17$
   A. To isolate x, first add 4 to both sides of the equation, then divide each side by 3. $x = 7$
   B. $\frac{z}{6} - 3 = 2$
   B. To isolate z, add 3 to both sides of the equation, then multiply each side by 6. $z = 30$
   C. $6x + 3 = 2x - 13$
   C. Possible answer: Subtract 3 from each side. Subtract $2x$ from each side. Divide both sides by 4 to isolate x. $x = -4$

2. A1-Rental charges $35 plus $6 per hour to rent a mini-loader, while BilJax charges $15 plus $8 per hour. Russell Casius needs to use a mini-loader for about 7 hours.
   A. For what number of hours would BilJax cost no more than A1-Rental? Write and solve a linear inequality.
   $8h + 15 \leq 6h + 35$, $h \leq 10$ hours
   B. If Russell can spend no more than $75 to excavate some land in his yard, what is the *maximum* amount of time he could use the rented mini-loader from A1? $6\frac{2}{3}$ hours
   C. What is the *maximum* amount of time he could use the mini-loader from BilJax? $7\frac{1}{2}$ hours
   D. A1 is 20 minutes from Russell's home, and BilJax is half an hour away. Taking drive time into consideration, where should Russell rent the mini-loader? Explain.
   If Russell rents from A1 he will need to deduct 40 minutes from his time for actual use. He could use it for approximately 6 hours. If he rents from BilJax he will need to deduct 1 hour from his time for actual use. He could use it for 6.5 hours. He should rent from BilJax, since he will get about an extra half an hour of usage.

CHAPTER 11 Performance Assessment
Graphing Lines

Complete the tasks given below.

1. Salina and Roberto are working on sectioning off a portion of the backyard for a rectangular garden. They have a total of 110 feet of fencing. Due to the shape of the yard, the garden has to be at least 25 feet long. 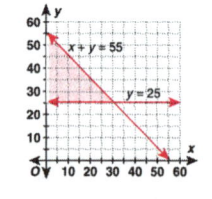
   a. Draw a graph that shows all the possible dimensions of the garden.
   b. Decide which dimensions would give the maximum gardening space. Explain your reasoning.
   27.5 feet by 27.5 feet, because for a given perimeter, a square has a bigger area than any rectangle.

2. Martin Norris is a land developer who plans to expand a subdivision. He will need to add three additional roads, the plans for which are shown in the graph. Washington Drive will be parallel to Kennedy Drive, and Lincoln Drive will be perpendicular to the other two roads. 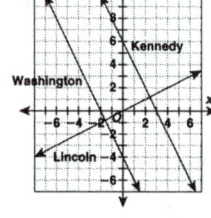
   a. Find the slope of each road.
   Slope of Washington and Kennedy is -2 and of Lincoln is $\frac{1}{2}$.
   b. Write an equation for each line.
   $y = -2x + 6$, $y = -2x - 4$, and $y = \frac{1}{2}x$
   c. What has to be true for Kennedy Drive and Washington Drive to be parallel?
   The slopes of the lines have to be the same.
   d. What has to be true for Lincoln Drive to be perpendicular to Washington Drive and Kennedy Drive?
   The slopes of the lines have to be negative reciprocals of each other. -2 and $\frac{1}{2}$ are negative reciprocals, so the lines are perpendicular to each other.

CHAPTER 12 Performance Assessment
Sequences and Functions

Complete the tasks given below.

1. Write a paragraph explaining the meaning of the terms function, domain and range. Give examples to support your answer.
   Possible answer: A function is a rule that relates two quantities. Each x-value corresponds to exactly one y-value. The domain is the set of all possible x-values and the range is the set of all possible y-values.

   Function
   3 → 6
   7 → 14
   9 → 18

   Not a function
   3 → 6
      → 8
   6 → 15

2. Identify the functions below as linear, exponential, or quadratic. Explain how you can classify each function. **Possible answers:**
   A. $f(x) = 9 + 8x$ linear; can be put in the form $f(x) = mx + b$.
   B. $f(x) = x^2 + 3x - 2$ quadratic; a quadratic function contains a squared term.
   C. $f(x) = \left(\frac{1}{2}\right)^x$ exponential; an exponential function contains a term raised to a power.
   D. $y = 4x - 9$ linear; of the form $y = mx + b$.
   E. $y = x^2 - 3$ quadratic; contains a squared term.

3. Explain the difference between direct and inverse variation. Explain how the y-values change.
   Possible answer: In a direct variation, y divided by x results in a constant of variation. A direct variation is a linear relationship. In an inverse variation, one variable increases in value as the other variable decreases in value. The product of the variables is a constant value.

Holt Pre-Algebra

Performance Assessment
Chapter 13 Polynomials

You have 34 meters of fencing to use in constructing the perimeter of a rectangular garden. The garden will have 4 sections as shown below. After you have used all the fencing to build the outside walls, you will need to buy more fencing to separate the sections.

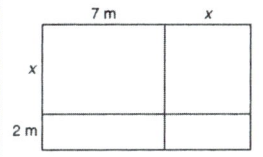

```
       7 m      x
   ┌────────┬──────┐
 x │        │      │
   ├────────┼──────┤
 2 m│        │      │
   └────────┴──────┘
```

- Write an expression in simplified form for the perimeter of the garden.

  $4x + 18$ meters

- Write as many expressions as you can for the area of the entire garden, using the variable x.

  Possible answer: $x^2 + 9x + 14$; $(x + 7)(x + 2)$

- What is the value of x in the drawing?

  $x = 4$ meters

- How much additional fencing will you need to build the inside walls of the garden?

  17 meters

- From least to greatest in area, the sections of the garden will be used for radishes, peppers, squash, and tomatoes. Find the area that will be used to plant each crop. Explain.

  radishes: 8 m², peppers: 14 m²; squash: 16 m², tomatoes: 28 m²; Possible answer: Substitute 4 for x to find the area of each section. Then order the areas from least to greatest.

Performance Assessment
Chapter 14 Analyzing a Graph

The graph represents a highway system connecting 6 towns: A, B, C, D, E, and F. A salesperson lives in one of the towns and travels between the towns on his route.

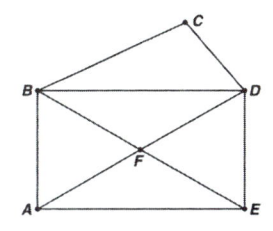

- Why would the salesperson be more interested in whether this graph is a Hamiltonian rather than an Euler circuit?

  because he needs to travel to each town, not over each road

- Name at least two Hamiltonian circuits on the graph.

  ABCDEFA, FEABCDF

- Another worker, a highway inspector, also lives in one of the towns. How can you tell without tracing the graph whether it would be possible for her to visit each town and return home without driving on the same road twice?

  If there are no vertices with odd degrees, it would be possible.

- Does the graph represent an Euler circuit? Explain.

  No; because there are 2 odd vertices.

- How could you convert this graph into an Euler circuit?

  Possible answer: Add another road connecting towns A and E. Then there would be no odd vertices.

Cumulative Test
Chapter 1 Form A

1. What is the median of the following data?
   12, 2, 6, 10, 8, 4, 9
   A 4 **C 8**
   B 6 D 10

2. Find $1.2 \cdot 10^3$.
   A 12 **C 1200**
   B 120 D 12,000

3. What is the greatest common factor of 8 and 32?
   A 2 **C 8**
   B 4 D 16

4. Simplify $8 - (-2)$.
   A 6 **C 10**
   B −6 D −10

5. Simplify $-4 \cdot 3$.
   A −12 C −7
   B 12 D 7

6. Which of the following is the best estimate of $4.9 \cdot 5.1$?
   A 20 **B 25**

7. Simplify $3\frac{3}{4} + 2\frac{1}{8}$.
   A $5\frac{7}{8}$ B $5\frac{1}{3}$

8. Which graph shows no correlation?
   A 
   B

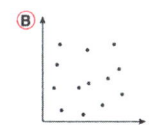

Cumulative Test
Chapter 1 Form A, continued

9. Which of the following pairs of figures are NOT similar?
   A 
   B

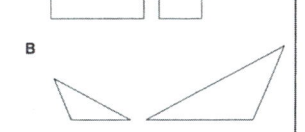

10. Solve the proportion $\frac{4}{3} = \frac{w}{21}$.
    A $w = 20$ C $w = 63$
    B $w = 28$ D $w = 82$

11. 25 is 10% of what number?
    A 2.5 **C 250**
    B 25 D 2500

12. Classify the triangle according to its sides and angles.

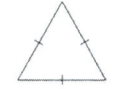

    A isosceles, acute
    B equilateral, equiangular
    C scalene, obtuse
    D isosceles, obtuse

13. Convert 25 m to cm.
    A 25 cm **C 2500 cm**
    B 250 cm D 25,000 cm

14. Find the volume of the prism. 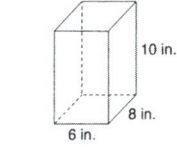

    A 24 in³ **C 480 in³**
    B 48 in³ D 960 in³

15. Use the Pythagorean Theorem to find the missing measure.

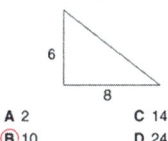

    A 2 C 14
    B 10 D 24

16. What is the slope and the y-intercept of $y = 2x + 1$?
    A 2, 1 B 2, −1

17. Maria wants to arrange four books on her shelf. How many ways could she arrange them?
    A 24 C 6
    B 12 D 1

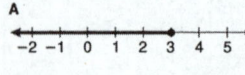

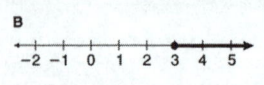

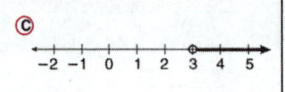

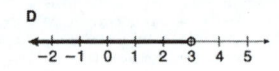

Cumulative Test
Form A, continued

18. What are the next two terms in this sequence?
3, 7, 11, 15 ...
A 19, 23
B 20, 25

19. Evaluate $3w$ for $w = 5$.
A 2
C 15
B 8
D 18

20. Use the formula $A = lw$ to find the area of the rectangle.

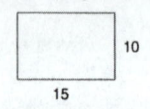

A 15
C 60
B 50
D 150

21. Which is an algebraic expression for 18 more than a number y.
A $18 - y$
C $18 \div y$
B $y + 18$
D $18y$

22. Sam read x books today; he had read 15 books before that. How many total books did Sam read? Which expression represents this problem?
A $x + 15$
C $x - 15$
B $15x$
D $15 - x$

23. Solve $x + 7 = 9$.
A $x = 1.27$
C $x = 16$
B $x = 2$
D $x = 63$

24. Solve $3x = 30$.
A $x = 10$
C $x = 60$
B $x = 27$
D $x = 90$

25. Which is the graph of $x + 2 > 5$?

26. Simplify $8w + 3w$.
A $5w$
C $11w$
B $11w$
D $24w$

27. Which ordered pair is a solution of $y = 6x$?
A (12, 2)
C (3, 18)
B (4, 18)
D (6, 1)

28. What are the coordinates of point B?

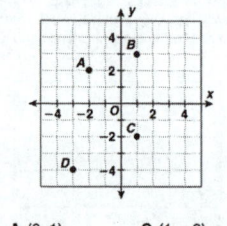

A (3, 1)
C (1, −2)
B (1, 3)
D (−3, −4)

29. Which graph matches the equation $y = 2x$?

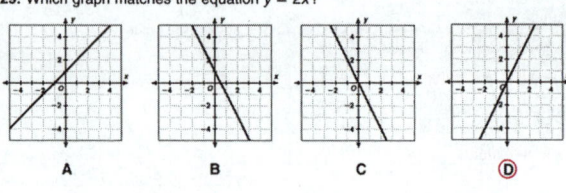

30. On a highway, a driver sets a car's cruise control for a constant speed of 55 mi/h. Which graph shows the speed of the car?

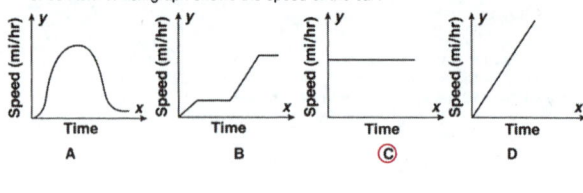

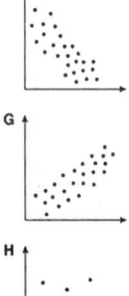

Cumulative Test
Form B

1. What is the median of the following data?
12, 26, 16, 14, 23, 28, 19, 20
A 17.5
C 19.5
B 19
D 20

2. Find $1.47 \cdot 10^4$.
F 14,700
H 147
G 1470
J 14.700

3. What is the greatest common factor of 18, 27, and 45?
A 3
C 6
B 5
D 9

4. Simplify $-7 - (-5)$.
F −12
H 2
G −2
J 12

5. Simplify $-3 \cdot 28$.
A 84
C −78
B 78
D −84

6. Which of the following is the best estimate of $4.12 \cdot 26.2$?
F 98
H 75
G 100
J 130

7. Simplify $5\frac{2}{5} + 3\frac{4}{15}$.
A $8\frac{1}{3}$
C $9\frac{2}{5}$
B $8\frac{2}{3}$
D $9\frac{1}{3}$

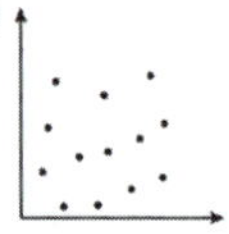

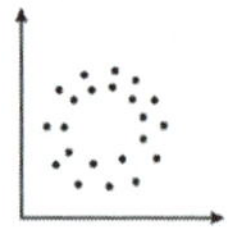

8. Which graph shows a negative correlation?

9. Which of the following pairs of figures are NOT similar?

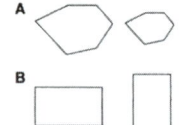

10. Solve the proportion $\frac{5}{w} = \frac{7.5}{15}$.
F $w = 2.5$
H $w = 9$
G $w = 6$
J $w = 10$

11. 35 is 7% of what number?
A 500
C 50
B 700
D 70

12. Classify the triangle according to its sides and angles.

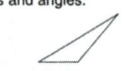

F isosceles, acute
G equilateral, equiangular
H scalene, obtuse
J isosceles, obtuse

13. Convert 350 m to km.
A 350,000 km
C 0.35 km
B 3500 km
D 0.035 km

14. Find the volume of the sphere to the nearest tenth. Use 3.14 for π.

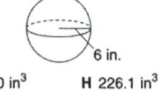

F 288.0 in^3
H 226.1 in^3
G 904.3 in^3
J 2713.0 in^3

15. Use the Pythagorean Theorem to find the missing measure.

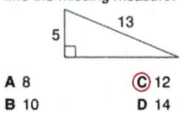

A 8
C 12
B 10
D 14

16. What is the slope and the y-intercept of $y = 3x - 2$?
F 2, −3
H 2, 3
G 3, −2
J −2, 3

17. Alana wants to arrange six trophies on her shelf. How many ways could she arrange them?
A 720
C 120
B 540
D 6

Cumulative Test
1 Form B, continued

18. What are the next three terms in this sequence?
    1, 4, 9, 16, …
    F 19, 22, 25   (H) 25, 36, 49
    G 23, 30, 37   J 24, 29, 36

19. Evaluate $3w - y$ for $w = 6$ and $y = 5$.
    (A) 13   C 4
    B 14   D 23

20. Use the formula $p = 2h + 2b$ to find the perimeter of the rectangle. 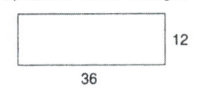

    12
    36

    F 48   (H) 96
    G 84   J 432

21. Which is an algebraic expression for 2 plus the product of 3 and w.
    A $3(2w)$ (C) $2 + 3w$
    B $2(3w)$ D $2w + 3$

22. To add to his 19-card collection, Kevin bought 2 baseball cards each week for x weeks. How many does he have now? Which expression represents this problem?
    (F) $19 + 2x$ H $2x - 19$
    G $2(3 + x)$ J $19 \div 2x$

23. Solve $w - 24 = 56$.
    A $w = 19$ C $w = 32$
    B $w = 2.3$ (D) $w = 80$

24. Solve $\frac{p}{5} = 450$.
    F $p = 9$ H $p = 445$
    G $p = 90$ (J) $p = 2250$

25. Which of the following is the graph of $4x + 9 > 21$?
    A
    B 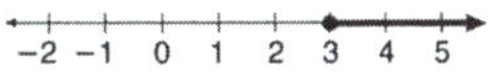
    (C)
    D

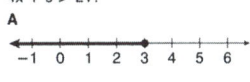

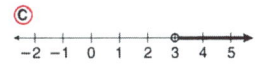

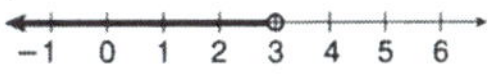

26. Simplify $14t - 4t + 7w + 3t$.
    F $13 + 7w$ (H) $13t + 7w$
    G $14wt$ J $10t + 7w$

27. Which ordered pair is a solution of $y = 2x - 7$?
    (A) $(-5, -17)$ C $(0, 7)$
    B $(-3, 1)$ D $(2, 11)$

28. What are the coordinates of point A? 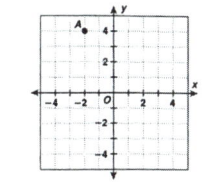

    F $(-2, -4)$ H $(2, -4)$
    (G) $(-2, 4)$ J $(2, 4)$

29. Which graph matches the equation $y = 2x + 1$?
    (A)   B   C   D

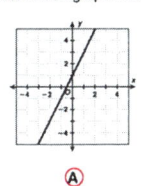

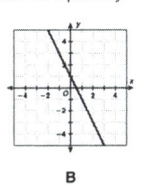

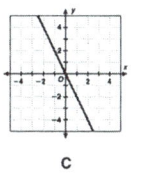

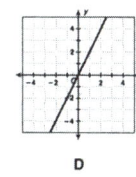

30. Which graph shows the speed of a scooter as you push for 2 minutes and then coast until coming to a rest?
    (F)   G   H   J

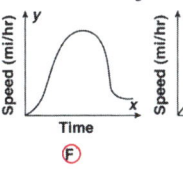

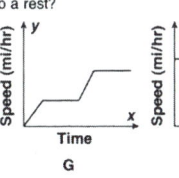

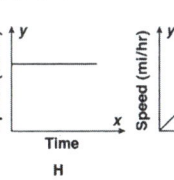

Cumulative Test
1 Form C

1. What is the median of the following data?
   212, 226, 216, 214, 223, 228, 219, 220
   A 217.5   (C) 219.5
   B 219   D 220

2. Find $2.482 \cdot 10^6$.
   F 24,820   (H) 2,482,000
   G 248,200   J 24,820,000

3. What is the greatest common factor of 16, 40, and 72?
   A 2   (C) 8
   B 4   D 16

4. Simplify $-32 - (-18)$.
   F 14   H 50
   (G) -14 J -50

5. Simplify $-9 \cdot -28$.
   A -37 C -252
   B 37   (D) 252

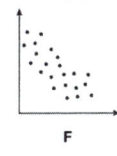

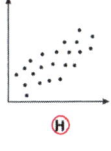

6. Which of the following is the best estimate of $8.725 \cdot 59.41$?
   F 72   H 480
   G 284   (J) 540

7. Simplify $18\frac{3}{16} + 122\frac{9}{40}$.
   (A) $140\frac{33}{80}$ C $140\frac{3}{14}$
   B $140\frac{12}{56}$ D 140

8. Which graph shows a positive correlation? 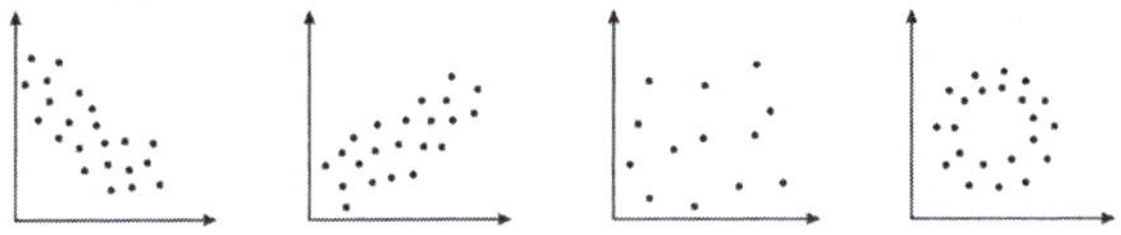
   F   (H)   G   J

Cumulative Test
1 Form C, continued

9. Which of the following pairs of figures are NOT similar?
   A 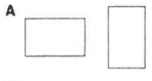
   (B) 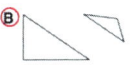
   C 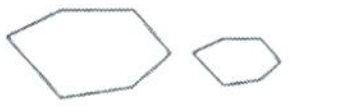
   D

10. Solve the proportion $\frac{26}{w} = \frac{19.5}{18}$.
    F 9   H 28.2
    G 15   (J) 24

11. 120 is 0.5% of what number?
    A 240   (C) 24,000
    B 2400   D 240,000

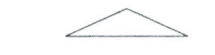

12. Classify the triangle according to its sides and angles.

    F isosceles, acute
    G equilateral, equiangular
    H scalene, obtuse
    (J) isosceles, obtuse

13. Convert 3725 cm to km.
    A 0.003725 km   C 0.3725 km
    B 3.725 km   (D) 0.03725 km

14. Find the volume of the sphere. 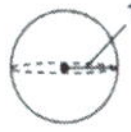
    18 in.
    (F) 24,416.64 in^3 H 6104.16 in^3
    G 18,321.48 in^3 J 1356.48 in^3

15. Use the Pythagorean Theorem to find the missing measure.
    41
    9
    A 50   (C) 40
    B 42   D 24

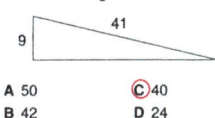

16. What is the slope and the y-intercept of $8x + 4y = 4$?
    F 8, 4   H 2, -1
    G 4, 8   (J) -2, 1

17. Marta wants to arrange eight bottles of perfume on her dresser. How many ways could she arrange them?
    A 52,430   C 1024
    (B) 40,320   D 36

---

351   **Holt Pre-Algebra**

CHAPTER 1 Cumulative Test
Form C, continued

18. What are the next three terms in this sequence?
1000, 500, 250, 125, ...
(F) 62.5, 31.25, 15.625
G 100, 75, 50
H 25, 36, 49
J 70, 50, 20

19. Evaluate $3(w - 2y + z)$ for $w = 14$, $y = 3$, and $z = 2$.
A 20
(C) 30
B 26
D 114

20. Use the formula $p = 2h + 2b$ to find the perimeter of the rectangle.

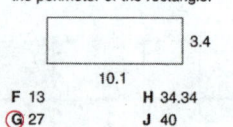

[rectangle: 10.1 by 3.4]

F 13
H 34.34
(G) 27
J 40

21. Which is an algebraic expression for 12 more than the product of 6 and a number.
A $12 + w$
C $12(6 + w)$
B $12(6 - w)$
(D) $6w + 12$

22. Renee bought a pounds of apples at $2 per pound and p pounds of peaches at $3 per pound. Which expression represents this problem?
F $2a - 3p$
H $2 + p$
(G) $2a + 3p$
J $2(a + p)$

23. Solve $142 = w - 85$.
A $w = 57$
(C) $w = 227$
B $w = 147$
D $w = 320$

24. Solve $5w - 4 = 21$.
(F) $w = 5$
H $w = 6$
G $w = 7$
J $w = 8$

25. Which of the following is the graph of $8y - 17 \leq 47$?
(A) [number line]
B [number line]
C [number line]
D [number line]

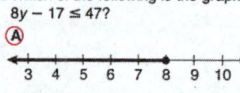

CHAPTER 1 Cumulative Test
Form C, continued

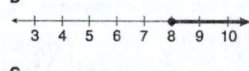

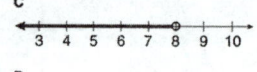

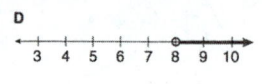

26. Simplify $4(2x + 3) - 3y + 5 + 2x$.
(F) $10x - 3y + 17$
H $10x + 17$
G $8x - 3y + 8$
J $17xy$

27. Which ordered pair is a solution of the equation $7x - 4y = 5$?
(A) (3, 4)
C (0, −4)
B (2, 3)
D (3, 24)

28. What are the coordinates of point A? 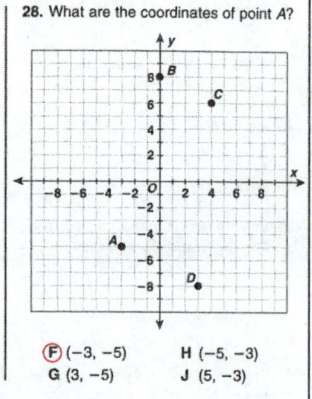
(F) (−3, −5)
H (−5, −3)
G (3, −5)
J (5, −3)

29. Which graph matches the equation $y = 2x - 1$?
(A) B C D

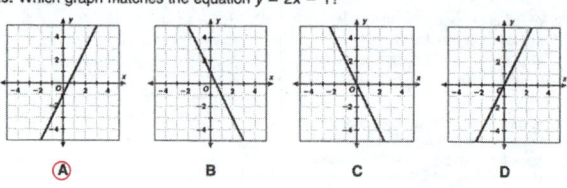

30. Which graph displays the speed of a car as it accelerates from a stopped position?
F G H (J)

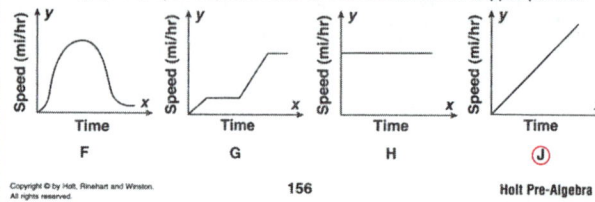

CHAPTER 2 Cumulative Test
Form A

1. Evaluate $a + 3$ for $a = 8$.
A 3
(B) 11

2. Evaluate $5m + 9n + 1$ for $m = 0$ and $n = 2$.
A 24
(B) 19

3. Evaluate $-8v - 2$ for $v = -2$.
A −18
(B) 14

4. Subtract $-11 - (-6)$.
A −17
(C) −5
B −12
D 5

5. What is the best estimate for the weight of an adult?
A 70 mg
(C) 70 kg
B 70 g
D 95 mL

6. Simplify $-5a + (-3a)$.
A $-2a$
(B) $-8a$

7. Simplify $-9m - 3n - 6n$.
A $-9m - 3n$
(B) $-9m - 9n$

8. Solve $8 - r = 5$.
A $r = -3$
C $r = 2$
B $r = -2$
(D) $r = 3$

9. Solve $s + 7 = 22$.
A $s = 29$
(B) $s = 15$

10. A straight path that extends without end in opposite directions is a _____.
A ray
C plane
B point
(D) line

11. Simplify $(-11) \cdot (-11)$.
A −121
(B) 121

12. Simplify $(-37) \cdot (0)$.
A 37
(C) 0
B 1
D −37

13. Evan drove 300 miles at the average rate of 60 miles per hour. How long did his trip take?
A 6 hr
(C) 5 hr
B $\frac{1}{5}$ hr
D 4 hr

14. Express 1.7×10^5 using standard notation.
(A) 170,000
B 1,700,000

15. Express 14,000,000 using scientific notation.
A 14×10^6
(B) 1.4×10^7

CHAPTER 2 Cumulative Test
Form A, continued

16. Find the median of the data in the stem and leaf plot.
(A) 64
B 65

5	0 0 4 7
6	1 3 4 5
7	1 4 6 8 9

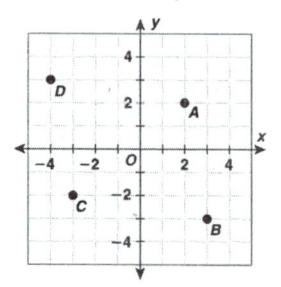

17. Give the coordinates of point A as shown on the graph above.
(A) (2, 2)
B (3, 3)

18. Give the coordinates of point B as shown on the graph above.
A (−3, −2)
(C) (3, −3)
B (−3, 3)
D (−4, 3)

19. Express $4 \cdot 4 \cdot 4 \cdot 4 \cdot 4 \cdot 4$ using exponents
(A) 4^6
B 4^5

Day	1	2	3	4	5	6	7
High	73°	78°	80°	84°	77°	69°	70°

20. Using the table above, determine the average high temperature to the nearest degree.
A 66°
(B) 76°

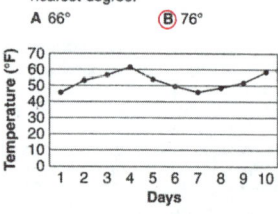

21. Use the graph above to determine the approximate temperature on day 7.
A 50 degrees
C 44 degrees
B 52 degrees
(D) 47 degrees

22. Simplify $-3 + (-16)$.
(A) −19
C 13
B −13
D 19

23. Express $\frac{m^9}{m^5}$ as one power.
(A) m^4
B $\frac{1}{m^4}$

Cumulative Test
Form A, continued

24. Which ordered pair is a solution of $3x = y$?
    A (3, 1)
    B (1, 3)

25. Which ordered pair is a solution of $y = \frac{1}{2}x$?
    A (1, 2)   **C (6, 3)**
    B (4, 1)   D (3, 6)

26. Natasha has $77 in her bank account when she writes a check for $48. She makes a deposit of $31. How much is now in the account?
    A $156
    B $60

27. Classes of 30 math students each met in the cafeteria to take achievement tests. If exactly 5 students sat at each table and 24 tables were used, how many classes took the tests?
    A 6 classes   **C 4 classes**
    B 17 classes   D 7 classes

28. Express the phrase, "11 is greater than a number n," as an algebraic expression.
    A $11 > n$
    B $n > 11$

29. A boy who is 5 feet tall casts a shadow 3 feet long. He is standing near a tree that casts a shadow of 24 feet. How tall is the tree?
    A 36 feet   C 44 feet
    B 40 feet   D 50 feet

30. Simplify 3^{-4}.
    A 81
    B $\frac{1}{81}$

31. Simplify $\frac{x^2}{x^4}$.
    A x C $2x$
    B x^2 **D $\frac{1}{x^2}$**

32. An acute angle measures ____.
    A less than 90°   C more than 90°
    B exactly 90°   D exactly 180°

33. Solve $\frac{x}{3} = 4$.
    A $x = 6$ **C $x = 12$**
    B $x = 7$ D $x = 1$

34. Express the phrase, "three times a number g," as an algebraic expression.
    A $3 + g$ **C $3g$**
    B $3 - g$ D $\frac{3}{g}$

35. Find the area of a parallelogram with base length 15 m and height 8 m.
    A 60 m²   C 48 m²
    B 44 m²   **D 120 m²**

36. Express the phrase, "the quotient of a number m and six," as an algebraic expression.
    A $6m$
    B $\frac{m}{6}$

37. Which expression could be used to find the approximate area of the base of a cylinder which has a height of 4 inches and a radius of 5 inches?
    A $5 \cdot \pi \cdot 3.14$
    B $5^2 \cdot 3.14$

38. Li has $1\frac{1}{2}$ pounds of tuna. She wants to serve each person a 6-oz serving. How many people can she serve?
    A 2 people
    B 4 people

39. Select the graph that is a solution to the inequality, $9x \leq 54$.
    A [number line graph]
    B [number line graph]

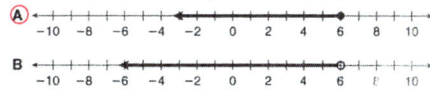

40. Select the graph that is a solution to the inequality, $x + 2 \leq -3$.
    A [number line graph]
    B [number line graph]

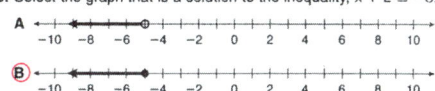

Cumulative Test
Form B

1. Evaluate $\frac{x+3}{3}$ for $x = 6$.
    A 9   C 1
    B 3   D 6

2. Evaluate $2m + 3n + 1$ for $m = -1$ and $n = 2$.
    F 9   **H 5**
    G 10   J 4

3. Evaluate $63 - (-4z)$ for $z = 12$.
    A 111   C 15
    B 59   D -15

4. Simplify $12x - (-17y) - 32y + (-43x)$.
    F $55x + 49y$ H $-31x - 60y$
    G $25x - 75y$ **J $-31x - 15y$**

5. What is the best estimate for the length of a baseball bat?
    A 1 m   C 3 cm
    B 10 m   D 30 km

6. Simplify $-6y - 4x - 2x$.
    F $-12y$ **H $-6y - 6x$**
    G $-12x$ J $-6y - 6x^2$

7. Simplify $-2 - (3 - 5t)$.
    A $5 + 5t$ C $1 + 5t$
    B $5t - 5$ D $-1 + 5t$

8. Solve $u + 47 = 238$.
    F 164   H 236
    G 191   J 285

9. Solve $v + 3 = 4$.
    A -7   C 7
    B -1   **D 1**

10. When two angles have the same measure, they are said to be ____.
    F complementary   H right
    G supplementary   **J congruent**

11. Simplify $(-4) \cdot (-4) \cdot (-4) \cdot (1)$.
    A -64   C 64
    B -65   D 65

12. Simplify $-\frac{143}{13}$.
    F 13   **H -11**
    G 11   J -13

13. A house worth $124,000 was assessed taxes based on $\frac{3}{4}$ of its value. What is the assessed value of the house?
    A $93,000   C $165,333
    B $124,000   D $930,000

14. Express 0.000086 using scientific notation.
    F 8.6×10^{-5}   H 8.6×10^{-4}
    G 86×10^{-4} J 8.6×10^4

15. Express 3.5×10^{-6} using standard notation.
    A 0.000035   **C 0.0000035**
    B 0.0035   D 0.00035

Cumulative Test
Form B, continued

16. Find the mode of the data in the stem and leaf plot.
    F 50   H 64.8
    G 64   J 65

    | 5 | 0 0 4 7 |
    | 6 | 1 3 4 5 |
    | 7 | 1 4 6 8 9 |

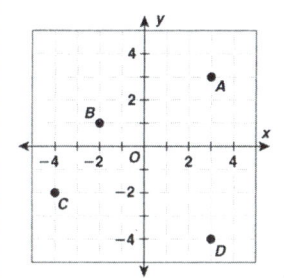

17. Give the coordinates of point D as shown on the graph above.
    A (3, 4)   C (-2, 1)
    B (3, -4)   D (-4, -1)

18. Give the coordinates of point C as shown on the graph above.
    F (4, 2)   H (2, 4)
    G (-2, 4)   **J (-4, -2)**

19. Simplify $(12 - 4^3)$.
    A 76   **C -52**
    B 64   D -64

Day	1	2	3	4	5	6	7
High	73°	85°	81°	85°	77°	69°	70°

20. Using the table above, determine the temperature difference between the warmest day and coolest day.
    F 3 degrees   **H 16 degrees**
    G 8 degrees   J 7.5 degrees

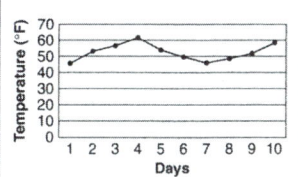

21. Use the graph above to determine the approximate change in temperature from day 1 to day 10.
    A 8 degrees   C 15 degrees
    B 12 degrees   D 10 degrees

22. Simplify $15 - (-4) + 0$.
    F 11   **H 19**
    G -11   J -19

23. Express the product $t^3 \cdot t^3 \cdot t^4$ as one power.
    A t^{36} C $3t^{10}$
    B t^{10}   D t^{18}

Cumulative Test
Form B, continued

24. Which ordered pair is a solution of $3x = 4y$?
    - **(F) (8, 6)**
    - H (3, 4)
    - G (6, 8)
    - J (1, 2)

25. Which ordered pair is a solution of $\frac{3x}{2} = y$?
    - A (−1, −2)
    - C (−2, 3)
    - **(B) (−1, −$\frac{3}{2}$)**
    - D (−$\frac{3}{2}$, −1)

26. Damon purchased CDs on sale for $2.45 each. If the total bill before sales tax was $34.30, how many CDs did he buy?
    - F 13
    - H 31
    - **(G) 14**
    - J 140

27. Winnie earned $2700 last summer. She spent $1200 to pay off her car debt, $540 for clothes, $360 for gifts, and put $600 in the bank. What fractional part of her income did she not spend?
    - **(A) $\frac{2}{9}$**
    - C $\frac{4}{9}$
    - B $\frac{2}{15}$
    - D $\frac{3}{15}$

28. Express the phrase, "8 more than three times a number is less than or equal to 34," as an algebraic expression.
    - F $3x + 8 > 34$
    - **(H) $3x + 8 \leq 34$**
    - G $3x - 8 \leq 34$
    - J $3x + 8 < 34$

29. How tall is a giraffe that casts a shadow 320 cm long, if a man standing nearby who is 180 cm tall casts a shadow 100 cm long?
    - A 177.78 cm
    - **(C) 576 cm**
    - B 232 cm
    - D 626 cm

30. Simplify 5^{-4}.
    - F −20
    - H $\frac{1}{20}$
    - G 625
    - **(J) $\frac{1}{625}$**

31. Simplify $\frac{b^{45}}{b^{62}}$.
    - **(A) $\frac{1}{b^{17}}$**
    - C $\frac{1}{b^{107}}$
    - B b^{17}
    - D $\frac{1}{b^{-17}}$

32. A right angle measures _____.
    - F less than 90°
    - H more than 90°
    - **(G) exactly 90°**
    - J exactly 180°

33. Solve $4r + 3 = 19$.
    - A $r = 12$
    - **(C) $r = 4$**
    - B $r = 1$
    - D $r = 16$

34. Express the phrase, "12 is less than twice a number x," as an algebraic expression.
    - **(F) $12 < 2x$**
    - H $2x < 12$
    - G $12 - 2x$
    - J $2x - 12$

35. Find the area of a parallelogram with base length 18 m and height 12 m.
    - A 30 m²
    - **(C) 216 m²**
    - B 60 m²
    - D 108 m²

36. Express the phrase, "2 times the quotient of a number y and nine," as an algebraic expression.
    - F $\frac{9}{2y}$
    - H $18y$
    - G $2y - 9$
    - **(J) $2\left(\frac{y}{9}\right)$**

37. The diameter of a bicycle wheel is 20 inches. How far does it travel in one complete revolution? Use 3.14 for π.
    - A 31.4 in.
    - C 314 in.
    - **(B) 62.8 in.**
    - D 1256 in.

38. How many one-cup servings of milk can be poured from 2 gallons of milk?
    - F 16
    - H 64
    - **(G) 32**
    - J 128

39. Select the graph that is the solution to the inequality $x \geq 4$.
    - A
    - **(B)**
    - C
    - D

40. Select the graph that is the solution to the inequality $x - 4 \geq -10$.
    - F
    - G
    - H
    - **(J)**

Cumulative Test
Form C

1. Evaluate $\frac{2x + 3}{10}$ for $x = 1$.
   - A 5
   - C 4
   - B $\frac{2}{5}$
   - **(D) $\frac{1}{2}$**

2. Evaluate $-4m + 2n + 1$ for $m = 3$ and $n = -2$.
   - **(F) −15**
   - H −7
   - G 16
   - J 9

3. Evaluate $22 - 6s$ for $s = -7$.
   - A 924
   - C −112
   - **(B) 64**
   - D −20

4. Simplify $157 - (-235)$.
   - **(F) 392**
   - H −78
   - G 78
   - J −392

5. What is the best estimate for the capacity of a large juice bottle?
   - A 1 mL
   - C 1 kL
   - **(B) 1 L**
   - D 1 kg

6. Simplify $5(v + 4) + 8v$.
   - F $20v$
   - **(H) $13v + 20$**
   - G $5v + 5$
   - J $13v + 4$

7. Simplify $-9m - 3n - 5n$.
   - A $-9m - 2n$
   - C $-9m + 2n$
   - **(B) $-9m - 8n$**
   - D $-17mn$

8. Together, Manuel and Fatima collected $342 in donations for new band uniforms. Their collections were $6539 less than the total collected amount. What was the total?
   - **(F) $6881**
   - H $3119
   - G $6197
   - J $342

9. Joshua has 248 more baseball cards than Sarah, who has 63 more than Calvin. If Joshua has 752 cards; how many cards does Calvin have?
   - A 689
   - **(C) 441**
   - B 504
   - D 248

10. Movement of a figure along a straight line is called _____.
    - **(F) translation**
    - H reflection
    - G rotation
    - J protraction

11. Divide $\frac{18(-5)}{3(15)}$.
    - A −90
    - **(C) −2**
    - B −45
    - D 45

12. Simplify $(-3) \cdot (-3) \cdot (3)$.
    - F −27
    - H −9
    - **(G) 27**
    - J 9

13. Elijah wishes to buy a car for $16,550 before tax. If the sales tax is 7%, what will be the total cost of the car?
    - A $1158.50
    - **(C) $17,708.50**
    - B $15,391.50
    - D $28,135.00

14. Express 4.6×10^6 in standard notation.
    - **(F) 4,600,000**
    - H 460,000
    - G 46,000,000
    - J 46,000

15. Express 0.00163 in scientific notation.
    - **(A) 1.63×10^{-3}**
    - C 1.63×10^{-4}
    - B 163×10^{-4}
    - D 1.63×10^4

16. Find the mean of the data in the stem and leaf plot to the nearest tenth.

    | 15 | 0 0 4 7 |
    | 16 | 1 3 4 5 |
    | 17 | 1 4 6 8 9 |

    - F 150.0
    - **(H) 164.8**
    - G 164.0
    - J 165.0

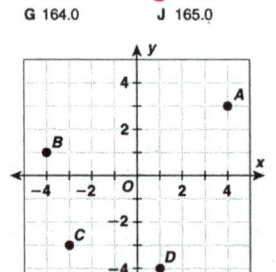

17. Give the coordinates of point B as shown on the graph above.
    - A (4, 1)
    - C (1, −4)
    - **(B) (−4, 1)**
    - D (−4, −1)

18. Give the coordinates of point C as shown on the graph above.
    - F (3, −4)
    - H (3, 4)
    - G (3, −3)
    - **(J) (−3, −3)**

19. Simplify 4^4.
    - **(A) 256**
    - C 3
    - B 44
    - D 16

20. What is the difference between the average temperatures for the first three days and the last three days?

Day	1	2	3	4	5	6	7
High	73°	78°	81°	85°	77°	69°	70°

    - F 77.3°
    - **(H) 5.3°**
    - G 72°
    - J 0.3°

21. Which statement is NOT true.
    - A On day 4, the temperature reached above 60°.
    - B The change in temperature from day 9 to day 10 was greater than from day 3 to day 4.
    - **(C) On most days, the temperature was below 50°.**
    - D The highest temperature reached on any day was approximately 61°.

22. Simplify $76 + (-15) + (-32)$.
    - F 123
    - H 44
    - G 59
    - **(J) 29**

23. Express the product $\frac{m^{12}}{m^0} \cdot m^2$ as one power.
    - A m^{24}
    - **(C) m^{14}**
    - B 1
    - D $\frac{1}{m^{12}}$

Cumulative Test
Chapter 2 Form C, continued

24. Which ordered pair is a solution of $2x = -y$?
    F (0, −6)   (H) (3, −6)
    G (−3, −6)   J (3, 6)

25. Which ordered pair is a solution of $5x + 1 = y$?
    (A) (−1, −4)   C (−5, −5)
    B (−5, −4)   D (1, −4)

26. Your bank account has $79 in it right now. You write checks for $41, $22, and $11. Then you make deposits of $58 and $24. How much is in the account after these transactions?
    F $71   (H) $87
    G $98   J $8

27. For Charmaine to get a B in math she needs to average 80 on four tests. Her scores on the first three tests were 78, 81, and 75. What is the lowest score she can receive on the next test and still get a B?
    A 78   C 80
    B 96   (D) 86

28. Express the phrase, "the quotient of 4 times a number, and the number times itself is greater than or equal to 12," as an algebraic expression.
    F $\frac{x^2}{4x} \geq 12$ (H) $\frac{4x}{x^2} \geq 12$
    G $\frac{4x}{x^2} \leq 12$ J $\frac{4x}{x^2} > 12$

29. On a particular scale drawing a measurement of 25 feet is represented by 2 inches. If two apartment buildings are actually 62.5 feet apart, what is the distance between them on the drawing?
    A 2.5 in.   C 11.4 in.
    (B) 5 in.   D 312.5 in.

30. What is the sum of 13 and −7, raised to the negative 4 power?
    F −1296   (H) $\frac{1}{1296}$
    G $\frac{1}{24}$ J 1296

31. Simplify $(16 − 11)^{-5}$.
    A 3125   C $\frac{1}{25}$
    (B) $\frac{1}{3125}$ D −3125

32. If two angles of a triangle measure 40° and 60°, how many degrees does the third angle measure?
    F 70°   H 90°
    (G) 80°   J 100°

33. Solve $\frac{x}{15} = 21$.
    A $x = 6$ (C) $x = 315$
    B $x = 36$ D $x = 1201$

34. Express the phrase, "the product of 5 and 9 more than a number y," as an algebraic expression.
    (F) $5(y + 9)$ H $(5 + 9)y$
    G $5 + 9 \cdot y$ J $5 \cdot 9 + y$

Cumulative Test
Chapter 2 Form C, continued

35. What is the area of a trapezoid with bases 15 m and 12 m and a height of 9 m.
    A 36 m²   C 54 m²
    (B) 121.5 m²   D 1620 m²

36. Express the phrase, "nine is equal to twelve minus three times a number n cubed," as an algebraic expression.
    F $12 = 9 − 3n^2$ (H) $9 = 12 − 3n^3$
    G $12 = 9 − 3n^3$ J $9 = 12 − 3n^2$

37. One bicycle wheel has a diameter of 25 inches, another wheel has a diameter of 27 inches. How much farther does the 27-inch wheel go than the 25-inch wheel in one rotation? Use 3.14 for π.
    A 2 inches   C 78.5 inches
    (B) 6.28 inches   D 84.78 inches

38. Tasha has ordered 2 pairs of jeans that weigh 12 oz each and 1 pair of running shoes that weigh 8 oz. If the cost to ship the items is $4.99 per pound, how much will it cost to ship the items to Tasha?
    F $4.99   (H) $9.98
    G $5.21   J $12.04

39. Select the graph that is a solution to $2x − 1 < 3$.
    A, (B), C, D

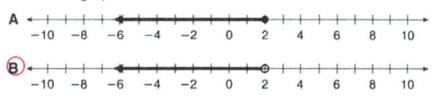

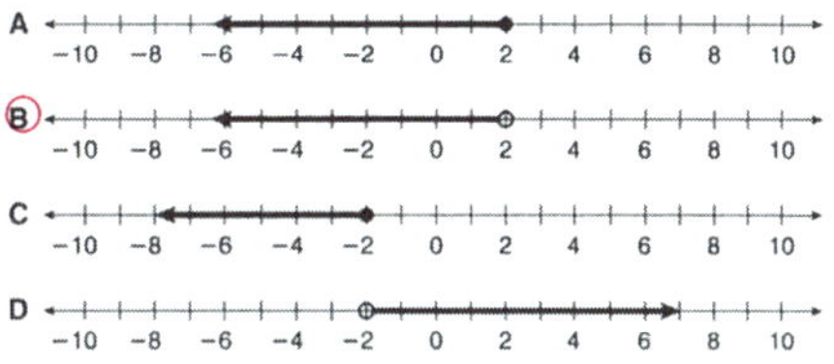

40. Select the graph that is a solution to $x \leq -3$.
    F, G, H, (J) 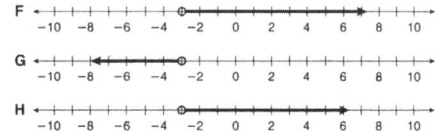 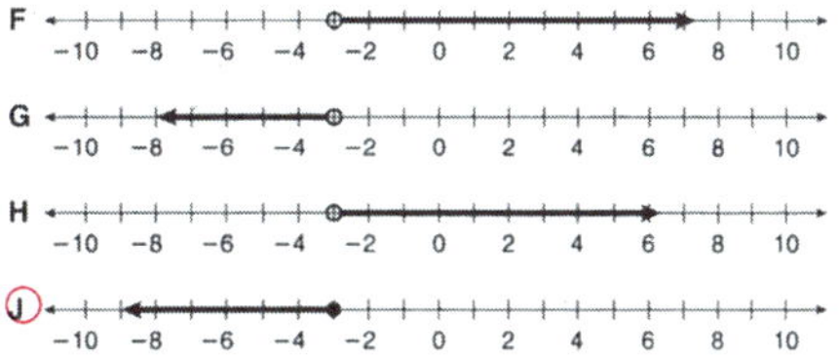

Cumulative Test
Chapter 3 Form A

1. Simplify $9 + (-5)$.
   A −4   (B) 4

2. Simplify $-7 + 21$.
   (A) 14   B −28

3. Simplify $3(10x + 7) + 2x$.
   A $30x + 21$ (B) $32x + 21$

4. Solve $11b − 3b = 32$.
   (A) $b = 4$ C $b = 8$
   B $b = 14$ D $b = 32$

5. How many days are there in 3 weeks?
   (A) 21   B 15

6. Simplify $\left(-\frac{1}{2}\right)\left(\frac{4}{19}\right)$.
   A $\frac{1}{19}$ (B) $-\frac{2}{19}$

7. Simplify $-\frac{18}{9}$.
   A 2   (B) −2

8. Which does not apply to the quadrilateral below? 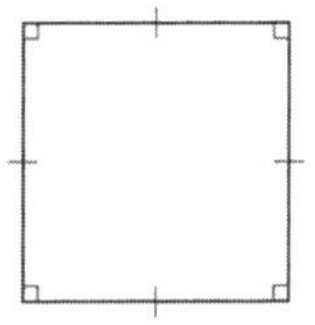
   A rhombus   C rectangle
   B parallelogram   (D) trapezoid

9. Evaluate $r + 3s$ for $r = 5$ and $s = 0$.
   (A) 5   B 15

10. Evaluate $\frac{c - 2}{2}$ for $c = -3$.
    A $-\frac{1}{2}$ (B) $-\frac{5}{2}$

11. Simplify $\frac{4}{11} - \left(-\frac{6}{11}\right)$.
    (A) $\frac{10}{11}$ B $-\frac{2}{11}$

12. Multiply $7\left(-\frac{6}{7}\right)$.
    A 6   C $\frac{13}{7}$
    B $-\frac{13}{7}$ (D) −6

13. Express 8.87×10^3 in standard notation.
    (A) 8870   B 887

14. Find the perimeter of the rectangle below. 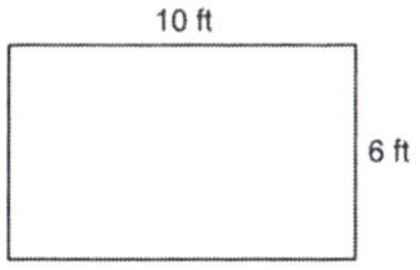
    10 ft
    6 ft
    A 60 ft   (B) 32 ft

Cumulative Test
Chapter 3 Form A, continued

15. Are the lines shown below parallel or perpendicular? 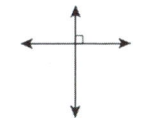
    A parallel   (B) perpendicular

16. Solve $4n = -8$.
    (A) $n = -2$ C $n = 0.5$
    B $n = 2$ D $n = -0.5$

17. Solve $-7 + t = 2$.
    A $t = -9$ (B) $t = 9$

18. Solve $3.2a − 4 = 4.2a − 13$.
    A $a = -6$ (B) $a = 9$

19. Express $x \cdot x \cdot x \cdot x \cdot x \cdot x \cdot x$ using exponents.
    (A) x^7 B $7x$

20. Solve $\frac{1}{2}m = 5$.
    (A) $m = 10$ C $m = 0.25$
    B $m = 2.5$ D $m = 1$

21. Simplify 10^3.
    A 30   (B) 1000

22. Choose the correct classification for the triangle below. 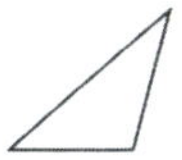
    A acute   (B) obtuse

23. Simplify 7^{-3}.
    (A) $\frac{1}{343}$ B $\frac{1}{49}$

24. Simplify $\frac{x^4}{x^3}$.
    A $\frac{1}{x}$ (C) x
    B x^7 D $7x$

25. Express 0.3 as a fraction.
    (A) $\frac{3}{10}$ B $\frac{3}{100}$

26. Simplify $\frac{8}{12}$.
    A $\frac{3}{4}$ (B) $\frac{2}{3}$

27. What is the phrase, "the product of 7 and b, minus the product of 3 and b," as an algebraic expression?
    A $3b - 7b$ C $3 \cdot b \cdot 7 \cdot b$
    (B) $7b - 3b$ D $(3 + 7)b$

28. Find the area of the triangle below.
    4 cm 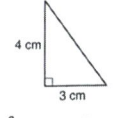
    3 cm
    A 12 cm²   C 7 cm
    B 10 cm²   (D) 6 cm²

Cumulative Test
Form A, continued

29. Divide $\frac{5}{12} \div 4$.
    A $\frac{20}{12}$
    (B) $\frac{5}{48}$

30. Find the volume of the prism below. 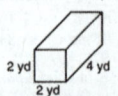

    A 8 yd³
    (B) 16 yd³

31. What is the quotient $\frac{b^3}{b^2}$ as one power?
    A b^5
    (B) b

32. An advertising company pays $275 a week plus a bonus of $23 for each new service contract. What is the total pay if 5 service contracts were sold in one week?
    (A) $390   C $115
    B $275   D $490

33. A sawmill trims a 2-inch board to be $1\frac{9}{16}$ inches wide. How much is trimmed off?
    A $\frac{11}{16}$ in.
    (B) $\frac{7}{16}$ in.

34. A rectangular scarf measures $2\frac{1}{2}$ feet by $3\frac{1}{3}$ feet. What is the distance around the scarf?
    A $5\frac{2}{5}$ ft
    (B) $11\frac{2}{3}$ ft

35. Venecia has $750 in her checking account. This month the bank charged $15 for checks, and she wrote checks for $95 and $55. What is the balance in the account?
    (A) $585   B $615

36. To the nearest tenth, find the volume of a sphere with a radius of 1 in. Use 3.14 for π.
    A 2.4 in³
    (B) 4.2 in³

37. Simplify $\sqrt{\frac{100}{25}}$.
    (A) 2
    B $\frac{5}{2}$

38. Find the surface area of the sphere below to the nearest tenth. Use 3.14 for π.

    r = 2 ft
    (A) 50.2 ft²   B 16.7 ft²

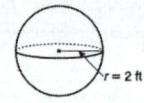

39. What kind of number is $\sqrt{-25}$?
    A rational   (C) not real
    B irrational   D negative

40. Translate $\frac{v}{2} + 1$ into words.
    (A) the quotient of v and 2, plus 1
    B the sum of two times v and 1
    C the sum of v and 1 divided by 2
    D the product of v and 2 and one

Cumulative Test
Form A, continued

41. Find the length of a side of a square with an area of 121 ft².
    A 44 ft.
    (B) 11 ft

42. To the nearest tenth, find the approximate distance around a square with an area of 110 cm².
    A 10.5 cm
    (B) 42.0 cm

43. Select the graph that is the solution of $x > 3$. 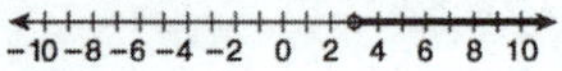
    (A)
    ◄++++++++++○++++►
    -10-8-6-4-2 0 2 4 6 8 10
    B
    ◄+++++++●++++++++►
    -10-8-6-4-2 0 2 4 6 8 10

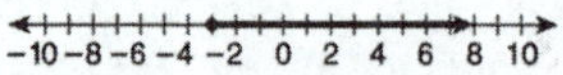

44. Select the graph that is the solution of $\frac{1}{2}x > 1$.
    A
    ◄+++++●++++++++++►
    -10-8-6-4-2 0 2 4 6 8 10
    (B)
    ◄+++++++++○+++++++►
    -10-8-6-4-2 0 2 4 6 8 10

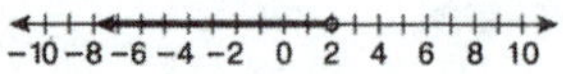

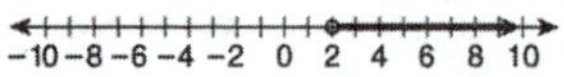

45. Solve $2.1x \leq -2.1$.
    A $x \geq -1$ C $x \leq 1$
    (B) $x \leq -1$ D $x \geq 1$

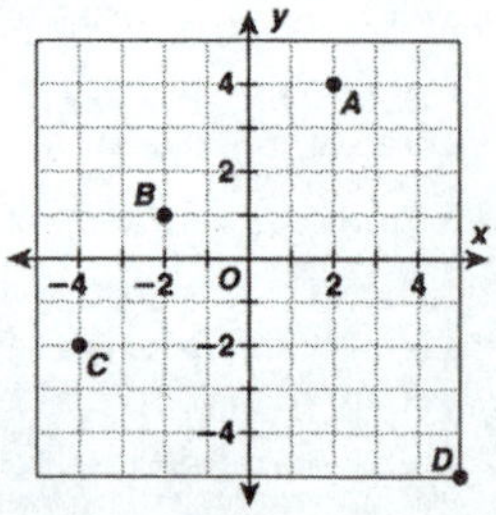

46. Use the graph above to give the coordinates of point B.
    (A) (-2, 1)   B (-2, -1)

47. Use the graph above to give the coordinates of point A.
    (A) (2, 4)   B (2, -4)

48. Which ordered pair is a solution of $2x + y = 5$?
    A (4, 1)   C (8, 1)
    B (16, 1)   (D) (1, 3)

49. Which ordered pair is a solution of $x + 3y = 18$?
    (A) (3, 5)   B (2, 3)

Cumulative Test
Form B

1. Simplify $-23 + (-14)$.
   A -9   (C) -37
   B 37   D 9

2. Simplify $10 + (-7)$.
   F 17   H -17
   (G) 3   J -3

3. Simplify $18v(4 - 2) - 25v + 72$.
   A $72 - 4v$ (C) $72 + 11v$
   B $18 + 2v$ D $18 + 11v$

4. Simplify $32x - (7x - 9)5$.
   F $68x - 45$ H $-45 + 3x$
   G $25x - 45$ (J) $45 - 3x$

5. How many minutes are there in 7.6 hours?
   A 402   (C) 456
   B 420   D 465

6. Simplify $\left(-\frac{4}{15}\right)\left(\frac{3}{7}\right)$.
   (F) $-\frac{4}{35}$ H $-\frac{24}{29}$
   G $\frac{2}{29}$ J $-\frac{2}{29}$

7. Simplify $\frac{42}{70}$.
   A $\frac{6}{12}$ (C) $\frac{3}{5}$
   B $\frac{6}{7}$ D $\frac{14}{15}$

8. Give all the names that apply to the quadrilateral below. 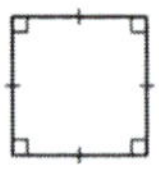

   F square, rhombus, kite, parallelogram
   (G) square, rhombus, rectangle, parallelogram
   H square, rhombus, trapezoid
   J square, rhombus, parallelogram

9. Evaluate $5r + 9s + 6$ for $r = 0$ and $s = 8$.
   (A) 78   C 83
   B 46   D 15

10. Evaluate $\frac{c + 4}{9}$ for $c = 4$.
    F $\frac{22}{9}$ H $\frac{9}{8}$
    G $\frac{16}{9}$ (J) $\frac{8}{9}$

11. Divide $\frac{3}{8} \div \frac{2}{5}$.
    (A) $\frac{15}{16}$ C $\frac{31}{40}$
    B $\frac{3}{20}$ D $\frac{1}{40}$

12. Multiply $\frac{5}{8}\left(-\frac{7}{12}\right)$.
    (F) $-\frac{35}{96}$ H $\frac{12}{20}$
    G $-\frac{60}{56}$ J $-\frac{2}{4}$

13. What is 5.6×10^6 in standard notation?
    A 56,000,000   C 56,000,000
    B 56,000   (D) 5,600,000

Cumulative Test
Form B, continued

14. Find the circumference of the circle to the nearest tenth. Use 3.14 for π.

    6 cm
    (F) 37.7 cm
    G 37.6 cm
    H 18.8 cm
    J 113.0 cm

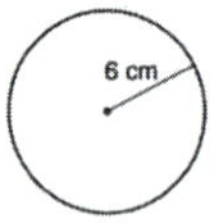

15. State whether the lines shown are parallel, perpendicular, skew, or none of these. 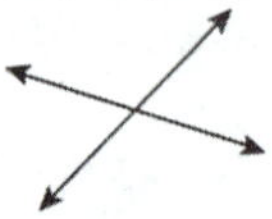
    A parallel
    B perpendicular
    C skew
    (D) none of these

16. Solve $\frac{1}{2}m = -4$.
    F $m = -2$ H $m = 1$
    (G) $m = -8$ J $m = 2$

17. Solve $-b - 2 = 18$.
    A $b = 16$ C $b = -16$
    B $b = 20$ (D) $b = -20$

18. Solve $3.7a + 4 = 2.7a - 8$.
    F $a = -11$ H $a = -13$
    G $a = -2$ (J) $a = -12$

19. Express $(-g) \cdot (-g) \cdot (-g) \cdot (-g) \cdot (-g)$ using exponents.
    A $-g^4$ (C) $(-g)^5$
    B $\frac{1}{g^5}$ D $-\frac{1}{g^5}$

20. Solve $-55.2 = -6.9c$.
    (F) $c = 8.0$ H $c = 48.3$
    G $c = 2.0$ J $c = -48.3$

21. Simplify $(-5)^4$.
    A -625   (C) 625
    B -125   D 3125

22. Classify the triangle according to its sides and angles. 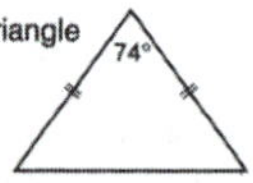

    74°
    F scalene acute triangle
    G right triangle
    (H) isosceles acute triangle
    J isosceles obtuse triangle

23. Simplify $\frac{4^7}{4^5}$.
    (A) $\frac{1}{16}$ C 16
    B $\frac{1}{8}$ D 8

24. Simplify $\frac{x^{17}}{x^5}$.
    F $\frac{1}{x^{12}}$ H x^{22}
    (G) x^{12} J x^{85}

25. Express 1.25 as a fraction.
    (A) $\frac{5}{4}$ C $\frac{1}{4}$
    B $1\frac{2}{5}$ D $1\frac{3}{4}$

26. Simplify $\frac{17}{51}$.
    F 3   (H) $\frac{1}{3}$
    G $\frac{1}{4}$ J 4

27. What is the phrase, "the difference of 8 times a number r and 12," as an algebraic expression?
    A $12 - 8r$ (C) $8r - 12$
    B $8 - r + 12$ D $12 + 8r$

Cumulative Test
Chapter 3 Form B, continued

28. Find the area of the triangle.
    F 56 ft²
    (G) 28 ft²
    H 15 ft
    J 112 ft²

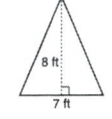

29. Divide $2 \div \frac{2}{15}$.
    A $\frac{4}{15}$
    C $\frac{1}{15}$
    (B) 15
    D $\frac{2}{15}$

30. To the nearest tenth, find the volume of the cylinder. Use 3.14 for π.
    F 63.6 in³
    G 763.0 in³
    (H) 254.3 in³
    J 84.8 in³

31. Express the quotient $\frac{m^5 m^3}{m^2}$ as one power.
    A m^{17}
    (C) m^6
    B m^{13}
    D m^4

32. Keisha earns $325 a week plus a bonus of $23 for each service contract she sells. What is her pay if she sells 3 service contracts in one week?
    F $325
    H $494
    (G) $394
    J $69

33. Martin has a 4-inch board that he wants to be $3\frac{11}{16}$ inches wide. How much does he need to trim off?
    A $7\frac{11}{16}$ in.
    C $\frac{11}{16}$ in.
    (B) $\frac{5}{16}$ in.
    D $4\frac{5}{16}$ in.

34. A rectangular garden measures $2\frac{2}{3}$ feet by $50\frac{1}{2}$ feet. What is the distance around the garden?
    (F) $106\frac{1}{3}$ ft
    H $134\frac{2}{3}$ ft
    G $53\frac{1}{6}$ ft
    J 109 ft

35. Raul has $506 in his checking account. This month the bank charged $6 for checks, and he wrote checks for $91 and $72. What is the balance in the account?
    A $531
    (C) $337
    B $675
    D $349

36. To the nearest tenth, find the volume of a sphere with a radius of 2 m. Use 3.14 for π.
    F 25.1 m³
    (H) 33.5 m³
    G 10.7 m³
    J 16.7 m³

37. Simplify $\sqrt{144}$.
    A 72
    (C) 12
    B 13
    D irrational

38. To the nearest tenth, find the surface area of the cylinder formed by the net. Use 3.14 for π.
    F 75.4 cm²
    (G) 150.7 cm²
    H 87.9 cm²
    J 138.2 cm²

39. What kind of number is $-\sqrt{17}$?
    A rational
    C not real
    (B) irrational
    D positive

Cumulative Test
Chapter 3 Form B, continued

40. Translate $\frac{v}{2} + 3v$ into words.
    (F) v divided by two, plus three times v
    G half a number plus 3
    H 2 times the quotient of v and 3
    J the product of half of v and 3 times v

41. To the nearest tenth, find the approximate length of a side of a square with an area of 652 cm².
    A 32.6 cm
    C 26.2 cm
    (B) 25.5 cm
    D 17.3 cm

42. To the nearest tenth, find the distance around a square with an area of 165 cm².
    (F) 51.4 cm
    H 52.1 cm
    G 12.8 cm
    J 50 cm

43. Which is the solution of $2x \le 6$?
    (D)

44. Which is the solution of $2x > -5$?
    (G)

45. Solve $-\frac{1}{3}x + 1 \ge 3$.
    A $x \ge 12$
    C $x \le 6$
    B $x \le 12$
    (D) $x \le -6$

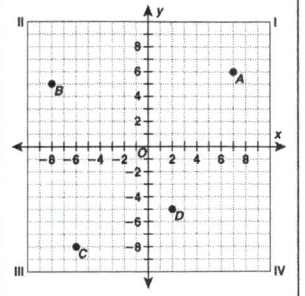

46. Use the graph above to give the coordinates of point B.
    F (2, −5)
    (H) (−8, 5)
    G (5, 8)
    J (−5, 8)

47. Use the graph above to give the coordinates of point C.
    A (8, 6)
    C (6, −8)
    B (−8, 6)
    (D) (−6, −8)

48. Which ordered pair is a solution of $5x - y = 6$?
    (F) (3, 9)
    H (3, 15)
    G (5, 15)
    J (45, 5)

49. Which ordered pair is a solution of $3y = 6x + 3$?
    A (7, 3)
    (C) (3, 7)
    B (−6, −3)
    D (−7, −12)

Cumulative Test
Chapter 3 Form C

1. Simplify $-9 + (4 - 9)$.
   A 4
   C −4
   B 14
   (D) −14

2. Simplify $26 + (-127)$.
   F 153
   (H) −101
   G 127
   J −153

3. Simplify $5(11 + 8y) - 2y$.
   A 55 − 42y
   C 55 − 38y
   (B) 55 + 38y
   D 55 + 42y

4. Simplify $7(21 - 9m) - 13m$.
   F 147 − 22m
   H 28 − 22m
   (G) 147 − 76m
   J 28 − 76m

5. How many minutes are there in 1 year?
   A 43,200
   C 8760
   (B) 525,600
   D 31,536,000

6. Simplify $\left(-\frac{7}{25}\right)\left(-\frac{1}{5}\right)$.
   F $\frac{7}{5}$
   H $\frac{7}{30}$
   (G) $\frac{7}{125}$
   J $\frac{7}{20}$

7. Simplify $-\frac{75}{15}$.
   A 15
   (C) −5
   B $\frac{-75}{8+7}$
   D 5

8. Give all the names that apply to the quadrilateral.
   (F) trapezoid
   G kite, trapezoid
   H trapezoid, parallelogram
   J trapezoid, rhombus

9. Evaluate $6(t - 6)$ for $t = -3$.
   A −18
   C −24
   (B) −54
   D −9

10. Evaluate $-5st$ for $s = -2$ and $t = -4$.
    F −20
    H −80
    G 40
    (J) −40

11. Simplify $-\frac{5}{13} - \left(-\frac{9}{13}\right)$.
    A $-\frac{14}{13}$
    C $\frac{14}{13}$
    (B) $\frac{4}{13}$
    D $-\frac{4}{13}$

12. Multiply $2\frac{8}{9}\left(\frac{3}{7}\right)$.
    F $\frac{21}{26}$
    H $\frac{16}{21}$
    (G) $\frac{26}{21}$
    J $\frac{21}{16}$

13. What is 2.321×10^{-6} in standard notation?
    A 0.0000002321
    (C) 0.000002321
    B 2,321,000
    D 0.002321

14. Find the perimeter of the polygon.
    (F) 29 cm
    G 28 cm
    H 30 cm
    J 31 cm

15. State whether the lines shown are parallel, perpendicular, skew, or none of these.
    A parallel
    B perpendicular
    (C) skew
    D none of these

Cumulative Test
Chapter 3 Form C, continued

16. Solve $-2(n - 6) = -4$.
    (F) $n = 8$
    H $n = 4$
    G $n = 10$
    J $n = 5$

17. Solve $-t + 20 = -54$.
    A $t = -74$
    C $t = 34$
    (B) $t = 74$
    D $t = -34$

18. Solve $-0.111c + 1 = 0.889c$.
    F $c = 10.0$
    (H) $c = 1$
    G $c = 0.778$
    J $c = -0.778$

19. Express $\frac{5}{6} \cdot \frac{5}{6} \cdot \frac{5}{6}$ using exponents.
    A $\frac{5^3}{6}$
    (C) $\left(\frac{5}{6}\right)^3$
    B $\left(\frac{5}{6}\right)^2$
    D $\frac{5}{6^3}$

20. Solve $\frac{1}{2}m = -\frac{5}{2}$.
    F $m = -2$
    H $m = 3$
    (G) $m = -5$
    J $m = 2$

21. Simplify $(-2)^9$.
    A $-\frac{5}{2}$
    C −256
    B $\frac{5}{2}$
    (D) −512

22. Classify the triangle according to its sides and angles.
    F scalene acute triangle
    H isosceles right triangle
    G isosceles acute triangle
    (J) scalene obtuse triangle

23. Simplify $(25 - 9)^{-2}$.
    A $\frac{1}{16}$
    C −16
    (B) $\frac{1}{256}$
    D −64

24. Simplify $y^8 \cdot y^3$.
    (F) y^{11}
    H y^{-5}
    G y^{-24}
    J y^{24}

25. What is 0.00625 as a fraction?
    A $\frac{625}{1000}$
    (C) $\frac{1}{160}$
    B $\frac{1}{625}$
    D $\frac{62}{100,000}$

26. Simplify $\frac{80}{104}$.
    F $\frac{1}{3}$
    (H) $\frac{10}{13}$
    G $\frac{40}{52}$
    J $\frac{8}{10}$

27. What is the phrase, "the product of two and six less than twice b," as an algebraic expression?
    A $2 - 6 \cdot b$
    C $2 \cdot 6 - b$
    (B) $2(2b - 6)$
    D $(2 - 6)b$

28. Find the area of the trapezoid below.
    (F) 13.5 m²
    G 12 m²
    H 27 m²
    J 18 m²

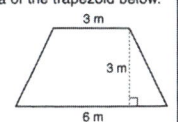

Cumulative Test
Chapter 3 Form C, continued

29. Divide $4\frac{4}{5} \div 12\frac{1}{5}$.
    A 17
    B $\frac{3}{8}$
    C $\frac{24}{61}$
    D $1\frac{8}{9}$

30. Find the volume of the prism.
    F 19 ft³
    G 50 ft³
    H 200 ft³
    (J) 100 ft³

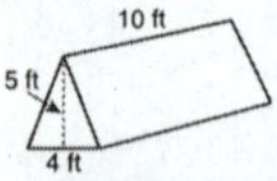

31. Express the quotient $\frac{m^5 m^3 n^2}{m^4 n^2}$ as one power.
    A m^{22}
    B m^{11}
    (C) m^4
    D $m^4 n$

32. Nela borrowed $430 from her sister. If it is paid back in 6 monthly payments of $95, how much is the sister charging for the loan?
    (F) $140
    G $570
    H $45
    J $670

33. Marla put $7\frac{1}{2}$ lb, $3\frac{1}{3}$ lb, and 2 lb of meat in the freezer. What is the total amount of meat?
    A $\frac{2}{7}$ lb
    (B) $12\frac{5}{6}$ lb
    C $12\frac{2}{5}$ lb
    D $3\frac{1}{2}$ lb

34. A rectangular scarf measures $5\frac{1}{3}$ inches by $49\frac{1}{2}$ inches. What is the distance around the scarf?
    F 264 in.
    G $54\frac{5}{6}$ in.
    (H) $109\frac{2}{3}$ in.
    J 115 in.

35. A football team gained 20 yards on the 1st play, lost 3 yards on the 2nd play, and then gained another 14 yards. What was the net gain or loss of yardage?
    (A) 31 yards
    B −3 yards
    C 37 yards
    D 3 yards

36. Find the volume of a cone that is 10 cm high and has a base with a radius of 3 cm. Use 3.14 for π.
    F 282.6 cm³
    (G) 94.2 cm³
    H 31.4 cm³
    J 141.3 cm³

37. Simplify $\sqrt{\frac{529}{49}}$.
    A 11
    B $\frac{23}{8}$
    C $\frac{23}{7}$
    D $\frac{24}{7}$

38. Find the surface area of the prism formed by the net below. 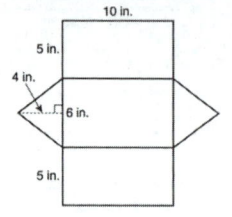
    F 172 in²
    G 174 in²
    H 294 in²
    J 184 in²

39. What kind of number is $\sqrt{\frac{25}{4}}$?
    (A) rational
    B irrational
    C not real
    D negative

Cumulative Test
Chapter 3 Form C, continued

40. Translate $\left(\frac{v}{2}\right)^2$ into words.
    F 2 times the quotient of v and 2
    G the square of the product of v and 2
    (H) the square of the quotient of v and 2
    J the product of half v and 2

41. A square has an area of 37 mm². To the nearest tenth, what is the length of one side?
    A 19.0 mm
    (B) 6.1 mm
    C 5.8 mm
    D 1396.0 mm

42. What is the distance around a square that has an area of 1500 cm²?
    F 38.7 cm
    G 50.0 cm
    (H) 154.9 cm
    J 375.0 cm

43. Which is the solution of $5x < -28 + x$?
    A 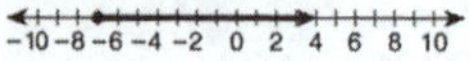
    B 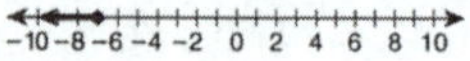
    (C) 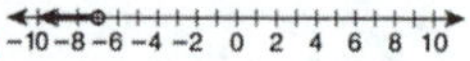
    D

44. Which is the solution of $-2 < \frac{2n}{7}$?
    (F) 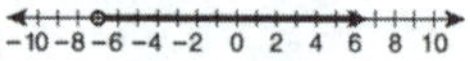
    G 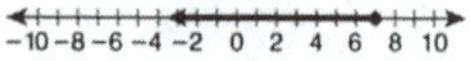
    H 
    J

45. Solve $6.8x \geq 8.5$.
    A $x \geq 0.8$
    B $x \leq 0.8$
    C $x \leq 1.25$
    (D) $x \geq 1.25$

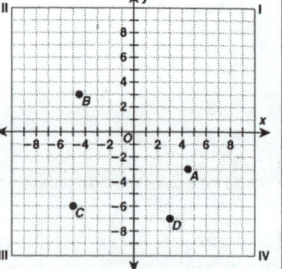

46. Use the graph above to give the coordinates of point D.
    (F) (3, −7)
    G (−7, 3)
    H (−3, 7)
    J (−3, −7)

47. Use the graph above to give the coordinates of point A.
    A (3, 4)
    B (4, −7)
    (C) (4.5, −3)
    D (−4.5, −3)

48. Which ordered pair is a solution of $0.25x + 1.5y = 4$?
    F (2, 4)
    G (0.5, 1.5)
    H (2, 1.5)
    (J) (4, 2)

49. Which ordered pair is a solution of $\frac{1}{3}x + \frac{2}{5}y = \frac{11}{15}$?
    A $\left(\frac{6}{5}, \frac{5}{3}\right)$
    B $\left(\frac{2}{5}, \frac{1}{3}\right)$
    (C) (1, 1)
    D $\left(\frac{1}{3}, \frac{2}{5}\right)$

Cumulative Test
Chapter 4 Form A

1. Simplify $\sqrt{-4}$.
   A −2
   (B) not a real number

2. Which is equivalent to 2^4?
   (A) $2 \cdot 2 \cdot 2 \cdot 2$
   B $2 \cdot 4$

3. Describe and give the value of $-\sqrt{9}$.
   A Irrational, −3
   (B) Rational, −3

4. Express 1900 in scientific notation.
   (A) 1.9×10^3
   B 10×1.9^5

5. Express 4.5×10^4 in standard notation.
   A 45
   B 450,000
   C 4500
   (D) 45,000

6. Find the two square roots of $\frac{9}{4}$.
   (A) $\frac{3}{2}, -\frac{3}{2}$
   B $\sqrt{\frac{3}{2}}, -\sqrt{\frac{3}{2}}$

7. Which of the following is the same as the phrase "the sum of three times a number and 1 is 13"?
   A $n + 3 = 13$
   (B) $3n + 1 = 13$

8. Solve $2x = 16$.
   A $x = -4$
   (B) $x = 8$
   C $x = 4$
   D $x = 32$

9. Simplify $5^8 \div 5^6$.
   A 5^{14}
   (B) 5^2

10. Which is equal to $(n)(n)(n)(n)$?
    A $4n$
    (B) n^4

11. Solve $5b = -10$.
    (A) $b = -2$
    B $b = 2$

12. This frequency table represents which data set?

Age When Learned to Catch	2	3	4	5
Frequency	4	3	2	1

    A 3 2 4 2 5 2 3 5 2 3
    B 4 2 4 2 5 2 3 4 2 3
    (C) 3 2 4 2 5 2 3 4 2 3
    D 2 2 4 2 5 2 3 4 2 3

13. Evaluate 3^{-2}.
    A −9
    (B) $\frac{1}{9}$

14. Simplify $5 - (-3)$.
    (A) 8
    B 2
    C −8
    D −2

15. Find the area of a rectangle that measures $2\frac{2}{3}$ in. by 3 in.
    A $6\frac{1}{3}$ in²
    (B) 8 in²

16. A 4-H club has 64 members. There are 20 more boys than girls. How many girls are there?
    (A) 22 girls
    B 44 girls
    C 20 girls
    D 64 girls

Cumulative Test
Chapter 4 Form A, continued

17. The area of the top of a square picnic table is 9 ft². What is the measurement of the top?
    (A) 3 feet × 3 feet
    B 2 feet × 7 feet

18. Luke biked 45 miles at an average rate of 15 miles per hour. How long did the trip take?
    (A) 3 hr
    B $\frac{1}{3}$ hr

19. Convert the following to an algebraic expression "Anna earned $20 a day at her job cleaning up at the fairgrounds. If she works one day a week, how much did she earn in n weeks?"
    A $20 + n$
    (B) $20n$

20. What is possibly misleading about the phrase, "The average school lunch price is $2.00"?
    (A) the average may be skewed by one high-priced item
    B school lunches are always the same price
    C the sample size may be too large
    D most students bring their lunches

21. Simplify $\frac{9}{-3}$.
    A 3
    (B) −3

22. Ten eighth graders said that hip-hop was their favorite kind of music, so Sarah states in her report that all eighth graders like hip-hop. What is possibly wrong with this statement?
    (A) the sample size is too small
    B too many people were asked

23. To make a fruit salad, Marla bought $2\frac{1}{4}$ pounds of apples, $1\frac{1}{3}$ pounds of peaches, and 1 pound of bananas. How much fruit salad did this make?
    (A) $4\frac{7}{12}$ lb
    B $4\frac{2}{7}$ lb

24. Simplify $\frac{2}{5} - \frac{3}{5}$.
    A $\frac{1}{5}$
    (B) $-\frac{1}{5}$

25. Which choice represents the data from the stem-and-leaf plot?

Stem	Leaves
2	7 8
3	5 8
4	7 8

    A 27, 21, 28, 38, 47, 48
    B 27, 28, 28, 35, 47, 48
    C 27, 28, 35, 38, 45, 48
    (D) 27, 28, 35, 38, 47, 48

26. Solve $\frac{1}{2}b > \frac{3}{4}$.
    (A) $b > 1\frac{1}{2}$
    B $b > \frac{3}{8}$

27. Find the range for the data set. 11, 12, 10, 14, 13.
    (A) 4
    B 1
    C 12
    D 2

Cumulative Test
Form A, continued

28. Use the data to find the median. May's normal monthly rainfall (in inches) for 5 different U.S. cities: 3.1, 1.2, 2.4, 3.2, 4.1.
    A 2.8
    (B) 3.1

29. Find the first quartile of the data set. 18, 14, 14, 23, 29, 10, 19
    A 12
    (B) 14

30. What type of correlation is shown in the scatter plot?

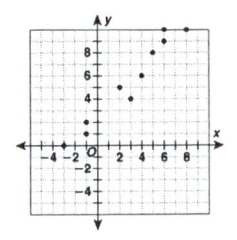

    (A) strong positive correlation
    B strong negative correlation
    C weak positive correlation
    D weak negative correlation

31. Divide $\frac{2}{3} \div \frac{1}{5}$.
    (A) $3\frac{1}{3}$
    B $\frac{3}{10}$

32. Kay's scores for her past six quizzes were 15, 10, 14, 12, 9, and 13. Find the average of her quiz scores to the nearest tenth.
    (A) 12.2
    B 14.6
    C 12.0
    D 11.5

33. If a waiter at Rudy's restaurant received the following tips for the last five nights, what is his range of tips? $30, $30, $10, $50, and $25
    A $20
    (B) $40

34. Solve $2x \geq -6$.
    A $x \leq -3$
    (B) $x \geq -3$

Cumulative Test
Form A, continued

35. Evaluate $\frac{3a - 2b}{9}$ when $a = 4$ and $b = 1$.
    (A) $\frac{10}{9}$
    B $\frac{1}{9}$

36. Simplify $6g - 8 + 6g$.
    A $4g$
    B $12g + 8$
    C $20g$
    (D) $12g - 8$

37. Express 1.2 as a fraction.
    (A) $1\frac{1}{5}$
    B $1\frac{2}{5}$

38. Solve $0.25a + 5 = 1.25a$.
    A $a = 7.5$
    (B) $a = 5$
    C $a = -5$
    D $a = 10$

39. Solve $-3 < \frac{b}{2}$.
    (A) $b > -6$
    B $b < -6$

40. Determine which ordered pair is a solution of the equation $3x + y = 8$.
    A (5, 0)
    (B) (3, −1)

41. On the graph, Point A represents which ordered pair?

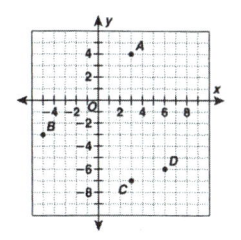

    A (3, −4)
    (B) (3, 4)

42. Simplify $-1 + (-5)$.
    A 4
    B 6
    (C) −6
    D −4

43. Simplify $2(5 - 3m)$.
    A $10 - 3m$
    (B) $10 - 6m$

44. Simplify $(2 + 7)^2$.
    A 51
    B −32
    (C) 81
    D 12

Cumulative Test
Form B

1. Simplify $\sqrt{-16}$.
   A −4
   B 8
   C 4
   (D) not real

2. Which is equivalent to 12^4?
   (F) $12 \cdot 12 \cdot 12 \cdot 12$
   G $12 \cdot 4$
   H $12 \cdot 12 \cdot 12 \cdot 12 \cdot 12$
   J $12 \div 4$

3. Describe and give the value of $-\sqrt{100}$.
   A Irrational, −50
   B Not a real number
   C Irrational, −10
   (D) Rational, −10

4. Express 1,900,000 in scientific notation.
   F 1.9×10^5
   G 10×1.9^5
   (H) 1.9×10^6
   J 10×1.9^6

5. Express 7.3×10^5 in standard notation.
   A 73
   B 7,300,000
   C 73,000
   (D) 730,000

6. Find the two square roots of $\frac{81}{25}$.
   (F) $\frac{9}{5}, -\frac{9}{5}$
   G 9, −9
   H $\sqrt{\frac{9}{5}}, -\sqrt{\frac{9}{5}}$
   J 9, 5

7. Which of the following is the same as the phrase, "The product of twice a number and 3 is 84"?
   A $6a = 84a$
   B $6a = 84$
   (C) $2a + 3 = 84$
   D $2a = 3(84)$

8. Solve $10x - 4 = 36$.
   F $x = -4$
   G $x = 6$
   (H) $x = 4$
   J $x = 32$

9. Simplify $6^3 \cdot 6^7$.
   A 6^{21}
   (B) 6^{10}
   C 6^4
   D 36^{21}

10. Which expression is equal to $(-3n)(-3n)(-3n)(-3n)$?
    F $-12n$
    G $-(3n)^4$
    H $-3n^4$
    (J) $(-3n)^4$

11. Solve $-3b = -18$.
    (A) $b = 6$
    B $b = -6$
    C $b < -6$
    D $b \leq -6$

12. This frequency table represents which data set?

Age When Learned to Ride a Bike	4	5	6	7
Frequency	2	9	5	2

    F 5 4 5 5 5 7 6 5 5 5 5 6 5 6 4 6 5 6
    G 3 4 5 5 5 7 6 5 5 5 5 6 6 7 4 6 5 6
    (H) 5 4 5 5 5 7 6 5 5 5 5 6 6 7 4 6 5 6
    J 5 4 5 5 5 7 6 5 4 4 5 6 6 7 4 6 5 6

13. Evaluate 4^{-3}.
    A −12
    (B) $\frac{1}{64}$
    C −81
    D $-\frac{3}{4}$

14. Simplify $12 - (-13)$.
    F −25
    G 1
    H −1
    (J) 25

Cumulative Test
Form B, continued

15. Find the area of a rectangle that measures $2\frac{2}{3}$ in. by $36\frac{1}{2}$ in.
    A $78\frac{1}{3}$ in^2
    B 81 in^2
    (C) $97\frac{1}{3}$ in^2
    D $39\frac{1}{6}$ in^2

16. The eighth grade at Byrnedale Junior High School has 464 students. There are 212 more boys than girls. How many boys are there?
    F 212 boys
    (G) 338 boys
    H 126 boys
    J 464 boys

17. The area of the top of a square table is 240 in^2. What are the dimensions of the top?
    A 12 inches × 20 inches
    (B) 15.49 inches × 15.49 inches
    C 24 inches × 10 inches
    D 61.97 inches × 4 inches

18. Jacob drove 450 miles at an average rate of 65 miles per hour. Rounded to the nearest tenth of an hour, how long did the trip take?
    (F) $6\frac{9}{10}$ hr
    G $\frac{1}{10}$ hr
    H $6\frac{93}{100}$ hr
    J $\frac{14}{100}$ hr

19. Convert the following to an algebraic expression "Leana earned $25 a day at her job. If she works five days a week, how much did she earn in n weeks?"
    A $25 + n$
    (B) $125n$
    C $25n$
    D $25 + 5n$

20. What is possibly misleading about the statistic, "Three out of four students prefer to ride the bus to school"?
    F The average is skewed by one outlier.
    G The measurements are not the same.
    (H) The sample size may be too small.
    J Students don't like to ride buses.

21. Simplify $\frac{10}{-5}$.
    A 5
    B 2
    C −5
    (D) −2

22. Fifty families who live near a freeway were asked whether the freeway should be expanded. Why might this sample be biased?
    F The average is skewed.
    G These people don't drive.
    (H) The people near the freeway are not likely to want it expanded.
    J Too many people were asked.

23. To obtain a party mix, Marla mixed $2\frac{1}{4}$ pounds of peanuts, $5\frac{1}{3}$ pounds of chocolate candies, and 5 pounds of raisins. What was the total weight?
    A $\frac{5}{18}$ lb
    B $3\frac{3}{5}$ lb
    (C) $12\frac{7}{12}$ lb
    D $\frac{12}{151}$ lb

24. Simplify $-\frac{2}{7} - \left(-\frac{4}{7}\right)$.
    F $-\frac{6}{7}$
    G $-\frac{2}{7}$
    (H) $\frac{2}{7}$
    J $\frac{6}{7}$

359
Holt Pre-Algebra

Cumulative Test
Form B, continued

25. Which choice represents the data from the stem-and-leaf plot?

Stem	Leaves
9	0 2 2
10	2 9
11	3 3 9

    A 90, 92, 99, 102, 109, 113, 113, 119
    B 90, 92, 92, 102, 109, 113, 113, 119
    C 92, 92, 102, 109, 113, 113, 119
    D 90, 92, 92, 102, 109, 112, 113, 119

26. Solve $4b \leq -\frac{2}{3}$.
    F $b \leq -\frac{1}{6}$ H $b \geq -\frac{8}{3}$
    G $b \leq -\frac{8}{3}$ J $b \geq -\frac{1}{6}$

27. Find the range for the data set. 326, 467, 588, 401, 326, 515
    A 189   C 121
    B 262   D 48

28. The average August monthly precipitation (in inches) for 10 different U.S. cities is given below. Find the median.
    3.5, 1.6, 2.4, 3.7, 4.1, 3.9, 1.0, 3.6, 4.2, 3.4.
    F 3.5 in.   H 3.6 in.
    G 3.7 in.   **J** 3.55 in.

29. Find the first quartile for the data set. 30, 38, 35, 38, 49, 38, 49, 39, 45
    A 46.5   **C** 36.5
    B 38   D 35

30. Which type of correlation is shown in the scatter plot?

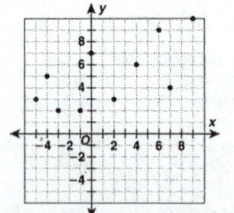

    F strong positive correlation
    G strong negative correlation
    H weak positive correlation
    J weak negative correlation

31. Divide $\frac{7}{8} \div \frac{5}{6}$.
    A $\frac{35}{42}$ C $\frac{20}{21}$
    B $1\frac{1}{20}$ D $\frac{6}{7}$

32. Jill's last grocery bills were $65.72, $55.82, $68.70, $78.19, $64.80, and $40.66. Find the average bill and round your answer to the nearest cent.
    F $93.47   H $62.78
    G $74.78   **J** $62.32

33. If you received the scores of 30, 37, 11, 50, and 53 on five math quizzes, what is the range of your scores?
    A 42   C 37
    B 20   D 36.2

34. Solve $10x \geq -70$.
    F $x > -7$ H $x = -7$
    G $x \geq -7$ J $x \leq -7$

35. Evaluate $\frac{7a - 5b}{6}$ when $a = 6$ and $b = 7$.
    A $\frac{19}{6}$ C $\frac{77}{6}$
    B $\frac{20}{3}$ **D** $\frac{7}{6}$

36. Simplify $6v - 8 - 6v + 16$.
    F 24   **H** 8
    G $v + 24$ J $12v + 8$

37. Express 2.125 as a fraction.
    A $2\frac{1}{8}$ C $2\frac{2}{5}$
    B $21\frac{1}{4}$ D $2\frac{1}{3}$

38. Solve $0.667a + 5 = 1.667a$.
    F $a = 1.667$ **H** $a = 5$
    G $a = -5$ J $a = 10$

39. Solve $-7 < \frac{b}{3}$.
    A $b < -21$ C $b < 21$
    B $b > -21$ D $b > 21$

40. Determine which ordered pair is a solution of the equation $y + 3x = 15$.
    F (5, 0)   H (5, 1)
    G (6, 3)   J (−6, 3)

41. On the graph, Point C represents which ordered pair?

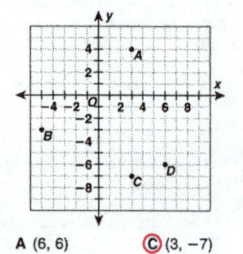

    A (6, 6)   **C** (3, −7)
    B (3, 7)   D (−3, 7)

42. Simplify $-4 + (-12)$.
    F −8   H 16
    G −16   J 8

43. Simplify $-1 + 9(5 - 3m)$.
    A $44 + 27m$ C $44 - 3m$
    B $45 - 27m$ **D** $44 - 27m$

44. Simplify $(2 - 6)^{-3}$.
    F $\frac{1}{64}$ **H** $-\frac{1}{64}$
    G 64   J 128

Cumulative Test
Form C

1. Simplify $-\sqrt{-10}$.
   A −5   C 3.16
   B −3.16   **D** not real

2. Which is equivalent to 9^5?
   F $9 \cdot 9 \cdot 9 \cdot 9$ **H** $9 \cdot 9 \cdot 9 \cdot 9 \cdot 9$
   G $9 \cdot 5$ J $9 \div 5$

3. Describe and give the value of $-\sqrt{200}$.
   A Irrational, −20
   B Not a real number
   C Irrational, −14.14
   D Rational, −20

4. Express 4,600,000,000 in scientific notation.
   F 4.6×10^9 H 4.6×10^8
   G 10×4.6^8 J 10×4.6^9

5. Express 9.5×10^{-4} in standard notation.
   A 0.0095   C 9500
   B 0.00095   D 95,000

6. Find the two square roots of $\frac{169}{196}$.
   F $\frac{13}{14}, -\frac{13}{14}$ H $\sqrt{\frac{13}{14}}, -\sqrt{\frac{13}{14}}$
   G 13, 14   J 14, 13

7. Which of the following is the same as the phrase, "The product of a number and 5 is less than the number cubed minus 47"?
   A $5a > a^3 - 47$ C $5a = a^3 - 47$
   B $5a < a^3 - 47$ D $5a < a^3 + 47$

8. Solve $9x - 27 - 6x + 3 = 0$.
   F $x = -8$ H $x = 6$
   G $x = 3$ **J** $x = 8$

9. Simplify $16^4 \cdot 16^7$.
   A 36^{21} C 16^3
   B 16^{-3} **D** 16^{11}

10. Which is equal to $(-12q)^3$?
    F $(-12q)(-12q)(-12q)$
    G $-[(-12q)(-12q)(-12q)]$
    H $12q^{-3}$
    J $\frac{1}{12q^3}$

11. Solve $-7d = 105$.
    A $d = -15$ C $d = 135$
    B $d = 15$ D $d = -135$

12. This frequency table represents which data set?

Age of students in karate class	6	7	8	9
Frequency	2	2	11	3

    F 7 9 8 8 7 8 6 8 6 9 7 6 8 8 9 8 8 8
    G 8 9 8 8 7 8 5 8 6 9 7 6 8 8 9 8 8 8
    H 8 9 8 8 7 8 8 8 6 9 7 6 8 8 9 8 8 8
    J 6 9 8 8 7 8 8 8 6 9 7 6 8 8 9 8 8 8

13. Evaluate 7^{-4}.
    A −2401   C −28
    B $\frac{1}{2401}$ D $-\frac{1}{28}$

14. Simplify $-12 - (-13) + 21 - 3^2$.
    F −10   **H** 13
    G 6   J 24

Cumulative Test
Form C, continued

15. Find the area of a table top that measures $2\frac{7}{9}$ in. by $4\frac{3}{4}$ in.
    A $7\frac{2}{5}$ in^2 **C** $13\frac{7}{36}$ in^2
    B 665 in^2 D $29\frac{1}{6}$ in^2

16. A store has an inventory of 1500 paint brushes. There are 372 more 4-inch paint brushes than 2-inch paint brushes. How many 2-inch paint brushes are there?
    F 225   H 1128
    G 564   J 1872

17. The area of a square game board is 245 in^2. What are the dimensions of the game board?
    A 15 × 5 inches
    B 16 × 16 inches
    C 2.45 × 100 inches
    D 15.65 × 15.65 inches

18. Jennifer rode a plane for 865 miles. The plane averaged 450 miles per hour. Rounded to the nearest tenth of an hour, how long did the trip take?
    F 1.9 hr   H 1.92 hr
    G 0.9 hr   J 0.92 hr

19. Convert the following to an algebraic expression "Lila earned $37.50 a day in tips. If she worked six days a week, how much did she earn in n weeks?"
    A $37.50 + n$ C $37.50n$
    B $225n$ D $37.50 + 6n$

20. What is possibly misleading about the statistic, "Maggie owns a surfing shop that made $12,000 from June to August and $500 from October to December."?
    F The average is skewed.
    G The sample size is too small.
    H The incomes are measured at different times.
    J The income amounts are too different.

21. Simplify $\frac{-125}{-5}$.
    A 25   C −25
    B 20   D −20

22. Twenty students were asked whether the school day should be lengthened. Why might this sample be biased?
    F The average is skewed.
    G Students can't vote.
    H Students won't likely want a longer school day.
    J Too many people were asked.

23. To obtain a trail mix, Marla mixed $3\frac{3}{4}$ pounds of candy pieces, $5\frac{1}{3}$ pounds of cashews, and $5\frac{1}{3}$ pounds of dried cherries. What was the total weight?
    A $12\frac{7}{12}$ lb   **C** $14\frac{5}{12}$ lb
    B $12\frac{5}{9}$ lb   D $13\frac{1}{9}$ lb

24. Simplify $-\frac{15}{13} - \left(-\frac{9}{13}\right)$.
    F $-\frac{6}{13}$ H $\frac{6}{13}$
    G $-\frac{24}{13}$ J $\frac{24}{13}$

Cumulative Test
Chapter 4 Form C, continued

25. Find the original data from the stem-and-leaf plot.

Stem	Leaves
8	0 3 7
9	4 6 6
10	1 1 4

A 80, 83, 86, 94, 96, 96, 101, 104, 101
B 84, 87, 93, 94, 96, 96, 101, 101, 104
C 80, 83, 87, 49, 96, 96, 101, 104, 101
D 80, 83, 87, 94, 96, 96, 101, 104, 101

26. Solve $-6c > -\frac{4}{3}$.
F $c < -\frac{2}{9}$
G $c < \frac{2}{9}$
H $c > \frac{2}{9}$
J $c > -\frac{2}{9}$

27. Find the range for the data set. 120, 116, 112, 117, 120
A 0
B 8
C 117
D 120

28. The average January monthly snowfall (in inches) for 10 different U.S. cities is given below. Find the median. 3.23, 1.67, 2.45, 3.45, 4.12, 3.61, 5.56, 3.68, 4.24, 3.40
F 3.53 in.
G 3.45 in.
H 3.89 in.
J 3.565 in.

29. Find the third quartile for the data set. 120, 138, 146, 138, 149, 138, 149, 123, 137
A 125
B 149
C 147.5
D 143.5

30. Which type of correlation is shown in the data set illustrated in the scatter plot?

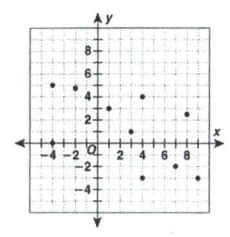

F strong positive correlation
G strong negative correlation
H weak positive correlation
J weak negative correlation

31. Divide $\frac{5}{11} \div \frac{2}{15}$.
A $\frac{75}{22}$
B $\frac{2}{33}$
C $\frac{22}{75}$
D $\frac{7}{15}$

32. Janelle's long distance bills for the last semester of college were $174.94, $154.72, $181.11, $183.12, $166.61 and $150.41. Find the average bill and round your answer to the nearest cent.
F $182.18
G $102.73
H $170.18
J $168.49

33. If the high temperatures for five days were 89.6°F, 87.3°F, 90.4°F, 88.9°F, and 86.1°F, what is the range of high temperatures for these five days?
A 88.46°F
B 4.3°F
C 3.5°F
D 88.9°F

34. Solve $18x + 30 < -60$.
F $x > -5$
G $x < -5$
H $x = -5$
J $x < -7$

35. Evaluate $\frac{13x - 19y}{6}$ when $x = -3$ and $y = -2$.
A $\frac{22}{6}$
B $\frac{4}{6}$
C $-\frac{77}{6}$
D $-\frac{1}{6}$

36. Simplify $14d - 18 - 12d + (-22)$.
F $22d + 4$
G $2d - 40$
H $22d - 40$
J $2d + 4$

37. Express 12.375 as a fraction.
A $12\frac{3}{8}$
B $12\frac{3}{4}$
C $12\frac{2}{5}$
D $12\frac{1}{3}$

38. Solve $-0.257b + 12 = 1.343b$.
F $b = 7.5$
G $b = -7.5$
H $b = 5$
J $b = 12$

39. Solve $-5 < \frac{b}{3} + 2$.
A $b < -21$
B $b > 21$
C $b < 21$
D $b > -21$

40. Determine which ordered pair is a solution of the equation $15 - x = 5y$.
F $(-5, 4)$
G $(20, 1)$
H $(5, 3)$
J $(2, 5)$

41. The equation $y = x + 1$, shown on the graph below, contains which ordered pair?

A $(-3, -2)$
B $(0, 0)$
C $(2, 2)$
D $(1, 4)$

42. Simplify $-3 + (-21) - 17$.
F -35
G 1
H 35
J -41

43. Simplify $-1 + 9(5 - 3m) + m$.
A $44 + 27m$
B $44 - 27m$
C $44 - 2m$
D $44 - 26m$

44. Simplify: $3(2 - 6)^3 + 3^0$.
F -191
G -1728
H -193
J 192

Cumulative Test
Chapter 5 Form A

Select the best answer.

1. Simplify $2 - (-1) \cdot (-4)$.
A 4
B -2

2. Solve $\frac{1}{2}v \le 30$.
A $v \le 60$
B $v \le 15$

3. What type of correlation probably exists between the amount of time you watch television and the amount of time you spend reading?
A negative
B neutral
C positive
D no correlation

4. Multiply $\frac{2}{3} \cdot \frac{5}{6}$.
A $\frac{7}{9}$
B $\frac{5}{9}$

5. Subtract $1\frac{1}{4} - \frac{1}{2}$.
A $\frac{5}{8}$
B $\frac{3}{4}$

6. Solve $c - 3 = 4$.
A $c = 7$
B $c = 1$

7. Solve $2n > 4.4$.
A $n > 8.8$
B $n < 2.2$
C $n > 2.2$
D $n < 8$

8. Simplify $-z - 3z$.
A $4z$
B $-4z$

9. Determine if the following number is rational, irrational, or not real and give the value if rational or irrational. $\sqrt{-9}$
A not real
B rational, -3

10. Find the algebraic expression for, "4 is less than 7 plus x."
A $4 < 7 + x$
B $7 < x + 4$

11. Find the slope of $\overline{AB}$.

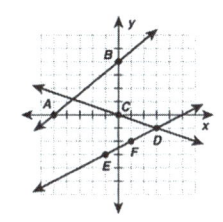

A $\frac{4}{5}$
B $-\frac{1}{3}$
C $\frac{1}{2}$
D $-\frac{5}{4}$

12. Express 1.3×10^{-4} in standard notation.
A 0.00013
B 0.0013

13. Find both square roots of $\sqrt{100}$.
A 5, 20
B 10, -10

14. Find the sum of the angle measures in a regular pentagon.
A 180°
B 540°
C 900°
D 720°

15. Find the original data from the stem-and-leaf plot.

Stem	Leaves
1	1 8
2	1 1 2
3	1 8

A 11, 18, 21, 21, 22, 31, 38
B 118, 211, 212, 318

16. Find the mean amount spent for snacks over the last four months. Round your answer to the nearest cent. $58.39, $89.53, $82.24, $70.94
A $75.28
B $76.59

17. Find the range for the data set. 11, 8, 11, 8, 15
A 9.5
B 7

18. Solve the equation $\frac{1}{3}x = 4$.
A $x = \frac{4}{3}$
B $x = 12$

19. Simplify $6^3 \cdot 6^2$ using positive exponents.
A 6^5
B 6^6
C 36^5
D 36^6

20. What is a quadrilateral with 4 congruent sides?
A rhombus
B rectangle

21. In the graph, how has figure DEF been transposed from figure ABC?

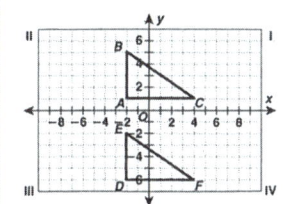

A reflection across the y-axis
B translation 7 units down

22. A piece of fabric is 18 inches wide. If $7\frac{1}{4}$ inches are trimmed off, how wide is the fabric?
A $10\frac{3}{4}$ in.
B $10\frac{1}{4}$ in.

361

Cumulative Test — Form A, continued

23. Which ordered pair is in the third quadrant?
- A (3, 2)
- B (−2, 3)
- C (2, 3)
- **D (−2, −3)** ✓

24. Which graph represents the inequality $x \leq 7$?
- A
- **B** ✓
- C
- D

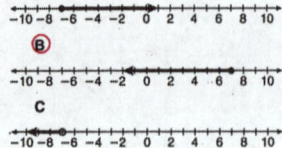

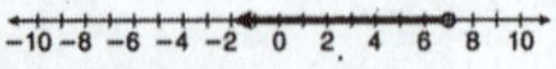

25. Simplify -2^2.
- A 4
- **B −4** ✓

26. Find the mode in the data set.
11, 8, 11, 15, 8, 15, 8
- A 11
- **B 8** ✓

27. Which is $\frac{7}{8}$ as a decimal?
- **A 0.875** ✓
- B 7.8

28. Determine which ordered pair is a solution of the equation $y = 1 + 3x$.
- **A (1, 4)** ✓
- B (4, 1)

29. Identify the type of triangle. 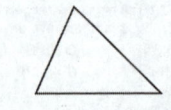
- A right
- **B acute** ✓
- C obtuse
- D isosceles

Refer to the following figure for problems 30–32.

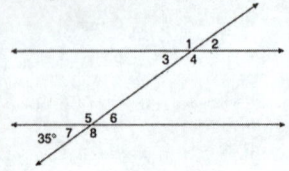

30. Find $m\angle 1$ in the figure above.
- A 35°
- **B 145°** ✓

31. In the figure above, which angle is vertical to $\angle 8$?
- **A ∠5** ✓
- B ∠6
- C ∠4
- D ∠7

32. In the figure above, name an angle that is supplementary to $\angle 2$.
- A ∠3
- **B ∠4** ✓

33. Express 42,000 in scientific notation.
- A 4.2×10^5
- **B 4.2×10^4** ✓

34. Polygon $ABCD \cong LMNO$. Find s. 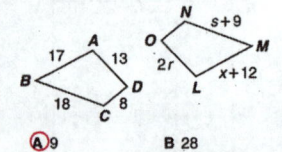
- **A 9** ✓
- B 28

35. Express 2.75 as a fraction.
- A $2\frac{7}{5}$
- **B $2\frac{3}{4}$** ✓

36. Line $a \parallel$ line b. If $m\angle 1 = 70°$, find $m\angle 4$. 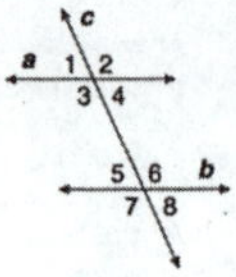
- A 110°
- **B 70°** ✓
- C 20°
- D 250°

37. At Peterson Junior High School an average of 100 students check out books in the school library every day. On Wednesday, the first 25 students who came into the library were asked what types of books they like to read. Identify the sample of the survey.
- A all students at the school
- **B the 25 that were surveyed** ✓

38. Which number is equivalent to 2^{-2}?
- A −4
- B $-\frac{1}{4}$
- **C $\frac{1}{4}$** ✓
- D 4

39. A recipe for cookies calls for 2 cups of sugar. How much sugar would be needed if the recipe is tripled?
- A $1\frac{1}{2}$
- B 3 cups
- C $\frac{1}{3}$ cup
- **D 6 cups** ✓

40. The surface area of the top of a square table is 81 cm². What are the dimensions of the top of the table?
- **A 9 cm × 9 cm** ✓
- B 8 cm × 11 cm

Cumulative Test — Form B

Select the best answer.

1. Simplify $10 - (-7) \cdot (-4) + 9$.
- A 15
- B −59
- C 12
- **D −9** ✓

2. Solve $\frac{5}{6}m > 12$.
- **F $m > 14\frac{2}{5}$** ✓
- G $m < 14\frac{2}{5}$
- H $m > 10$
- J $m < 10$

3. What type of correlation exists between the amount of time you exercise and the number of calories you use?
- A negative
- B neutral
- **C positive** ✓
- D no correlation

4. Multiply $\frac{20}{23} \cdot \frac{15}{16}$.
- F $\frac{5}{7}$
- G $\frac{35}{39}$
- **H $\frac{75}{92}$** ✓
- J $\frac{301}{368}$

5. Subtract $\frac{11}{15} - 1\frac{1}{5}$.
- A $\frac{23}{25}$
- **B $-\frac{7}{15}$** ✓
- C $\frac{11}{18}$
- D $\frac{11}{180}$

6. Solve $m - 13 = -16$.
- F $m = 3$
- **G $m = -3$** ✓
- H $m = -29$
- J $m = 29$

7. Solve $0.25n > 2$.
- **A $n > 8$** ✓
- B $n < 0.125$
- C $n > 0.5$
- D $n < 4$

8. Simplify $-7z - 3z - 7z$.
- F $17z$
- G $-17z^2$
- **H $-17z$** ✓
- J $17z^2$

9. Determine if the following number is rational, irrational, or not real and give the value if rational or irrational. $\sqrt{64}$
- **A rational, 8** ✓
- B not real
- C rational, 32
- D irrational, 32

10. Find the algebraic expression for "fourteen is less than or equal to seven subtracted from x."
- F $14 < 7 - x$
- G $14 \leq 7 - x$
- **H $14 \leq x - 7$** ✓
- J $x - 7 < 14$

11. Find the slope of CD. 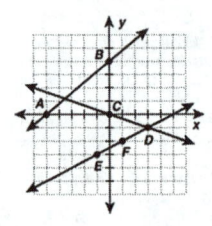
- A $\frac{4}{3}$
- **B $-\frac{1}{3}$** ✓
- C $\frac{1}{2}$
- D 3

12. Express 4.45×10^{-5} in standard notation.
- **F 0.0000445** ✓
- G 0.00445
- H 0.000445
- J 44,500

13. Find both square roots of $\sqrt{169}$.
- **A 13, −13** ✓
- B 16, −16
- C 15, −15
- D 26, −26

14. Find the sum of the angle measures in a regular hexagon.
- F 540°
- **G 720°** ✓
- H 900°
- J 1080°

15. Find the original data from the stem-and-leaf plot.

Stem	Leaves
4	1 8
5	1 1 2 7
6	1 8 8

- A 418, 51127, 6188
- **B 41, 48, 51, 51, 52, 57, 61, 68, 68** ✓
- C 41, 48, 51, 52, 57, 61, 68
- D 418, 511, 527, 618, 618

16. Find the mean amount spent for groceries over the last eight weeks. Round your answer to the nearest cent.
$58.39, $89.53, $82.24, $70.94, $58.39, $87.74, $58.17, $61.42
- F $72.63
- G $66.18
- H $70.86
- **J $70.85** ✓

17. Find the range for the data set.
11, 25, 11, 28, 28, 15
- **A 17** ✓
- B 19.5
- C 4
- D 11

18. Solve the equation $\frac{3}{5}x + 2 = \frac{2}{5}x - 4$.
- F $x = \frac{26}{5}$
- G $x = -\frac{7}{5}$
- H $x = \frac{1}{5}$
- **J $x = -30$** ✓

19. Simplify $3^5 \cdot 3^3 \cdot 3^2$ using positive exponents.
- **A 3^{10}** ✓
- B 27^{10}
- C 3^{30}
- D 27^{30}

20. What is a quadrilateral with 4 congruent angles and 4 congruent sides?
- F trapezoid
- G rhombus
- **H square** ✓
- J rectangle

21. In the graph, how has figure ADE been transposed from figure ABC? 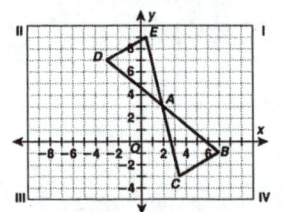
- A reflection across the x-axis
- **B a 180° rotation about (2, 3)** ✓
- C reflection across the y-axis
- D translation 2 units down

22. A piece of fabric is 2 yards long and 8 inches wide. If the width is trimmed to $7\frac{11}{16}$ inches, how much was trimmed off the width?
- F $\frac{123}{16}$ in.
- G $3\frac{3}{5}$ in.
- **H $\frac{5}{16}$ in.** ✓
- J $\frac{11}{16}$ in.

Cumulative Test — Form B, continued

23. Which ordered pair is in the fourth quadrant?
    A (3, −5)   C (3, 5)
    B (5, 3)   D (−5, 3)

24. Which graph represents the inequality $4x < -28$?
    F 
    G 
    H 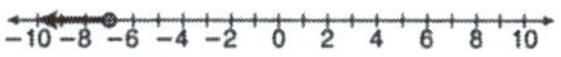
    J

25. Simplify -5^2.
    A −25   C 10
    B 25   D −10

26. Find the mode in the data set.
    11, 28, 11, 15, 28, 15, 28
    F 11   H 17
    G 28   J 15

27. Which is $\frac{7}{25}$ as a decimal?
    A 0.28   C 3.5
    B 1.4   D 7.25

28. Determine which ordered pair is a solution of the equation $2y = 6 + 3x$.
    F $\left(-1, 1\frac{1}{2}\right)$ H $\left(-1, \frac{9}{2}\right)$
    G $(-1, 3)$ J $(3, 2)$

29. Identify the type of triangle. 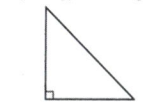
    A right   C obtuse
    B acute   D equilateral

Refer to the following figure for problems 30–32.

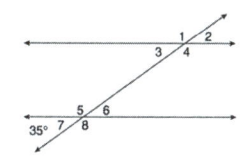

30. Find $m\angle 1$ in the figure above.
    F 35°   H 90°
    G 145°   J 180°

31. In the figure above, which angle is vertical to $\angle 3$?
    A $\angle 2$ C $\angle 4$
    B $\angle 5$ D $\angle 1$

32. In the figure above, name the angles that are supplementary to $\angle 8$.
    F $\angle 5$ and $\angle 6$ H $\angle 6$ and $\angle 7$
    G $\angle 3$ and $\angle 4$ J $\angle 1$ and $\angle 4$

33. Express 2,230,000 in scientific notation.
    A 2.23×10^6 C 0.223×10^{-7}
    B 2.23×10^7 D 22.3×10^6

34. Polygon $ABCD \cong LMNO$. Find x. 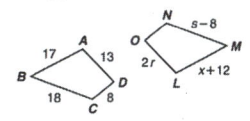
    F 1.667   H 5
    G −5   J 10

35. Express 0.225 as a fraction.
    A $\frac{9}{40}$ C $2\frac{1}{4}$
    B $22\frac{1}{2}$ D $2\frac{2}{5}$

36. Line $a \parallel$ line b. If $m\angle 2 = 128°$, find $m\angle 8$.
    F 128°   H 38°
    G 52°   J 218°

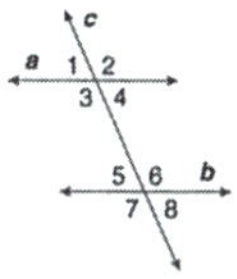

37. Of the 350 students at Byrnedale Junior High School, 250 buy their lunch in the school cafeteria. On Wednesday, the first 100 students eating in the cafeteria were asked their favorite food that was served in the cafeteria that day. Identify the population of the survey.
    A all students at the school
    B all students in school on Wednesday
    C 100 that were surveyed
    D 250 that buy lunch in cafeteria

38. Which number is equivalent to 6^{-2}?
    F −36   H $\frac{1}{36}$
    G $-\frac{1}{36}$ J $\frac{1}{12}$

39. A recipe for cookies calls for $1\frac{1}{2}$ cups of sugar. How much sugar would be needed if the recipe is tripled?
    A $4\frac{1}{2}$ cups   C $\frac{1}{2}$ cup
    B 3 cups   D 6 cups

40. The surface area of the top of a square table is 110.25 in.². What are the dimensions of the top of the table?
    F $10\frac{1}{2}'' \times 1''$ H $10\frac{1}{2}'' \times 10\frac{1}{2}''$
    G $10\frac{1}{4}'' \times 11\frac{1}{4}''$ J $11\frac{1}{2}'' \times 10''$

Cumulative Test — Form C

Select the best answer.

1. Simplify $15 - (-5) \div (-5) \cdot 2$.
   A −2   C 13
   B −8   D −4

2. Solve $\frac{7}{12}m + 2 \leq 16$.
   F $m \leq 24$ H $m \leq \frac{18}{7}$
   G $m \geq 24$ J $m \geq \frac{18}{7}$

3. What type of correlation likely exists between the amount of time you spend playing video games and the amount of time you spend studying?
   A negative   C positive
   B neutral   D no correlation

4. Simplify $2\frac{11}{15} \cdot 1\frac{4}{5}$.
   F $5\frac{7}{15}$ H $4\frac{4}{5}$
   G $3\frac{23}{15}$ J $4\frac{23}{25}$

5. Simplify $1\frac{23}{24} - 1\frac{17}{20}$.
   A $3\frac{97}{120}$ C $\frac{13}{120}$
   B $2\frac{1}{2}$ D $-\frac{13}{120}$

6. Solve $3m - 10 = -14 - m$.
   F $m = -1$ H $m = -12$
   G $m = -6$ J $m = -2$

7. Solve $1.25n \geq -2$.
   A $n > -3.25$ C $n \geq 3.25$
   B $n \geq -1.6$ D $n > -2.5$

8. Simplify $-6z - 3z - 6z^2$.
   F $-15z$ H $-9z - 6z^2$
   G $-15z^2$ J $9z^2$

9. Determine if the following number is rational, irrational, or not real and give the value if rational or irrational.
   $-\sqrt{30}$
   A rational, −5.48
   B not real
   C irrational, 5.48
   D irrational, −5.48

10. Find the algebraic expression for the phrase "twice x is less than or equal to three less than three times x."
    F $2x \leq 3 - 3x$ H $2x \geq 3 - 3x$
    G $2x \leq 3x - 3$ J $2x < 3x - 3$

11. Find the slope of EF. 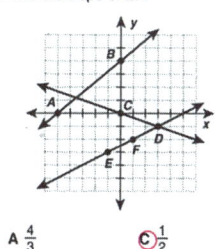
    A $\frac{4}{3}$ C $\frac{1}{2}$
    B $-\frac{1}{3}$ D 2

12. Express 1.05×10^{-6} in standard notation.
    F 0.0000105   H 0.000105
    G 0.00105   J 0.00000105

13. Find both square roots of $\sqrt{484}$.
    A 24, −24   C 12, −12
    B 22, −22   D 44, −44

Cumulative Test — Form C, continued

14. Find the sum of the angle measures in a nonagon.
    F 540°   H 1260°
    G 720°   J 1620°

15. Find the original data from the stem-and-leaf plot.

Stem	Leaves
4.	1 8
5.	1 1 2 7
6.	1 8

    A 4.18, 5.1127, 6.188
    B 41, 48, 51, 51, 52, 57, 61, 68, 68
    C 41, 48, 51, 52, 57, 61, 68
    D 4.1, 4.8, 5.1, 5.1, 5.2, 5.7, 6.1, 6.8

16. Find the mean amount that Jean spent on long distance telephone calls over the past 8 months. Round your answer to the nearest cent.
    $161.42, $158.39, $158.39, $170.94, $189.53, $182.24, $187.74, $158.17.
    F $455.61   H $170.85
    G $116.18   J $172.63

17. Find the range for the data set.
    185, 246, 372, 801, 675, 212
    A 415   C 309
    B 616   D 110

18. Solve $\frac{3}{5}x + 2 = \frac{2}{5}\left(x + \frac{9}{4}\right)$.
    F $-1\frac{1}{10}$ H $-5\frac{1}{2}$
    G $1\frac{1}{4}$ J $14\frac{1}{2}$

19. Simplify using positive exponents.
    $(3a)^5 \cdot (3a)^3 \cdot (3a)^2 \cdot (3a)^0$
    A $3a^{10}$ C $(3a)^{30}$
    B $(3a)^{10}$ D $27a^{30}$

20. What is a quadrilateral with 4 congruent sides where, $\angle A \cong \angle C$ and $\angle B \cong \angle D$, but $\angle A \not\cong \angle B$?
    F rhombus   H trapezoid
    G square   J rectangle

21. In the graph, how has figure DEF been transposed from figure ABC? 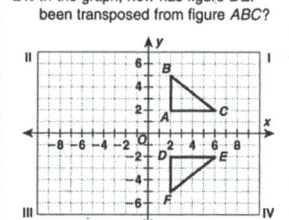
    A reflection across the x-axis
    B a 180° rotation about (6, 2)
    C reflection across the y-axis
    D translation 2 units down

22. A piece of fabric is 2 feet long and 18 inches wide. If both width and length are trimmed by $7\frac{11}{16}$ in., what are the new dimensions?
    F 2 ft × $10\frac{5}{6}$ in.
    G 2 ft × $7\frac{11}{16}$ in.
    H $16\frac{5}{16}$ in. × $10\frac{5}{16}$ in.
    J $16\frac{5}{16}$ in. × 18 in.

Cumulative Test
5 Form C, continued

23. Which ordered pair is not in the fourth quadrant?
    A $(2\frac{1}{2}, -6)$
    C $(3\frac{1}{2}, 6)$
    B $(6, -3\frac{1}{2})$
    D $(6, -2\frac{1}{2})$

24. Which graph represents the inequality $3x \geq -28 - x$? 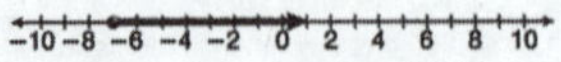
    F
    G 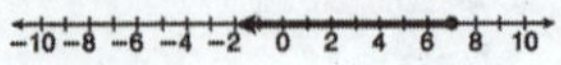
    H 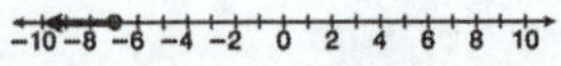
    J

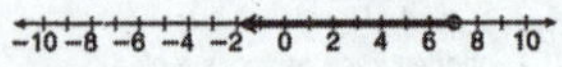

25. Simplify $-5^2 \cdot 6^2$.
    A 1
    C 900
    B −900
    D 810,000

26. Find the mode in the data set.
    110, 280, 110, 150, 280, 150, 280
    F 110
    H 170
    G 150
    J 280

27. Which is $\frac{7}{11}$ as a decimal?
    A 0.63
    C 1.3
    B $0.\overline{63}$
    D $1.\overline{3}$

28. Determine which ordered pair is a solution of the equation $3y + 1 = 2x$.
    F $(1, 1\frac{1}{2})$
    H $(\frac{3}{2}, \frac{2}{3})$
    G $(-1, 3)$
    J $(2, 3)$

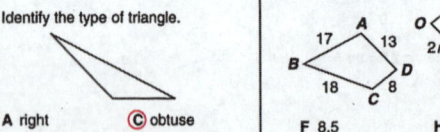

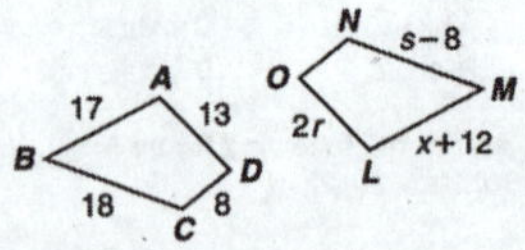

29. Identify the type of triangle.
    A right
    C obtuse
    B acute
    D isosceles

Refer to the following figure for problems 30–32.

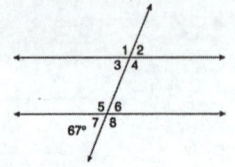

30. Find $m\angle 3$ in the figure above.
    F 67°
    H 77°
    G 113°
    J 180°

31. In the figure above, which angle is vertical to $\angle 4$?
    A $\angle 2$
    C $\angle 7$
    B $\angle 6$
    D $\angle 1$

32. In the figure above, name the angles that are supplementary to $\angle 3$.
    F $\angle 5$ and $\angle 6$
    H $\angle 1$ and $\angle 2$
    G $\angle 1$ and $\angle 4$
    J $\angle 2$ and $\angle 4$

Cumulative Test
5 Form C, continued

33. Express 343,000,000 in scientific notation.
    A 3.43×10^8
    C 343×10^8
    B 3.43×10^6
    D 34.3×10^8

34. Polygon $ABCD \cong LMNO$. Find r.
    F 8.5
    H 4
    G 9.5
    J 6.5

35. Express 2.325 as a fraction.
    A $2\frac{13}{40}$
    C $2\frac{1}{4}$
    B $23\frac{1}{2}$
    D $2\frac{3}{5}$

36. Line $a \parallel$ line b. If $m\angle 5 = 79°$, find $m\angle 2$. 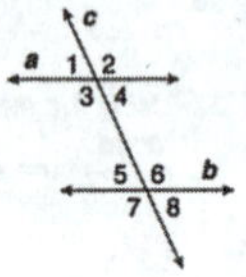
    F 79°
    G 101°
    H 11°
    J 259°

37. Of the 350 students at Burnet Junior High School, an average of 20 visit the nurse's office every day. On Wednesday, the first 5 students who visited the nurse's office were asked to describe their symptoms. Identify the sample in the survey.
    A the 5 who were surveyed
    B all students at the school
    C all students in school on that day
    D the 20 who visit the nurse's office

38. Which number is equivalent to 5^{-3}?
    F −125
    H $\frac{1}{125}$
    G $-\frac{1}{15}$
    J $\frac{1}{15}$

39. A recipe for cookies calls for $1\frac{1}{2}$ cups of sugar. How much sugar would be needed if the recipe is increased $1\frac{1}{2}$ times?
    A $\frac{1}{2}$ cup
    C 3 cups
    B $2\frac{1}{4}$ cups
    D 6 cups

40. The surface area of the top of a square table is 150.0625 in². What are the dimensions of the top of the table?
    F 10.1″ × 15″
    H 10.5″ × 10.6″
    G $10\frac{1}{4}$″ × $15\frac{1}{4}$″
    J $12\frac{1}{4}$″ × $12\frac{1}{4}$″

Cumulative Test
6 Form A

1. Simplify $-3 + 7 - 1 - 7$.
   A −4
   C 11
   B 17
   D −8

2. Simplify $-2 - (-3 + 3)$.
   A 0
   C −8
   B −4
   D −2

3. A cone has a radius of 2.3 cm and a height of 4.0 cm. Find its volume to nearest tenth. Use 3.14 for π.
   A 22.1 cm³
   B 66.4 cm³

4. Find the square root of 8 to the nearest hundredth.
   A 4.00
   C 1.14
   B 2.83
   D 64.00

5. Simplify $-6 + (-3)$.
   A −9
   B −3

6. Simplify $mn - 2mn$.
   A $-mn$
   B −1

7. Simplify $-2(d + 1)$.
   A $-2d - 1$
   B $-2d - 2$

8. Evaluate $\frac{c+4}{4}$ for $c = -1$.
   A $\frac{3}{4}$
   B −1

9. Solve $p - 4 = 4$.
   A $p = 0$
   C $p = -8$
   B $p = 4$
   D $p = 8$

10. Clark received the following scores on biology lab assignments: 12, 30, 15, 45, 5. Find the range of his scores.
    A 21.4
    C 40
    B 7
    D 15

11. Express $m \cdot m \cdot m$ using exponents.
    A $3m$
    C $3m^6$
    B m^3
    D 3^m

12. Find the measure of an angle whose supplement is 2 times the angle.
    A 60°
    B 45°

13. Express 1.2×10^{-3} in standard notation.
    A 0.00012
    C 0.0012
    B 0.012
    D 1.2000

14. Express 360,000 in scientific notation.
    A 3.6×10^3
    C 36×10^5
    B 36×10^3
    D 3.6×10^5

15. A cabinetmaker installs $\frac{1}{4}$-in. thick laminate material on top of $\frac{3}{8}$-in. plywood. What is the thickness of the completed countertop?
    A $\frac{5}{8}$ in.
    B $\frac{6}{12}$ in.

Cumulative Test
6 Form A, continued

16. Find the perimeter of the polygon. 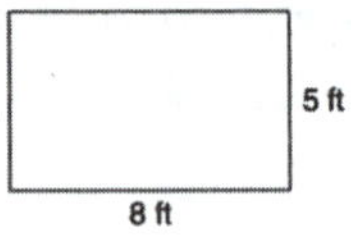
    A 13 ft
    C 26 ft
    B 40 ft
    D 80 ft

17. Find the average number of e-mails that Joshua received each day if the number of e-mails he received each day for the past week were 20, 5, 10, 30, 4, 4, and 5. Round to the nearest tenth.
    A 10
    C 10.6
    B 11.1
    D 5

18. Find the missing measurement in the triangle to the nearest tenth. 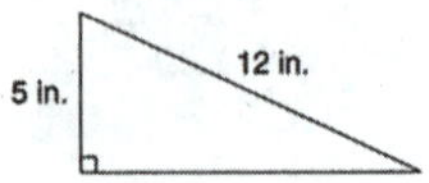
    A 8.5 in.
    C 12.5 in.
    B 10.9 in.
    D 13.0 in.

19. Solve the equation $1.5 = s + 2.8$.
    A $s = -1.3$
    C $s = 4.3$
    B $s = 1.3$
    D $s = -4.3$

20. Find the circumference of a circle with $d = 10$ mi. Use 3.14 for π.
    A 78.5 mi
    C 62.8 mi
    B 31.4 mi
    D 314 mi

21. Anissa writes one page in $\frac{1}{2}$ of an hour. At this rate, how many pages can she write in 3 hours?
    A 6 pages
    B $1\frac{1}{2}$ pages

22. Find the sum of the angles of a regular quadrilateral.
    A 360°
    C 1080°
    B 720°
    D 1620°

23. Find the volume of the figure to the nearest hundredth. Use 3.14 for π. 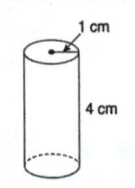
    A 25.12 cm²
    B 12.56 cm²

24. Find the area of the triangle. 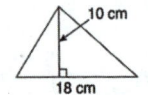
    A 90 cm²
    B 180 cm²

Cumulative Test
Form A, continued

25. Simplify $-\frac{16}{8}$.
    A 2
    (B) −2

26. Find the value of 10^2.
    A 20
    B 200
    (C) 100
    D 10

27. Which equation means "five minus a number d is -4"?
    A $d - 5 = -4$
    (B) $5 - d = -4$

28. Solve for the hypotenuse in the right triangle ABC with leg $a = 3$ and leg $b = 4$.
    (A) $c = 5$
    B $c = 7$

29. Express $2^4 \cdot 2^3$ using positive exponents.
    (A) 2^7
    B 2^{12}
    C 4^7
    D 4^{12}

30. Solve $\frac{3}{5}y = \frac{5}{8} + \frac{2}{5}y + \frac{5}{8}$.
    A $y = \frac{5}{4}$
    (B) $y = \frac{25}{4}$

31. Which of the following inequalities means "two is less than or equal to x plus 2"?
    A $2 < x + 2$
    (B) $2 \leq x + 2$
    C $x - 2 \leq 2$
    D $2 = x - 2$

32. In $\triangle ABC$, $\angle A$ and $\angle B$ have the same measure. $\angle C$ is 30° larger than each of the other angles. Find $m\angle C$.
    A 50°
    B 30°
    (C) 80°
    D 150°

33. To the nearest tenth, find the volume of a sphere with a radius of 10 cm. Use 3.14 for π.
    A 3140 cm³
    (B) 4186.7 cm³
    C 1046.7 cm³
    D 2355 cm³

34. What is the name for a quadrilateral with all sides having equal length and angles having equal measurement?
    A parallelogram
    (B) square

35. The perimeter of a square is 100 inches. Find the length of one side.
    A 10 in.
    (B) 25 in.

36. Find the area of a circle with $d = 10$ mi. Use 3.14 for π.
    A 31.4 mi²
    B 785 mi²
    C 314 mi²
    (D) 78.5 mi²

37. Using the stem-and-leaf plot, find the mode.

Stem	Leaves
3	1 8
4	1 1 1 5
5	1 2 2 8 9

    A 46.27
    (B) 41
    C 50
    D no mode

38. How much will it cost to carpet a 15 ft by 12 ft room if carpeting costs $10 per square yard?
    A $1800
    (B) $200

39. Find the volume of the figure.

    5 in. × 5 in. × 5 in.

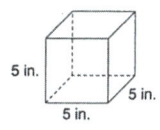

    A 25 in³
    (B) 125 in³

40. Express 5.25 as a fraction.
    A $7\frac{1}{8}$
    (B) $5\frac{1}{4}$

41. Find the surface area of the figure.

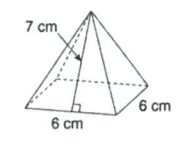

    7 cm, 6 cm, 6 cm

    (A) 120 cm²
    B 252 cm²
    C 84 cm²
    D 168 cm²

42. Name the figure.

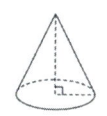

    A circle
    B pyramid
    (C) cone
    D prism

43. Find the surface area of the figure. Use 3.14 for π.

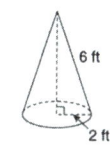

    6 ft, 2 ft

    (A) 50.2 ft²
    B 25.1 ft²

44. Find the area of the figure.

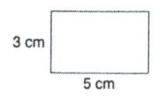

    3 cm × 5 cm

    A 16 cm²
    (B) 15 cm²

45. Find all roots of $\sqrt{\frac{1}{16}}$.
    A $4, -4$
    (B) $\frac{1}{4}, -\frac{1}{4}$

Cumulative Test
Form B

1. Simplify $-3 + 17 - 1 - 7$.
   (A) 6
   B 28
   C −22
   D −3

2. Simplify $-1 - (-4 + 3) + (-1)$.
   F 1
   G −5
   H 5
   (J) −1

3. An ice cream cone has a diameter of 7.7 cm and a height of 11.0 cm. Use 3.14 for π to find the volume to the nearest tenth.
   (A) 170.7 cm³
   B 657.0 cm³
   C 682.6 cm³
   D 1971.0 cm³

4. Find $\sqrt{808}$ to the nearest tenth.
   F 808.0
   (G) 28.4
   H 28.3
   J 28.422

5. Simplify $-36 \div (-3) + (-4)$.
   A −43
   B $\frac{36}{7}$
   (C) 8
   D 16

6. Simplify $13mn - 13mn$.
   F $26mn$
   (G) 0
   H $-mn$
   J mn

7. Simplify $-2(d + 3)$.
   A $-2d - 3$
   (B) $-2d - 6$
   C $-2d + 6$
   D $-2d + 3$

8. Evaluate $\frac{c + 4}{9}$ for $c = -9$.
   F $\frac{13}{9}$
   G 4
   H $-\frac{13}{9}$
   (J) $-\frac{5}{9}$

9. Solve $p - 8 = -4 + 12$.
   A $p = 0$
   B $p = 24$
   C $p = -16$
   (D) $p = 16$

10. The prices for five meal choices in a school cafeteria are the following: $5.20, $4.49, $4.11, $3.80, $5.05. Find the range of these prices.
    F $4.53
    (G) $1.40
    H $1.25
    J $0.15

11. Express $m \cdot m \cdot m \cdot m \cdot m \cdot m$ using exponents.
    A $6m$
    B $6m^6$
    (C) m^6
    D 6^m

12. Find the measure of an angle whose supplement is 3 times the angle.
    F 60°
    (G) 45°
    H 135°
    J 90°

13. Express 4.23×10^{-5} in standard notation.
    A 0.000423
    B 0.00423
    (C) 0.0000423
    D 4.230000

14. Express 5,360,000 in scientific notation.
    F 53.6×10^3
    G 5.36×10^3
    H 5.36×10^5
    (J) 5.36×10^6

15. A cabinetmaker installs $\frac{3}{16}$-in. thick laminate material on top of $\frac{3}{4}$-in. plywood. What is the thickness of the completed countertop?
    (A) $\frac{15}{16}$ in.
    B $\frac{3}{10}$ in.
    C $1\frac{1}{2}$ in.
    D $\frac{1}{16}$ in.

16. Find the perimeter of the polygon.

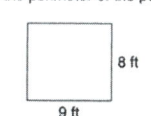

    8 ft × 9 ft

    F 72 ft
    (G) 34 ft
    H 32 ft
    J 17 ft

17. Find the average number of e-mails that Joshua received during the past week: 22, 4, 11, 38, 43, 47, 5. Round to the nearest tenth.
    A 22.5
    (B) 24.3
    C 42
    D 22

18. Find the missing measurement in the triangle to the nearest tenth.

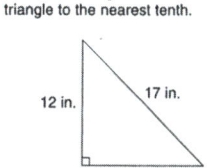

    12 in., 17 in.

    F 14.5 in.
    G 20.8 in.
    H 72.5 in.
    (J) 12.0 in.

19. Solve the equation $4.3 = s + 7.2$.
    (A) $s = -2.9$
    B $s = 11.5$
    C $s = 0.60$
    D $s = 2.9$

20. Find the circumference of a circle with $d = 19$ mi to the nearest tenth. Use 3.14 for π.
    F 29.8 mi
    (G) 59.7 mi
    H 119.3 mi
    J 1133.5 mi

21. Ann reads $\frac{3}{4}$ of a book in 2 days. At this rate, how many books can she read in $4\frac{1}{3}$ days?
    A $3\frac{1}{4}$ books
    B 2 books
    C $1\frac{1}{2}$ books
    (D) $1\frac{5}{8}$ books

22. Find the sum of the angles of a regular 9-gon.
    F 900°
    (G) 1260°
    H 1080°
    J 1620°

23. Find the volume of the figure to the nearest tenth. Use 3.14 for π.

    2 cm, 6 cm

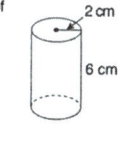

    A 301.4 cm³
    (B) 75.4 cm³
    C 37.7 cm³
    D 25.1 cm³

24. Find the area of the triangle.

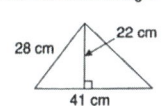

    28 cm, 22 cm, 41 cm

    F 242 cm²
    G 902 cm²
    H 308 cm²
    (J) 451 cm²

365

Holt Pre-Algebra

Cumulative Test
Form B, continued

25. Simplify $\frac{56}{-8}$.
  - A 7
  - B 8
  - C −8
  - (D) −7

26. Find the value of 10^4.
  - F 40
  - (G) 10,000
  - H 1000
  - J 100,000

27. Which of the following is the same as "five less than a number d is −40"?
  - A $5 - d = -40$
  - B $-5d = -40$
  - (C) $d - 5 = -40$
  - D $5 + d = -40$

28. Solve for the unknown side in the right triangle ABC: $b = 4$, $c = 5$.
  - (F) $a = 3$
  - G $a = 9$
  - H $a = 6.4$
  - J $a = 4.5$

29. Express $9^4 \cdot 9^9$ using positive exponents.
  - A 9^{36}
  - (B) 9^{13}
  - C 81^{36}
  - D 81^{13}

30. Solve $\frac{3}{5}y + \frac{2}{9} = \frac{5}{8} - \frac{2}{5}y + \frac{5}{8}$.
  - F $y = \frac{53}{36}$
  - G $y = \frac{29}{72}$
  - (H) $y = \frac{37}{36}$
  - J $y = \frac{53}{36}$

31. Choose an inequality that means "14 is less than or equal to x minus 2."
  - A $14 < x - 2$
  - (B) $14 \leq x - 2$
  - C $x - 2 \leq 14$
  - D $14 = x - 2$

32. In $\triangle ABC$, $\angle A$ and $\angle B$ have the same measure. $\angle C$ is 42° larger than each of the other angles. Find m$\angle C$.
  - F 92°
  - (G) 88°
  - H 46°
  - J 134°

33. To the nearest tenth, find the volume of a sphere with a diameter of 16 cm. Use 3.14 for π.
  - A 539.9 cm³
  - (B) 2143.6 cm³
  - C 1607.7 cm³
  - D 17,148.6 cm³

34. What is the name for a quadrilateral that is a parallelogram with all sides having equal length?
  - F equilateral
  - G rectangle
  - (H) rhombus
  - J trapezoid

35. The perimeter of a square is 88 inches. Find the length of one side.
  - (A) 22 in.
  - B 9.4 in.
  - C 44 in.
  - D 88 in.

36. To the nearest tenth find the area of a circle with $d = 19$ mi. Use 3.14 for π.
  - F 1133.5 mi²
  - G 119.3 mi²
  - (H) 283.4 mi²
  - J 59.7 mi²

37. Using the stem-and-leaf plot, find the mode.

Stem	Leaves
7	1 8
8	1 1 1 3 5
9	1 3 3 8 9 9
10	3 5

  - (A) 81
  - B 1
  - C 91
  - D no mode

38. How much will it cost to paint a 15.5 ft by 12.5 ft ceiling if the painter charges $23 per square yard?
  - (F) $495.14
  - G $193.75
  - H $4456.25
  - J $1485.42

39. Find the volume of the figure. 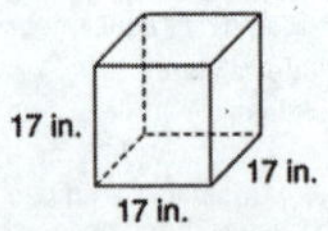
  - A 51 in³
  - B 578 in³
  - C 289 in³
  - (D) 4913 in³

40. Express 7.125 as a fraction.
  - F $7\frac{1}{4}$
  - G $7\frac{1}{12}$
  - (H) $7\frac{1}{8}$
  - J $7\frac{1}{16}$

41. Find the surface area of the figure. 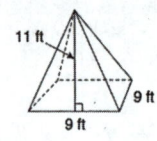
  - (A) 279 ft²
  - B 180 ft²
  - C 234 ft²
  - D 477 ft²

42. Name the figure. 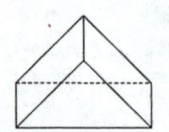
  - F triangle
  - G pyramid
  - (H) prism
  - J quadrilateral

43. Find the surface area of the figure to the nearest tenth. Use 3.14 for π. 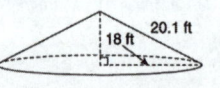
  - (A) 2153.4 ft²
  - B 3052.1 ft²
  - C 2289.1 ft²
  - D 1526 ft²

44. Find the area of the figure. 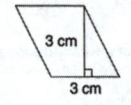
  - F 6 cm²
  - G 4.5 cm²
  - (H) 9 cm²
  - J 12 cm²

45. Find all roots of $\sqrt{\frac{121}{16}}$.
  - A 12, 11, 4
  - (B) $\frac{11}{4}, -\frac{11}{4}$
  - C 11, 4
  - D $\frac{13}{4}, -\frac{13}{4}$

Cumulative Test
Form C

1. Simplify $-3 + 17 - (-2)$.
  - A 18
  - B 12
  - C −22
  - (D) 16

2. Simplify $-1 - 2(-4 + 3)$.
  - (F) 1
  - G −4
  - H −4
  - J −1

3. An ice cream cone has a diameter of 7.2 cm and a height of 11.2 cm. Use 3.14 for π to find the volume to the nearest tenth.
  - (A) 151.9 cm³
  - B 455.8 cm³
  - C 506.4 cm³
  - D 607.7 cm³

4. Find $\sqrt{1808}$ to the nearest tenth.
  - F 1808
  - G 32,400
  - H 904
  - (J) 42.5

5. Simplify $-36 + (-12) \cdot (-4)$.
  - A $-\frac{3}{4}$
  - (B) −12
  - C $\frac{3}{4}$
  - D 12

6. Simplify $13mn - 13mn^2$.
  - F $26mn$
  - G 0
  - H $-mn^2$
  - (J) $13mn - 13mn^2$

7. Simplify $-2(-d^2 - 3)$.
  - (A) $2d^2 + 6$
  - B $2d - 6$
  - C $-2d^2 + 6$
  - D $-2d + 3$

8. Evaluate $\frac{2c + 5}{18}$ for $c = -9$.
  - F $\frac{5}{8}$
  - G 5
  - H $-\frac{5}{9}$
  - (J) $-\frac{13}{18}$

9. Solve $2p - 8 = -3p + 12$.
  - (A) $p = 4$
  - B $p = -4$
  - C $p = -20$
  - D $p = 20$

10. Jack records the average temperature for five days during the winter and obtains the following: 23.8°F, 10°F, 0.7°F, −5.3°F, 15.5°F. Find the range of these temperatures.
  - F 8.3°F
  - (G) 29.1°F
  - H 8.94°F
  - J 18.5°F

11. Express $m \cdot m \cdot m \cdot n \cdot n$ using exponents.
  - A $6m$
  - B $3mn$
  - (C) m^3n^2
  - D $3m^n$

12. Find the measure of an angle whose complement is 3 times the angle.
  - (F) 22.5°
  - G 67.5°
  - H 45°
  - J 135°

13. Express 1.413×10^{-5} in standard notation.
  - A 0.0001413
  - B 0.001413
  - (C) 0.00001413
  - D 14.130000

14. Express 50,600,000 in scientific notation.
  - F 50.6×10^6
  - G 5.06×10^3
  - H 5.06×10^5
  - (J) 5.06×10^7

15. A chef spreads a $\frac{1}{4}$-in. thick layer of icing on top of a cake that is $2\frac{2}{3}$-in. in height. What is the height of the completed cake?
  - (A) $2\frac{11}{12}$ in.
  - B $3\frac{1}{4}$ in.
  - C $\frac{11}{12}$ in.
  - D $2\frac{7}{8}$ in.

Cumulative Test
Form C, continued

16. Find the perimeter of the polygon. 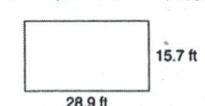
  - F 44.6 ft
  - (G) 89.2 ft
  - H 112.3 ft
  - J 453.7 ft

17. To the nearest tenth find the average number of e-mails that Joshua received during the past two weeks: 22, 4, 0, 11, 38, 43, 47, 0, 5, 10, 10, 0, 10, 11. Round to the nearest tenth.
  - A 19.2
  - (B) 15.1
  - C 10
  - D 105.5

18. Find the missing measurement in the triangle to nearest tenth. 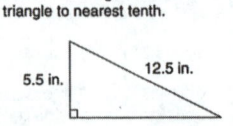
  - F 9.0 in.
  - G 34.4 in.
  - (H) 11.2 in.
  - J 13.7 in.

19. Solve the equation $8.62 = s + 14.5$.
  - (A) $s = -5.88$
  - B $s = 23.12$
  - C $s = 0.60$
  - D $s = 5.88$

20. Find the circumference of a circle with $d = 20.2$ mi to the nearest tenth. Use 3.14 for π.
  - F 126.9 mi
  - (G) 63.4 mi
  - H 320.3 mi
  - J 31.7 mi

21. Mark runs $3\frac{1}{4}$ miles in $\frac{1}{3}$ of an hour. At this rate, how long would it take him to run 6 miles?
  - (A) $\frac{8}{13}$ hr
  - B $6\frac{1}{2}$ hr
  - C $1\frac{1}{12}$ hr
  - D $\frac{2}{3}$ hr

22. Find the sum of the angles of a regular 11-gon.
  - F 1980°
  - G 1260°
  - H 1080°
  - (J) 1620°

23. Find the volume of the figure to the nearest tenth. Use 3.14 for π. 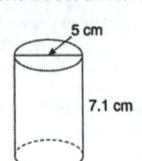
  - A 557.4 cm³
  - (B) 139.3 cm³
  - C 791.4 cm³
  - D 137.4 cm³

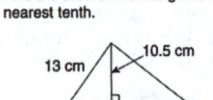

24. Find the area of the triangle to the nearest tenth.
  - F 64.5 cm²
  - G 225.8 cm²
  - (H) 112.9 cm²
  - J 451.9 cm²

Cumulative Test — Form C, continued (Chapter 6)

25. Simplify $\frac{-56}{-16}$.
    A 5
    B $\frac{56}{16}$
    C -5
    D $3\frac{1}{2}$ ⊙

26. Find the value of 10^6.
    F 60
    G 100,000
    H 1000
    J 1,000,000 ⊙

27. Which of the following is the same as "five less than twice a number d is negative 4"?
    A $5 - 2d = -4$
    B $2d - 5 = -4$ ⊙
    C $2d - 5 = 4$
    D $5 + 2d = -4$

28. To the nearest tenth, solve for the unknown side in the right triangle ABC: $a = 9$, $c = 11.4$.
    F $b = 7.0$ ⊙
    G $b = 9$
    H $b = 14.5$
    J $b = 10.2$

29. Express $9^4 \cdot 9^9 \cdot 9^0$ using positive exponents.
    A 9^{36}
    B 9^{13} ⊙
    C 0
    D 81^{13}

30. Solve $\frac{3}{5}y + \frac{2}{9} = \frac{5}{8} - \frac{2}{5}y + \frac{5}{9}$.
    F $y = \frac{115}{24}$
    G $y = -\frac{11}{72}$
    H $y = \frac{23}{24}$ ⊙
    J $y = -\frac{55}{72}$

31. Choose the inequality for "four is greater than two times x minus 2."
    A $4 > 2x - 2$ ⊙
    B $4 \leq 2x - 2$
    C $2x - 2 \leq 4$
    D $4 = 2x - 2$

32. In $\triangle ABC$, $\angle A$, $\angle B$, and $\angle C$ have a 1:2:3 ratio to each other. Find $m\angle C$.
    F 30°
    G 90° ⊙
    H 60°
    J 135°

33. To the nearest tenth, find the volume of a sphere with a diameter of 21 cm. Use 3.14 for π.
    A 3247.2 cm³
    B 4846.6 cm³ ⊙
    C 3634.9 cm³
    D 38,772.7 cm³

34. What is the name for a quadrilateral that has exactly one pair of parallel sides?
    F equilateral
    G rectangle
    H rhombus
    J trapezoid ⊙

35. The perimeter of a square is 188 inches. Find the length of one side.
    A 13.7 in.
    B 94 in.
    C 47 in. ⊙
    D 3.7 in.

36. To the nearest tenth, find the area of a circle with $d = 20.2$ mi. Use 3.14 for π.
    F 63.4 mi²
    G 126.9 mi²
    H 320.3 mi² ⊙
    J 1281.2 mi²

37. Using the stem-and-leaf plot, find the mode.

Stem	Leaves
7	1 8
8	1 2 3 5 6
9	1 3 8 9
10	3 5

    A 86
    B 1
    C 90
    D no mode ⊙

38. How much will it cost to build a 57.5 ft by 72.25-ft skating rink if the builders charge $27.10 per square yard?
    F $37,527.85
    G $12,509.28 ⊙
    H $112,583.56
    J $4154.38

39. Find the volume of the figure. 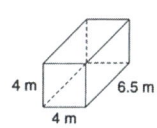
    (4 m × 4 m × 6.5 m box)
    A 22.5 m³
    B 136 m²
    C 26 m³
    D 104 m³ ⊙

40. Express 17.125 as a fraction.
    F $7\frac{1}{8}$
    G $17\frac{1}{12}$
    H $17\frac{1}{8}$ ⊙
    J $17\frac{1}{16}$

41. Find the surface area of the figure. 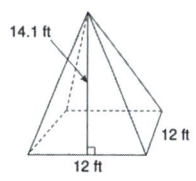
    (pyramid, 14.1 ft, 12 ft, 12 ft)
    A 482.4 ft² ⊙
    B 388.8 ft²
    C 314.4 ft²
    D 338.4 ft²

42. Name the figure.
    (rectangular prism)
    F triangle
    G pyramid
    H prism ⊙
    J quadrilateral

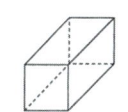

43. Find the surface area of the figure. Use 3.14 for π. 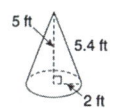
    (cone, 5 ft, 5.4 ft, 2 ft)
    A 46.5 ft² ⊙
    B 54 ft²
    C 20.9 ft³
    D 102.4 ft³

44. Find the area of the figure to the nearest tenth. 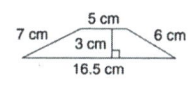
    (trapezoid: 5 cm, 7 cm, 6 cm, 3 cm height, 16.5 cm base)
    F 32.3 cm²
    G 32.5 cm²
    H 46.5 cm²
    J 93.0 cm²

45. Find all roots of $\sqrt{\frac{169}{81}}$.
    A 13, 16, 9
    B $\frac{13}{9}, -\frac{13}{9}$ ⊙
    C 16, 9
    D $\frac{16}{9}, -\frac{16}{9}$

Cumulative Test — Form A (Chapter 7)

1. Evaluate $2ab$ for $a = 3$ and $b = -1$.
   A 7
   B 5
   C -6 ⊙
   D 6

2. Solve: $m + 4 = -4$.
   A $m = 0$
   B $m = -16$
   C $m = 8$
   D $m = -8$ ⊙

3. A length on a map is 4 in. The scale is 1 in:1.5 mi. What is the actual distance?
   A 5.5 mi
   B 6 mi ⊙

4. Solve $2(s - 3) = 6$.
   A $s = 6$ ⊙
   B $s = 0$

5. A photograph 5 in. wide by 7 in. high is to be enlarged to 12.5 in. wide. How high will the enlarged photo be?
   A 14.5 in.
   B 17.5 in. ⊙
   C 19.5 in.
   D 8.9 in.

6. Which algebraic expression means "73 times a number"?
   A $73n$ ⊙
   B $\frac{73}{n}$

7. Find the area of a rectangular poster measuring $\frac{1}{2}$ ft by $1\frac{1}{4}$ ft.
   A $\frac{3}{4}$ ft²
   B $1\frac{3}{4}$ ft²
   C $3\frac{1}{2}$ ft²
   D $\frac{5}{8}$ ft² ⊙

8. You received science test scores of 70, 40, 70, 90, and 40. What was your mean score?
   A 62 ⊙
   B 70
   C 66.7
   D 40

9. A block of wood has dimensions of 4 in. × 2 in. × 3 in. What is its volume?
   A 9 in³
   B 24 in³ ⊙

10. A poultry farm hatched the following number of chicks during the past week: 50, 10, 10, 40, 30, 25, 20. What was the median number of chicks hatched?
    A 25 ⊙
    B 26.4

11. Find $\sqrt{16}$.
    A 4 ⊙
    B 8

12. Solve for the unknown side in the right triangle. 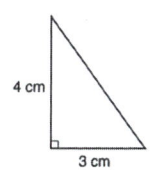
    (4 cm, 3 cm)
    A 5 cm ⊙
    B 12 cm
    C 3 cm
    D 4 cm

13. Solve $2.5 = n + 6.5$.
    A $n = 4$
    B $n = -4$ ⊙

14. Marla's pay for a week was $84.12. If she worked 12 hr, what was her rate of pay?
    A $7.01 per hr ⊙
    B $8.59 per hr

15. Find the unknown number in the proportion: $\frac{4}{7} = \frac{10}{8}$.
    A $f = 3.2$ ⊙
    B $f = 8$

16. A 9-in. cube is built from small cubes, each 3 in. on a side. The volume of the large cube is how many times that of a small cube?
    A 3
    B 27 ⊙

17. Convert 12 cups to pints.
    A 6 pt ⊙
    B 24 pt
    C 4 pt
    D 2 pt

18. Express $3 \cdot 3 \cdot 3 \cdot 3$ in simplest form using exponents.
    A 3^{27}
    B 9^4
    C 24
    D 3^4 ⊙

19. Simplify $(2)^{-3}$.
    A $-\frac{1}{16}$
    B $\frac{1}{8}$ ⊙

20. The population of a small town is 45,000. Express this number using scientific notation.
    A 4.5×10^4 ⊙
    B 4.5×10^3

21. You received scores of 70, 40, 70, 90, and 40. Find the first quartile.
    A 40 ⊙
    B 70

22. In the figure, which angle is a supplement to $\angle ABE$? 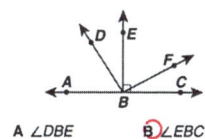
    A $\angle DBE$
    B $\angle EBC$ ⊙

23. Find the unknown length in the pair of similar triangles.
    (3, 4, 5 and 6, x, 10) 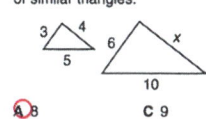
    A 8 ⊙
    B 4
    C 9
    D 12

24. Find the volume of the figure. 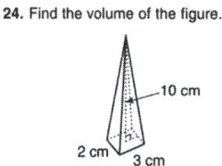
    (pyramid, 10 cm, 2 cm, 3 cm)
    A 60 cm³
    B 20 cm³ ⊙

25. Which angle is congruent to $\angle AOD$? 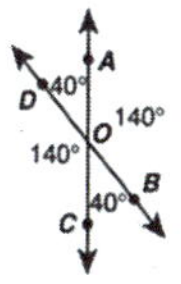
    (intersecting lines with 40° and 140°)
    A $\angle BOC$ ⊙
    B $\angle AOB$

CHAPTER 7 Cumulative Test
Form A, continued

26. Simplify $\dfrac{d^5}{d^5}$.
   A d^{10}
   (B) 1

27. What are the coordinates of point A on the graph?

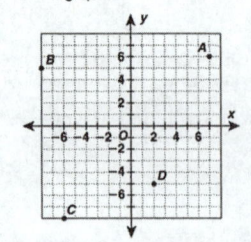

   A (5, −8) (C) (7, 6)
   B (−7, 6) D (6, 7)

28. Simplify $1\dfrac{1}{3} - \dfrac{5}{12}$.
   (A) $\dfrac{11}{12}$ B $\dfrac{5}{12}$

29. Triangle ABC is similar to triangle DEF. Find x.

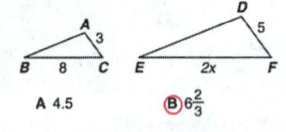

   A 4.5   (B) $6\dfrac{2}{3}$

30. The graph represents which inequality? 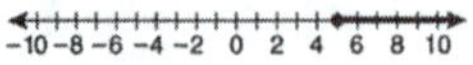
   A $x \geq -5$ (B) $x \geq 5$

31. Simplify $2\dfrac{3}{8} + \dfrac{1}{4}$.
   A $2\dfrac{1}{3}$ (C) $2\dfrac{5}{8}$
   B $2\dfrac{2}{3}$ D $2\dfrac{3}{32}$

32. In △ABC, ∠A and ∠B have the same measure. m∠C is 90°. Find m∠A.
   (A) 45°   B 90°

33. What is the term for part of a line that is between two points?
   A point   C line
   B ray   (D) line segment

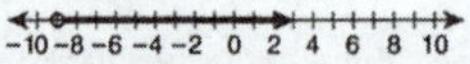

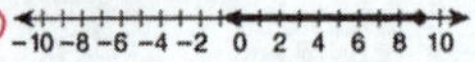

34. Which of the following graphs represents $\dfrac{x}{3} \leq 3$?

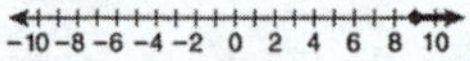

35. In a town of 4500 voters, a pollster asks 400 people if they will support the next school levy. Identify the population.
   (A) 4500 voters
   B All people in the town
   C All school children
   D 400 people surveyed

36. Name the figure.

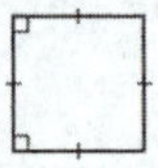

   A pentagon   C hexagon
   B octagon   (D) square

37. Which of the following pairs of ratios forms a proportion?
   (A) $\dfrac{3}{4}$ and $\dfrac{6}{8}$
   B $\dfrac{4}{6}$ and $\dfrac{5}{8}$

38. Classify $-\sqrt{11}$ and give the approximate value.
   A rational; −3.3   (B) irrational; −3.3

39. If a drawing has a scale 1 cm:1 mi, is the drawing a reduction or an enlargement of the actual object?
   A enlargement   (B) reduction

40. Express 1.5 as a fraction.
   A $1\dfrac{1}{5}$ (C) $1\dfrac{1}{2}$
   B $1\dfrac{1}{4}$ D $1\dfrac{1}{10}$

41. Find $\sqrt{20}$ to the nearest hundredth.
   (A) 4.47   B 4.52

42. Solve $3c = -9$.
   A $c = 3$ C $c = 27$
   (B) $c = -3$ D $c = -27$

43. Solve $2(s - 2) + 2s$.
   A $4s - 10$ (C) $4s - 4$
   B $8s$ D $4s - 2$

44. Determine which of the following is a solution to $y = x - 3$.
   A (−1, −1)   C (2, 3)
   (B) (4, 1)   D (1, 2)

45. Find $(2 + 1)^2$.
   A 5   (B) 9

46. When dilating a triangle, using a scale factor greater than 1 will:
   (A) enlarge the triangle.
   B shrink the triangle.

CHAPTER 7 Cumulative Test
Form B

1. Evaluate $6xy + 3$ for $x = -5$ and $y = 3$.
   A −90   C 87
   B 24   (D) −87

2. Solve $m + 5 = 11$.
   F $m = 16$ (H) $m = 6$
   G $m = -16$ J $m = -6$

3. The scale on a map is 1 in.:2.5 mi. What is the length of 42 miles on the map?
   A 105 in.   C 44.5 in.
   (B) 16.8 in.   D 19.3 in.

4. Solve $8(s - 3) = 24$.
   (F) $s = 6$ H $s = 3$
   G $s = 0$ J $s = -3$

5. A picture 6 in. wide by 9 in. high is enlarged to 15 in. wide. What is the scale factor for the enlargement?
   A 6   (C) 2.5
   B 2   D 9

6. Which algebraic expression means "73 minus a number"?
   F $n - 73$ H $-73 - n$
   G $-n + 73$ (J) $73 - n$

7. Find the area of a table top measuring $\dfrac{1}{3}$ ft by $\dfrac{2}{7}$ ft.
   A $\dfrac{3}{7}$ ft²   C $\dfrac{3}{21}$ ft²
   B $\dfrac{3}{10}$ ft²   (D) $\dfrac{2}{21}$ ft²

8. Bill received tips of $73, $48, $73, $92, and $48 over the past 5 nights. What were his mean nightly tips?
   (F) $66.80   H $70
   G $73   J $71

9. A small box has dimensions of 5 in. × 2 in. × 9 in. What is its volume?
   A 225 in³   C 20 in³
   B 32 in³   (D) 90 in³

10. A restaurant served the following number of steaks during the past week: 50, 9, 7, 42, 33, 25, 20. What was the median number of steaks served?
   F 33   H 42
   G 26.6   (J) 25

11. Find $\sqrt{43}$ and round your answer to the nearest hundredth.
   A 43.0   (C) 6.56
   B 6.5   D 6.55

12. Solve for the unknown side in the right triangle. 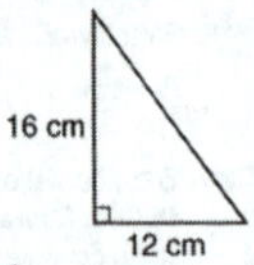
   (F) 20 cm
   G 19 cm
   H 14 cm
   J 16 cm

13. Solve $7.8 = n + 6.6$.
   A $n = 14.4$ (C) $n = 1.2$
   B $n = 1.18$ D $n = -1.2$

14. Jen's pay for a week was $82.20. If she worked 12 hr, what was her rate of pay?
   (F) $6.85 per hr   H $6.98 per hr
   G $6.73 per hr   J $7.29 per hr

15. Find the unknown number in the proportion: $\dfrac{5}{f} = \dfrac{20}{8}$.
   A $f = 0.08$ (C) $f = 2$
   B $f = 20$ D $f = 12.5$

16. A 15 cm cube is built from small cubes, each 3 cm on a side. The surface area of the large cube is how many times that of a small cube?
   F 5   H 9
   G 12   (J) 25

17. Convert 24 pints to cups.
   A 6 c   C 3 c
   (B) 48 c   D 12 c

18. Express $4 \cdot 4 \cdot 4 \cdot 4 \cdot 4 \cdot 4$ using exponents.
   (F) 4^6 H 24
   G 6^4 J 4^5

19. Find $(-2)^{-4}$.
   A 16   C −16
   B $-\dfrac{1}{16}$ (D) $\dfrac{1}{16}$

20. The population of a small country is 5,740,000. Express this number using scientific notation.
   (F) 5.74×10^6 H 5.74×10^4
   G 5.74×10^5 J 57.4×10^4

21. You received scores of 73, 48, 73, 92, and 48. Find the first quartile.
   A 44   C 60.5
   (B) 48   D 73

22. In the figure, which angle is a complement to ∠EBF? 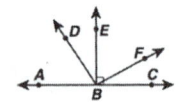
   F ∠DBE   (H) ∠CBF
   G ∠EBA   J ∠ABD

23. Find the unknown length in the pair of similar triangles. 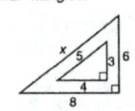
   A 11   C 15
   B 5   (D) 10

24. Find the volume of the figure. 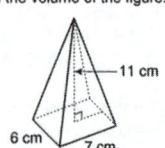
   F 185 cm³   H 90 cm³
   (G) 154 cm³   J 462 cm³

Holt Pre-Algebra

Cumulative Test
Form B, continued

25. Identify a pair of congruent angles.

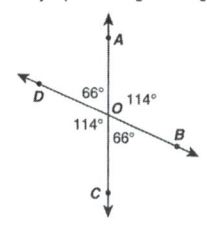

    (A) ∠AOD and ∠BOC
    B ∠AOD and ∠AOB
    C ∠BOC and ∠COD
    D ∠AOB and ∠BOC

26. Simplify $\frac{d^3}{d^5}$.

    F d^2
    (G) $\frac{1}{d^2}$
    H d^{-5}
    J $\frac{d^3}{d^2}$

27. What are the coordinates of point B on the graph?

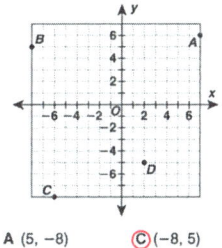

    A (5, −8)
    B (5, 8)
    (C) (−8, 5)
    D (8, 5)

28. Simplify $1\frac{2}{3} - \frac{11}{18}$.

    (F) $\frac{19}{18}$
    G $\frac{3}{7}$
    H $\frac{10}{21}$
    J $1\frac{4}{21}$

29. Triangle ABC ≅ triangle DEF. Find z.

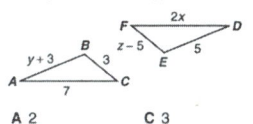

    A 2
    (B) 8
    C 3
    D 5

30. The graph represents which inequality?

    F $x \geq -5$
    G $x > -5$
    H $x < 5$
    (J) $x < -5$

31. Simplify $2\frac{5}{6} + \frac{1}{3}$.

    A $3\frac{7}{12}$
    (B) $3\frac{1}{6}$
    C $3\frac{1}{3}$
    D $2\frac{1}{3}$

32. In △ABC, ∠A and ∠B have the same measure. m∠C is 92°. Find m∠A.

    (F) 44°
    G 88°
    H 46°
    J 134°

33. What is the term for part of a line that starts at one point and extends infinitely in one direction?

    A point
    (B) ray
    C line
    D line segment

34. Which of the following graphs represents: $\frac{x}{3} \geq -3$?

    F
    (G)
    H
    J

35. In a town of 5500 voters, a pollster asks 500 people if they will support the challenger in the mayoral race. Identify the population.

    A 500 people surveyed
    B All people in the town
    C All supporters
    (D) 5500 voters

36. Name the figure.

    F pentagon
    (G) octagon
    H hexagon
    J heptagon

37. Which of the following pairs of ratios forms a proportion?

    (A) $\frac{3}{8}$ and $\frac{6}{16}$
    B $\frac{4}{9}$ and $\frac{5}{10}$
    C $\frac{1}{2}$ and $\frac{2}{3}$
    D $\frac{11}{20}$ and $\frac{1}{2}$

38. Classify $-\sqrt{25}$ as rational, irrational, or not real. If rational or irrational, give the approximate equivalent value.

    (F) rational; −5
    G irrational; −5
    H irrational; 5
    J not real

39. Choose which scale preserves the size of the actual object.

    A 12 in.:6 ft
    B 1 cm:1 mi
    C 1 m:100 mm
    (D) 36 in.:1 yd

40. Express 2.25 as a fraction.

    F $2\frac{2}{5}$
    G $2\frac{1}{25}$
    (H) $2\frac{1}{4}$
    J $2\frac{1}{5}$

41. Find $-\sqrt{21}$ to the nearest tenth.

    A 2.1
    (B) −4.6
    C 10.5
    D −441.0

42. Solve $3(3n) = -9$.

    F $n = 1$
    (G) $n = -1$
    H $n = 9$
    J $n = -9$

43. Solve $5(s - 2) - s$.

    A $5s - 10$
    B $4s - 2$
    (C) $4s - 10$
    D $6s - 2$

44. Determine which of the following is a solution to $y = 2x - 3$.

    F (−1, −1)
    (G) (1, −1)
    H (2, 3)
    J (1, 2)

45. Find $(-2)^{-2}$.

    A 8
    B $-\frac{1}{4}$
    C 4
    (D) $\frac{1}{4}$

46. When dilating a triangle, to enlarge the triangle you must use a scale factor that is:

    (F) > 1
    G < 1
    H = 1
    J = 0

Cumulative Test
Form C

1. Evaluate $12m + 15n$, for $m = -2$ and $n = -3$.

    A −21
    B −66
    C 54
    (D) −69

2. Solve $p + 5 = 11 - p$.

    F $p = 8$
    (G) $p = 3$
    H $p = 6$
    J $p = -16$

3. If a scale of 1 cm:3.5 m is used, how long is an object 4.8 cm long in a drawing?

    A 8.3 m
    (B) 16.8 m
    C 1.37 m
    D 20.3 m

4. Solve: $5(3s - 3) = 25$.

    F $s = 1.9$
    (G) $s = 2\frac{2}{3}$
    H $s = 5$
    J $s = 25$

5. A scale factor of 3.2 is used to enlarge a 15 cm by 30 cm picture. What are the dimensions of the enlargement?

    A 18.2 cm × 33.2 cm
    B 45 cm × 95 cm
    (C) 48 cm × 96 cm
    D 4.7 cm × 9.4 cm

6. Which algebraic expression means "44 less than a number"?

    F $44 - n$
    G $-n + 44$
    H $-44 - n$
    (J) $n - 44$

7. Find the area of a mirror measuring $1\frac{1}{3}$ ft by $\frac{7}{12}$ ft.

    (A) $\frac{7}{9}$ ft²
    B $1\frac{3}{4}$ ft²
    C $1\frac{11}{12}$ ft²
    D $2\frac{2}{7}$ ft²

8. You received quiz scores of 7.3, 4.8, 7.3, 9.2, and 4.8. What was your mean score rounded to the nearest tenth?

    (F) 6.7
    G 7.3
    H 7.1
    J 33.4

9. A locker has dimensions of 5.2 in. × 2.5 in. × 9 in. What is its volume?

    A 16.7 in³
    B 22.5 in³
    (C) 117 in³
    D 243.4 in³

10. A dentist treated the following number of patients during the past week: 50, 19, 17, 32, 33, 25, 20. What was the median number of patients treated?

    (F) 25
    G 28
    H 32
    J 33

11. Find $-\sqrt{142}$ and round your answer to the nearest tenth.

    A −3.5
    B −11.8
    (C) −11.9
    D 12.0

12. Solve for the unknown side in the right triangle. Round to the nearest tenth.

    F 14.0 cm
    G 9.2 cm
    (H) 19.9 cm
    J 28.0 cm

13. Solve $17.8 - n = n + 6.6$.

    A $n = -12.2$
    (B) $n = 5.6$
    C $n = 11.2$
    D $n = -11.2$

14. Sarah's pay for two weeks was $158.16. If she worked 8 hours the first week and 4 hours the second, what was her rate of pay?

    F $3.23 per hr
    (G) $13.18 per hr
    H $22.60 per hr
    J $31.64 per hr

15. Find the unknown number in the proportion $\frac{5}{f} = \frac{20}{4.4}$.

    (A) $f = 1.1$
    B $f = 8$
    C $f = 11$
    D $f = 22.7$

16. A 2-m cube is built from small cubes, each 5 cm on a side. The volume of the large cube is how many times that of a small cube?

    F 40,000
    G 1600
    (H) 64,000
    J 16,000

17. Convert 24 cups to quarts.

    A 3 qt
    (B) 6 qt
    C 8 qt
    D 24 qt

18. Express $(5 \cdot 5 \cdot 5 \cdot 5) \cdot (2 \cdot 2 \cdot 2)$ using exponents.

    F 24^7
    G $(2 \cdot 5)^7$
    H $5^6 \cdot 2^6$
    (J) $5^4 \cdot 2^3$

19. Find $(-3)^{-3}$.

    A $\frac{1}{27}$
    B 27
    (C) $-\frac{1}{27}$
    D −27

20. The population of a large city is 3,040,000. Express this number using scientific notation.

    (F) 3.04×10^6
    G 3.04×10^5
    H 3.04×10^4
    J 30.4×10^7

21. What is the third quartile for 730, 480, 730, 920, and 480?

    A 440
    B 480
    C 668
    (D) 825

22. In the figure, which angle is a complement to ∠EBD?

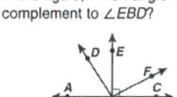

    F ∠DBE
    G ∠EBC
    H ∠CBF
    (J) ∠ABD

23. Find the unknown length in the pair of similar triangles.

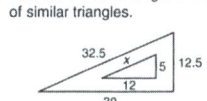

    A 15
    B 20
    (C) 13
    D 11

24. Find the height of the figure, if the volume is 154 cm³.

    F 33 cm
    (G) 11 cm
    H 3.7 cm
    J 9 cm

CHAPTER 7 Cumulative Test
Form C, continued

25. Identify a pair of congruent angles.

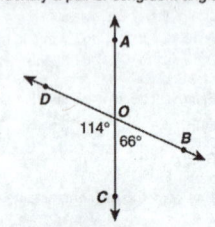

- A ∠AOD and ∠COD
- B ∠AOD and ∠AOB
- C ∠BOC and ∠COD
- **D ∠AOB and ∠COD** ○

26. Simplify $\frac{x^2 d}{d^6 x}$.

- F 0
- **G $\frac{x}{d^5}$** ○
- H $d^5 x$
- J 1

27. What are the coordinates of point C on the graph?

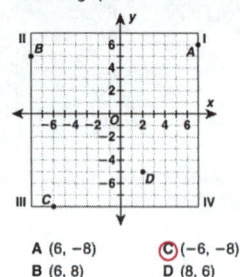

- A (6, −8)
- B (6, 8)
- **C (−6, −8)** ○
- D (8, 6)

28. Simplify $2\frac{2}{3} - 1\frac{5}{8}$.

- F $\frac{8}{17}$
- **G $\frac{25}{24}$** ○
- H $1\frac{8}{9}$
- J $4\frac{1}{2}$

29. Triangle ABC ≅ triangle DEF. Find y.

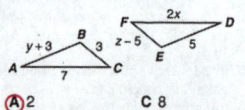

- **A 2** ○
- B 3
- C 8
- D 15

30. The graph represents which inequality?

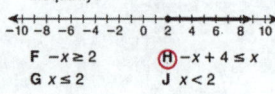

- F $-x \geq 2$
- G $x \leq 2$
- **H $-x + 4 \leq x$** ○
- J $x < 2$

31. Simplify $2\frac{5}{7} + 1\frac{1}{3}$.

- A $2\frac{5}{21}$
- **B $4\frac{1}{21}$** ○
- C $3\frac{7}{10}$
- D $3\frac{9}{10}$

32. In △ABC, ∠A and ∠B have the same measure. m∠C is 15° more than the other angles. Find m∠A.

- **F 55°** ○
- G 82.5°
- H 60°
- J 90°

33. Which is the term for a flat surface that extends infinitely in all directions?

- A ray
- **B plane** ○
- C line
- D line segment

34. Which of the following graphs represents: $\frac{-x}{9} \geq -1$

- F 
- **G** ○ 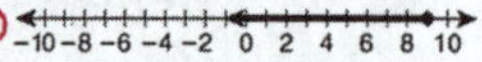
- H 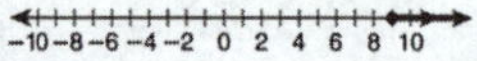
- J

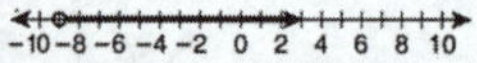

35. In a town of 15,500 voters, a pollster asks 1500 people if they will support construction of a new landfill. Identify the sample.

- A 15,500 voters
- B All trash producers
- C All people in the town
- **D 1500 people surveyed** ○

36. Name the figure.

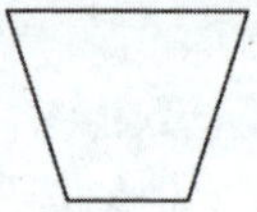

- **F trapezoid** ○
- G octagon
- H hexagon
- J rectangle

37. Which of the following pairs of ratios forms a proportion?

- **A $\frac{3.5}{5.5}$ and $\frac{7}{11}$** ○
- B $\frac{4}{9}$ and $\frac{2.5}{5}$
- C $\frac{2}{5}$ and $\frac{2}{3}$
- D $\frac{11}{20}$ and $\frac{1.1}{20}$

38. Classify $-\sqrt{-25}$ as rational, irrational, or not real. If rational or irrational, give the approximate equivalent value.

- F rational; −5
- G irrational; −5
- H irrational; 5
- **J Not real** ○

39. Choose the one which enlarges the most.

- A 1 cm:1 m
- B 2 m:15.5 km
- **C 1.3 in.:0.8 mi** ○
- D 4 in.:2 mi

40. Express 12.125 as a fraction.

- F $12\frac{2}{5}$
- G $12\frac{1}{25}$
- **H $12\frac{1}{8}$** ○
- J $12\frac{1}{5}$

41. Find $-\sqrt{110}$ to the nearest tenth.

- A −11
- **B −10.5** ○
- C not real
- D −12

42. Solve 3.5(3g) = −19.95.

- **F $g = -1.9$** ○
- G $g = 5.7$
- H $g = 7.7$
- J $g = 6.65$

43. Simplify $2.5(s - 2) - 2.5s$.

- **A −5** ○
- B −2
- C $5s - 5$
- D $5s - 2$

44. Determine which of the following is a solution to $2y = 2x - 3$.

- F (−1, −1)
- **G $(1, -\frac{1}{2})$** ○
- H (2, 3)
- J (1, 2)

45. Find $(-3)^{-2} + (3)^2$.

- A 0
- B 1
- C −1
- **D $9\frac{1}{9}$** ○

46. When dilating a triangle, which of the following scale factors will reduce the image of the initial triangle?

- F $s > 1$
- G $s < 0$
- H $s = 1$
- **J $0 < s < 1$** ○

CHAPTER 8 Cumulative Test
Form A

1. What is 9.8% as a decimal?

- **A 0.098** ○
- B 9.8
- C 0.98
- D 98

2. Allison needs 15 feet of ribbon for a craft project. How many yards is this?

- **A 5 yd** ○
- B 45 yd

3. Simplify using exponents: $5^5 \cdot 5^4$.

- **A 5^9** ○
- B 25^{20}

4. Determine which of the following is a solution to $y - 2 = 2x$.

- **A (1, 4)** ○
- B (0, 1)

5. In the figure, △ABC is similar to △GHJ. Find length x.

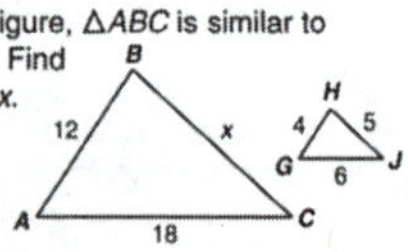

- A 5
- **B 15** ○

6. Solve $f - 10 = 40 - f$.

- A $f = 15$
- B $f = 30$
- **C $f = 25$** ○
- D $f = 50$

7. Find the mode of these numbers.
7, 7, 2, 7, 13, 5, 9, 6

- **A 7** ○
- B 13

8. Simplify the expression $\frac{v^2}{v^3}$.

- **A $\frac{1}{v}$** ○
- B $\frac{1}{v^5}$

9. Construct a box-and-whisker plot for the data set:
15 24 34 18 20 48
26 18 24 19 39 32
20 28 25 24 35 23

- A 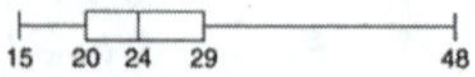
- **B** ○

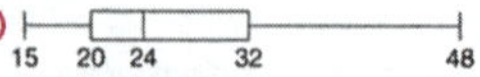

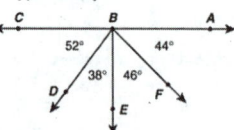

10. In the figure, which two angles are supplementary?

- **A ∠CBE and ∠EBA** ○
- B ∠CBD and ∠DBE

11. Find all square roots of $\sqrt{36}$.

- A 6.6
- B 6, 6
- **C 6, −6** ○
- D 36, −36

12. Evaluate $-3vw$ for $v = 3$ and $w = -1$.

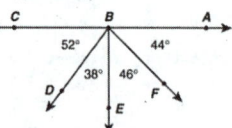

- A −9
- **B 9** ○
- C −10
- D 10

13. Sarah is enclosing a square flower garden with timbers. The perimeter of the flower garden is 36 feet. Find the length of one side of the garden.

- A 4 ft
- B 6 ft
- **C 9 ft** ○
- D 12 ft

14. Find an algebraic expression for "2 plus the quotient of 13 and z."

- **A $2 + \frac{13}{z}$** ○
- B $\frac{2 + 13}{z}$

15. Find the original data from the stem-and-leaf plot.

Stem	Leaves
8	1 7
9	1 1 7
10	1 4 9

- A 8, 1, 7, 9, 11, 17, 10, 14, 19
- **B 81, 87, 91, 91, 97, 101, 104, 109** ○

16. Find the area of the figure.

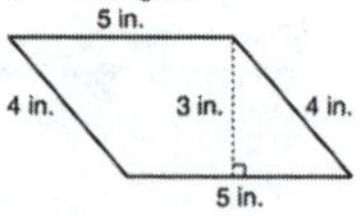

- A 7.5 in²
- B 12 in²
- **C 15 in²** ○
- D 20 in²

17. What is the value of x for the equivalent fraction? $\frac{12}{9} = \frac{x}{18}$

- A $x = 2$
- B $x = 9$
- C $x = 12$
- **D $x = 24$** ○

18. A circle has a radius of 2 inches. If the radius is doubled, what is the circumference to the nearest tenth? Use 3.14 for π.

- A 6.3 in.
- B 12.6 in.
- **C 25.1 in.** ○
- D 50.2 in.

19. A bin has 250 lag and toggle bolts. There are 80 more toggle bolts than lag bolts. How many lag bolts are in the bin?

- A 80
- **B 85** ○
- C 165
- D 170

20. A reception hall holds 192 people. If at a wedding reception the hall is filled to $\frac{3}{4}$ of the maximum capacity, how many people are at the reception?

- A 48
- **B 144** ○

21. Describe the radical as "rational," "irrational," or "not a real number." If a real number, give the value: $-\sqrt{81}$.

- A rational, 9
- **B rational, −9** ○
- C irrational, −9
- D not a real number

22. Find the best buy for a box of cereal and give the cost per ounce.
Fruity Pops: 12 oz for $2.34
Raisin Nut Crunch: 16 oz for $3.06

- A Fruity Pops, $0.19 per oz
- **B Raisin Nut Crunch, $0.19 per oz** ○
- C Fruity Pops, $0.20 per oz
- D Raisin Nut Crunch, $0.21 per oz

23. Find the area of the figure.

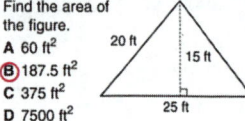

- A 60 ft²
- **B 187.5 ft²** ○
- C 375 ft²
- D 7500 ft²

Copyright © by Holt, Rinehart and Winston.
All rights reserved.

Holt Pre-Algebra

Cumulative Test
Form A, continued

24. Find the volume of the figure. 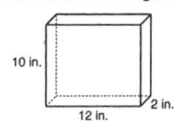
 (10 in. × 12 in. × 2 in.)
 A 20 in^3
 B 24 in^3
 C 120 in^3
 (D) 240 in^3

25. What is the mean of the following temperatures?

Temperature Highs			
Mon.	55°	Fri.	55°
Tues.	65°	Sat.	60°
Wed.	55°	Sun.	70°
Thurs.	60°		

 A 55°
 (B) 60°
 C 65.5°
 D 70°

26. Find the surface area of the figure. (7 ft, 4 ft, 4 ft)
 A 13 ft^2
 B 27 ft^2
 C 42 ft^2
 (D) 72 ft^2

27. Below is a diagram of how a set of bracing leans against a wall of a house under construction. About how far from the house is the base of the bracing to the nearest tenth? (house, 20 ft, 30 ft)
 (A) 22.4 ft
 B 36.1 ft

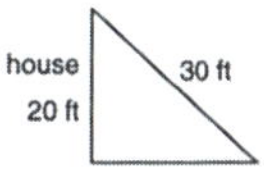

28. Simplify $1\frac{2}{3} \cdot \frac{5}{6}$.
 (A) $1\frac{7}{18}$
 B $1\frac{7}{9}$
 C $1\frac{5}{9}$
 D 3

29. Find the sum of the angle measures in a regular pentagon.
 A 360°
 (B) 540°
 C 720°
 D 900°

30. Solve $3x \le -30$.
 (A) $x \le -10$
 B $x < 10$
 C $x \ge -10$
 D $x > -10$

31. A recipe calls for $2\frac{1}{6}$ cups of flour and $1\frac{1}{3}$ cups of sugar. How many cups of dry ingredients are in the recipe?
 A $2\frac{1}{18}$ cups
 B $3\frac{2}{9}$ cups
 C $3\frac{1}{9}$ cups
 (D) $3\frac{1}{2}$ cups

32. In $\triangle ABC$, $m\angle A = m\angle B$ and $m\angle C$ is 30°. What is $m\angle B$ to the nearest tenth of a degree?
 (A) 75°
 B 150°

33. Seventy-five out of 100 students are going to the school dance. What is this number as a reduced fraction and as a percent?
 A $\frac{3}{4}$, 7.5%
 B $\frac{3}{4}$, 750%
 (C) $\frac{3}{4}$, 75%
 D $\frac{7.5}{10}$, 7.5%

34. An investment broker invests $60,000 in bonds. If he earns 5% per year on the investment, how much money is earned in just one year?
 (A) $3000
 B $6000
 C $63,000
 D $66,000

35. On a science test, a student got 25 questions correct. On a second test the student got 30 questions correct. What was the percent of increase in the score?
 A 0.2%
 (B) 20%

36. Find 5^{-2}.
 A $\frac{1}{10}$
 (B) $\frac{1}{25}$

37. Express 1.6×10^5 miles per hour in standard notation.
 A 0.000016 mi/h
 B 16,000 mi/h
 (C) 160,000 mi/h
 D 1,600,000 mi/h

38. Simplify $-10 - (-4) + 10$.
 (A) 4
 B -4
 C 24
 D 30

39. The Deerfield cross-country team ran 5.25 miles to prepare for a meet. What is this distance as a fraction?
 A $5\frac{1}{5}$ mi
 B $5\frac{2}{5}$ mi
 (C) $5\frac{1}{4}$ mi
 D $52\frac{1}{2}$ mi

40. The ratio of boys to girls on a baseball team is 5 to 2. If there are 4 girls on the team, how many boys are on the team?
 A 12
 (B) 10
 C 9
 D 8

Cumulative Test
Form B

1. Express 79.8% as a decimal.
 A 0.0798
 (C) 0.798
 B 7.98
 D 79.8

2. Brad's new kite has 66 ft of string. How many yards is this?
 F 7.33 yd
 (G) 22 yd
 H 198 yd
 J 2376 yd

3. Simplify using exponents: $9^5 \cdot 9^4 \cdot 9^8$.
 (A) 9^{17}
 B 9^{160}
 C 729^{17}
 D 729^{160}

4. Determine which of the following is a solution to $2y - 2 = 2x$.
 F $(-1, -1)$
 G $(1, -1)$
 (H) $(0, 1)$
 J $(1, 3)$

5. In the figure, $\triangle ABC$ is similar to $\triangle GHJ$. Find length x. (12, 15, 4, 5, 6)
 A 6
 B 20
 (C) 18
 D 24

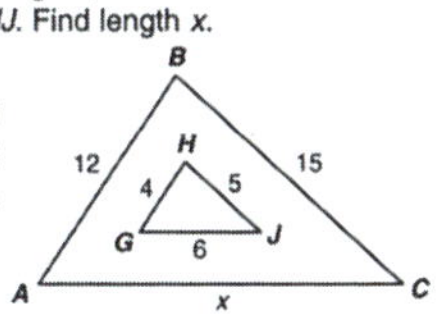

6. Solve $f - 18 = 48 - 10f$.
 (F) $f = 6$
 G $f = 12$
 H $f = -6$
 J $f = -12$

7. The following were temperatures of seven different cities in the United States on March 1, 2002: 77°, 25°, 77°, 13°, 25°, 29°, 56°, 77°. What was the mode of the temperatures?
 A 25°
 B 42.5°
 C 47.4°
 (D) 77°

8. Simplify the expression: $\frac{v^2}{v^8}$.
 F $\frac{v^2}{v^4}$
 G $\frac{v}{v^{12}}$
 (H) $\frac{1}{v^4}$
 J v^4

9. Construct a box-and-whisker plot for the data set:
 15 20 18 34 18 24 48
 29 39 23 24 26 18 29
 33 35 24 25 19 28 23
 (A) (15, 19.5, 24, 31, 48)
 B (15, 20, 24, 29, 48)
 C (15, 19, 24, 33, 48)
 D (15, 20, 24, 33, 48)

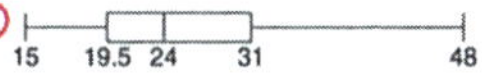

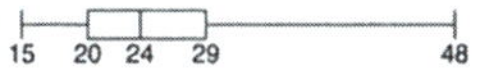

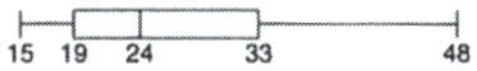

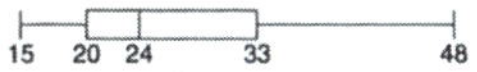

10. In the figure, which two angles are complementary? (52°, 44°, 38°, 46°)
 (F) $\angle ABF$ and $\angle FBE$
 G $\angle CBE$ and $\angle EBA$
 H $\angle CBD$ and $\angle FBE$
 J $\angle FBE$ and $\angle EBD$

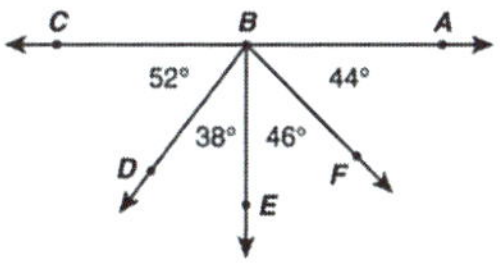

11. Find all square roots of $\sqrt{25}$.
 A 5.5
 B 5, 5
 (C) 5, -5
 D 25, -25

12. Evaluate $-5v^2w$ for $v = 4$ and $w = -1$.
 F 20
 (G) 80
 H -20
 J -80

13. The perimeter of a square pavilion at a park is 128 feet. Find the length of one side of the pavilion.
 A 5.7 ft
 B 11.3 ft
 (C) 32 ft
 D 256 ft

14. Find an algebraic expression for "the quotient of 3 and the sum of z and 5."
 (F) $\frac{3}{z+5}$
 G $\frac{3}{z} \div 5$
 H $\frac{z+5}{3}$
 J $\frac{3}{5z}$

15. Find the original data from the stem-and-leaf plot.

Stem	Leaves
8	1 7
9	1 1 7
10	1 4 9
11	4

 A 81, 17, 19, 19, 11, 14, 17, 19, 14
 B 8, 81, 84, 91, 91, 97, 94, 101, 101, 114
 (C) 81, 87, 91, 91, 97, 101, 104, 109, 114
 D 81, 87, 91, 94, 97, 104, 109, 114

16. Find the area of the figure. (7 in., 5 in., 4 in., 5 in., 7 in.)
 F 14 in^2
 (G) 28 in^2
 H 24 in^2
 J 35 in^2

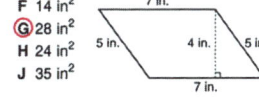

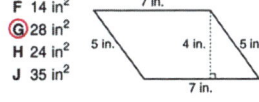

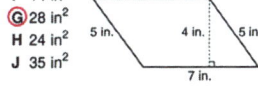

17. What is the value of x for the equivalent fraction $\frac{12}{9} = \frac{x}{45}$?
 A $x = 3$
 B $x = 5$
 C $x = 30$
 (D) $x = 60$

18. A circle has a radius of 6 inches. If the radius is doubled, what is the circumference to the nearest tenth? Use 3.14 for π.
 F 37.7 in.
 (G) 75.4 in.
 H 113.0 in.
 J 452.2 in.

19. A jar has 415 red and blue balls. There are 87 more blue balls than red balls. How many blue balls are there in the jar?
 A 77
 B 164
 (C) 251
 D 338

20. A restaurant has a seating capacity of 187 customers. If the restaurant is $\frac{5}{11}$ full, how many customers are in the restaurant?
 F 17
 G 80
 (H) 85
 J 90

21. Describe the radical as "rational," "irrational," or "not a real number." If a real number, give the approximate value: $-\sqrt{191}$.
 (A) irrational, -13.8
 B irrational, 19.1
 C rational, 19.1
 D not a real number

---

371 Holt Pre-Algebra

Cumulative Test
Form B, continued

22. Find the best buy and give the cost per ounce for the can of juice concentrate.
    Libby: 12 oz for $1.32
    Motts: 16 oz for $2.08
    (F) Libby, $0.11 per oz
    G Motts, $0.11 per oz
    H Libby, $0.13 per oz
    J Motts, $0.13 per oz

23. Find the area of the figure. 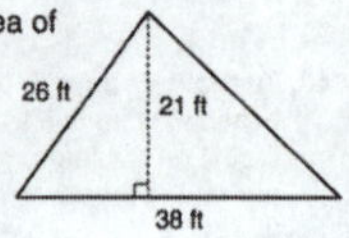
    A 84 ft^2
    (B) 399 ft^2
    C 494 ft^2
    D 798 ft^2

24. Find the volume of the figure. 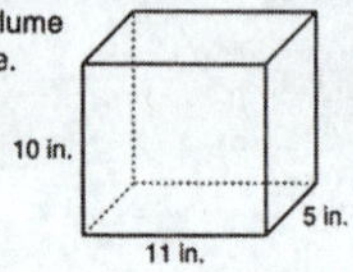
    F 26 in^3
    G 115 in^3
    H 550 in^2
    (J) 550 in^3

25. Matthew has a birthday party at a local bowling alley. The following are the scores of the children's first game. What is the mean score?
    55, 68, 69, 55, 69, 70, 55?
    A 55
    (B) 63
    C 65.5
    D 68

26. Find the surface area of the figure. 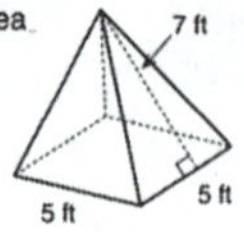
    F 29 ft^2
    G 43 ft^2
    (H) 95 ft^2
    J 105 ft^2

27. Below is a diagram of how a ladder leans against one wall of a house. How far from the house is the base of the ladder to the nearest tenth? 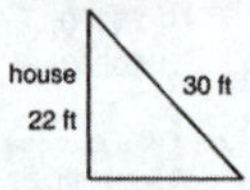
    A 8 ft
    B 37.2 ft
    (C) 20.4 ft
    D 416 ft

28. Simplify $2\frac{2}{3} \cdot \frac{15}{18}$.
    (F) $2\frac{2}{9}$
    G $2\frac{9}{21}$
    H $3\frac{1}{5}$
    J 10

29. Find the sum of the angle measures in a regular heptagon.
    A 720°
    (B) 900°
    C 1080°
    D 1260°

30. Solve $5x \le -40$.
    (F) $x \le -8$
    G $x < -8$
    H $x \ge -8$
    J $x > -8$

31. Kendra makes $2\frac{5}{6}$ quarts of fruit salad. She then decides to add $\frac{1}{3}$ quart of blueberries. How many quarts of fruit does she now have?
    A $3\frac{7}{12}$ qt
    (B) $3\frac{1}{6}$ qt
    C $3\frac{1}{3}$ qt
    D $2\frac{1}{3}$ qt

32. In $\triangle ABC$, $\angle A$ is 10° more than $\angle B$ and $\angle C$ is 30° more than $\angle B$. What is $m\angle A$ to the nearest tenth of a degree?
    F 46.7°
    (G) 56.7°
    H 60.0°
    J 76.7°

Cumulative Test
Form B, continued

33. For a fishing derby a pond is stocked with 100 trout. If 80 of the trout are caught, what is this ratio as a reduced fraction and a percent?
    A $\frac{4}{5}$, 8%
    (C) $\frac{4}{5}$, 80%
    B $\frac{4}{5}$, 800%
    D $\frac{8}{10}$, 8%

34. Marty receives $74,000 from her grandmother's estate. If she invests the money in an IRA and earns 14% per year on the investment, how much money is earned in one year?
    (F) $10,360
    G $103,600
    H $52,857
    J $528,571

35. On a two-part history project, Joel got 25 points on the first part and 34 points on the second part. What was the percent of increase in the score?
    A 9%
    (B) 36%
    C 26.5%
    D 64%

36. Find 3^{-5}.
    F $\frac{1}{15}$
    (G) $\frac{1}{243}$
    H $\frac{1}{81}$
    J -15

37. The speed of light is 1.86×10^5 miles per second. What is this number in standard notation?
    A 0.0000186 mi/s
    (B) 186,000 mi/s
    C 1,860,000 mi/s
    D 18,600,000 mi/s

38. Simplify $7 - 15 - (-4) + 15$.
    F 3
    (G) 11
    H 30
    J -30

39. A carpenter cuts 15.3 inches off of a board of wood. What is this length as a fraction?
    A $15\frac{1}{3}$ in.
    (C) $15\frac{3}{10}$ in.
    B $15\frac{2}{3}$ in.
    D $15\frac{3}{5}$ in.

40. The ratio of juniors to seniors in the Lakewood High School Honor Society is 2:7. If there are 56 seniors, how many juniors are there in the Honor Society?
    F 14
    (G) 16
    H 18
    J 19

Cumulative Test
Form C

1. Express 179.8% as a decimal.
   A 0.1798
   (C) 1.798
   B 17.98
   D 179.8

2. A house in the country has a driveway that is 600 feet long. How many yards is this?
   F 1800 yd
   H 100 yd
   G 90 yd
   (J) 200 yd

3. Simplify using exponents: $9^5 \cdot 9^4 \cdot 9^4 \cdot 9^2$.
   A 9^{11}
   C 9^{160}
   (B) 9^{15}
   D 160^9

4. Determine which of the following is a solution to $-3y = 2x + 3$.
   F (1, −1)
   H (2, 3)
   (G) (0, −1)
   J (1, 2)

5. In the figure, $\triangle ABC$ is similar to $\triangle GHJ$. Find length y. 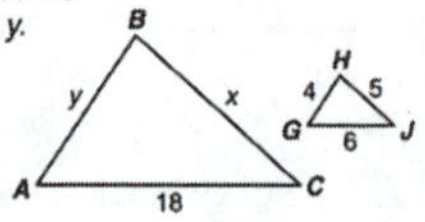
   (A) 12
   B 20
   C 16
   D 24

6. Solve $2f - 10 = 38 - 10f$.
   F $f = 3.5$
   H $f = 6$
   (G) $f = 4$
   J $f = -6$

7. A plumber has a bunch of scrap pipe in the following lengths. What is the mode for these lengths?
   7.7 ft, 2.5 ft, 7.7 ft, 1.3 ft, 2.5 ft, 2.9 ft, 5.6 ft, 7.7 ft
   A 2.5 ft
   C 4.7 ft
   B 4.3 ft
   (D) 7.7 ft

8. Simplify the expression $\frac{v^2 w^3}{v^5 w}$.
   F $\frac{1}{v^2}$
   (H) $\frac{w^2}{v^4}$
   G $\frac{w}{v^2}$
   J $\frac{(vw)^5}{(vw)^7}$

9. Construct a box-and-whisker plot for the data set:
   150 190 180 240 470 180 480
   290 390 260 180 240 230 290
   330 390 190 280 250 240 230
   (A) 150 | 240 | 480 (190, 310)
   B 150 | 240 | 480 (190, 300)
   C 150 | 240 | 480 (200, 310)
   D 150 | 250 | 480 (190, 310)

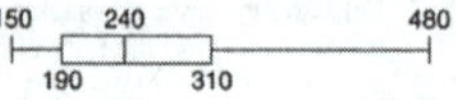

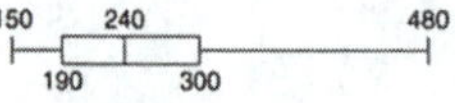

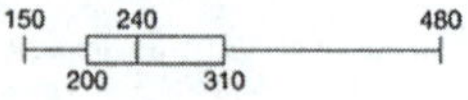

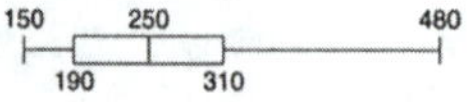

10. In the figure, give all combinations of angles that are complementary.
    F $\angle ABF$ and $\angle FBE$ 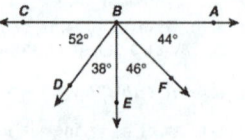
    G $\angle CBE$ and $\angle EBA$
    (H) $\angle CBD$ and $\angle DBE$; $\angle FBE$ and $\angle ABF$
    J $\angle FBE$ and $\angle EBD$; $\angle ABF$ and $\angle CBD$

Cumulative Test
Form C, continued

11. Find all square roots of $\sqrt{225}$.
    A 5, −5
    (C) 15, −15
    B 5
    D 25, −25

12. Evaluate $-6v^2 w^2$ for $v = 4$ and $w = -1$.
    F 24
    H 96
    G -24
    (J) -96

13. The perimeter of a floor in a square shed is 222 feet. Find the length of one side of the floor.
    A 14.9 ft
    (C) 55.5 ft
    B 22 ft
    D 111 ft

14. Find an algebraic expression for "5 times the quotient of 2 and the sum of z and 5."
    (F) $5\left(\frac{2}{z+5}\right)$
    H $\frac{2z+5}{3}$
    G $2\left(\frac{3}{z} \div 5\right)$
    J $2\left(\frac{3}{5z}\right)$

15. Find the original data from the stem-and-leaf plot.

Stem	Leaves
8.	1 7
9.	1 1 4 7
10.	1 4 4 7 9
11.	4 5

    A 9, 15, 9, 9, 12, 15, 11, 11, 14, 17, 19, 15
    (B) 8.1, 8.7, 9.1, 9.1, 9.4, 9.7, 10.1, 10.4, 10.4, 10.4, 10.7, 10.9, 11.4, 11.5
    C 81, 87, 91, 91, 94, 97, 101, 104, 104, 107, 109, 114, 115
    D 8.1, 8.7, 9.4, 9.4, 9.7, 10.4, 10.5, 10.7, 10.9, 11.4, 11.5

16. Find the area of the figure. 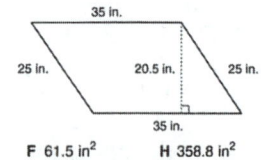
    F 61.5 in^2
    H 358.8 in^2
    G 80.5 in^2
    (J) 717.5 in^2

17. Solve $\frac{12}{5} = \frac{x}{18}$.
    A $x = 3.6$
    C $x = 15$
    B $x = 12$
    (D) $x = 43.2$

18. A circle has a diameter of 4.8 inches. If the diameter is doubled, what is the circumference to the nearest tenth? Use 3.14 for π.
    F 15.1 in.
    H 72.3 in.
    (G) 30.1 in.
    J 289.4 in.

19. A jar has 413 pink, purple, and yellow beads. There are 30 more purple beads than pink beads, and 20 more yellow than pink, and 10 more purple than yellow. How many purple beads are there in the jar?
    A 121
    (C) 151
    B 141
    D 292

20. A water dispenser holds 312 ounces of fluid. If the water dispenser is $\frac{7}{12}$ full, how many ounces of water are in the dispenser?
    F 26 oz
    (H) 182 oz
    G 130 oz
    J 312 oz

Holt Pre-Algebra

Cumulative Test
Chapter 8 Form C, continued

21. Describe the radical as "rational," "irrational," or "not a real number." If a real number, give the approximate value: $-\sqrt{189}$.
   A irrational, 13.7
   B irrational, −13.7
   C rational, 13.7
   D not a real number

22. Find the best buy for laundry soap and give the cost per ounce.
   Brand A: 112 oz for $31.13
   Brand B: 116 oz for $37.58
   F Brand A, $0.278 per oz
   G Brand B, $0.278 per oz
   H Brand A, $0.324 per oz
   J Brand B, $0.324 per oz

23. Find the area of the figure. 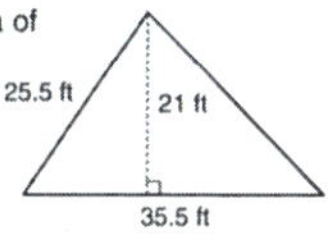
   A 91 ft^2
   B 372.8 ft^2
   C 452.6 ft^2
   D 745.5 ft^2

24. Find the volume of the figure. 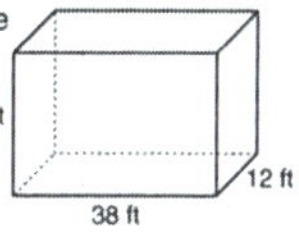
   F 76 ft^3
   G 456 ft^3
   H 988 ft^3
   J 11,856 ft^3

25. Find the mean of these numbers.
   55, 72, 69, 55, 79, 70, 55
   A 49.3 bags
   B 65 bags
   C 69 bags
   D 69.1 bags

26. Find the surface area of the figure. 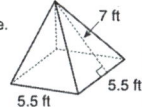
   F 16 ft^2
   G 44 ft^2
   H 107.25 ft^2
   J 134.8 ft^2

27. Below is a diagram of how a rope comes off the top of one tent pole. To the nearest tenth, how far from the tent is the rope staked? 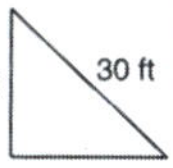
   A 3.1 ft
   B 9.5 ft
   C 21.9 ft
   D 36.3 ft

28. Simplify $12\frac{2}{3} \cdot 1\frac{5}{8}$.
   F $7\frac{31}{39}$
   G $13\frac{5}{12}$
   H $20\frac{7}{12}$
   J $47\frac{1}{2}$

29. Find the sum of the angle measures in a regular 10-gon.
   A 900°
   B 1080°
   C 1440°
   D 1800°

30. Solve for x in the inequality $-4x \le -36$.
   F $x \ge 9$
   G $x \le -9$
   H $x \ge -9$
   J $x \le 9$

31. Peggy poured $1\frac{2}{5}$ liters of pop and $2\frac{1}{3}$ liters of juice into a punch bowl for a luncheon. How many liters of punch did she make?
   A $3\frac{2}{15}$ L
   B $3\frac{1}{4}$ L
   C $3\frac{3}{8}$ L
   D $3\frac{11}{15}$ L

32. In $\triangle ABC$, $m\angle A$ is 15° more than $m\angle B$ which is 10° more than $m\angle C$. What is $m\angle A$ to the nearest tenth of a degree?
   F 48.3°
   G 58.3°
   H 63.3°
   J 73.3°

33. Carla has completed 18 out of the 27 questions on her science homework. What fraction, and what percent, of her science homework has she completed?
   A $\frac{2}{3}$, 0.667%
   B $\frac{2}{3}$, 667%
   C $\frac{2}{3}$, $66\frac{2}{3}$%
   D $\frac{18}{27}$, $6\frac{2}{3}$%

34. Mr. Harrington invests $174,500 in an IRA. If he earns 12.5% per year on the investment, how much money is earned in one year?
   F $218.13
   G $2181.25
   H $21,812.50
   J $152,687.50

35. The Bulldogs fifth-grade basketball team scored 23 points at their first game of the season. At the second game they scored 38 points. What was the percent of increase in the score to the nearest tenth of a percent?
   A 15.0%
   B 39.5%
   C 65.2%
   D 60.5%

36. Find $(-4)^{-4}$.
   F $\frac{1}{16}$
   G $\frac{1}{256}$
   H $-\frac{1}{16}$
   J $-\frac{1}{256}$

37. Express 2.06×10^{-6} in standard notation.
   A 0.00000206
   B 2,060,000
   C 0.000000206
   D 0.000206

38. Simplify $7 - 15 - (-4) \div (-2)$.
   F −2
   G −6
   H 8
   J −10

39. A cement truck carries 20.125 yards of concrete to a job. What is this amount as a fraction?
   A $20\frac{1}{8}$ yd
   B $20\frac{1}{5}$ yd
   C $20\frac{1}{4}$ yd
   D $20\frac{2}{5}$ yd

40. The ratio of cars to trucks in a parking lot is 12 to 5. If there are 90 trucks in the parking lot, how many cars are there?
   F 38
   G 142
   H 216
   J 324

Cumulative Test
Chapter 9 Form A

1. A multiple-choice history test has questions with 4 possible choices. If you were to randomly guess on the first question, what is the probability that you would guess correctly?
   A 0.10
   B 0.25
   C 0.798
   D 79.8

2. Simplify 1^3.
   A 0
   B $\frac{1}{3}$
   C 1
   D 3

3. Which is $\frac{9}{4}$ as a percent?
   A $2\frac{1}{4}$%
   B $20\frac{1}{4}$%
   C 22.5%
   D 225%

4. Tasha rode her bike for 300 minutes. How many hours did she bike?
   A 5 hr
   B 60 hr

5. Simplify the 3^{-2}.
   A $\frac{1}{9}$
   B −9

6. Find the perimeter of the figure. 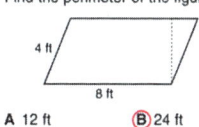
   A 12 ft
   B 24 ft

7. Ryan's chess club had a meeting every day for 5 days to practice for an upcoming tournament. What was the mean attendance at the meetings?

Chess Club	
Monday	5
Tuesday	12
Wednesday	8
Thursday	10
Friday	12

   A 9.4
   B 10

8. Roger has a garden in the shape of a triangle. What is the area? 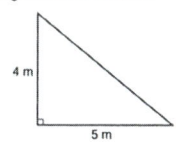
   A 9 m^2
   B 10 m^2
   C 18 m^2
   D 20 m^2

9. Find the square root to the nearest hundredth. $\sqrt{5}$
   A 2.24
   B 2.5

10. Solve the equation $1.2z = 2.4$.
    A $z = 2$
    B $z = 3.6$

11. A rectangular package is 6 in. high, 3 in. wide, and 2 in. deep. What is the volume of the package if the width is doubled?
    A 36 in^3
    B 72 in^3

12. Simplify $-2(a + 3) + 3a$.
    A $a - 6$
    B $a + 6$

13. Evaluate ab^3 for $a = 2$ and $b = -1$.
    A −2
    B 2

14. What are the coordinates of point B? 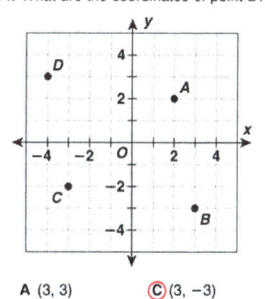
    A (3, 3)
    B (−3, −3)
    C (3, −3)
    D (−3, 3)

15. Tom can type one letter in $\frac{1}{2}$ an hour. How long do 7 letters take?
    A 1 hr
    B 14 hr
    C 2 hr
    D $3\frac{1}{2}$ hr

16. Deena has a circular flower garden with a diameter of 10 feet. Deena would like to edge the garden. How many feet of edging material would she need? Use 3.14 for π.
    A 31.4 ft
    B 62.8 ft

17. Solve the proportion $\frac{10}{9} = \frac{5}{x}$.
    A $x = 4.5$
    B $x = 18$

18. Sam took a survey of the number of hours of sleep that several family members received over the weekend. What is the mode?

Sleep Survey	
Member	Hours
Mom	8
Dad	8
Kim	13
Roger	14
Grandma	8
Grandpa	12

    A 10
    B 8

19. Simplify using positive exponents: $\left(\frac{1}{6}\right)\left(\frac{1}{6}\right)\left(\frac{1}{6}\right)$.
    A $\left(\frac{1}{6}\right)^3$
    B $\left(\frac{3}{6}\right)^3$
    C $6^{\frac{1}{6}}$
    D $6\frac{1}{6}$

20. An auditorium currently has a seating capacity of 200 people. If the capacity is increased to 300 people, what is the percent of increase?
    A $33\frac{1}{3}$%
    B 50%

21. Which statement says that 3 minus a number equals twice the number?
    A $3 - n = 2n$
    B $n - 3 = 2 + n$

22. Find all square roots of the number 81.
    A 1, 8
    B 9, −9
    C −81
    D 6561

Cumulative Test
Chapter 9 Form A, continued

23. Express 2.9 × 10³ in standard notation.
    A 0.0029
    B 0.029
    C 290
    D) 2900

24. Solve $3c = -15$.
    A) $c = -5$
    B $c = -\frac{1}{5}$

25. Simplify $4 - (8) \cdot (2) + 5$.
    A -3
    B) -7
    C -28
    D -92

26. Find the surface area of a sphere that has a radius of 3 m. Use 3.14 for π. Recall that $S = 4\pi r^2$.
    A) 113.04 m²
    B 37.68 m²

27. Find the measure of an angle whose supplement is three times the measure of the angle.
    A 22.5°
    B) 45°

28. A dressmaker purchased $2\frac{1}{2}$ yd of denim, $3\frac{1}{2}$ yd of corduroy, and $1\frac{1}{4}$ yd of cotton. How much total fabric did the dressmaker purchase?
    A) $7\frac{1}{4}$ yd
    B $6\frac{1}{2}$ yd

29. Which inequality is represented by the graph? 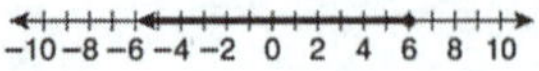
    A) $x \leq 6$
    B $x \geq 6$

30. How many sides does a kite have?
    A) 4
    B 6

31. Two angles whose sum is 90° are known as what type of angles?
    A) complementary angles
    B supplementary angles
    C congruent angles
    D obtuse angles

32. Carla had $400 in a checking account on Monday. On Tuesday she wrote a check for $200. What was the percent decrease in her account balance?
    A $33\frac{1}{3}$%
    B) 50%

33. Identify the figure. 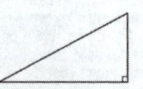
    A) right triangle
    B obtuse triangle
    C isosceles triangle
    D equilateral triangle

34. A radio with an original price of $90 was marked down to $60. What is the percent decrease?
    A 50%
    B) $33\frac{1}{3}$%

35. Larissa has read 12 out of 60 books that are required reading for her English class. What fraction and percent of the books has she read?
    A) $\frac{1}{5}$, 20%
    B $\frac{1}{5}$, 2.0%

36. What is the measure of ∠EBD? 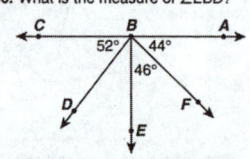
    A) 38°
    B 52°

37. Find the volume of the figure. Use 3.14 for π.
    A) 1177.5 m³
    B 4710 m³

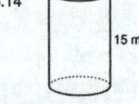

38. 300 people at a local mall were asked about their favorite cookie. Chocolate chip was the favorite of 125 people. Identify the sample.
    A 125 preferring chocolate chip
    B) 300 asked their favorite cookie
    C all people at the mall
    D cannot be determined

39. If $P(A) = \frac{2}{3}$, find the odds in favor of A happening.
    A) 2:1
    B 2:3

40. Lucinda has purchased $9\frac{3}{4}$ pounds of peaches. Which represents the decimal amount?
    A 9.7
    B) 9.75
    C 97
    D 97.5

41. Simplify $\frac{5!}{3!}$.
    A) 20
    B $8\frac{1}{3}$
    C $\frac{5}{3}$
    D 2

42. Kelly has 200 red and green beads in a bag. If there are 70 red beads, what is the probability of randomly selecting a red bead?
    A 70%
    B) 35%

43. You have 10 problems on a math test. The teacher is allowing you to select 4 problems to complete. How many different possible combinations are there?
    A) 210
    B 5040

44. Find $_4P_3$.
    A 12
    B) 24

Cumulative Test
Chapter 9 Form B

1. Margaret is playing a carnival game. There are 5 possible prizes she could win by spinning a spinner (tiger, giraffe, bear, koala, or bird) divided into 5 equal sections. Every contestant wins a prize. What is the probability she will win the tiger?
    A 0.15
    B) 0.20
    C 0.50
    D 0.80

2. Simplify -1^6.
    F $\frac{1}{6}$
    G) -1
    H 1
    J -6

3. Express $\frac{10}{3}$ as a percent.
    A $3\frac{1}{3}$%
    B $33\frac{1}{3}$%
    C) $333\frac{1}{3}$%
    D 3000%

4. Becky drove 480 minutes. How many hours did she drive?
    F 4 hr
    G) 8 hr
    H 20 hr
    J 192 hr

5. Simplify $(-4)^{-2}$.
    A) $\frac{1}{16}$
    B 16
    C $-\frac{1}{16}$
    D -16

6. Find the perimeter. 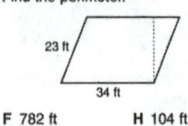
    F 782 ft
    G) 114 ft
    H 104 ft
    J 57 ft

7. For the past 5 years Tony has had the following bonuses. What was Tony's mean bonus?

Year	Bonus
2002	$7830
2001	$4320
2000	$1570
1999	$2370
1998	$3600

    A $1570
    B $3600
    C) $3938
    D $4320

8. Find the area. 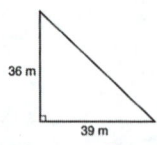
    F 53 m²
    G 75 m²
    H) 702 m²
    J 1404 m²

9. What is $\sqrt{46}$?
    A 6.77
    B) 6.78
    C 6.79
    D 46.00

10. Solve the equation $4.5z = 31.5$.
    F $z = \frac{1}{7}$
    G) $z = 7$
    H $z = 24.5$
    J $z = 27$

11. A small planter measures 12 in. high by 6 in. wide by 3 in. deep. What is the volume of the box if the width is doubled?
    A 1728 in³
    B) 432 in³
    C 216 in³
    D 42 in³

12. Simplify $-4(a + 7) + 10a$.
    F) $6a - 28$
    G $-14a - 28$
    H $6a + 28$
    J $-6a + 28$

13. Evaluate ab^3 for $a = 5$ and $b = -3$.
    A -22
    B 45
    C -45
    D) -135

14. What are the coordinates of point D? 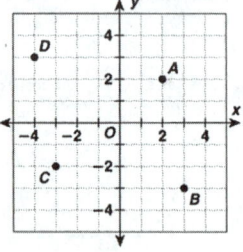
    F (3, -4)
    G (-4, -3)
    H (4, 3)
    J) (-4, 3)

15. If you can paint one box in $\frac{2}{3}$ of an hour, how long do 5 boxes take?
    A 2 hr
    B 4 hr
    C) $3\frac{1}{3}$ hr
    D $4\frac{2}{3}$ hr

16. A circular fountain has a diameter of 15 ft. What is the distance around the fountain? Use 3.14 for π.
    F) 47.1 ft
    G 60 ft
    H 94.2 ft
    J 176.6 ft

17. Solve the proportion $\frac{25}{9} = \frac{15}{x}$.
    A $x = 1\frac{2}{3}$
    B $x = 5$
    C $x = -5$
    D) $x = 5.4$

18. A group of fishermen on a charter boat have the following ages: 89, 49, 52, 49, 79, 39. What is the mode?
    F 59.5
    G) 49
    H 60
    J 99

19. Simplify using positive exponents $\left(\frac{5}{6}\right)\left(\frac{5}{6}\right)\left(\frac{5}{6}\right)\left(\frac{5}{6}\right)\left(\frac{5}{6}\right)\left(\frac{5}{6}\right)$.
    A) $\left(\frac{5}{6}\right)^6$
    B $\left(\frac{5}{6}\right)^5$
    C $6^{\frac{5}{6}}$
    D $6\frac{5}{6}$

20. Lindsay currently deposits $250 each month into her savings account. If she increases the amount to $350, what is the percent increase in her monthly deposit?
    F 140%
    G 100%
    H) 40%
    J 28.6%

21. Which statement says that 56 minus twice a number equals the number?
    A $56 - 2n = 2n$
    B $2n - 56 = n$
    C $56 - 2 = n$
    D) $56 - 2n = n$

22. Find all square roots of the number 225.
    F) 15, -15
    G 16, -16
    H 20, -25
    J 25, -25

374

Holt Pre-Algebra

Cumulative Test
Form B, continued

23. Express 6.6×10^5 in standard notation.
    A 330
    (C) 660,000
    B 66,000
    D 6,600,000

24. Solve the equation $9c = -135$.
    F $c = 1$
    H $c = -144$
    (G) $c = -15$
    J $c = 144$

25. Simplify $9 - (-12) \cdot (-6) + 10$.
    A -116
    C -29
    (B) -53
    D -92

26. Find the surface area to the nearest tenth, of a sphere that has a radius of 4.5 m. Use 3.14 for π.
    F 56.5 m²
    (H) 254.3 m²
    G 28.26 m²
    J 317.9 m²

27. Find the measure of an angle whose complement is two times the measure of the angle.
    A 22.5°
    C 60°
    (B) 30°
    D 67.5°

28. The camp cook purchased $2\frac{1}{6}$ lb of hamburger, $6\frac{1}{2}$ lb of chicken, and $1\frac{3}{5}$ lb of pork. How many total pounds of meat did he purchase?
    F $9\frac{1}{5}$ lb
    H $9\frac{2}{3}$ lb
    G $9\frac{5}{13}$ lb
    (J) $10\frac{4}{15}$ lb

29. Which inequality is represented by the graph? 
    A $2x - 3 \leq 6$
    C $x \leq 6 - x$
    (B) $-x + 3 \geq -3$
    D $x < 6$

30. How many sides does a rhombus have?
    (F) 4
    H 8
    G 6
    J 12

31. Two angles whose sum is 180° are known as what type of angles?
    A complementary angles
    (B) supplementary angles
    C congruent angles
    D acute angles

32. A submarine on maneuvers starts at a depth of 300 feet, rises 50 feet, then dives an additional 100 feet. What is the overall percent change from the original depth?
    F 10%
    H 50%
    (G) $16\frac{2}{3}$%
    J 110%

33. Identify the figure. 
    A right triangle
    B obtuse triangle
    C isosceles triangle
    (D) equilateral triangle

34. Selena was looking at a coat, originally priced at $89.00. It was marked down to $56.99. What is the percent decrease in price?
    F 32.01%
    (H) 35.97%
    G 35.90%
    J 56.20%

35. Ayna sent out 80 resumes. As of Friday, she had received 15 replies. What fraction and percent of replies did she receive?
    (A) $\frac{3}{16}, 18\frac{3}{4}$%
    C $\frac{15}{18}, 18$%
    B $\frac{3}{16}, 5\frac{1}{3}$%
    D $\frac{3}{16}, 533\frac{1}{3}$%

36. Find the measure of $\angle ABD$. 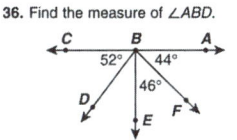
    F 38°
    G 90°
    (H) 128°
    J not enough information

37. Find the volume of the cylinder. Use 3.14 for π. 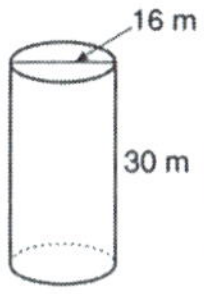
    A 480 m³
    B 753.6 m³
    (C) 6028.8 m³
    D 24,115.2 m³

38. 350 people at a local carnival were asked about their favorite flavor of ice cream. Vanilla was the favorite of 115 people. Identify the sample.
    F 115 preferring vanilla
    (G) 350 asked their favorite ice cream
    H all people at the carnival
    J cannot be determined

39. If $P(A) = \frac{5}{6}$, what are the odds in favor of A happening?
    A 1:5
    (C) 5:1
    B 5:6
    D 6:5

40. A tentmaker uses $15\frac{3}{5}$ yards of fabric. What is $15\frac{3}{5}$ as a decimal?
    F 15.3
    H 15.8
    (G) 15.6
    J 15.9

41. Simplify $\frac{9!}{5!}$.
    A 2!
    (C) 3024
    B $\frac{4}{5}$
    D 45,000

42. A decorated tree has 415 red and blue balls. If there are 87 blue balls, what is the probability of randomly selecting a red ball to the nearest percent?
    (F) 79%
    H 21%
    G 50%
    J 20.9%

43. You are looking at new cars. The car lot has 20 cars on it. You want to test drive 5 of the cars. How many different possible combinations are there?
    A 1
    C 120
    B 5
    (D) 15,504

44. Find $_5P_4$.
    (F) 120
    H 5
    G 24
    J 1

Cumulative Test
Form C

1. A contestant on a game show will win the grand prize if he or she selects the correct door. The contestant has 7 possible doors to choose from. If the contestant randomly guesses, what is the probability that the contestant will select the correct door?
   (A) 0.14
   C 0.83
   B 0.17
   D 0.92

2. Simplify -5^4.
   F $-\frac{4}{5}$
   (H) -625
   G 5
   J $\frac{1}{625}$

3. Express $\frac{33}{4}$ as a percent.
   A $8\frac{1}{4}$%
   C $33\frac{1}{4}$%
   B $8\frac{3}{4}$%
   (D) 825%

4. A lot in the country has 450 feet of frontage. How many yards is this?
   (F) 150 yd
   H 37.5 yd
   G 50 yd
   J 20 yd

5. Simplify $(-3)^{-3}$.
   (A) $-\frac{1}{27}$
   C $\frac{1}{27}$
   B -27
   D 27

6. Find the perimeter of the figure. 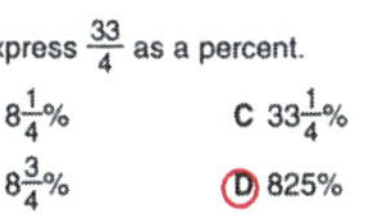
   F 810.8 ft
   H 58 ft
   (G) 116 ft
   J 57 ft

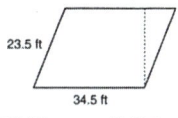

7. During the past month, $708.34, $430.11, $150.72, $230.75, and $360.00 were spent on repairs. What is the median amount spent on repairs?
   A $150.72
   C $390.38
   (B) $360.00
   D $375.98

8. Find the area of the figure. 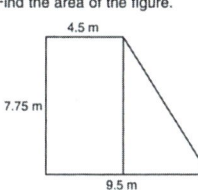
   F 34.875 m²
   (H) 54.25 m²
   G 73.6 m²
   J 36.8125 m²

9. Find $\sqrt{111}$ to the nearest hundredth.
   A 10.52
   (C) 10.54
   B 10.53
   D 10.55

10. Solve the equation $3.5z = 31.5 - z$.
    F $z = \frac{1}{7}$
    H $z = 24.5$
    (G) $z = 7$
    J $z = 27$

11. A shipping carton measures 12 in. high by 6 in. wide by 4.5 in. deep. What is the volume of the carton if the width is tripled?
    A 8748 in³
    (B) 972 in³
    C 324 in³
    D 108 in³

12. Simplify $-5(-a - 7) + 10a$.
    F $5a - 35$
    H $15a - 35$
    G $-5a + 35$
    (J) $15a + 35$

13. Evaluate $5ab^3$ for $a = 2$ and $b = -3$.
    (A) -270
    C 30
    B 270
    D -30

14. What are the coordinates of point C? 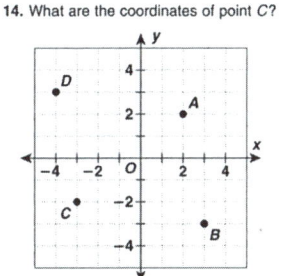
    F $(-2, -3)$
    H $(4, 3)$
    (G) $(-3, -2)$
    J $(-4, 3)$

15. Josh can walk one mile in $\frac{1}{3}$ of an hour. If he kept up this rate, how many hours would he take to walk 7 miles?
    A $7\frac{1}{3}$ hr
    C 3 hr
    B 4 hr
    (D) $2\frac{1}{3}$ hr

16. A small stepping stone has a diameter of 1.5 ft. What is the distance around the stepping stone? Use 3.14 for π.
    (F) 4.71 ft
    H 7.17 ft
    G 6 ft
    J 9.42 ft

17. Solve the proportion $\frac{x}{9} = \frac{4.1}{36.9}$.
    A $x = \frac{1}{4}$
    C $x = 4.1$
    (B) $x = 1$
    D $x = 9$

18. Your bowling scores for the last six games are: 119, 104, 113, 119, 119, and 129. What is the mode?
    F 117
    H 188
    G 165
    (J) 119

19. Simplify using positive exponents: $\left(\frac{5}{6}\right)\left(\frac{5}{6}\right)\left(\frac{5}{6}\right)\left(\frac{1}{5}\right)\left(\frac{1}{5}\right)\left(\frac{1}{5}\right)$.
    (A) $\left(\frac{5}{6}\right)^3 \left(\frac{1}{5}\right)^3$
    C $6^{\frac{5}{6}}$
    B $\left(\frac{1}{5}\right)\left(\frac{5}{6}\right)^6$
    D $6^{\frac{5}{6}}$

20. A farmer currently farms 2000 acres of land. If he increases his acreage to 2700, what is the percent of increase in acreage?
    F 25.9%
    H 125.9%
    (G) 35%
    J 135%

21. Which statement says that 6 less than three times a number equals twice the number?
    A $6 - 3n = 2n$
    C $6 - 3n = n$
    (B) $3n - 6 = 2n$
    D $3(n - 6) = 2n$

22. Find all square roots of the number 576.
    (F) 24, -24
    H 18, 32
    G 28.8
    J 288

Cumulative Test 9 — Form C, continued

23. Express 3.16×10^7 in standard notation.
    A 31,600
    (C) 31,600,000
    B 3,160,000
    D 316,000,000

24. Solve $55c + 30 = -135$.
    (F) $c = -3$
    H $c = -50$
    G $c = 3$
    J $c = 110$

25. Simplify $9 - (-12) \div (-6) + 9$.
    A -2
    C -12
    B 12.5
    (D) 16

26. Find the surface area of a sphere that has a diameter of 1 m. Use 3.14 for π.
    F 1.05 m^2
    H 9.86 m^2
    (G) 3.14 m^2
    J 12.56 m^2

27. Find the measure of an angle whose supplement is 3 more than 5 times the measure of its complement.
    A 18°
    C 67.5°
    B 33.8°
    (D) 68.25°

28. Susan is making trail mix. She purchased $3\frac{3}{5}$ lb of almonds, $5\frac{1}{2}$ lb of peanuts, and $1\frac{1}{6}$ lb of cashews. How many pounds of nuts did she purchase?
    F $9\frac{1}{10}$ lb
    H $9\frac{5}{6}$ lb
    G $9\frac{1}{2}$ lb
    (J) $9\frac{13}{15}$ lb

29. Which inequality is shown? 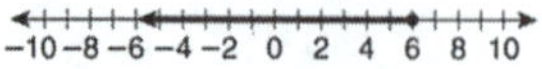
    A $4x - 6 \le 12$
    C $2x \le 12 - x$
    (B) $-2x + 5 \ge -7$
    D $-2x < 12$

30. How many sides does a trapezoid have?
    (F) 4
    H 8
    G 6
    J 10

31. What are two angles that both measure 30° called?
    A complementary angles
    B supplementary angles
    (C) congruent angles
    D right angles

32. A hot air balloon started at 800 feet, rose 50 feet, and then dropped 250 feet. What is the percent change from its original height?
    F $33\frac{1}{3}$%
    H $33\frac{1}{4}$%
    (G) -25%
    J 20%

33. Identify the figure.
    A acute triangle
    (B) obtuse triangle
    C isosceles triangle
    D equilateral triangle

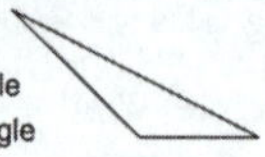

34. Ms. Whipple is looking at a camera for her husband. When she looked at the camera originally, it had a price of $189.00. Two weeks later it was marked down to $126.99. What is the percent of markdown to the nearest tenth?
    (F) 32.8%
    H 62.01%
    G 48.8%
    J 67.2%

35. Rich and Christine sent out 180 wedding invitations. 25 people responded a week later. What fraction and percent, to the nearest tenth of a percent, of responses had they received?
    A $\frac{5}{12}$, 41.6%
    C $\frac{1}{9}$, 7.2%
    B $\frac{15}{38}$, 39.4%
    (D) $\frac{5}{36}$, 13.9%

36. Find the measure of $\angle CBF$. 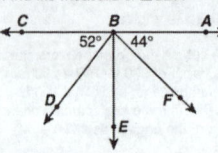
    F 44°
    G 108°
    (H) 136°
    J not enough information

37. Find the volume of the figure to the nearest tenth. Use 3.14 for π.
    A 594.0 m^3
    B 1865.2 m^3
    C 2564.6 m^3
    (D) 7693.8 m^3

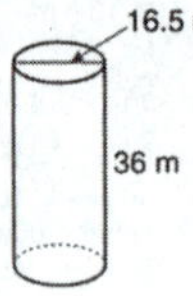

38. 120 people were asked about their favorite color. Blue was the favorite of 58 people. Identify the sample.
    F 58 preferring blue
    (G) 120 asked their favorite color
    H all people at the school
    J cannot be determined

39. If $P(A) = \frac{4}{5}$, what are the odds in favor of A not happening?
    A 4:1
    (C) 1:4
    B 4:5
    D 5:4

40. Sandra bought $150\frac{3}{4}$ feet of ribbon. What is $150\frac{3}{4}$ as a decimal?
    F 15.75
    H 150.3
    G 1.575
    (J) 150.75

41. Simplify $\frac{9!}{5!(9-5)!}$.
    A 1
    C 3024
    (B) 126
    D 15,120

42. By mistake someone has mixed up some nuts, bolts, and screws. You know there are a total of 400 in the bin. If there are 170 bolts and 200 screws, what is the probability of randomly selecting a nut?
    (F) 0.075
    H 7.5
    G 0.75
    J 75

43. You are re-hanging some posters in your room. You have 15 different posters, but you only want to hang 4. How many different possible combinations are there?
    A 273
    C 32,760
    (B) 1365
    D 5.45×10^{10}

44. Find $_7P_4$.
    F 6
    H 210
    G 35
    (J) 840

Cumulative Test 10 — Form A

1. Solve the system:
   $4x - 3y = -8$
   $x + 3y = 13$
   (A) (1, 4)
   B (−1, 4)

2. A sample space consists of 18 separate events that are equally likely. What is the probability of each?
   A 0
   C 1
   (B) $\frac{1}{18}$
   D 18

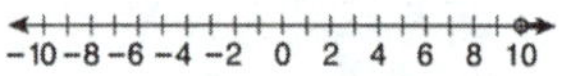

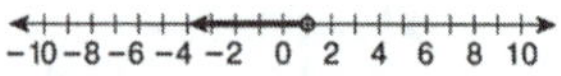

3. Select the graph of the solution set for $0.2x > 0.2$.
   (C)

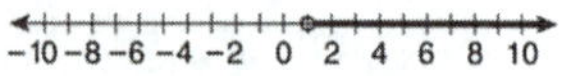

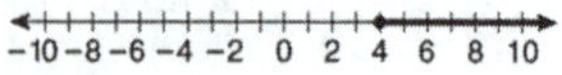

4. 2^3 means:
   (A) $2 \cdot 2 \cdot 2$
   B $2 \cdot 3$

5. Express 0.15 as a percent.
   A 150%
   C 0.15%
   (B) 15%
   D 0.0015%

6. If your window measures 60 inches, what is its length in feet?
   (A) 5 ft
   B 6.6 ft

7. A file contains 3300 bytes of data. What is this number in scientific notation?
   A 3.3×10^4
   (C) 3.3×10^3
   B 3.3×10^2
   D 33×10^{-3}

8. Find the missing length in the right triangle. 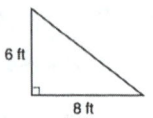
   (A) 10 ft
   B 14 ft

9. Simplify $-4 - (-14)$.
   (A) 10
   C 18
   B -10
   D -18

10. A scale model of a statue is 14 inches high by 3 inches wide. If the height of the actual statue is 14 feet, how wide is it?
    A 3 in.
    (C) 3 ft
    B 42 in.
    D 14 ft

11. A particular size battery is available in packages of 3 for $1.25 or packages of 4 for $1.75. Which is the better buy, and what is its price per battery?
    (A) 3 for $1.25; $0.42
    B 3 for $1.25; $0.44
    C 4 for $1.75; $0.42
    D 4 for $1.75; $0.44

12. Which angles in the figure are complementary? 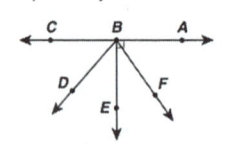
    (A) $\angle ABF$ and $\angle FBE$
    B $\angle ABE$ and $\angle EBC$

13. Solve $-10 + h = 10$.
    A $h = 0$
    C $h = 1$
    (B) $h = 20$
    D $h = 10$

14. The perimeter of a square flower bed is 8 feet. What is the length of one side?
    (A) 2 ft
    B 32 ft

15. What is "2.4 times a number" written as an algebraic expression?
    (A) $2.4x$
    B $2.4 + x$

16. Solve $\frac{3}{4}x = -\frac{3}{8}$.
    A $x = -\frac{9}{32}$
    (B) $x = -\frac{1}{2}$

17. Find the surface area of the figure. 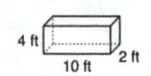
    A 16 ft^2
    C 80 ft^2
    B 68 ft^2
    (D) 136 ft^2

18. Louis surveyed his friends to see how long, in minutes, it takes them to walk to school. What is the mean length of time of the data 7, 5, 11, 4, 8, and 13?
    A 7.5 min
    (C) 8 min
    B 9.6 min
    D 9 min

19. Find the square root and round to the nearest thousandth, if necessary. $\sqrt{13}$
    A 3.605
    (B) 3.606

20. Find the value of -3^2.
    A 9
    (B) -9

21. You are making a quilt out of rectangles. One piece has dimensions of $1\frac{1}{3}$ in. by $2\frac{1}{6}$ in. What is the perimeter of this rectangle?
    A $3\frac{2}{9}$ in.
    C $4\frac{1}{3}$ in.
    B $3\frac{1}{2}$ in.
    (D) 7 in.

22. Choose the box-and-whisker plot that matches the data: 3, 5, 5, 5, 6, 9, 7, 7, 9, 8.
    (A)

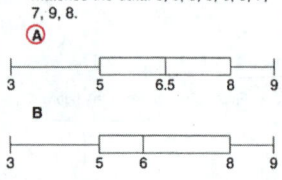

23. Solve $\frac{2}{5} + 2a = -\frac{2}{5}$.
    A $a = -\frac{4}{5}$
    (B) $a = -\frac{2}{5}$

Cumulative Test
Form A, continued

24. What is the surface area of the figure? Use 3.14 for π. (2 m, 6 m)
    A 18.8 m²
    B 25.12 m²
    C 37.7 m²
    D 43.96 m²

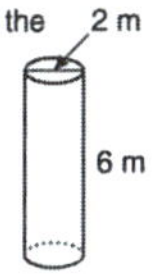

25. Find the measure of ∠x. 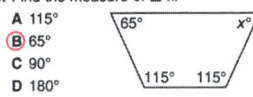
    A 115°
    B 65°
    C 90°
    D 180°

26. Identify the angle that is congruent to ∠2. 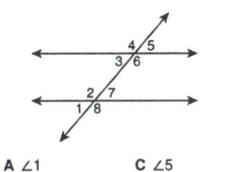
    A ∠1
    B ∠7
    C ∠5
    D ∠8

27. Evaluate the expression $2mn$ when $m = -3$ and $n = 1$.
    A −7
    B −6
    C 7
    D 6

28. Which name does not apply to the figure? 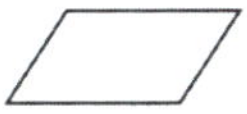
    A square
    B quadrilateral
    C polygon
    D parallelogram

29. Solve the inequality $-y + 10 < y - 10$.
    A $y \leq 0$
    B $y > 0$
    C $y < 1$
    D $y > 10$

30. If $P(A) = \frac{1}{5}$, what are the odds against A happening?
    A 1:6
    B 1:5
    C 4:1
    D 5:1

31. Bagel Emporium conducted a survey and found that five out of every ten customers prefer plain bagels. What is this statistic as a percent?
    A 50%
    B 10%
    C 5%
    D 25%

32. On a map, $\frac{1}{4}$ in. equals 10 miles. How many miles does a map distance of $2\frac{1}{4}$ in. represent?
    A $2\frac{1}{4}$ mi
    B $18\frac{1}{4}$ mi
    C 20 mi
    D 90 mi

33. A mixture of nuts and candy contains 2 pounds of peanuts for every pound of candy. In 30 pounds of mix, how many pounds of candy are there?
    A 10 lb
    B 15 lb
    C 20 lb
    D 30 lb

34. Simplify the expression using positive exponents: $\frac{h^4}{h^3}$.
    A $\frac{1}{h}$
    B $4h$
    C $\frac{1}{2h}$
    D h

35. Find the value of 3^{-2}.
    A $\frac{1}{9}$
    B −9
    C $-\frac{1}{9}$
    D 9

36. Combine like terms: $a + 2a - 5a$.
    A $-3a$
    B $-3a^3$
    C $-2a$
    D $-2a^3$

37. Sean wants to put a braid trim around a box that measures 14 inches by 12 inches. If the braid costs $0.10 per linear inch, how much will the decoration cost?
    A $2.60
    B $5.20

38. Find the area of the figure. (3 in., 5 in.)
    A 7.5 in²
    B 15 in²

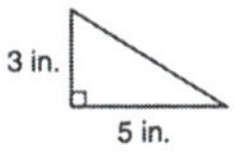

39. A camera that costs $200 wholesale is sold in a store for $233. What is the percent increase in price?
    A 33%
    B 16.5%
    C 14.2%
    D 66%

40. Find the total interest on a simple interest loan if the amount borrowed is $1000 at 5% for 3 years. Assume one payment is made at the end of the 3 years.
    A $1150.00
    B $150.00

Cumulative Test
Form B

1. Solve the system:
   $2x + 5y = 0$
   $x = -3y + 1$
   A (−5, 2)
   B (0, 0)
   C (2, 5)
   D (−2, 0)

2. A sample space consists of 84 separate events that are equally likely. What is the probability of each?
   F 0
   G $\frac{1}{84}$
   H 1
   J 84

3. Select the graph of the solution set for $0.6x + 2 > 8$.
   A 
   B 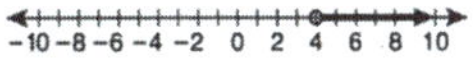
   C 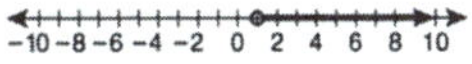
   D

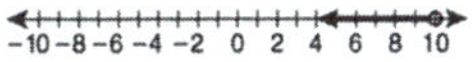

4. 16^4 means:
   F $16 \div 4$
   G $16 \cdot 4$
   H $16 \cdot 16 \cdot 16 \cdot 16$
   J $16 \cdot 16 \cdot 16 \cdot 16 \cdot 16$

5. Express 7.1 as a percent.
   A 710%
   B 71%
   C 0.71%
   D 0.0071%

6. Dennis measured a board to be 102 inches. How many feet long is the board?
   F 34 ft
   G 10.2 ft
   H 8.5 ft
   J 51 ft

7. A company produces 443,000 small appliances per year. What is this number in scientific notation?
   A 4.43×10^6
   B 44.3×10^4
   C 4.43×10^5
   D 4.43×10^{-5}

8. Find the missing length in the right triangle to the nearest tenth. (5 m, 14 m)
   F 9.5 m
   G 14.9 m
   H 110.5 m
   J 221 m

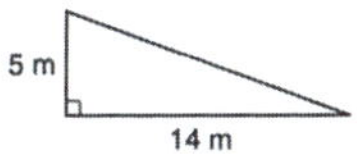

9. Simplify $-3 + (-15) - (-7)$.
   A 5
   B −5
   C 11
   D −11

10. A photograph that is 4 inches high by 6 inches wide is to be scaled to 12 inches wide. To keep the picture in proportion, how high will it need to be?
    F 18 in.
    G 5.3 in.
    H 6 in.
    J 8 in.

11. Foodmart has your favorite cereal on sale. A 28-ounce box sells for $1.40. Shop-n-Save has a 32-ounce box for $1.75. Which would be the better buy and price?
    A the 28-oz box at $0.050 per oz
    B the 32-oz box at $0.050 per oz
    C the 28-oz box at $0.055 per oz
    D the 32-oz box at $0.055 per oz

Cumulative Test
Form B, continued

12. Which angles in the figure are supplementary? 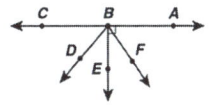
    F ∠ABE and ∠CBE
    G ∠ABE and ∠FBE
    H ∠ABF and ∠EBF
    J ∠ABF and ∠CBD

13. Solve $-14 + 3h = 13$.
    A $h = 27$
    B $h = 9$
    C $h = -\frac{1}{3}$
    D $h = 3$

14. The perimeter of the square addition to Mrs. Weagley's house is 100 feet. What is the length of one side?
    F 15 ft
    G 20 ft
    H 25 ft
    J 400 ft

15. What is "0.8 less than a number" written as an algebraic expression?
    A $0.8 - x$
    B $0.8 + x$
    C $x - 0.8$
    D $-0.8 - x$

16. Solve $\frac{16}{45}x = -\frac{4}{9}$.
    F $x = -\frac{5}{16}$
    G $x = -\frac{5}{4}$
    H $x = -\frac{20}{9}$
    J $x = -\frac{16}{5}$

17. Find the surface area of the figure. (5 ft, 4 ft, 14 ft)
    A 280 ft²
    B 118 ft²
    C 292 ft²
    D 380 ft²

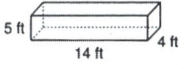

18. What is the mean amount spent on taxi fares for a month in which you spent $25, $45, $10, $45, and $30?
    F $30
    G $35
    H $31
    J $45

19. Find the square root and round to the nearest thousandth, if necessary. $\sqrt{34}$
    A 5.828
    B 5.831
    C 5.836
    D 34.000

20. Find the value of -5^3.
    F −125
    G 125
    H 625
    J −15,625

21. Find the perimeter of a rectangle measuring $2\frac{2}{3}$ in. by $2\frac{1}{5}$ in.
    A $4\frac{3}{8}$ in.
    B $4\frac{13}{15}$ in.
    C $5\frac{13}{15}$ in.
    D $9\frac{11}{15}$ in.

Cumulative Test
Form B, continued

22. Choose the box-and-whisker plot that matches the data: 43, 51, 52, 52, 69, 69, 71, 87, 65, 83.
- (F) 43.0 … 52.0 … 67.0 … 71.0 … 87.0
- G 43.0 … 52.0 … 69.0 71.0 … 70.0
- H 43.0 … 52.0 … 69.0 … 71.0 … 87.0
- J 43.0 … 52.0 … 69.0 … 70.0 … 83.0

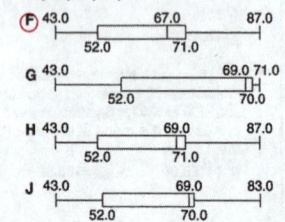

23. Solve $\frac{2}{5} + \frac{3a}{4} = \frac{5}{4} - \frac{2a}{5}$.
- A $a = \frac{33}{4}$
- C $a = \frac{33}{20}$
- B $a = \frac{17}{4}$
- (D) $a = \frac{17}{20}$

24. What is the surface area of the figure? Use 3.14 for π. 7 ft, 10 ft
- F 384.65 ft^2
- G 219.8 ft^2
- (H) 296.73 ft^2
- J 747.32 ft^2

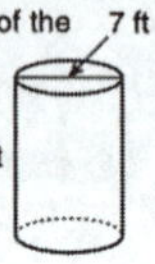

25. Find the measure of $\angle x$.
- A 116°
- B 113°
- C 90°
- (D) 85°
(64°, 116°, 95°, x°)

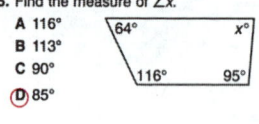

26. Identify one angle that is congruent to $\angle 8$. 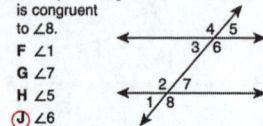
- F $\angle 1$
- G $\angle 7$
- H $\angle 5$
- (J) $\angle 6$

27. Evaluate the expression $5ab$ when $a = -5$ and $b = 3$.
- A 3
- C 75
- B −60
- (D) −75

28. Which name does not apply to the figure? 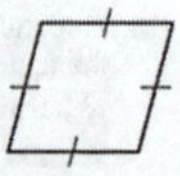
- (F) trapezoid
- G square
- H rectangle
- J parallelogram

29. Solve the inequality $-10y + 15 < -15y + 20$.
- A $y \le 1$
- (C) $y < 1$
- B $y > 1$
- D $y \ge 1$

30. If $P(A) = \frac{1}{6}$, what are the odds against A happening?
- F 1:6
- (H) 5:1
- G 1:5
- J 6:1

31. In a recent survey it was found that three out of every five pre-school children prefer singing over doing arts and crafts. Express this statistic as a percent.
- (A) 60%
- C 5%
- B 30%
- D 3%

32. The scale on a blueprint drawing is such that $\frac{1}{4}$ in. equals 1 ft. If on the blueprint a living room measures $3\frac{3}{4}$ in. wide, then what is the actual width of the room?
- F $3\frac{3}{4}$ ft
- H $12\frac{3}{4}$ ft
- G 4 ft
- (J) 15 ft

33. A paint mixture contains 9 gallons of base for every gallon of color. In 230 gallons of paint, how many gallons of color are there?
- (A) 23 gal
- C 115 gal
- B 76 gal
- D 207 gal

34. Simplify the expression using positive exponents: $\frac{t^4}{t^8}$.
- F $\frac{t}{t^2}$
- H $\frac{1}{2t}$
- (G) $\frac{1}{t^4}$
- J t^4

35. Find the value of 9^{-2}.
- (A) $\frac{1}{81}$
- C $-\frac{1}{81}$
- B −81
- D $\frac{1}{18}$

36. Combine like terms: $4a^2 + 2a^2 - 9a^2$.
- (F) $-3a^2$
- H $-72a^2$
- G $-3a^6$
- J $-72a^6$

37. What will it cost to buy ceiling molding to go around a rectangular room that measures 11 feet by 10 feet, if the molding costs $3.31 per linear foot?
- A $66.20
- C $72.82
- B $69.51
- (D) $139.02

38. Find the area of the figure. 15 ft, 18 ft
- F 83 ft^2
- (G) 135 ft^2
- H 67.5 ft^2
- J 270 ft^2

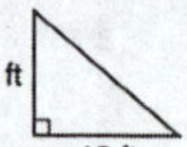

39. Last year Marta earned $356 per week. This year her salary is $371 per week. What is the percent of increase to the nearest tenth of a percent?
- A 4.0%
- C 95.8%
- (B) 4.2%
- D 96.0%

40. Find the total amount due on a simple interest loan if the principal is $800 with a rate of 8% for 5 years. Assume one payment is made at the end of the 5 years.
- F $320.00
- (H) $1120.00
- G $864.00
- J $2080.00

Cumulative Test
Form C

1. Solve the system: $x + 4y = 37$, $6x + 3y = 75$
- A (8, 8)
- (C) (9, 7)
- B (−7, 8)
- D (−9, 8)

2. A sample space consists of 124 separate events that are equally likely. What is the probability of each?
- F 0
- H 1
- (G) $\frac{1}{124}$
- J 124

3. Select the graph of the solution set for $0.5x - 8 > -4 - 3.5x$.
- A, B,  (C),  D (number line graphs)

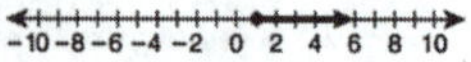

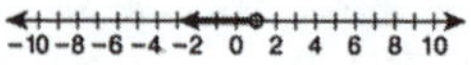

4. $\left(\frac{2}{3}\right)^4$ means:
- F $\frac{2}{3} + \frac{2}{3}$
- G $\frac{2}{3} \cdot \frac{2}{3}$
- (H) $\frac{2}{3} \cdot \frac{2}{3} \cdot \frac{2}{3} \cdot \frac{2}{3}$
- J $\frac{2}{3} \cdot \frac{2}{3} \cdot \frac{2}{3} \cdot \frac{2}{3} \cdot \frac{2}{3}$

5. Express 0.00047 as a percent.
- A 4.7%
- C 0.47%
- (B) 0.047%
- D 0.0047%

6. You measured the height of a barn to be 234 inches. How many yards tall is the barn?
- F 78 yd
- H 19.5 yd
- G 23.4 yd
- (J) 6.5 yd

7. A science experiment produced 10,230,000 cells. What is this number in scientific notation?
- (A) 1.023×10^7
- C 1.23×10^7
- B 1.023×10^6
- D 1.23×10^6

8. Find the missing length in the right triangle to the nearest tenth. 14 ft, 8 ft
- F 6.0 ft
- (G) 11.5 ft
- H 16.1 ft
- J 132.0 ft

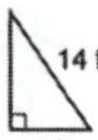

9. Simplify $-5(-15) - (-5)$.
- A −15
- C 70
- B −35
- (D) 80

10. A sketch of a decorative wall panel is 8.5 inches wide by 11 inches tall. If the finished panel is to be scaled to 5 feet wide, how high, to the nearest tenth of a foot, will it need to be to keep the same proportions?
- F 5.5 ft
- H 42.5 ft
- (G) 6.5 ft
- J 18.7 ft

11. Walgreen's has bottled water advertised a $0.69 for a 20-ounce bottle and $1.20 for a 32-ounce bottle. Which would be the better buy and price?
- (A) the 20-oz bottle at $0.035 per oz
- B the 32-oz bottle at $0.035 per oz
- C the 20-oz bottle at $0.038 per oz
- D the 32-oz bottle at $0.038 per oz

12. Which angles in the figure are NOT supplementary? 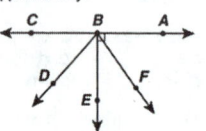
- F $\angle CBE$ and $\angle ABE$
- G $\angle CBF$ and $\angle ABF$
- H $\angle CBD$ and $\angle ABD$
- (J) $\angle CBD$ and $\angle EBD$

13. Solve $-10 + 3h = 10 - h$.
- A $h = 0$
- C $h = -10$
- (B) $h = 5$
- D $h = 10$

14. The area of a square outside patio is 49 square feet. What is the perimeter of the patio?
- F 7 ft
- (H) 28 ft
- G 14 ft
- J 49 ft

15. What is "0.8 less than the product of a number and 5" written as an algebraic expression?
- A $0.8 - 5x$
- (C) $5x - 0.8$
- B $0.8 + x + 5$
- D $0.8x - 5$

16. Solve $-\frac{27}{48}x = -\frac{3}{5}$.
- F $\frac{15}{16}$
- H $-\frac{15}{16}$
- (G) $\frac{16}{15}$
- J $-\frac{16}{15}$

17. Find the surface area of the figure. 13.5 ft, 3.5 ft, 2.5 ft
- A 19.5 ft^2
- C 118.13 ft^2
- B 89.75 ft^2
- (D) 179.5 ft^2

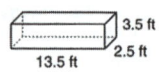

18. What is the mean cost of your lunch in a week in which you spend $7.50, $5.00, $5.61, $4.50, and $3.55?
- F $3.95
- H $5.00
- (G) $5.23
- J $7.50

19. Find the square root and round to the nearest hundredth, if necessary. $\sqrt{222}$
- A 14.89
- C 14.91
- (B) 14.90
- D 15.0

20. Find the value of -8^4.
- F −32
- H 4096
- G 32
- (J) −4096

21. What is the perimeter of a rectangle measuring $3\frac{1}{6}$ in. by $1\frac{3}{5}$ in.?
- A $1\frac{47}{48}$ in.
- C $5\frac{1}{15}$ in.
- B $4\frac{4}{11}$ in.
- (D) $9\frac{8}{15}$ in.

Cumulative Test
Chapter 10 Form C, continued

22. Choose the box-and-whisker plot that matches the data: 53, 61, 62, 62, 79, 79, 81, 97, 75, 93.
    - F
    - G
    - H
    - (J)

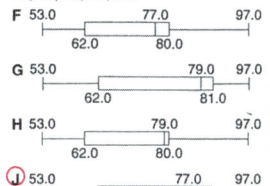

23. Solve $\frac{2}{9} + \frac{3}{5}b = \frac{5}{8} - \frac{2}{5}b + \frac{5}{8}$.
    - A $b = \frac{53}{36}$
    - (C) $b = \frac{37}{36}$
    - B $b = -\frac{53}{36}$
    - D $b = \frac{36}{72}$

24. What is the surface area of the figure? Use 3.14 for π and round to the nearest hundredth. 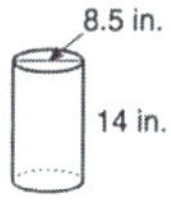
    - F 794.03 in^2
    - G 1201.05 in^2
    - (H) 487.09 in^2
    - J 747.32 in^2

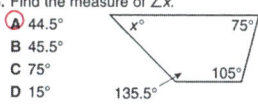

25. Find the measure of $\angle x$.
    - (A) 44.5°
    - B 45.5°
    - C 75°
    - D 15°

26. Identify all angles congruent to $\angle 1$. 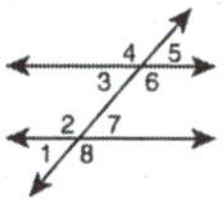
    - F $\angle 2, \angle 7, \angle 8$
    - (H) $\angle 3, \angle 5, \angle 7$
    - G $\angle 2, \angle 3, \angle 4$
    - J $\angle 5, \angle 6, \angle 7$

27. Evaluate the expression $6ab^2$ when $a = 4$ and $b = -3$.
    - (A) 216
    - C 72
    - B −216
    - D −72

28. Which name does not apply to the figure? 
    - (F) prism
    - G polygon
    - H trapezoid
    - J quadrilateral

29. Solve the inequality $-9.5y + 14 < -6y - 7.3$. Round to the nearest tenth.
    - A $y \geq 6.1$
    - (C) $y > 6.1$
    - B $y \leq 6.1$
    - D $y < 6.1$

30. If $P(A) = \frac{1}{3}$, what are the odds against A happening?
    - F 4:3
    - H 2:3
    - G 3:2
    - (J) 2:1

31. A study showed that seven out of every eight people regularly brush their teeth before going to bed. Express this statistic as a percent.
    - A 62.5%
    - (C) 87.5%
    - B 70%
    - D 92%

32. On a scale model of a garden $\frac{1}{4}$ inch equals 1 yard. If the model is $5\frac{3}{4}$ inch wide, then what is the actual width of the garden?
    - F $1\frac{3}{16}$ yd
    - H $5\frac{3}{4}$ yd
    - G 6 yd
    - (J) 23 yd

33. A punch recipe calls for 5 quarts of fruit juice for every quart of sparkling water. In 36 quarts of punch, how many quarts of fruit juice are there?
    - A 6 qt
    - C 36 qt
    - (B) 30 qt
    - D 41 qt

34. Simplify the expression using positive exponents: $\frac{t^4 u^5}{t^8 u}$.
    - F $\frac{u}{t}$
    - H $\frac{u^5}{u^2}$
    - (G) $\frac{u^4}{t^4}$
    - J ut

35. Find the value of $(-3)^{-3}$.
    - (A) $-\frac{1}{27}$
    - C $\frac{1}{27}$
    - B −27
    - D 27

36. Combine like terms: $3a^2 + 2b^2 - 3a^2$.
    - F $-2ab$
    - H $7ab^2$
    - G $2a^4$
    - (J) $2b^2$

37. A border for a flower bed costs $3.50 per foot for the first 10 feet and $2.75 per foot for every additional foot. How much will the border cost for a flower bed that measures 12 feet by 10 feet?
    - A $156.00
    - C $121.00
    - (B) $128.50
    - D $345.00

38. Find the area of the figure. 
    - F 4.75 ft^2
    - G 9.5 ft^2
    - (H) 9.75 ft^2
    - J 19.5 ft^2

39. Dianne made 230 chocolate chip cookies and 125 butterscotch cookies last week. This week, she made 380 cinnamon raisin cookies. What is the percent increase in the number of cookies she baked to the nearest tenth of a percent?
    - A 6.6%
    - C 39.5%
    - (B) 7.0%
    - D 45.7%

40. Find the total amount due on a simple interest loan if the principal is $20,800 with a rate of 5.5% for 10 years. Assume one payment is made at the end of the 10 years.
    - F $9360
    - (H) $32,240
    - G $11,440
    - J $21,944

Cumulative Test
Chapter 11 Form A

Select the best answer.

1. Which of the following equations is not linear?
    - (A) $x + y^2 = 7$
    - B $5x = y + 2$

2. Solve the system of equations:
    $-3x + y = -3$
    $y = x - 3$.
    - (A) (0, −3)
    - B (−3, 0)

3. Solve $4y + 1 = 8 + 2y$.
    - A $y = -\frac{2}{7}$
    - (B) $y = \frac{7}{2}$

4. In Mrs. Trista's class there are 15 girls and 20 boys. Mrs. Trista randomly selects one student. What is the probability it will be a girl?
    - A $\frac{3}{4}$
    - (B) $\frac{3}{7}$

5. Express 55% as a decimal.
    - A 0.0055
    - (C) 0.55
    - B 0.055
    - D 5.5

6. Which statement explains why the graph is misleading? 
    - A The numbers are not realistic.
    - (B) The area of the books distorts the comparison.

7. Solve for l: $P = 2w + 2l$.
    - (A) $l = \frac{P - 2w}{2}$
    - B $l = P - w$

8. Solve $\frac{1}{4}x = 1$.
    - (A) $x = 4$
    - B $x = 0.25$

9. Dilation of a figure changes _____.
    - A size and shape
    - (B) size only

10. $10 is 25% of what amount?
    - A $2.50
    - (C) $40
    - B $25
    - D $100

11. To determine an experimental probability, do _____.
    - (A) multiple trials
    - B a tree diagram

12. What scale factor relates a 5-inch scale model to a 30-foot statue?
    - (A) 1: 72
    - C 6: 5
    - B 1: 6
    - D 1 in. to 30 ft

13. Sue left a $2.00 tip for a meal that cost $15. To the nearest tenth, at what rate did she tip?
    - A 22.5%
    - (C) 13.3%
    - B 20%
    - D 6.6%

14. Tessellations are patterns that _____.
    - A are rotated
    - B are shifted and rotated
    - (C) cover a plane with no gaps
    - D are changed in size and shape

15. Find the slope of the line that passes through (1, 2) and (2, 4).
    - (A) 2
    - B $\frac{1}{2}$

16. Calculate the means of the x- and y-coordinates for the data:

x	3	3	5	4	5
y	4	3	7	5	6

    - (A) 4; 5
    - B 4; 6

17. If a 3-inch cube is built from 1-inch cubes, what is the ratio of the corresponding volumes?
    - A 3:1
    - (B) 27:1

18. What is the slope-intercept form of the line passing through the points (−2, 5) and (6, 1)?
    - (A) $y = -\frac{1}{2}x + 4$
    - B $y = x - 2$

19. Express 220% as a decimal.
    - A 0.22
    - C 22.0
    - (B) 2.2
    - D 220.0

20. Choose the equation which means "y varies directly with x, and when y is 10, x is 5."
    - A $10y = 5x$
    - (C) $y = 2x$
    - B $y = 5x$
    - D $2 = x$

21. 50% of 201 is about what number?
    - (A) 100
    - B 50

22. The graph represents which inequality? 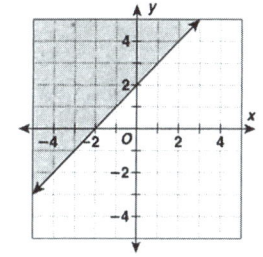
    - A $y \leq -x + 3$
    - C $y > -x + 2$
    - B $2y < -2x - 5$
    - (D) $y \geq x + 2$

23. Simplify $\frac{\sqrt{36}}{10}$.
    - (A) $\frac{3}{5}$
    - B $\frac{\sqrt{18}}{5}$

24. A child's large wooden cube measures 12 inches by 12 inches by 12 inches. Find the surface area of the cube.
    - A 1728 in^2
    - (B) 864 in^2

25. Find the unknown number in the proportion $\frac{6}{x} = \frac{24}{5}$.
    - (A) $x = 1.25$
    - B $x = 4$

26. In a three-dimensional figure, a(n) _____ is a flat surface.
    - A vertex
    - (C) face
    - B edge
    - D line

379

Holt Pre-Algebra

Cumulative Test
Chapter 11 Form A, continued

27. Find the measure of the smaller angle.

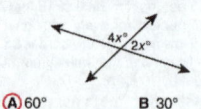

- (A) 60°
- B 30°

28. What is the point-slope form of the line with slope 3 that passes through (2, −1)?
- (A) $y + 1 = 3(x - 2)$
- B $y - 1 = -3(x + 2)$
- C $y + 1 = 3(x + 2)$
- D $y - 1 = -3(x - 2)$

29. The two figures shown are in proportion. Find the value of x.
- (A) $6\frac{2}{3}$ ft
- B 60 ft

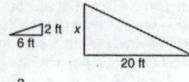

30. The perimeter of a garden is to be at least 90 feet. If the width is 20 feet, then which statement correctly indicates the length?
- A $L \geq 70$ ft
- B $L \geq 50$ ft
- C $L \geq 30$ ft
- (D) $L \geq 25$ ft

31. Simplify $2 + 6 \cdot (-2)$.
- A −16
- (B) −10

32. Express the ratio of 5 pieces of red candy to 20 total pieces of candy as a fraction in lowest terms.
- (A) $\frac{1 \text{ red}}{4 \text{ total}}$
- B $\frac{3 \text{ red}}{4 \text{ total}}$

33. Find the perimeter of the parallelogram.
- A 40 ft
- B 20 ft
- (C) 18 ft
- D 9 ft

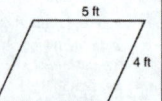

34. Find the average number of customer complaints over the past 8 days at Clara's Coffee Cart: 8, 2, 15, 15, 20, 15, 3, and 7.
- (A) 10.6
- B 12.1

35. Find the surface area of the given figure. Use 3.14 for π.
- A 251.2 ft²
- (B) 113.04 ft²

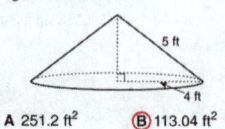

36. Find the value of x in the parallelogram.
- (A) $x = 90$
- B $x = 180$
- C $x = 360$
- D not enough information

37. Express $(x)(x)(x)(x)(x)(x)(x)(x)$ using exponents.
- (A) x^8
- B $x + 8$
- C $8x$
- D 8^x

38. What is an equation for the line that is perpendicular to $x - y = 5$ and passes through (2, 2)?
- (A) $x + y = 4$
- B $y - x = 4$

39. Find the measure of ∠CBD.
- (A) 67°
- B 23°

40. Rey flips a coin and spins a number spinner, then records the outcomes. These two events are _____.
- (A) independent events
- B dependent events
- C random numbers
- D permutations

41. Find the value of $_5P_4$.
- (A) 120
- B 1.25
- C 5
- D 1

42. Which of the graphs represents the inequality $x \geq -2$?
- A
- (B)
- C
- D

Cumulative Test
Chapter 11 Form B

Select the best answer.

1. Which of the following equations is not linear?
- A $6x + 3y = 75$
- B $0.5x = y$
- C $y = 3$
- (D) $y = x^2$

2. Solve the system of equations:
$x - 2y = 7$
$-x + 3y = -14$.
- (F) $(-7, -7)$
- G no solution
- H $(-7, -1)$
- J $(7, 0)$

3. Solve $-3y + 9 = -7 + 9y$.
- A $y = -\frac{3}{4}$
- (B) $y = \frac{4}{3}$
- C $y = \frac{3}{4}$
- D $y = 3$

4. There are 87 green balls and 41 red balls in a bag. If one is randomly selected, what is the probability that it will be a green ball?
- F $\frac{87}{41}$
- (G) $\frac{87}{128}$
- H $\frac{41}{128}$
- J $\frac{1}{128}$

5. Express 3.2% as a decimal.
- A 0.0032
- (B) 0.032
- C 0.32
- D 32

6. Which statement explains why the graph is misleading?
- F The hands distract from the data.
- G The numbers are not realistic.
- H Nothing is wrong with the graph.
- (J) The area changes more than the amount.

7. Solve for h: $V = \frac{1}{3}Bh$.
- A $h = \frac{V}{3B}$
- B $h = \frac{3B}{V}$
- C $h = \frac{B}{3V}$
- (D) $h = \frac{3V}{B}$

8. Solve $-\frac{1}{8}x = 18$.
- (F) $x = -144$
- G $x = 144$
- H $x = -2.25$
- J $x = 2.25$

9. Changing the size but not the shape of a figure is called _____.
- (A) dilation
- B rotation
- C reflection
- D translation

10. $75 is 20% of what amount?
- F $15
- G $90
- (H) $375
- J $1500

11. Repeating trials many times to try to determine the likelihood of an event is related to _____.
- A dependent events
- B theoretical probability
- (C) experimental probability
- D the Fundamental Counting Principle

12. What scale factor relates a 10-inch scale model to a 50-foot actual model?
- F 1:5
- (G) 1:60
- H 5:1
- J 1:6

13. The sales tax is $44.75 on a TV that sells for $895. What is the sales tax rate?
- A 40%
- B 20%
- (C) 5%
- D 4.5%

14. What is the name given to a repeating pattern of figures that completely covers a plane with no gaps or overlaps?
- (F) tessellation
- G transformation
- H correspondence
- J rotational symmetry

15. Find the slope of the line that passes through (3, −1) and (2, 3).
- A 2
- B −2
- C 4
- (D) −4

16. Calculate the means of the x- and y-coordinates for the data:

x	14	15	16	16	18
y	13	14	17	15	16

- (F) 15.8; 15
- G 5; 15
- H 16; 15
- J 10.8; 7.5

17. If a 4-inch cube is built from 1-inch cubes, what is the ratio of the corresponding volumes?
- A 4:1
- B 16:1
- C 24:1
- (D) 64:1

18. What is the slope-intercept form of the line passing through the points (−5, 0) and (2, −2)?
- F $y = \frac{5}{7}x - \frac{10}{7}$
- G $y = \frac{5}{7}x - \frac{10}{7}$
- H $y = \frac{2}{7}x - \frac{10}{7}$
- (J) $y = -\frac{2}{7}x - \frac{10}{7}$

19. Express 4% as a decimal.
- (A) 0.04
- B 0.40
- C 4.0
- D 40.0

20. Choose the equation which means "y varies directly with x, and y is 12 when x is 4."
- (F) $y = 3x$
- G $y = 12x$
- H $y = 12$
- J $12 = 3x$

21. 19 is about what percent of 82?
- (A) 25%
- B 35%
- C 40%
- D 50%

Cumulative Test
11 Form B, continued

22. The graph represents which inequality?

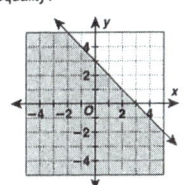

   (F) $y \le -x + 3$ H $y > -x + 3$
   G $y < -x + 3$ J $y \ge -x + 3$

23. Simplify $\frac{\sqrt{125}}{20}$.

   A $\frac{\sqrt{25}}{5}$ C $\frac{5}{2}$
   B $\frac{25}{4}$ (D) $\frac{\sqrt{5}}{4}$

24. A briefcase measures 10 inches by 4.9 inches by 13.7 inches. Find the surface area of the briefcase.
   F 671.3 in²   H 253.1 in²
   (G) 506.3 in²   J 28.6 in²

25. Find the unknown number in the proportion $\frac{5}{x} = \frac{20}{16}$.
   A $x = 40$ (C) $x = 4$
   B $x = 6.2$ D $x = 0.16$

26. In a three-dimensional figure, an edge is where two _____ meet.
   F vertices   H points
   G curves   (J) faces

27. Find the measure of the smaller angle. 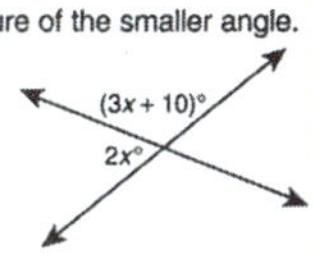
   A 112°
   B 90°
   (C) 68°
   D 34°

28. What is the point-slope form of the line with slope −2 that passes through (3, −1)?
   (F) $y + 1 = -2(x - 3)$
   G $y - 1 = -2(x - 3)$
   H $y + 1 = -2(x + 3)$
   J $y - 1 = -2(x + 3)$

29. The wheelchair ramp shown below is going to be rebuilt, with the proportions staying the same. How many feet will it rise if the new length is going to be 9 feet? 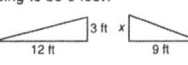
   A 36 ft   C 3 ft
   B 4 ft   (D) 2.25 ft

30. Andrea wants the area of her rectangular garden to be at least 36 square feet. If the garden is 4 feet wide, what must the length be?
   (F) $L \ge 9$ ft   H $L \ge 14$ ft
   G $L > 9$ ft   J $L \le 14$ ft

31. Simplify $72 \div 8 + 8 \cdot (-4)$.
   (A) −23   C 41
   B −68   D −18

32. Express the ratio of 11 cars to 55 people as a fraction in lowest terms.
   (F) $\frac{1 \text{ car}}{5 \text{ people}}$ H $\frac{5 \text{ cars}}{1 \text{ person}}$
   G $\frac{11 \text{ cars}}{55 \text{ people}}$ J $\frac{44 \text{ cars}}{55 \text{ people}}$

33. Find the perimeter of the parallelogram. 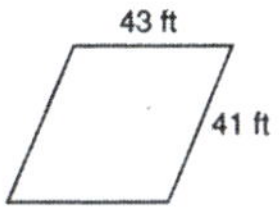
   (A) 168 ft
   B 127 ft
   C 125 ft
   D 84 ft

34. Find the average number of phone calls made by a salesperson over the past 8 hours: 8, 21, 3, 19, 27, 41, 35, 32.
   F 24   H 24.5
   (G) 23.25   J 27

35. Find the surface area of the given figure. Use 3.14 for π.
   A 3815.1 ft²
   B 1271.7 ft²
   (C) 678.24 ft²
   D 423.9 ft²

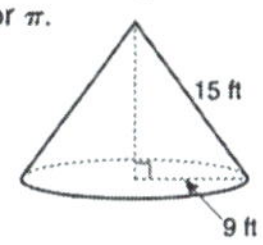

36. Find the value of x in the parallelogram. 
   F $x = 65$
   G $x = 90$
   (H) $x = 115$
   J $x = 145$

37. Express $(-6x)(-6x)(-6x)(-6x)$ using exponents.
   A $-6x^4$ (C) $(-6x)^4$
   B $-24x$ D $-(-6x)^4$

38. What is an equation for the line that is parallel to $3x - 2y = 6$ and passes through (3, −1)?
   (F) $2y - 3x = -11$ H $-3x + 2y = 7$
   G $2y + 3x = -7$ J $2x + 3y = 3$

39. Find the measure of $\angle DBF$. 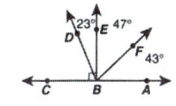
   A 203°
   (B) 70°
   C 43°
   D 47°

40. Claire draws a card from a deck, puts it back, and draws another card from the same deck. These two events are _____.
   (F) independent events
   G dependent events
   H random numbers
   J permutations

41. Find the value of $_8C_6$.
   A 20,160   (C) 28
   B 56   D 48

42. Which of the graphs represents the inequality $3x \ge 6$?
   F
   G
   H
   (J)

Cumulative Test
11 Form C

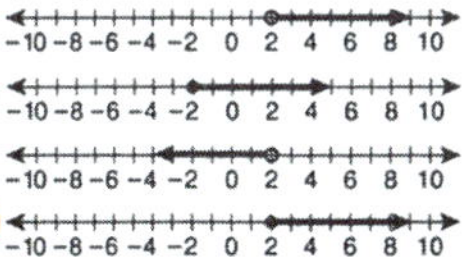

Select the best answer.

1. Which of the following equations is not linear?
   A $6x = 7$ C $y = x + 4$
   (B) $0.5x^2 = y$ D $y = x$

2. Solve the system of equations:
   $5x - 4y = 15$
   $-3x + 6y = -9$.
   F (−1, 1)   H (−1, 3)
   G no solution   (J) (3, 0)

3. Solve $-4y + 7 = -7 + 4y - 4$.
   A $y = -\frac{9}{4}$ C $y = 2$
   (B) $y = \frac{9}{4}$ D $y = \frac{4}{9}$

4. A regular deck of cards has four suits with 13 cards per suit. The deck is shuffled several times. If one card is randomly selected, what is the probability that it will be a heart or a four?
   (F) $\frac{4}{13}$ H $\frac{1}{13}$
   G $\frac{17}{52}$ J $\frac{1}{4}$

5. Express 0.82% as a decimal.
   (A) 0.0082   C 0.82
   B 0.082   D 8.2

6. Which statement explains why the graph is misleading? 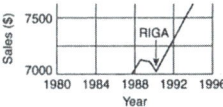
   (F) The scale is misleading.
   G It does not show this year's data.
   H The numbers are not realistic.
   J Nothing is wrong with the graph.

7. Solve for C: $F = \frac{9}{5}C + 32$.
   A $C = \frac{5}{9}(F + 32)$ C $C = \frac{9}{5}F + 32$
   (B) $C = \frac{5}{9}(F - 32)$ D $C = \frac{5}{9}F + 32$

8. Solve $-\frac{4}{3}x = 12$.
   F $x = -16$ (H) $x = -9$
   G $x = 16$ J $x = 9$

9. When a figure is dilated, it _____.
   (A) changes size but not shape
   B changes size and shape
   C is translated
   D is rotated

10. $3000 is 150% of what amount?
   F $1500   H $3000
   (G) $2000   J $4500

11. In experimental probability, the probability is determined by _____.
   A tree diagrams
   (B) repeated trials
   C mathematical calculations
   D Fundamental Counting Principle

12. What scale factor relates a 12-inch scale model to a 72-foot actual model?
   F 1:6   H 6:1
   (G) 1:72   J $\frac{1}{12}$ in. to 1 ft

13. A softball bat, with tax, costs $37.17. If the sales tax is $2.17, what is the sales tax rate?
   A 5.8%   (C) 6.2%
   B 7%   D 4.5%

14. Which of the following figures will not form a tessellation?
   F
   G
   H
   (J)

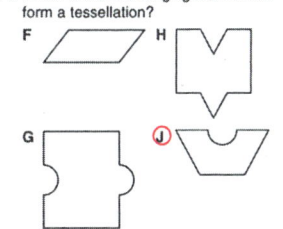

15. Find the slope of the line that passes through (−1, −3) and (−2, 2).
   (A) −5   C −2
   B $-\frac{1}{5}$ D $\frac{1}{3}$

16. Calculate the means of the x- and y-coordinates for the data:

x	114	115	116	116	118
y	113	114	117	115	116

   (F) 115.8; 115   H 116; 115
   G 115; 115.8   J 115.8; 117

17. If a 6-inch cube is built from 1-inch cubes, what is the ratio of the corresponding volumes?
   A 6:1   C 36:1
   B 18:1   (D) 216:1

18. What is the slope-intercept form of the line passing through the points (−2, 1) and (5, −2)?
   F $y = -\frac{3}{7}x + 2$ H $y = -\frac{3}{7}x - \frac{1}{7}$
   G $y = \frac{3}{7}x - \frac{1}{7}$ (J) $y = -\frac{3}{7}x + \frac{1}{7}$

19. Express 0.304% as a decimal.
   A 0.304   C 0.0034
   (B) 0.00304   D 0.0304

20. Choose the equation which means "y varies directly with x, and y is 6.5 when x is 2.5."
   F $y = 4x$ H $y = 6.5$
   G $y = 8x$ (J) $y = 2.6x$

21. 40% of 118.7 is about what number?
   (A) 48   C 57
   B 40   D 30

22. The graph represents which inequality?

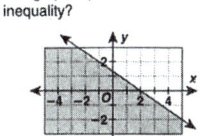

   F $y \le -3x + 2$ (H) $4y \le -3x + 6$
   G $4y < -3x + 6$ J $4y \ge -3x + 6$

CHAPTER 11 Cumulative Test
Form C, continued

23. Simplify $\frac{\sqrt{49}}{14}$.
    A $\frac{\sqrt{7}}{2}$
    B $\frac{\sqrt{7}}{14}$
    C $\frac{7}{2}$
    D $\frac{1}{2}$ ✓

24. Serena is wrapping a gift for her brother's birthday. The package measures 20 inches × 14.5 inches × 23.7 inches. How much wrapping paper does she need to wrap the gift?
    F 6873.0 in²
    G 2215.3 in² ✓
    H 1107.7 in²
    J 58.2 in²

25. Find the unknown number in the proportion $\frac{5}{x} = \frac{40}{15}$.
    A $x = 1.875$ ✓
    B $x = 3$
    C $x = 5$
    D $x = 8$

26. In a three-dimensional figure, a vertex is where _____ edges meet.
    F exactly two
    G exactly three
    H two or more
    J three or more ✓

27. Find the measure of the angle.
    A 220°
    B 90°
    C 110° ✓
    D 30°

    $(3x + 20)°$ $(4x - 10)°$

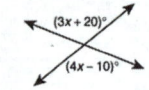

28. What is the point-slope form of the line with slope −3 that passes through (−4, −1)?
    F $y - 1 = -3(x - 4)$
    G $y + 1 = -3(x + 4)$ ✓
    H $y - 1 = -3(x + 4)$
    J $y - 1 = 3(x + 3)$

29. For the incline shown below, what measurement in the horizontal direction corresponds to 12 feet in the vertical direction?

    3 ft, 15 ft

    A 2.4 ft
    B 3 ft
    C 15 ft
    D 60 ft ✓

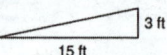

30. The area of a triangle must be no greater than 75 cm². If the base is 20 cm, what must the height be?
    F $h ≤ 3.25$ cm
    G $h ≥ 3.25$ cm
    H $h ≤ 7.5$ cm ✓
    J $h < 7.5$ cm

31. Simplify $18 + 8 ÷ (-4) · 7 - 3^2$.
    A 36.5
    B −19
    C −8.07
    D −5 ✓

32. Express the ratio of 24 balls for 80 cats as a fraction in lowest terms.
    F $\frac{3 \text{ balls}}{10 \text{ cats}}$ ✓
    G $\frac{24 \text{ balls}}{80 \text{ cats}}$
    H $\frac{10 \text{ balls}}{3 \text{ cats}}$
    J $\frac{80 \text{ balls}}{24 \text{ cats}}$

33. Find the perimeter of the figure created by the triangle and the parallelogram.

    30 ft, 25.5 ft, 25.5 ft, 25.5 ft

    A 765.0 ft
    B 249.8 ft
    C 111.0 ft
    D 136.5 ft ✓

34. Over the past 9 days, 39, 41, 39, 39, 47, 61, 55, 46, and 52 people entered the hiking trail. Find the mean of the data set to the nearest hundredth.
    F 42.71
    G 46.56 ✓
    H 46.00
    J 22.00

35. Find the surface area of the given figure. Use 3.14 for π and round to the nearest tenth.

    6 ft, 8 ft

    A 351.7 ft²
    B 401.9 ft²
    C 452.2 ft² ✓
    D 480.0 ft²

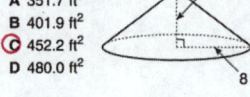

36. Find the value of x in the regular hexagon.
    F $x = 120$ ✓
    G $x = 86.7$
    H $x = 60$
    J $x = 30$

37. Simplify $(-ab)^3 (-ab)^2$.
    A $(-ab)^6$
    B $(-ab)^5$ ✓
    C $(ab)^5$
    D $(ab)^6$

38. What is an equation for the line that is perpendicular to $4x - 5y = 20$ and passes through $(-2, -3)$?
    F $4y - 5x = 7$
    G $4y - 5x = -2$
    H $5x + 4y = -22$ ✓
    J $-4x + 5y = 2$

39. Find the measure of ∠CBF.
    A 137° ✓
    B 70°
    C 67°
    D 43°

    23°, 47°

40. A bag contains 10 marbles, 3 of which are blue. Two random picks without replacement constitute _____.
    F independent events
    G dependent events ✓
    H random numbers
    J permutations

41. Find the value of $_8P_6$.
    A 1.33
    B 48
    C 56
    D 20,160 ✓

42. Choose the graph which represents the inequality $2x + 4 < 3x + 2$.
    F ✓
    G
    H
    J

CHAPTER 12 Cumulative Test
Form A

Select the best answer.

1. Each morning at Park Elementary, teachers take a count of how many students will be buying lunch. On Monday, the cook was told to prepare 340 lunches. If the 1st grade teacher counted 17 lunch buyers in her classroom, what percent of the lunches prepared were for first graders?
   A 5% ✓
   B 20%

2. Identify the angle that is complementary to ∠ABF.

   23°, 67°, 47°, 43°

   A ∠CBE
   B ∠CBD
   C ∠EBF ✓
   D ∠FBC

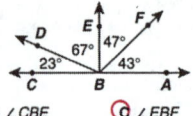

3. The diameter of a plate is 8.0 inches. What is the area of the flat upper surface? Use 3.14 for π.
   A 50.24 in² ✓
   B 25.12 in²

4. A scuba diver is swimming 3 meters below sea level then descends another 6 meters. What is the diver's new depth?
   A −3 m
   B −9 m ✓

5. Find the slope of the line $2x - 3y = 16$.
   A $\frac{3}{2}$
   B $\frac{2}{3}$ ✓

6. Find the value of $_8C_3$.
   A 5
   B 56 ✓
   C 120
   D 360

7. Solve the system of equations:
   $x + y = -3$
   $x - y = -5$.
   A (4, 2)
   B (−5, 2)
   C (−4, 1) ✓
   D no solution

8. Solve $6c - (5c - 1) = 2$.
   A $c = \frac{1}{11}$
   B $c = 1$ ✓

9. Company A rents copy machines for $300 plus $0.05 per copy. Company B charges of $600 plus $0.01 per copy. For which number of copies is Company B's rate higher?
   A 7500 copies
   B 8000 copies
   C 7000 copies ✓
   D 8500 copies

10. Solve the equation for s:
    $a + b = s + r$.
    A $s = a + b - r$ ✓
    B $s = r(a + b)$
    C $s = \frac{a}{r} + b$
    D $s = \frac{a + b}{r}$

11. Find the mean for the data set 49, 52, 52, 52, 74, 67, 55, 55.
    A 52
    B 57 ✓

12. Find $f(1)$ for $f(x) = 2(2x + 1)$.
    A 6 ✓
    B 5

13. Express $3^4 3^4$ using positive exponents.
    A 9^{16}
    B 9^8
    C 3^{16}
    D 3^8 ✓

14. Simplify $-4b - 4b$.
    A $-8b$ ✓
    B 0
    C $8b$
    D $8b^2$

15. Find the quotient $\frac{2}{3} ÷ \frac{1}{6}$.
    A $\frac{1}{9}$
    B 4 ✓

16. Find the rule for the linear function.

    A $f(x) = -(2x + 1)$
    B $f(x) = 2x + 1$ ✓

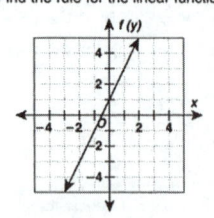

17. What is 10% of 500?
    A 500
    B 50 ✓
    C 5
    D 0.5

18. Convert 6 hours to seconds.
    A 360 s
    B 5400 s
    C 10,800 s
    D 21,600 s ✓

19. Solve $-4(-2y) = -2$.
    A $y = -\frac{1}{4}$ ✓
    B $y = -4$
    C $y = \frac{1}{4}$
    D $y = 4$

20. The sum of 6 and four times a number is 38. What is the number?
    A 7
    B 8 ✓
    C 11
    D 16

21. Find the missing value in the proportion $\frac{1}{12} = \frac{x}{60}$.
    A $x = 60$
    B $x = 6$
    C $x = 5$ ✓
    D $x = 1$

22. Solve the inequality $5x < 45$.
    A $x < 9$ ✓
    B $x > 9$

23. A particular substance has a half-life of 60 minutes. To the nearest thousandth, how much of a 200 mg sample is left after 10 hours?
    A 0.391 mg
    B 0.195 mg ✓

Cumulative Test
Chapter 12 Form A, continued

24. Which angle is supplementary to ∠7?

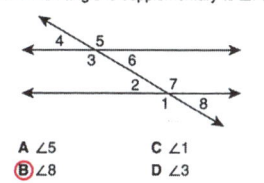

A ∠5   C ∠1
(B) ∠8   D ∠3

25. Find the perimeter of the figure.

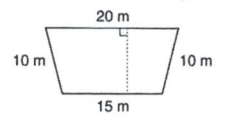

(A) 55 m   B 175 m

26. Find the missing length in the right triangle to the nearest tenth.

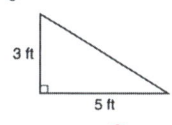

A 4.0 ft   **(B) 5.8 ft**

27. If a costume maker takes 18 hours to sew 8 costumes, at what rate per hour are the costumes being made?

(A) $\frac{4}{9}$ costume/hour
B $\frac{4}{18}$ costume/hour
C $\frac{18}{8}$ costume/hour
D $\frac{9}{4}$ costume/hour

28. Find the y-intercept for $x + y = 5$.

A (2, 5)   C (1, 0)
B (0, 3)   **(D) (0, 5)**

29. Use the data in the table to find the inverse variation equation.

x	1	5	10	15
y	5	1	$\frac{1}{2}$	$\frac{1}{3}$

(A) $y = \frac{5}{x}$ B $y = \frac{1}{5x}$

30. Which ordered pair is a solution to the equation $y = -x + 1$?

A (−1, −3)   C (−1, −2)
B (−1, 3)   **(D) (−1, 2)**

31. What is the point-slope form equation for the line with a slope of 5 that passes through (2, 0)?

(A) $y = 5(x - 2)$ B $y = -5(x + 2)$

Chapter 12 Form A, continued

32. How many sides does a regular hexagon have?

A 4   **(C) 6**
B 7   D 8

33. On a multiple-choice test, each question has 7 possible answers. What is the probability that a random guess on the first question will be correct?

A 7   **(C) $\frac{1}{7}$**
B 1   D 0

34. 60 is 80% of what number?

A 750   C 48
(B) 75   D 7.5

35. Simplify $\left(\frac{3}{7}\right)^2$.

A $\frac{6}{14}$ **(B) $\frac{9}{49}$**

36. Solve $-\frac{1}{4}x = -23$.

A $x = 27$ C $x = -88$
B $x = -69$ **(D) $x = 92$**

37. Find the 5th term of the sequence defined by $a_n = 2(n + 1)$.

(A) 12   B 10

38. Find the 8th term in the arithmetic sequence 3, 5, 7, 9, ...

A 19   **(B) 17**

39. The price of a couch is reduced from $800 to $675. Find the rate of discount to the nearest tenth of a percent.

(A) 15.6%   B 18.5%

40. A large carton in the shape of a cube measures 9 feet on a side. What is the volume of the carton?

A 18 ft³   C 720 ft³
B 81 ft³   **(D) 729 ft³**

41. Find the area of the triangle.

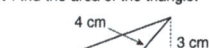

A 12 cm²   **(B) 9 cm²**

42. Express 51,000 in scientific notation.

A 5.1×10^6 C 5.1×10^3
B 5.1×10^5 **(D) 5.1×10^4**

Cumulative Test
Chapter 12 Form B

Select the best answer.

1. On Saturday, the Henry County Fair had an attendance of 5625 people. The gates admitted 2645 adults, 2154 children, and the remainder senior citizens. About what percent were senior citizens?

A 7%   C 24%
(B) 15%   D 31%

2. Which angle is complementary to ∠CBD?

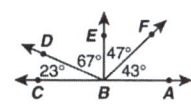

F ∠ABF
G ∠FBA
H ∠ABE
(J) ∠DBE

3. The diameter of a circle is 6 inches. What is the area? Use 3.14 for π.

A 3.5 in²   **(C) 28.26 in²**
B 18.84 in²   D 28.3 in²

4. An airplane flies at a cruising altitude of 2900 feet. It descends 1200 feet as it begins to reach its destination. As it approaches the airport, it descends an additional 1550 feet. What is the new altitude of the airplane?

F 1700 ft   H 300 ft
G 1500 ft   **(J) 150 ft**

5. Find the slope of the line $4x - 5y = 32$.

A $-\frac{5}{4}$ C $-\frac{4}{5}$
B $\frac{5}{4}$ **(D) $\frac{4}{5}$**

6. Find the value of $_8C_2$.

(F) 28   H 40,320
G 56   J 80,640

7. Solve the system of equations:
$3x + 3y = 18$
$2x - 2y = 4$

A (−4, 2)   C (7, 2)
(B) (4, 2)   D (1, 11)

8. Solve $7c + 6 = 2 + 3c$.

F $c = -\frac{2}{5}$ **(H) $c = -1$**
G $c = 1$ J $c = \frac{4}{5}$

9. A salesperson has two job offers. Company A is offering $200 weekly plus 10% commission on sales. Company B is offering $400 weekly plus 5% commission on sales. For what level of sales does Company A have the better offer?

A $2000   **(C) $4100**
B $4000   D $3900

Chapter 12 Form B, continued

10. Solve the equation for h:
$V = \frac{1}{3}\pi r^2 h$.

F $h = \frac{V}{3}\pi r^2$ H $h = \frac{V}{\pi r}$
(G) $h = \frac{3V}{\pi r^2}$ J $h = \frac{V}{\pi}r^2$

11. Find the median weight of a group of apples with individual weights of 4.7, 4.0, 6.2, 6.5, 6.1, 4.7, 4.0, 6.2, 6.3, 6.5, 4.7, 6.2, 6.5, and 6.0 ounces.

A 4.7 oz   C 5.6 oz
B 6.5 oz   **(D) 6.15 oz**

12. Find $f(3)$ for $f(x) = 2(x + 1)^2$.

F $2x^2 + 4x - 1$ H 22
(G) 32   J 20

13. Express $(-4)^4(-4)^9$ using positive exponents.

A $(16)^{36}$ C $(16)^{13}$
(B) $(-4)^{13}$ D $(4)^{13}$

14. Simplify $-5n + 3n$.

F $2n$ **(H) $-2n$**
G $8n$ J $-8n$

15. Find the quotient $\frac{8}{15} \div \frac{4}{5}$.

(A) $\frac{2}{3}$ C $\frac{2}{15}$
B $\frac{32}{75}$ D $\frac{4}{15}$

16. Find the rule for the linear function.

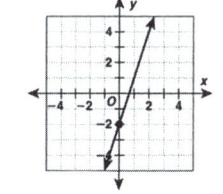

F $f(x) = -3x - 2$ **(H) $f(x) = 3x - 2$**
G $f(x) = 3x + 2$ J $f(x) = x - 2$

17. What is $6\frac{1}{5}\%$ of 80,000?

A 50   **(C) 4960**
B 496   D 49,600

18. Convert 4 weeks to minutes.

F 80,640 min   **(H) 40,320 min**
G 57,600 min   J 10,080 min

19. Solve $-12 = -3(-3y)$.

A $y = \frac{4}{3}$ **(C) $y = -\frac{4}{3}$**
B $y = 9$ D $y = -9$

20. The sum of twice a number and 28 equals 36 plus the number.

F 21   **(H) 8**
G 20   J −8

21. Find the missing value in the proportion $\frac{4}{8} = \frac{x}{24}$.

A $x = 4$ C $x = 32$
(B) $x = 12$ D $x = 256$

---

383   Holt Pre-Algebra

Cumulative Test
Form B, continued

22. Solve the inequality $-5x > 45$.
    F $x > 9$
    G $x < -5$
    H $x < 9$
    J $x < -9$

23. An investment of $2000 will double every 10 years. Which exponential function could be used to calculate the balance in x years?
    A $f(x) = 2000(2)^{x/10}$
    B $f(x) = 2000(2)^x$
    C $f(x) = 2000(x)^{10}$
    D $f(x) = x(10)^2$

24. Which angle is congruent to $\angle 4$? 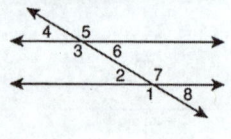
    F $\angle 6$
    G $\angle 8$
    H $\angle 7$
    J $\angle 2$

25. Find the perimeter of the figure. 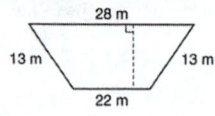
    A 286 m
    B 76 m
    C 41 m
    D 28 m

26. Find the missing length in the right triangle. 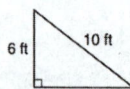
    F 7 ft
    G 8 ft
    H 9 ft
    J 10 ft

27. If you can type 1080 words in 40 minutes, then how many words per minute can you type?
    A 180 words/min
    B 31 words/min
    C 27 words/min
    D 9 words/min

28. Find the y-intercept for $3x + y = 0$.
    F (0, 0)
    G (−9, 0)
    H (3, 0)
    J (0, −9)

29. Use the data in the table to find the inverse variation equation.

x	15	30	45	60
y	10	5	$3\frac{1}{3}$	$2\frac{1}{2}$

    A $y = \frac{150}{x}$
    B $y = 2x$
    C $y = \frac{x}{2}$
    D $y = x + 15$

30. Which ordered pair is a solution to the equation $y = -x + 7$?
    F (4, 12)
    G (4, −3)
    H (4, 4)
    J (4, 3)

31. What is the point-slope form equation for the line with a slope of $-\frac{2}{3}$ passes through (0, −2)?
    A $y - 2 = \frac{2}{3}x$
    B $y + 2 = \frac{2}{3}x$
    C $y - 3 = \frac{3}{2}x$
    D $y + 2 = -\frac{2}{3}x$

32. How many sides does a regular octagon have?
    F 4
    G 7
    H 6
    J 8

33. A six-sided die is rolled. What is the probability of rolling a number less than 5?
    A 4
    B $\frac{5}{6}$
    C $\frac{2}{3}$
    D $\frac{1}{6}$

34. 560 is 140% of what number?
    F 40
    G 78.4
    H 400
    J 19,600

35. Simplify $\left(\frac{8}{3}\right)^2$.
    A $21\frac{1}{3}$
    B $\frac{64}{9}$
    C $\frac{8}{9}$
    D $\frac{3}{8}$

36. Solve $\frac{16}{45}x = -\frac{4}{9}$.
    F $x = \frac{5}{16}$
    G $x = -\frac{16}{5}$
    H $x = -\frac{20}{9}$
    J $x = -\frac{5}{4}$

37. Find the 5th term of the sequence defined by $a_n = \frac{n+1}{n+2}$.
    A $\frac{1}{2}$
    B $\frac{5}{6}$
    C $\frac{6}{7}$
    D $\frac{7}{8}$

38. Find the 13th term in the arithmetic sequence 2, 7, 12, 17, ...
    F 62
    G 60
    H 57
    J 55

39. A computer, priced at $1380, is marked down to $971.52. What is the rate of markdown to the nearest tenth of a percent?
    A 22.7%
    B 23.4%
    C 29.6%
    D 30.6%

40. Find the volume of a rectangular solid measuring 11 feet by 14 feet by 4 feet.
    F 1331 ft³
    G 616 ft³
    H 154 ft³
    J 29 ft³

41. Find the area of the triangle. 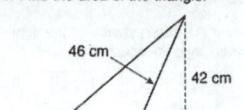
    A 1344 cm²
    B 966 cm²
    C 736 cm²
    D 672 cm²

42. Express 680,000 in scientific notation.
    F 6.8×10^5
    G 6.8×10^4
    H 6.8×10^{-5}
    J 6.8×10^{-4}

Cumulative Test
Form C

Select the best answer.

1. Every morning Thrush's bakery makes 15 dozen cookies. The baker chooses to make peanut butter, chocolate chip, and oatmeal cookies. If she makes 42 peanut butter, 66 chocolate chip, and the rest oatmeal, what percentage of the cookies are oatmeal?
    A 22%
    B 40%
    C 54%
    D $63\frac{2}{3}$%

2. Identify the pair of angles that are supplementary. 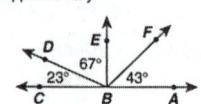
    F $\angle CBD$ & $\angle ABD$
    G $\angle ABF$ & $\angle CBD$
    H $\angle DBE$ & $\angle EBF$
    J $\angle ABF$ & $\angle FBE$

3. The diameter of a discus is about 8.6 inches. To the nearest hundredth, what is the area of one flat surface of the discus? Use 3.14 for π.
    A 27.00 in²
    B 29.03 in²
    C 58.06 in²
    D 232.23 in²

4. On a cold winter night, the temperature at midnight is 6°F below zero. By five o'clock in the morning the temperature has dropped another four degrees. What is the temperature at this time?
    F 10°F
    G 2°F
    H −10°F
    J −12°F

5. Find the slope of the line $2x - 5y = -16$.
    A $-\frac{5}{2}$
    B $\frac{5}{2}$
    C $-\frac{2}{5}$
    D $\frac{2}{5}$

6. Find the value of $_{11}C_3$.
    F 990
    G 165
    H 120,960
    J 40,320

7. Solve the system of equations:
    $x + y = 14$
    $\frac{1}{3}x = \frac{1}{3}y + \frac{2}{3}$
    A (−8, 7)
    B (7, 7)
    C (8, 6)
    D no solution

8. Solve $-6c + 9 = 3 + 4c + 3c$.
    F $c = -\frac{2}{15}$
    G $c = \frac{6}{13}$
    H $c = -\frac{13}{6}$
    J $c = \frac{13}{6}$

9. A car rental company has two rental rates. The first rate is $45 per day plus $0.15 per mile. The second rate is $90 per day plus $0.07 per mile. If you plan to rent for one day, what is the least number of miles you would need to drive to pay less by taking the second rate?
    A 501 mi
    B 282 mi
    C 563 mi
    D 205 mi

10. Solve the equation for B: $V = \frac{1}{3}Bh$.
    F $B = \frac{V}{3h}$
    G $B = \frac{3h}{V}$
    H $B = \frac{3V}{h}$
    J $B = \frac{h}{3V}$

11. Find the mean speed of cars measured by radar to the nearest tenth. The individual speeds are 41.0, 43.5, 40.9, 44.3, 40.6, 43.4, 41.6, 40.7, 44.2, 41.0, 41.6, 44.3, 41.7, and 43.5 miles per hour.
    A 42.3 mph
    B 42.9 mph
    C 43.5 mph
    D 41.7 mph

12. Find $f(-1)$ for $f(x) = x(x^2 + 1)$.
    F 0
    G 2
    H −2
    J 3

13. Express $(-3)^7(-3)^8$ using positive exponents.
    A $(-3)^{15}$
    B $(3)^{15}$
    C $(9)^{56}$
    D $(9)^{15}$

14. Simplify $-14mn - (-7mn - 7mn)$.
    F $-7mn$
    G $-14mn$
    H $-28mn$
    J 0

15. Find the quotient $\frac{2}{7} \div \frac{1}{8}$.
    A $\frac{1}{28}$
    B $\frac{2}{7}$
    C $\frac{1}{5}$
    D $\frac{16}{7}$

16. Find the rule for the linear function. 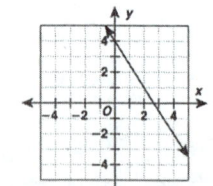
    F $f(x) = -2x + 4$
    G $f(x) = -\frac{3}{2}x + 4$
    H $f(x) = 3x + 4$
    J $f(x) = \frac{3}{2}x + 4$

17. What is 190% of 3490?
    A 663
    B 6631
    C 66,310
    D 663,100

18. Convert 10,800 seconds to days.
    F 10 days
    G $\frac{1}{4}$ day
    H $\frac{1}{8}$ day
    J $\frac{1}{16}$ day

19. Solve $-18 = -2(-5y)$.
    A $y = -\frac{9}{5}$
    B $y = -10$
    C $y = \frac{9}{5}$
    D $y = 10$

20. A tree 7 feet tall grows at the rate of 2 feet per year. How many years will it take for it to be 15 feet tall?
    F 14 yr
    G 9 yr
    H 11 yr
    J 4 yr

Cumulative Test
Chapter 12 Form C, continued

21. Find the missing value in the proportion $\frac{x}{72} = \frac{3}{12}$.
    A $x = 3$
    C $x = 18$
    B $x = 9$
    D $x = 216$

22. Solve the inequality $-10x < -100$.
    F $x < 10$
    H $x > -10$
    G $x > 10$
    J $-x > -10$

23. The half-life of caffeine in children is 3 hours. If a child consumes 40 mg of caffeine, how much is present after 4 hours? Round to the nearest tenth.
    A 15.9 mg
    C 5.0 mg
    B 15.0 mg
    D 2.5 mg

24. Which angles are congruent to ∠1? 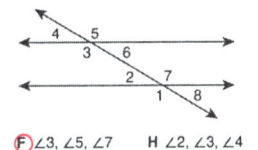
    F ∠3, ∠5, ∠7
    H ∠2, ∠3, ∠4
    G ∠4, ∠5, ∠8
    J ∠3, ∠6, ∠7

25. Find the perimeter of the trapezoid. 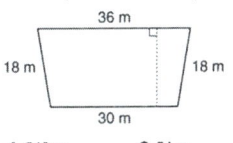
    A 540 m
    C 54 m
    B 102 m
    D 84 m

26. To the nearest tenth, find the missing length in the right triangle. 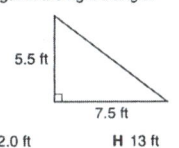
    F 2.0 ft
    H 13 ft
    G 4.0 ft
    J 9.3 ft

27. A machine can fill 4818 bags of candy in 0.6 hour. How many bags can be filled in an hour?
    A 2891 bags
    C 6883 bags
    B 4810 bags
    D 8030 bags

28. Find the y-intercept for $-2x + 3y = 6$.
    F $(-2, 3)$
    H $(3, 0)$
    G $(-2, 0)$
    J $(0, 2)$

29. Use the data in the table to find the inverse variation equation.

x	10	5	4	2
y	0.01	0.02	0.025	0.05

    A $y = \frac{1}{10x}$
    C $y = \frac{1}{2x}$
    B $y = 10x$
    D $y = 0.01x$

30. Which ordered pair is a solution to the equation $y = -7x - 33$?
    F $(0, -6)$
    H $(0, -40)$
    G $(0, -33)$
    J $(-6, 0)$

31. What is the point-slope form equation for the line with a slope of $-\frac{2}{3}$ that passes through $(5, 2)$?
    A $y - 5 = -\frac{2}{3}(x - 2)$
    B $y - 2 = -\frac{2}{3}(x - 5)$
    C $y + 5 = \frac{2}{3}(x - 2)$
    D $y + 2 = \frac{2}{3}(x - 5)$

32. How many sides does a regular heptagon have?
    F 5
    H 7
    G 6
    J 9

33. A bag contains four red marbles, three blue marbles, and five green marbles. What is the probability that a randomly selected marble is not blue?
    A $\frac{1}{6}$
    C $\frac{3}{4}$
    B $\frac{1}{4}$
    D $\frac{5}{6}$

34. $2\frac{1}{2}\%$ of what number is 68?
    F 27,200
    H 272
    G 2720
    J 1.7

35. Simplify $\left(\frac{5}{8}\right)^3$.
    A $\frac{5}{512}$
    C $1\frac{3}{5}$
    B $\frac{125}{512}$
    D $15\frac{5}{8}$

36. Solve $-\frac{16}{21}x = -\frac{12}{35}$.
    F $x = \frac{9}{20}$
    H $x = \frac{45}{16}$
    G $x = \frac{16}{45}$
    J $x = \frac{36}{5}$

37. Find the 5th term of the sequence defined by $a_n = 3\left(\frac{n}{n+1}\right)$.
    A $\frac{12}{5}$
    C $\frac{5}{6}$
    B $\frac{5}{2}$
    D $\frac{4}{5}$

38. Find the 25th term in the arithmetic sequence 1, 7, 13, 19, ...
    F 151
    H 145
    G 150
    J 144

39. A TV set costs $345 before tax. The sales tax is $17.25. What is the sales tax rate to the nearest tenth of a percent?
    A 4.8%
    C 5.0%
    B 5.3%
    D 0.5%

40. A rectangular carton measures 15 feet by 19 feet by 13 feet. What is the volume of the carton?
    F 247 ft³
    H 3705 ft³
    G 285 ft³
    J 4275 ft³

41. Find the area of the triangle. 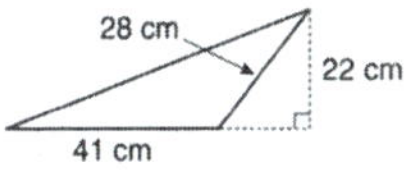
    A 1148 cm²
    C 451 cm²
    B 574 cm²
    D 902 cm²

42. Express 29,000,000 in scientific notation.
    F 2.9×10^6
    H 2.9×10^7
    G 2.9×10^8
    J 2.9×10^9

Cumulative Test
Chapter 13 Form A

Choose the best answer.

1. Evaluate the expression $5a - 3b$ for $a = 6$ and $b = 8$.
   A 18
   B 6

2. Solve $6x = 84$.
   A $x = 14$
   B $x = 78$

3. Simplify $(-2)^3 + (-3)^2$.
   A -17
   B 1

4. What is $x^5 \cdot x^2 \cdot x$ written as one power?
   A x
   C x^7
   B x^2
   D x^8

5. What is 3.62×10^{-2} in standard notation?
   F 0.00362
   H 0.362
   G 0.0362
   J 362

6. Simplify $\frac{18}{24}$.
   A $\frac{2}{3}$
   B $\frac{3}{4}$

7. Add $\frac{5}{12} + \frac{1}{12}$. Give the answer in simplest form.
   A $\frac{1}{2}$
   B $\frac{1}{4}$

8. Multiply $3.7(-2.4)$.
   A -8.88
   B 1.3

9. Solve $h + 8.2 = 4.9$.
   A $h = 13.1$
   B $h = 3.7$
   C $h = -3.3$
   D $h = -4.7$

10. Evaluate $\sqrt{9 + 27}$.
    A 6
    B 30

11. Mr. Jetter's class made a stem-and-leaf plot showing the daily high temperatures in °F. How many stems does the stem-and-leaf plot have?

    Daily High Temperatures (°F)

Stem	Leaves
5	4 4 8 9 9 9 9
6	1 1 3 4 7 7 8
7	0 2 4 5 5 6 6 6

    Key: 6|3 means 63

    A 25 stems
    B 22 stems
    C 13 stems
    D 3 stems

12. Marisa's quiz scores were 7, 9, 10, 9, 8, and 10. Find the first quartile.
    A 8
    B 9

13. What type of relationship is shown by the scatter plot? 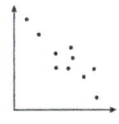
    A positive correlation
    B negative correlation

Use the figure for 14 and 15.

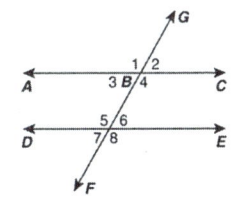

14. What type of angle is ∠ABC?
    A acute
    C right
    B obtuse
    D straight

15. Line AC is parallel to line DE. If $m\angle 3$ is 70°, what is $m\angle 6$?
    F 110°
    H 70°
    G 90°
    J 20°

16. Which quadrilateral could have its vertices at coordinates (4, 2), (7, 2), (7, 5), (4, 5)?
    A trapezoid
    B square

17. Which figure is symmetric?
    A
    B

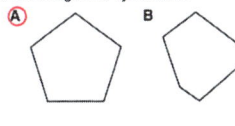

18. What is the length of the hypotenuse of the triangle, to the nearest tenth?
    A 4.9 in. 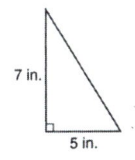
    B 8.6 in.

19. A round table has a diameter of 36 inches. What is the circumference of the top of the table? Use 3.14 for π.
    A 226.08 in.
    B 113.04 in.

20. A rectangular prism is 7 inches long, 5 inches wide, and 3 inches high. What is the volume of the prism?
    A 15 in³
    B 105 in³

21. Find the surface area of the pyramid. 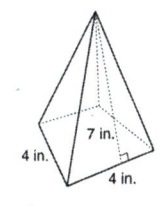
    A 72 in²
    B 128 in²

385

Holt Pre-Algebra

Cumulative Test — Form A, continued

22. Which pair of ratios is proportional?
A $\frac{6}{10}$ and $\frac{10}{15}$
B $\frac{12}{18}$ and $\frac{8}{12}$

23. A car is traveling at the rate of 60 miles per hour. How far does the car travel in 1 minute?
A 3,600 mi
B 1 mi

24. On a scale drawing of a room, 1 inch represents 2 feet. In the scale drawing, the room is 7 inches long. How long is the actual room?
A 3.5 ft
B 7 ft
C 14 ft
D 84 ft

25. In a survey of 75 people, 60 people said they own a car. What percent of the people surveyed own a car?
F 8%
G 60%
H 75%
J 80%

26. Brandy and 2 friends went out to dinner. They left a 15% tip on a dinner check of $45. How much did they leave as a tip?
A $6.75
B $20.25

27. What is the probability of the spinner landing on a number less than 4?
A $\frac{2}{3}$
B $\frac{1}{2}$

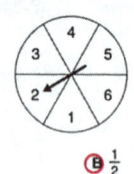

28. In how many ways can the letters of the word WORD be arranged?
A 24 ways
B 12 ways

29. A bag contains 5 green candles, 2 yellow candles, and 3 red candles. What are the odds of drawing a red candle from the bag?
A 3 to 10
B 3 to 7
C 10 to 3
D 7 to 3

30. Solve $6d - 24 = 3d - 3$.
A $d = 7$
B $d = -7$

31. Which ordered pair is a solution to this system of equations?
$x + y = 14$
$x - y = 2$
A (6, 8)
B (8, 6)

32. What is the slope of the line that passes through the points $(-3, 5)$ and $(1, 8)$?
A $\frac{4}{3}$
B $\frac{3}{4}$

33. What is the y-intercept of the line shown by the equation $-3x + y = 6$?
A -3
B 6

34. Multiply $(y - 5)(y + 2)$.
F $y^2 - 3y - 7$
G $y^2 - 7y - 10$
H $y^2 - 3y + 10$
J $y^2 - 3y - 10$

35. Which inequality is shown on the graph?
A $y > x + 1$
B $y < x + 1$

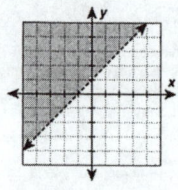

36. What are the next three terms in the sequence 1, 3, 6, 10, . . .?
A 15, 21, 28
B 11, 13, 16

37. What is the rule for the function table?

x	0	1	2	3
y	5	6	7	8

A $y = 5x$
B $y = x + 5$

38. Find $f(2)$ for $f(x) = x^2 - 3x + 7$.
A $f(2) = 5$
B $f(2) = -21$

Cumulative Test — Form B

Choose the best answer.

1. Evaluate the expression $x(3 + y) - 4$ for $x = 7$ and $y = 1$.
A 70
B 24
C 18
D 0

2. Solve $\frac{a}{13} = 15$.
F $a = 205$
G $a = 195$
H $a = 185$
J $a = 28$

3. Simplify $(1 + 2 \cdot 4)^2$.
A 18
B 33
C 81
D 144

4. What is $7^0 \cdot 7^4 \cdot 7^2 \cdot 7^3$ written as one power?
F 7^0
G 7^4
H 7^9
J 7^{24}

5. What is 5.09×10^{-4} in standard notation?
A 50,900
B 5,090
C 0.00509
D 0.000509

6. Which decimal is equivalent to $\frac{9}{37}$?
F 0.937
G $0.\overline{243}$
H 0.243
J $0.\overline{234}$

7. Add $\frac{13}{15} + \frac{8}{15}$. Give the answer in simplest form.
A $\frac{7}{10}$
B $1\frac{2}{5}$
C $1\frac{2}{3}$
D $2\frac{1}{15}$

8. Evaluate $\left(2\frac{3}{4}\right)p$ for $p = \frac{2}{5}$.
F $1\frac{1}{10}$
G $2\frac{3}{10}$
H $3\frac{3}{20}$
J $6\frac{7}{8}$

9. Solve $2.3n = 0.92$.
A $n = 0.4$
B $n = 2.5$
C $n = 4$
D $n = 25$

10. Evaluate $\sqrt{81} - \sqrt{25}$.
F $\sqrt{56}$
G 16
H 14
J 4

11. Mr. Jetter's class made a stem-and-leaf plot showing the daily high temperatures in °F. How many leaves does the stem-and-leaf plot have?

Daily High Temperatures (°F)

Stem	Leaves
5	4 4 8 9 9 9 9
6	1 1 3 4 7 7 8
7	0 2 4 5 5 6 6 6

Key: 6|3 means 63

A 3 leaves
B 13 leaves
C 22 leaves
D 25 leaves

12. Armando's quiz scores were 7, 9, 10, 9, 8, 10, 9, and 8. Find the third quartile.
F 8
G 8.5
H 9
J 9.5

Cumulative Test — Form B, continued

13. What type of relationship is shown by the scatter plot?
A positive correlation
B negative correlation
C constant correlation
D no correlation

Use the figure for 14 and 15.

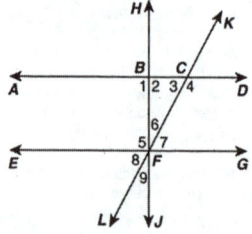

14. What type of angle is $\angle GFC$?
F acute
G obtuse
H right
J straight

15. Line AD is parallel to line EG. If $m\angle 3$ is 70°, what is $m\angle 7$?
A 20°
B 70°
C 90°
D 110°

16. Which quadrilateral could have its vertices at coordinates (3, 2), (7, 2), (1, 5), (5, 5)?
F rectangle
G square
H parallelogram
J trapezoid

17. Which of the following figures has line symmetry?
A
B
C
D

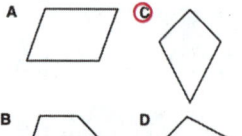

18. What is the length of the missing side of the triangle, to the nearest tenth?
F 5.0 ft
G 8.6 ft
H 8.7 ft
J 11.2 ft

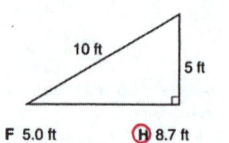

19. A round table has a diameter of 36 inches. What is the area of the top of the table? Use 3.14 for π.
A 1,017.4 in^2
B 508.7 in^2
C 226.1 in^2
D 113.0 in^2

Cumulative Test
Chapter 13 Form B, continued

20. A rectangular prism is 2 inches long, 4.5 inches wide, and 8 inches long. What is the volume of the prism?
    F 9 in^3
    G 14.5 in^3
    H 36 in^3
    J 72 in^3 ✓

21. Find the surface area of the pyramid. 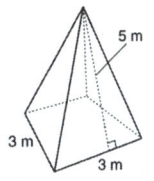
    A 15 m^2
    B 31.5 m^2
    C 39 m^2 ✓
    D 69 m^2

22. Which pair of ratios is proportional?
    F $\frac{9}{15}$ and $\frac{6}{9}$
    G $\frac{8}{12}$ and $\frac{6}{8}$
    H $\frac{15}{20}$ and $\frac{10}{16}$
    J $\frac{16}{24}$ and $\frac{10}{15}$ ✓

23. A painter can paint 1 foot of molding in 1 minute. How many inches of molding can he paint in one hour?
    A 720 in. ✓
    B 144 in.
    C 60 in.
    D 12 in.

24. The distance between two cities on a map is 4 centimeters. The scale on the map says 1 cm = 45 km. What is the actual distance between the cities?
    F 11.25 km
    G 90 km
    H 162 km
    J 180 km ✓

25. In a survey of 65 students, 80% said they make their bed every morning. How many of the students surveyed make their bed every morning?
    A 80 students
    B 65 students
    C 52 students ✓
    D 15 students

26. Consuelo bought a sweater on sale at 15% off the regular price. If the regular price was $45.00, how much did she pay for the sweater?
    F $6.75
    G $38.25 ✓
    H $42.00
    J $51.75

27. What is the probability of the spinner landing on an odd number less than 6? 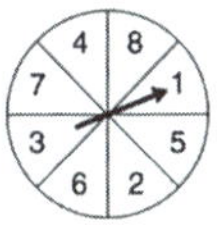
    A $\frac{5}{6}$
    B $\frac{5}{8}$
    C $\frac{1}{2}$
    D $\frac{3}{8}$ ✓

28. A pizza store offers a choice of 8 toppings. How many different pizzas are possible if you choose a pizza with 3 toppings?
    F 24
    G 48
    H 56 ✓
    J 336

29. A 1–6 number cube is rolled. What are the odds that the outcome is a multiple of 3?
    A 1 to 1
    B 1 to 2 ✓
    C 2 to 1
    D 1 to 3

30. Solve $7(b + 4) = 5b - 2$.
    F $b = 15$
    G $b = 13$
    H $b = -15$ ✓
    J no solution

31. Which ordered pair is a solution to this system of equations?
    $2x + y = 14$
    $3x - 2y = 7$
    A (4, 6)
    B (5, 8)
    C (5, 4) ✓
    D (4, 5)

32. What is the slope of the line that passes through the points (−2, −5) and (1, 4)?
    F $\frac{1}{3}$
    G 3 ✓
    H $-\frac{1}{3}$
    J −3

33. What is the y-intercept of the line represented by $-3x + 2y = 6$?
    A 6
    B 3 ✓
    C 2
    D $\frac{3}{2}$

34. Multiply $(3y - 5)(y - 1)$.
    F $3y^2 - 8y + 5$ ✓
    G $3y^2 - 2y + 5$
    H $3y^2 - 8y - 6$
    J $3y^2 - 8y - 5$

35. Which inequality is shown on the graph? 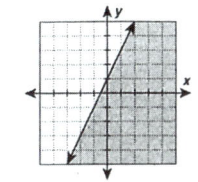
    A $y > 2x + 1$
    B $y < 2x + 1$
    C $y \geq 2x + 1$
    D $y \leq 2x + 1$ ✓

36. What are the next three terms in the sequence 5, 8, 12, 17, ... ?
    F 23, 30, 38 ✓
    G 18, 19, 20
    H 23, 24, 25
    J 22, 27, 32

37. Which is the rule for the function table?

x	−1	0	1	2
y	4	7	10	13

    A $y = x + 5$
    B $y = 2x + 8$
    C $y = x + 11$
    D $y = 3x + 7$ ✓

38. Find $f(-2)$ for $f(x) = 3x^2 - x + 7$.
    F $f(-2) = 45$
    G $f(-2) = 21$ ✓
    H $f(-2) = 15$
    J $f(-2) = -3$

Cumulative Test
Chapter 13 Form C

Choose the best answer.

1. Evaluate the expression $8ab - 3ac$ for $a = 2$, $b = 6$, and $c = 3$.
    A 15
    B 78 ✓
    C 139
    D 503

2. Solve $26n = 117$.
    F $n = 4.05$
    G $n = 4.13$
    H $n = 4.5$ ✓
    J $n = 91$

3. Simplify $(9 + 5)^3 - (9 - 5)^3$.
    A 0
    B 30
    C 1,000
    D 2,680 ✓

4. What is $\frac{(8^6 \cdot 8^4)}{8^2}$ written as one power?
    F 8^8 ✓
    G 8^5
    H 8^2
    J 8^1

5. What is 9.812×10^5 in standard notation?
    A 0.00009812
    B 9,812
    C 98,120
    D 981,200 ✓

6. Which number does not have the same value as the others?
    F $\frac{75}{37}$
    G $2.0\overline{27}$
    H $2\frac{1}{37}$
    J $2.1\overline{37}$ ✓

7. Evaluate $1 - y - \frac{17}{20}$ for $y = \frac{7}{20}$.
    A $-\frac{1}{5}$ ✓
    B $\frac{1}{10}$
    C $\frac{1}{5}$
    D $1\frac{1}{2}$

8. Multiply $\left(\frac{4}{5}\right)\left(1\frac{2}{3}\right)\left(\frac{15}{16}\right)$. Give the answer in simplest form.
    F $\frac{1}{2}$
    G $1\frac{1}{4}$ ✓
    H $1\frac{1}{2}$
    J $3\frac{1}{2}$

9. Solve $b - 0.39 = 1.61 - 2.04$.
    A $b = 0.82$
    B $b = 0.04$
    C $b = -0.04$ ✓
    D $b = -0.82$

10. Evaluate $\sqrt{121} - \sqrt{36 + 64}$.
    F −3
    G 1 ✓
    H 3
    J 13

11. Mrs. Wexler's class made a back-to-back stem-and-leaf plot showing the daily high and low temperatures in °F. How many leaves does the stem-and-leaf plot have?

    Daily High and Low Temperatures (°F)

Leaves	Stem	Leaves
1 7 8 9	4	
0 0 1 2 4 4 6	5	4 4 8 9 9 9 9
9	6	1 1 3 4 7 7 8
0 1 3 5 5 5 8	7	0 2 4 5 5 6 6 6
1 2 3		

    Key: 6|3 means 63

    A 4 leaves
    B 31 leaves
    C 44 leaves ✓
    D 48 leaves

12. Hope's math test scores are 85, 95, 100, 90, 85, 100, 95, and 80. Find the third quartile.
    F 85
    G 95
    H 97.5 ✓
    J 100

13. What type of relationship is shown by the scatter plot? 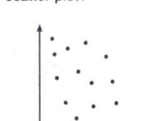
    A positive correlation
    B negative correlation
    C constant correlation
    D no correlation ✓

Use the figure for 14 and 15.

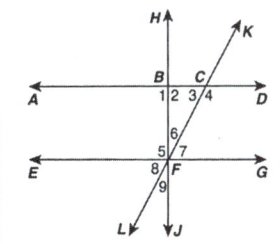

14. What type of angle is $\angle KFE$?
    F obtuse ✓
    G straight
    H right
    J acute

15. Line AD is parallel to line EG. If $m\angle 4$ is 110°, what is $m\angle 8$?
    A 20°
    B 70° ✓
    C 90°
    D 110°

16. Which quadrilateral could have its vertices at coordinates (4, 2), (9, 2), (2, 5), (9, 5)?
    F rectangle
    G square
    H parallelogram
    J trapezoid ✓

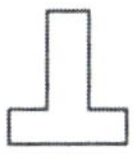

17. Which of the following figures is not symmetric?
    A M
    B G ✓
    C T
    D D

18. What is the length of the missing side of the triangle, to the nearest tenth?
    F 7.0 ft 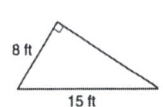
    G 12.6 ft
    H 12.7 ft ✓
    J 17.0 ft

19. A round table has a diameter of 3 feet. What is the area of the top of the table? Use 3.14 for π.
    A 7.1 ft² ✓
    B 9.4 ft²
    C 18.8 ft²
    D 28.3 ft²

Cumulative Test
Chapter 13 Test C, continued

20. A fish tank is 20.5 inches long, 12 inches wide, and 14 inches high. What is the volume of the fish tank?
    F 279 in³
    G 287 in³
    H 1,066 in³
    J 3,444 in³

21. Find the surface area of the cone to the nearest tenth. Use 3.14 for π.

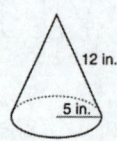

    12 in.
    5 in.

    A 188.4 in²
    B 219.8 in²
    (C) 266.9 in²
    D 314 in²

22. Which pair of ratios is not proportional?
    (F) $\frac{15}{24}$ and $\frac{21}{32}$
    G $\frac{30}{45}$ and $\frac{14}{21}$
    H $\frac{24}{40}$ and $\frac{9}{15}$
    J $\frac{16}{20}$ and $\frac{20}{25}$

23. A runner can jog 6 miles in 1 hour. How many feet does she jog in 1 minute?
    A 52.8 ft
    (B) 528 ft
    C 5,280 ft
    D 52,800 ft

24. The distance between two cities on a map is 1.25 inches. The scale on the map is 1 in. = 60 mi. What is the actual distance between the cities?
    F 48 mi
    G 61.25 mi
    H 65 mi
    (J) 75 mi

25. In a survey of 225 students, 175 of them said they drink a glass of milk every morning. To the nearest tenth of a percent, what percent of the students surveyed drink a glass of milk every morning?
    A 17.5%
    B 22.5%
    C 77.7%
    (D) 77.8%

26. The rate for defective light bulbs at a light bulb plant is 1.5%. If 3,000 light bulbs are produced per hour, how many defective light bulbs are produced in 24 hours?
    F 108 bulbs
    G 450 bulbs
    (H) 1,080 bulbs
    J 1,800 bulbs

27. What is the probability of the spinner landing on a prime number?

    A $\frac{1}{4}$
    (B) $\frac{1}{2}$
    C $\frac{3}{8}$
    D $\frac{5}{8}$

28. A pizza store offers a choice of 2 crusts and 8 toppings. How many different pizzas are possible if you choose a pizza with 3 toppings?
    F 16 pizzas
    G 48 pizzas
    H 56 pizzas
    (J) 112 pizzas

29. Two 1–6 number cubes are rolled. What are the odds that the sum of the 2 cubes is 7?
    (A) 1 to 5
    B 1 to 6
    C 5 to 1
    D 6 to 1

30. Solve $2.2p + 5.9 = 8.3 + 1.6p$.
    (F) $p = 4$
    G $p = 2$
    H $p = -4$
    J no solution

31. Which ordered pair is a solution to this system of equations?
    $y = 3x + 9$
    $2x + 5y = 11$
    (A) $(-2, 3)$
    B $(3, 0)$
    C $(3, 1)$
    D $(-2, 15)$

32. Which set of points represents a line that does not have a slope of -1?
    F $(3, -1), (-7, 9)$
    G $(-2, 2), (5, -5)$
    (H) $(-1, 0), (3, 4)$
    J $(-6, 3), (0, -3)$

33. What is the y-intercept of the line shown by the equation $4(x - 2y) = 3$?
    A $-\frac{1}{2}$
    (B) $-\frac{3}{8}$
    C $\frac{3}{8}$
    D $\frac{1}{2}$

34. Multiply $(3a - 5b)(3a + 5b)$.
    F $9a^2$
    G $9a^2 - 30ab + 25b^2$
    (H) $9a^2 - 25b^2$
    J $9a^2 + 25b^2$

35. Which inequality is shown on the graph?

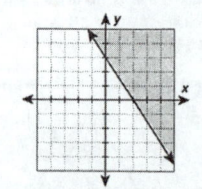

    A $y \leq -\left(\frac{3}{2}\right)x + 3$
    (B) $y \geq -\left(\frac{3}{2}\right)x + 3$
    C $y < -\left(\frac{3}{2}\right)x + 3$
    D $y > -\left(\frac{3}{2}\right)x + 3$

36. What are the next 3 terms in the sequence 13, 12, 10, 7, 3, …?
    (F) $-2, -8, -15$
    G 8, 14, 21
    H 2, 1, 0
    J 0, -4, -9

37. What is the rule for the function table?

x	-5	-1	4	8
y	4	-4	-14	-22

    A $y = -3x + 2$
    (B) $y = -2x - 6$
    C $y = -5x + 1$
    D $y = -3x - 11$

38. Find $f(8)$ for $f(x) = \left(\frac{1}{2}\right)x^2 - \left(\frac{1}{4}\right)x + 8$.
    F $f(8) = 10$
    G $f(8) = 22$
    H $f(8) = 35$
    (J) $f(8) = 38$

Cumulative Test
Chapter 14 Form A

Choose the best answer.

1. Which is the algebraic expression for the word phrase *6 more than twice a number* n?
   A $6n + 2$
   (B) $2n + 6$

2. Find the solution for $x - 7 = 24$.
   (A) $x = 31$
   B $x = 17$

3. Evaluate $9 - k$ for $k = -20$.
   A -29
   B -11
   C 11
   (D) 29

4. Solve $g + 18 \leq -45$.
   F $g \leq -27$
   G $g \leq 27$
   (H) $g \leq -63$
   J $g \leq 63$

5. Evaluate 2^4.
   A 8
   (B) 16

6. What is 0.45 expressed as a fraction in simplest form?
   (A) $\frac{9}{20}$
   B $\frac{9}{25}$

7. Add $4.32 + (-7.19)$.
   A -3.13
   (B) -2.87

8. Divide $\frac{9}{10} \div \frac{3}{5}$.
   (A) $1\frac{1}{2}$
   B $\frac{27}{50}$

9. Solve $d + \frac{1}{3} < \frac{5}{6}$.
   A $d > \frac{1}{2}$
   B $d > 1\frac{1}{6}$
   (C) $d < \frac{1}{2}$
   D $d < 1\frac{1}{6}$

10. What is $\sqrt{44}$ rounded to the nearest tenth?
    (A) 6.6
    B 6.7

For 11–12, use the stem-and-leaf plot, which shows the heights in inches of the boys on a middle school basketball team.

Basketball Team Heights

Stems	Leaves
6	4 5 7 8 9
7	0 1 1 3

11. What is the mode of the heights?
    A 69
    B 69.5
    C 70
    (D) 71

12. What is the first quartile?
    (A) 66
    B 67

Cumulative Test
Chapter 14 Form A, continued

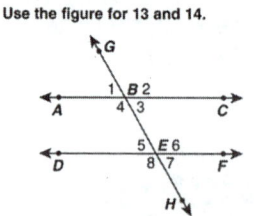

Use the figure for 13 and 14.

13. Which angle is obtuse?
    (A) $\angle ABE$
    B $\angle ABG$

14. Line AC is parallel to line DF. If $m\angle 5$ is 50°, what is $m\angle 3$?
    A 130°
    B 90°
    (C) 50°
    D 40°

15. What is the sum of the measures of the angles of a pentagon?
    F 900°
    (G) 540°
    H 360°
    J 180°

16. Identify the type of translation.

    (A) translation
    B rotation
    C reflection
    D tessellation

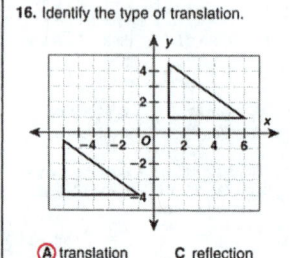

17. What is the area of a figure with vertices (3, 1), (3, 5), and (5, 5)?
    A 8 square units
    (B) 4 square units

18. A round table has a radius of 18 inches. What is the area of the top of the table to the nearest tenth? Use 3.14 for π.
    A 113.0 in²
    (B) 1,017.4 in²

19. A rectangular pyramid has a base with an area of 36 square centimeters and a height of 7 centimeters. What is the volume of the pyramid?
    A 252 cm³
    (B) 84 cm³

20. Which is the better buy?
    (A) a 28-oz bottle of ketchup for $1.29
    B a 40-oz bottle of ketchup for $1.89

21. Solve the proportion $\frac{5}{m} = \frac{15}{24}$.
    A $m = 3$
    (B) $m = 8$

22. Which scale reduces the size of the actual object?
    A 1 ft to 6 in.
    (B) 1 in. to 6 ft

23. Michael bought a CD on sale for $12. The sale price was 75% of the regular price. What is the regular price of the CD?
    A $9.00
    B $12.25
    C $12.75
    (D) $16.00

Cumulative Test
Chapter 14 Form A, continued

24. Jamie borrowed $3,000 from a bank for 2 years at a simple interest rate of 6% per year. How much interest will she pay on the loan?
    A) $360
    B $3,360

25. A baseball team won 14 of its last 21 games. What is the probability that the team will win its next game?
    A $\frac{1}{2}$
    B) $\frac{2}{3}$

26. How many combinations of 3 different flowers can you choose from 5 different flowers?
    A) 10 different combinations
    B 15 different combinations

27. A bag contains 5 green candles, 2 yellow candles, and 3 red candles. What is the probability of picking a red candle and then a green candle from the bag without replacement?
    A $\frac{3}{20}$
    B) $\frac{1}{6}$

28. Which graph represents the solution to the inequality, $4x - 9 > 7$?
    A) 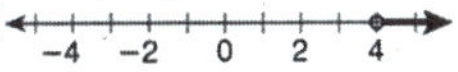 (number line)
    B (number line)

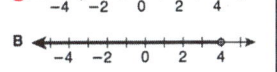

29. Solve $P = 2a + b$ for b.
    A) $b = P - 2a$
    B $b = 2a - P$

30. What is the equation of the line that passes through the points (3, 5) and (1, 1)?
    A $y = -2x + 11$
    B) $y = 2x - 1$

31. Which equation gives the direct variation if $y = 8$ when $x = 2$?
    A) $y = 4x$
    B $y = \left(\frac{1}{4}\right)x$

32. Which inequality is shown on the graph? 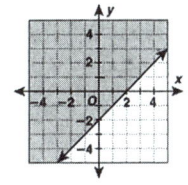
    A) $y \geq x - 2$
    B $y \leq x - 2$

33. What is the ninth term in the sequence, 3, 6, 9, ...?
    A) 27
    B 30

34. What is the rule for the function table?

x	-1	0	1	2
y	3	5	7	9

    A $y = 2x - 1$
    B) $y = 2x + 5$

35. Find $f(3)$ for $f(x) = 4x^2 - 1$.
    A) $f(3) = 35$
    B $f(3) = 143$

36. What is the degree of the polynomial $x + 3x^2 - 9x$?
    A) 2
    B 3

37. Add $(7m + 3) + (2m^2 - 3m - 5)$.
    A $6m^2 + 2$
    C) $2m^2 + 4m - 2$
    B $2m^2 + 7m - 5$
    D $4m^2$

38. Multiply $(x^4y^2)(2x^2y^3)$.
    F $2(xy)^{48}$
    H $2x^8y^6$
    G $(2xy)^{48}$
    J) $2x^6y^5$

39. Which set is a finite set?
    A {whole numbers greater than 50}
    B) {whole numbers less than 50}

40. Which statement is true about the conjunction P and Q?
    A) P and Q must both be true if the conjunction is true.
    B P and Q must both be false if the conjunction is false.

41. What is the degree of vertex B? 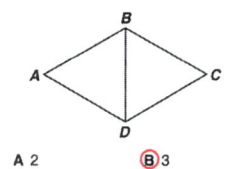
    A 2
    B) 3

Cumulative Test
Chapter 14 Form B

Choose the best answer.

1. Which is the algebraic expression for the word phrase *5 less than the product of 8 and p*?
    A $5 - 8p$
    C $5p - 8$
    B) $8p - 5$
    D $p + 8 - 5$

2. Find the solution for $x - 26 = 78$.
    F $x = 3$
    H) $x = 104$
    G $x = 52$
    J $x = 2,028$

3. Evaluate $w - (-17)$ for $w = 8$.
    A -25
    C 9
    B -9
    D) 25

4. Solve $29 \geq b - 17$.
    F) $b \leq 46$
    H $b \leq 12$
    G $b \geq 46$
    J $b \geq 12$

5. Evaluate $(-8)^2$.
    A -64
    C 16
    B -16
    D) 64

6. What is 0.125 as a fraction in simplest form?
    F $\frac{1}{125}$
    H) $\frac{1}{8}$
    G $\frac{1}{80}$
    J $1\frac{1}{4}$

7. Add $6.85 + 3.3$.
    A) 10.15
    C 8.88
    B 9.88
    D 7.18

8. Divide $16 \div 2\frac{2}{3}$.
    F) 6
    H $13\frac{1}{3}$
    G 12
    J $42\frac{2}{3}$

9. Solve $3n \leq -\frac{7}{9}$.
    A $n \leq -3\frac{6}{7}$
    C $n \geq -\frac{7}{27}$
    B) $n \leq -\frac{7}{27}$
    D $n \geq -3\frac{6}{7}$

10. What is $\sqrt{176}$ rounded to the nearest tenth?
    F 13
    H) 13.3
    G 13.2
    J 14

For 11–12, use the stem-and-leaf plot, which shows the points scored by a middle-school football team during the season.

Football Points Scored

Stems	Leaves
0	3 7 7 7
1	2 3 4 4 9
2	1 4

11. What is the mode of the scores?
    A) 7
    C 12.8
    B 11
    D 13

12. What is the third quartile?
    F 14
    H 21
    G) 19
    J 24

Use the figure for 13 and 14.

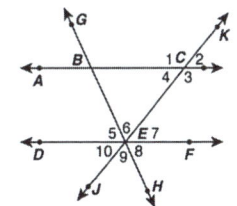

13. Which angle is supplementary to $\angle ABE$?
    A $\angle GBC$
    C) $\angle EBC$
    B $\angle BCE$
    D $\angle DEJ$

14. Line AC is parallel to line DF. If $m\angle DEC$ is 150°, what is $m\angle 1$?
    F 210°
    H) 150°
    G 180°
    J 30°

15. What is the sum of the measures of the angles of an octagon?
    A 180°
    C 1,440°
    B) 1,080°
    D 1,800°

16. Identify the type of translation. 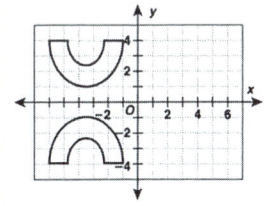
    F translation
    H) reflection
    G dilation
    J tessellation

17. What is the area of a figure with vertices (1, 1), (8, 1), and (5, 5)?
    A 28 square units
    B 21 square units
    C) 14 square units
    D 11 square units

18. A tree trunk has a radius of 11 inches. What is the circumference of the tree trunk to the nearest tenth? Use 3.14 for π.
    F 34.5 in.
    H 138.2 in.
    G) 69.1 in.
    J 379.9 in.

19. The base of a cone has a radius of 6 centimeters. The cone is 7 centimeters tall. What is the volume of the cone to the nearest tenth? Use 3.14 for π.
    A 260 cm³
    C) 263.8 cm³
    B 263.7 cm³
    D 264.0 cm³

CHAPTER 14 Cumulative Test
Form B, continued

20. Which is the best buy?
- F an 8-oz can of peaches for $0.59
- G a 12-oz can of peaches for $0.89
- H a 20-oz can of peaches for $1.49
- (J) a 32-oz can of peaches for $2.29

21. Solve the proportion $\frac{5}{8} = \frac{a}{20}$.
- A $a = 2.5$
- C $a = 15$
- (B) $a = 12.5$
- D $a = 17$

22. Which scale enlarges the size of the actual object?
- F 1 m to 100 cm
- H 1 cm to 10 m
- G 10 mm to 1 cm
- (J) 10 m to 1 cm

23. Alyssa is selling bags of peanuts to raise money for a fundraiser. So far, she has sold 36 bags of peanuts. This is 45% of her goal. What is Alyssa's goal?
- A 16 bags of peanuts
- B 52 bags of peanuts
- (C) 80 bags of peanuts
- D 81 bags of peanuts

24. Noah borrowed $7,500 for 2 years at a simple interest rate of 6.5% per year. How much interest will he pay on the loan?
- F $97.50
- (H) $975.00
- G $487.50
- J $8,475.00

25. Mario tossed a coin 40 times, and it landed heads up 24 times. What is the experimental probability that the coin will land heads up on the next toss?
- A $\frac{2}{3}$
- C $\frac{1}{2}$
- (B) $\frac{3}{5}$
- D $\frac{2}{5}$

26. How many different 3-topping pizzas are possible if there are 10 toppings from which to choose?
- F 30 pizzas
- (H) 120 pizzas
- G 60 pizzas
- J 720 pizzas

27. A bag contains 5 green marbles, 4 yellow marbles, and 3 red marbles. What is the probability of picking a red marble and then a yellow marble from the bag without replacement?
- (A) $\frac{1}{11}$
- C $\frac{7}{12}$
- B $\frac{1}{12}$
- D $\frac{20}{33}$

28. Which graph represents the solution for the inequality $3x + 9 < 6$?
- (F)
- G
- H
- J

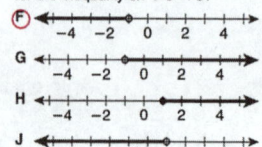

29. Solve $y = 2x + 5$ for x.
- A $x = 5 - 2y$
- (C) $x = \frac{(y-5)}{2}$
- B $x = 2y - 5$
- D $x = 2(y - 5)$

30. What is the equation of the line that passes through the points $(-2, 5)$ and $(1, -1)$?
- F $y = 3x + 11$
- (H) $y = -2x + 1$
- G $y = 2x + 9$
- J $y = 3x - 4$

31. Which equation gives the direct variation if $y = 8$ when $x = -2$?
- (A) $y = -4x$
- C $y = 4x$
- B $y = \left(\frac{1}{4}\right)x$
- D $y = -\left(\frac{1}{4}\right)x$

32. Which inequality is shown on the graph?
- F $y > 2x + 1$
- H $y \geq 2x + 1$
- G $y < 2x + 1$
- (J) $y \leq 2x + 1$

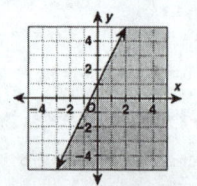

33. What is the tenth term in the sequence 5, 8, 11, 14, ...?
- A 29
- C 35
- (B) 32
- D 38

34. What is the rule for the function table?

x	-1	0	1	2
y	7	5	3	1

- F $y = -7x$
- H $y = x + 5$
- (G) $y = -2x + 5$
- J $y = -3x + 7$

35. Find $f(-2)$ for $f(x) = 2x^2 + 4x + 1$.
- A $f(-2) = -15$
- C $f(-2) = 17$
- (B) $f(-2) = 1$
- D $f(-2) = 25$

36. What is the degree of the polynomial $2x + 3x^2 - 4x^5$?
- F 2
- H 4
- G 3
- (J) 5

37. Add $(3m^2 - 5m) + (3m - 5m^2)$.
- A $-4m^2$
- C $-2m^2 - 2m$
- B 0
- D $-4m$

38. Multiply $3z(2z^4 - 4z^3)$.
- (F) $6z^5 - 12z^4$
- H $6z^4 - 12z^3$
- G $5z^5 - z^4$
- J $5z^4 - z^3$

39. Which set is an infinite set?
- (A) {integers less than 5}
- B {positive integers less than 5}
- C {integers between -5 and 5}
- D {integers with absolute values less than 5}

40. Which statement is **not** true about the conjunction P and Q?
- F Both P and Q must be true if the conjunction is true.
- (G) Only P must be true if the conjunction is true.
- H If both P and Q are false, the conjunction is false.
- J The conjunction is false if P is false.

41. What is the degree of vertex A?
- A 3
- (C) 5
- B 4
- D 6

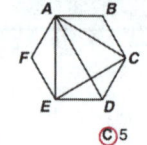

CHAPTER 14 Cumulative Test
Form C

Choose the best answer.

1. Which is the algebraic expression for the word phrase *half the difference of n and 15*?
- A $\left(\frac{1}{2}\right)(n + 15)$
- C $\left(\frac{1}{2}\right)(15 - n)$
- (B) $\left(\frac{1}{2}\right)(n - 15)$
- D $\left(\frac{1}{2}\right)n - 15$

2. Find the solution for $3.9 + y = 11.7$.
- F $y = 3$
- H $y = 8.2$
- (G) $y = 7.8$
- J $y = 15.6$

3. Evaluate $t - (-17) - 9$ for $t = -24$.
- A -50
- C -2
- (B) -16
- D 2

4. Solve $3k + 11 > -55$.
- F $k < -15$
- (H) $k > -22$
- G $k < 22$
- J $k > 55$

5. Evaluate $-(-9)^3$.
- (A) 729
- C -27
- B 27
- D -729

6. What is -2.325 expressed as a fraction in simplest form?
- F $-2\frac{13}{400}$
- H $-2\frac{3}{8}$
- G $-2\frac{3}{25}$
- (J) $-2\frac{13}{40}$

7. Add $-8.95 + (-4.093) + 6.43$.
- A 1.573
- (C) -6.613
- B 11.287
- D -19.473

8. Divide $-12\frac{2}{3} \div 9\frac{1}{2}$.
- F $-\frac{3}{4}$
- (H) $-1\frac{1}{3}$
- G $-1\frac{1}{6}$
- J $-120\frac{1}{3}$

9. Solve $5n \leq \frac{1}{4} - \frac{7}{8}$.
- A $n \leq -3\frac{1}{8}$
- (C) $n \leq -\frac{1}{8}$
- B $n \leq -\frac{9}{40}$
- D $n \leq 3\frac{1}{8}$

10. What is $\sqrt{276.8}$ rounded to the nearest tenth?
- F 16.0
- H 16.7
- (G) 16.6
- J 17.0

For 11 and 12, use the stem-and-leaf plot, which shows the daily high temperatures (in °F) for 10 days.

Daily High Temperatures (in °F)

Stems	Leaves
8	8 9
9	2 3 4 6 9
10	0 1 4

11. What is the mode of the temperatures?
- A 95
- C 104
- B 95.6
- (D) no mode

12. What is the third quartile?
- F 92
- (H) 100
- G 95
- J 101.5

CHAPTER 14 Cumulative Test
Form C, continued

Use the figure for 13 and 14.

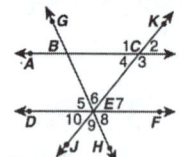

13. Line AC is parallel to line DF. Which angle is **not** supplementary to $\angle DEB$?
- A $\angle GEF$
- C $\angle DEH$
- B $\angle ABE$
- (D) $\angle JEF$

14. Line AC is parallel to line DF. If $m\angle CEF = 45°$, what is $m\angle 1$?
- F 325°
- (H) 135°
- G 180°
- J 45°

15. What is the measure of each angle of a regular decagon?
- A 100°
- C 180°
- (B) 144°
- D 1,440°

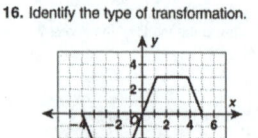

16. Identify the type of transformation.
- F translation
- G 90° clockwise rotation
- H reflection
- (J) 180° rotation

17. What is the area of a figure with vertices $(-5, -3)$, $(2, -3)$, $(-3, 1)$, and $(0, 1)$?
- A 10 square units
- B 14 square units
- (C) 20 square units
- D 40 square units

18. The wheel of a bicycle has a diameter of 24 inches. About how far does the bicycle go if the wheel revolves 20 times? Use 3.14 for π.
- (F) 126 ft
- H 480 ft
- G 252 ft
- J 1,507 ft

19. The base of a rectangular pyramid has a length of 6.3 meters and a width of 5.9 meters. The height of the pyramid is 7.2 meters. What is the volume of the pyramid to the nearest tenth?
- (A) 89.2 m³
- C 267.6 m³
- B 90 m³
- D 270.0 m³

20. Which is the best buy?
- F a 9.5-oz box of crackers for $1.59
- (G) a 14.5-oz box of crackers for $2.39
- H a 22.5-oz box of crackers for $3.79
- J a 33.5-oz box of crackers for $5.59

21. Solve the proportion $\frac{2.5}{1.8} = \frac{15}{m}$.
- A $m = 10.5$
- C $m = 27$
- (B) $m = 10.8$
- D $m = 67.5$

Cumulative Test
Test C, continued

22. Which scale does **not** preserve the size of the actual object?
    F 18 ft to 6 yd
    H 24 in. to 2 ft
    G 6 yd to 2 ft ✓
    J 72 in. to 2 yd

23. Maggie has saved $84 for a new racing bicycle. This is 37.5% of the price. How much does the bicycle cost?
    A $31.50
    C $224 ✓
    B $121.50
    D $252

24. Tomás took out a new car loan with an annual simple interest rate of 2.9% for 36 months. At the end of the loan he had paid $1,305 in interest. What was the principal of the loan?
    F $13,050
    H $15,000 ✓
    G $13,695
    J $18,305

25. Marcia tossed 2 number cubes 75 times and found that 30 times the sum was less than or equal to 7. What is the probability that the sum will be less than or equal to 7 on the next toss?
    A $\frac{1}{3}$
    C $\frac{1}{2}$
    B $\frac{2}{5}$ ✓
    D $\frac{3}{5}$

26. How many 5-player starting squads can be formed from a basketball team of 15 players?
    F 75 squads
    H 2,250 squads
    G 225 squads
    J 3,003 squads ✓

27. A bag contains 5 green marbles, 3 yellow marbles, and 2 red marbles. What is the probability of picking three green marbles from the bag without replacement?
    A $\frac{1}{8}$
    C $\frac{3}{50}$
    B $\frac{1}{12}$ ✓
    D $\frac{3}{100}$

28. Which graph represents the solution of the inequality $3x + 9 \geq 1 - x$?
    F ✓
    G
    H
    J

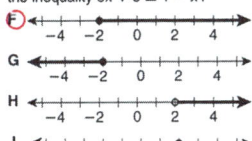

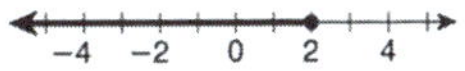

29. Solve $a^2 + b^2 = c^2$ for b.
    A $b = \sqrt{c^2 - a^2}$ ✓
    C $b = c^2 - a^2$
    B $b = a^2 - c^2$
    D $b = \sqrt{a^2 - c^2}$

30. What is the equation of the line that passes through the points (3, −5) and (−5, 11)?
    F $y = 3x + 11$
    H $y = 3x - 4$
    G $y = 2x + 9$
    J $y = -2x + 1$ ✓

31. What equation gives the direct variation if $y = -2$ when $x = 8$?
    A $y = -4x$
    C $y = 4x$
    B $y = \left(\frac{1}{4}\right)x$
    D $y = -\left(\frac{1}{4}\right)x$ ✓

32. What inequality is shown on the graph? 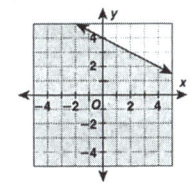
    F $y \leq -\left(\frac{1}{2}\right)x + 4$ ✓
    H $y \leq \left(\frac{1}{2}\right)x + 4$
    G $y \leq \left(\frac{1}{2}\right)x - 4$
    J $y \leq -\left(\frac{1}{2}\right)x + 4$

33. What is the eleventh term in the sequence 50, 46, 42, 38, …?
    A 10 ✓
    C 0
    B 6
    D −10

34. What is the rule for the function table?

x	−4	−1	2	5
y	−18	−9	0	9

    F $y = x - 2$
    H $y = 3(x - 2)$ ✓
    G $y = 3(2 - x)$
    J $y = 3x - 2$

35. Find $f(-5)$ for $f(x) = \left(\frac{1}{2}\right)x^2 - \left(\frac{1}{2}\right)x + 5$.
    A $f(-5) = 12$
    C $f(-5) = 13\frac{3}{4}$
    B $f(-5) = 12\frac{1}{2}$
    D $f(-5) = 20$ ✓

36. What is the degree of the polynomial $\left(\frac{1}{2}\right)x^2 + 5x^3 - 4x$?
    F $\frac{1}{2}$
    H 2
    G 1
    J 3 ✓

37. Add $(3m^2 + mn) + (3mn - m^2)$.
    A $4m^2 + 4mn$
    C $2m^2 - 4mn$
    B $2m^2 + 4mn$ ✓
    D $m^2 + 4mn$

38. Multiply $(-3abc^2)(-2a^2b + 4ac^2)$.
    F $6a^3b^2c^2 - a^2bc^4$
    G $6a^3b^2c^2 - 12a^2bc^4$ ✓
    H $6a^3b^2c^2 + 12a^2bc^4$
    J $-6a^5b^3c^6$

39. Which set is an infinite set?
    A {rational numbers between −10 and 10} ✓
    B {whole numbers between −10 and 10}
    C {integers between −10 and 10}
    D {natural numbers between −10 and 10}

40. What conditions are required to make the disjunction P or Q false?
    F Both P and Q are true.
    G P is true and Q is false.
    H P is false and Q is true.
    J Both P and Q are false. ✓

41. What is the degree of vertex D? 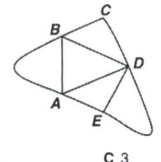
    A 5 ✓
    C 3
    B 4
    D 2

End of Course Assessment

Select the best answer.

1. Which of the following is a correct statement?
    A 0.03 > 30%
    C $0.30 < \frac{3}{10}$
    B $0.03 = \frac{3}{10}$
    D $30\% = \frac{3}{10}$ ✓

2. Valleywood Golf Course offers classes Monday through Saturday. The Saturday classes last 75 minutes while the weekday classes last 45 minutes. If they offered 705 minutes of classes last week, what was the maximum number of weekday classes?
    F 9
    H 12
    G 14 ✓
    J 17

3. Find the area of a trapezoid with the dimensions $b_1 = 9$, $b_2 = 16$, $h = 13.4$.
    A 335 sq units
    B 223.4 sq units
    C 189.2 sq units
    D 167.5 sq units ✓

4. A car salesman makes $445 per week plus commission. If during the week he sells a new van for $34,960 and earns a 2.5% commission on the sale, how much money did he earn for the week?
    F $429
    H $885
    G $874
    J $1319 ✓

5. What is the value of $f(-3)$ for the function $f(x) = 4.7x + 1.6$?
    A −18.3
    C −12.5 ✓
    B −15.7
    D 15.7

6. If $\angle A$ and $\angle B$ are supplementary, and $m\angle A = 57°$, what is $m\angle B$?
    F 33°
    H 213°
    G 123° ✓
    J 303°

7. At a school festival, a colored chip is randomly drawn out of a bag, and replaced. The table below shows the results of 50 draws. Estimate the probability of choosing a purple chip.

Outcome	Blue	Red	Green	Purple	Gold
Draws	7	11	6	12	14

    A 12%
    C 24% ✓
    B 18%
    D 88%

8. What are the coordinates of point K? 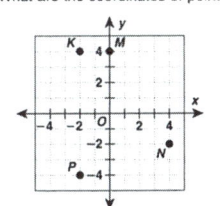
    F (−2, 4) ✓
    H (2, −4)
    G (−2, 4)
    J (2, −4)

9. Solve for k, $\frac{4}{5}k + \frac{7}{10} = \frac{13}{15}k - \frac{3}{5}$.
    A $k = \frac{1}{10}$
    C $k = 19\frac{1}{2}$ ✓
    B $k = 1\frac{1}{2}$
    D $k = 21\frac{2}{5}$

10. What is the slope of a line that is perpendicular to the line that passes through the points (4, 7) and (9, 3)?
    F $-\frac{4}{5}$
    H $\frac{4}{5}$
    G $-\frac{5}{4}$
    J $\frac{5}{4}$ ✓

11. Evaluate $\frac{3 + k + 9}{2}$ for $k = -16$.
    A −2 ✓
    C −4
    B 2
    D 14

12. In a school survey, three homerooms are chosen and ten students from each homeroom are randomly chosen to complete the survey. Identify the sampling method.
    F Population
    H Systematic
    G Random
    J Stratified ✓

13. Give the 8th term in the sequence the numerator of each fraction is 1. $6\frac{1}{6}, 6\frac{1}{3}, 6\frac{1}{2}, \ldots$
    A $6\frac{2}{3}$
    C 7
    B $6\frac{5}{6}$
    D $7\frac{1}{3}$ ✓

14. On a drawing with a scale of $\frac{1}{8}$ in.: 1 ft, a window is $\frac{3}{4}$ in. wide. What is the actual width of the window?
    F 5.5 ft
    H 6 ft ✓
    G 5.75 ft
    J 6.25 ft

15. What is the equation of direct variation given that y is 16 when x is −2?
    A $y = -8x$ ✓
    C $y = 8x$
    B $y = -\frac{1}{8}x$
    D $x = -8y$

16. The graph represents which inequality? 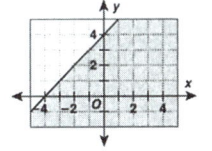
    F $y \leq x + 4$ ✓
    H $y < x + 4$
    G $y = 4x$
    J $y > x + 4$

17. Combine like terms $7a + 4b - 3a - 2b$.
    A $4a + 2b$ ✓
    C $4a - 2b$
    B $11a - b$
    D $7ab$

18. Find the mean, median, and mode of the data set. 5, 8, 6, 5, 8, 4, 8, 4
    F 6, 5.5, 8 ✓
    H 8, 5.5, 8
    G 5.5, 6, 6
    J 6, 5.5, 5

19. Simplify $19 + (3 \cdot 2^4)$.
    A 38
    C 912
    B 67 ✓
    D 1315

20. What is the square root of 429 to the nearest tenth?
    F 8.6
    H 214.5
    G 20.7 ✓
    J 184,041

21. When starting a family vacation, the Singler's odometer in their van read 15,674.7. At the end of the trip the odometer read 16,495.2, how far did the Singler's travel?
    A 32,169.9 mi
    C 820.5 mi ✓
    B 1117.2 mi
    D −820.5 mi

End of Course Assessment

22. Caleb and Drew are playing a game with a pair of dice. Caleb needs a sum of 5 or greater to win. What is his probability of winning on his next turn?
    F) $\frac{5}{6}$
    G) $\frac{2}{5}$
    H) $\frac{1}{6}$
    J) $\frac{2}{3}$

23. Wesley invested $6500 in a mutual fund at a yearly rate of $3\frac{3}{4}$%. If he has earned $975 in interest, how long has the money been invested?
    A) 2.5 years
    C) 4 years
    B) 3 years
    D) 4.5 years

24. If a pool table measures 4 ft by 8 ft, what is the length from the back edge of the top left pocket to the bottom right pocket to the nearest tenth?
    F) 80 ft
    H) 6.9 ft
    G) 24.3 ft
    J) 8.9 ft

25. What is the sum of the interior angles of the polygon?

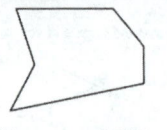

    A) 180°
    C) 540°
    B) 360°
    D) 720°

26. Evaluate $2x + 5y$ for $x = 12$ and $y = 6$.
    F) 72
    H) 54
    G) 25
    J) 42

27. Solve $\frac{3}{18} = \frac{a}{30}$.
    A) 6
    C) 15
    B) 5
    D) 90

28. Your favorite brand of cereal comes in four different sized boxes. Which size is the best deal?

Toasted Almond Crunch	
15 oz	$2.39
18 oz	$2.87
24 oz	$3.79
32 oz	$5.10

    F) 15 oz
    H) 24 oz
    G) 18 oz
    J) 32 oz

29. The line has what type of slope?

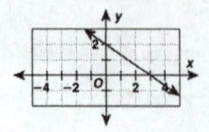

    A) positive
    C) zero
    B) negative
    D) undefined

30. Which is the number 0.0000042 in scientific notation?
    F) 0.42×10^{-7}
    H) 4.2×10^{-6}
    G) 4.2×10^{6}
    J) 4.2×10^{-5}

31. Which ordered pair is a solution of the system of equations?
    $y = 3x + 1$ $y = 5x - 3$
    A) (2, 3)
    C) (1, 2)
    B) (0, 1)
    D) (2, 7)

32. For a graduation party Mrs. Kennedy prepares a meat tray. If she buys 3 pounds each of ham and turkey, 2 pounds of roast beef, and 1 pound of salami, how much will the meat cost?

Today's Specials	
Roast Beef	$6.29 lb
Ham	$3.29 lb
Salami	$3.89 lb
Turkey	$4.79 lb

    F) $18.26
    H) $40.71
    G) $31.13
    J) $47.26

33. If Hannah purchased a 5-day parking pass for $36.25, how much did she pay per day?
    A) $6.75
    C) $7.00
    B) $6.85
    D) $7.25

34. A school festival has sold 760 tickets. If this is 95% of their goal, how many more tickets do they need to sell to reach 100% of their goal?
    F) 22
    H) 58
    G) 40
    J) 800

35. Find the perimeter of the figure.

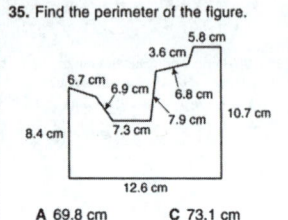

    A) 69.8 cm
    C) 73.1 cm
    B) 70 cm
    D) 76.7 cm

36. Jasmin is walking a group of 6 dogs. In how many different orders can the dogs enter her house in single file to get a drink of water?
    F) 21
    H) 480
    G) 40
    J) 720

37. How many times did the temperature rise one degree per hour?

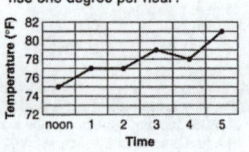

    A) none
    C) 2
    B) 1
    D) 3

38. Give the 9th term in the sequence 2, −6, 18, −54 . . .
    F) −39,366
    H) 1459
    G) −4374
    J) 13,122

39. What is the range and the first and third quartiles of the data set?
    14, 16, 12, 15, 22, 18, 16, 10, 12
    A) 22, 12, 17
    C) 12, 12, 17
    B) 22, 15, 17
    D) 12, 12, 15

40. Find the volume. Use 3.14 for π. 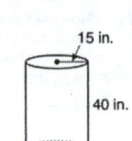
    F) 1200 in^3
    G) 1884 in^3
    H) 3768 in^3
    J) 28,260 in^3